AF352870

Emerging and Advanced Green Energy Technologies for Sustainable and Resilient Future Grid

Emerging and Advanced Green Energy Technologies for Sustainable and Resilient Future Grid

Editor

Surender Reddy Salkuti

MDPI • Basel • Beijing • Wuhan • Barcelona • Belgrade • Manchester • Tokyo • Cluj • Tianjin

Editor
Surender Reddy Salkuti
Department of Railroad and Electrical Engineering,
Woosong University
Korea

Editorial Office
MDPI
St. Alban-Anlage 66
4052 Basel, Switzerland

This is a reprint of articles from the Special Issue published online in the open access journal *Energies* (ISSN 1996-1073) (available at: https://www.mdpi.com/journal/energies/special_issues/Future_Grid).

For citation purposes, cite each article independently as indicated on the article page online and as indicated below:

LastName, A.A.; LastName, B.B.; LastName, C.C. Article Title. *Journal Name* **Year**, *Volume Number*, Page Range.

ISBN 978-3-0365-5769-4 (Hbk)
ISBN 978-3-0365-5770-0 (PDF)

Contents

About the Editor

Surender Reddy Salkuti

Surender Reddy Salkuti has been working at Woosong University, South Korea, as an Associate Professor in the Department of Railroad and Electrical Engineering since April 2014. He received his Ph.D. degree in Electrical Engineering from the Indian Institute of Technology Delhi (IITD), India, in 2013. He was a Postdoctoral Researcher at Howard University, Washington, DC, USA, from 2013 to 2014. His research interests include power system restructuring issues, smart grid development with the integration of wind and solar photovoltaic energy sources, battery storage and electric vehicles, demand response, power system analysis and optimization, soft computing techniques application in power systems and renewable energy.

He has published one edited volume with Springer (LNEE) and more than 200 research articles in peer-reviewed international journals and conference proceedings. He has served as or is serving as Guest Editor for various international journals. He is also an editorial board member for many journals. He is the recipient of the 2016 Distinguished Researcher Award from Woosong University Educational Foundation, South Korea, and POSOCO Power System Award (PPSA) 2013, India. He is a Member of IEEE and IEEE Power and Energy Society.

Editorial

Emerging and Advanced Green Energy Technologies for Sustainable and Resilient Future Grid

Surender Reddy Salkuti

Department of Railroad and Electrical Engineering, Woosong University, Daejeon 34606, Korea;
surender@wsu.ac.kr; Tel.: +82-10-9674-1985

Citation: Salkuti, S.R. Emerging and Advanced Green Energy Technologies for Sustainable and Resilient Future Grid. *Energies* **2022**, *15*, 6667. https://doi.org/10.3390/en15186667

Received: 6 September 2022
Accepted: 12 September 2022
Published: 13 September 2022

Publisher's Note: MDPI stays neutral with regard to jurisdictional claims in published maps and institutional affiliations.

1. Introduction

Future grid refers to the next generation of the electrical grid, which will enable smart integration of conventional, renewable, and distributed power generation, energy storage, transmission and distribution, and demand management. Renewable energy is crucial in transitioning to a less carbon-intensive economy and a more sustainable energy system. The high penetration and uncertain power outputs of renewable energy poses a great challenge to the stable operation of energy systems. The deployment of the smart grid is revolutionary and also imperative around the world. It involves and deals with multidisciplinary fields such as energy sources, control systems, communications, computational, generation, transmission, distribution, customer, operations, markets, and service provider. Smart grids are emerging in both developed and developing countries, with the aim of achieving a reliable and secure electricity supply. Smart grids will eventually need standards, policy, and a regulatory framework for successful implementation.

This Special Issue invited original contributions on, but not limited to, themes and topics related to: renewable power and clean energy technologies; design and operation of sustainable energy systems; smart grid architectures and cyber and physical security; smart grid and green energy integration; operation and control of renewable energy sources; smart grid and smart cities modeling; forecasting techniques for renewable energy sources and loads; electric vehicle systems for smart grid; distributed generation and distributed storage; agent-based smart grid simulation; decision support approaches for smart grids; electricity market modeling and simulation for the integration of renewable sources; intelligent approaches for smart grid management; multi-agent applications for smart grids; smart grid energy management system; computational intelligence technologies for sustainable energy; machine learning, IoT, and big data applications for energy systems; and demand side management.

2. Special Issue Content

This Special Issue on "Emerging and Advanced Green Energy Technologies for Sustainable and Resilient Future Grid" includes 18 papers presenting various advanced technologies related to the future grid. The first paper, by Battula et al. [1], provides an in-depth review on energy management in microgrid; its components, control techniques, communication technologies in the use, architecture, auxiliary services, structure, and standards available for implementation and future scope in the area of energy management system (EMS). The prime aspects that are covered in this review are on the prospects, solutions, and opportunities of the objective functions of the EMS using efficient strategies. This article introduces the microgrid, its architecture composing of an AC microgrid, a DC microgrid, and a hybrid microgrid. The microgrid components composing of distributed generators, storage elements, loads and their composition, integration of electrical vehicles (EVs) and their applications, demand response, and electricity market pricing. The main focus of the review is on the different techniques adopted, which are classified as classical methods, meta-heuristic methods, and intelligent methods used to solve different objective functions considering cost, active power loss, voltage stability, and emission of greenhouse gasses with usage on non-renewable generators. The review further broadened with the structure of control such as centralized, decentralized, and distributed control with multiple distribution generation sources, integration of EVs, and utilization of demand response and their mode of operation which is either in islanded mode or grid connected mode. The areas such as customer confidentiality regulations, management of communication systems, and reliability studies with depth of discharge of the batteries, effect of the conventional grid on greenhouse gas emissions, and demand response, active and reactive losses along with resilience and customer management for effective and efficient operation of microgrids have scope to emphasize in future studies.

Soliman et al. [2] proposed a robust tracking (servomechanism) controller for linear time-invariant (LTI) islanded (autonomous, isolated) microgrid voltage control. The studied microgrid (MG) consists of many distributed energy resources (DERs) units, each using a voltage-sourced converter (VSC) for the interface. The optimal tracker design uses the ellipsoidal approximation to the invariant sets. The MG system is decomposed into different subsystems (DERs). Each subsystem is affected by the rest of the system that is considered as a disturbance to be rejected by the controller. The proposed tracker (state feedback integral control) rejects bounded external disturbances by minimizing the invariant ellipsoids of the MG dynamics.

Malik et al. [3] designed and fabricated a novel thermal energy storage system using phase change material. The concept of utilizing phase change materials (PCMs) has attracted wide attention in recent years. This is due to their ability to extract thermal energy when used in collaboration with photovoltaic (PV), thus improving the photoelectric conversion efficiency. The aim of this study was to design and study a latent heat thermal energy storage system which can store solar energy for a reasonable amount of time. This stored energy is environmentally friendly and can be used for any indoor heating purpose such as cooking, water, and room heating applications in the absence of sunlight. The experimental results of the designed system show that the system is capable of storing enough energy during sunshine hours that can later be used for any heating application in the absence of sunlight. In the current study, potash alum was identified as a phase change material combined with renewable energy sources that can be efficiently and effectively used in storing thermal energy at comparatively lower temperatures that can later be used in daily life heating requirements. A parabolic dish which acts of a heat collector is used to track and reflects solar radiation at a single point on a receiver tank.

Swain et al. [4] presented a smart city energy model along with a comprehensive state-of-the-art overview of blockchain-powered energy systems with an objective to transform traditional cities into smart communities. It aims to increase the energy efficiency and network security of smart grids with the help of nature-inspired algorithms and blockchain technology. This work also identifies some of the major issues associated with the energy grids and provides solution to overcome these challenges. Focusing mainly on the intelligent management of energy systems, it introduced wireless sensor networks, microgrid, prosumers, and blockchain technology in this energy model to generate and consume clean energy on a decentralized trading system. The smart grid network consists of several wireless sensor nodes that come with various challenges such as limited computational capabilities, huge consumption of energy by the sensors in longer communication distances resulting in decreased network lifespan, and increased communication costs. This led to the introduction of nature-inspired algorithms such as particle swarm optimization (PSO) and genetic algorithm which provided optimal solutions for these challenges. This work presents the properties and mechanisms of biological systems and deployed PSO and genetic algorithm for energy-efficient selection of cluster head in a group of sensor nodes and achieved optimal routing in the communication system, respectively, for faster transmission of data in less processing time. In order to improve energy distribution capability in the smart community and maintain security of the systems, the blockchain-based smart microgrid technology is deployed using a decentralized application to enable peer-to-peer energy trading. This is done by implementing Web3 technologies and Ethereum smart contract to maintain energy records in a decentralized, transparent, and cost-effective way.

Kim et al. [5] proposed a photovoltaic (PV) string-level isolated DC–DC power optimizer with wide voltage range. A hybrid control scheme in which pulse frequency modulation (PFM) control and pulse width modulation (PWM) control are combined with a variable switching frequency is employed to regulate the wide PV voltage range. By adjusting the switching frequency during the PWM control process, the circulating current period can be eliminated and the turn-on period of the bidirectional switch of the dual-bridge LLC (DBLLC) resonant converter is reduced compared to that with a conventional PWM control scheme with a fixed switching frequency, resulting in better switching and conduction loss. Soft start-up control under a no-load condition is proposed to charge the DC-link electrolytic capacitor from 0 V. A laboratory prototype of a 6.25 kW DBLLC resonant converter with a transformer, including integrated resonant inductance, is built and tested in order to verify the performance and theoretical claims.

Yoo et al. [6] presented a demonstration and validation of primary frequency control (PFC) and secondary frequency control (SFC) with a vehicle-to-grid (V2G)-capable Nissan leaf EV using a commercially available supervisory control and data acquisition (SCADA) and MATLAB/Simulink, based on droop characteristics designed for PFC and the dispatch of area control error (ACE) signals for SFC at a V2G testing facility of the National Research Council (NRC), Canada. The simulation models derived, implemented, and described in this work are based on time-domain dynamic analysis using the phasor modeling technique. Then, model-based simulations of PFC and SFC with an intelligent optimal control algorithm, including a coordinated control of the state of charge (SOC) and charging schedule for five aggregated EVs with different departure times and SOC management profiles, are presented to validate the effectiveness of frequency control using the integrated fleet of V2G-capable EVs and to verify its relevant technical issues.

Veeramsetty et al. [7] uses a stochastic gradient descent optimizer to update the parameters in the regression models. Simple linear regression and polynomial regression models were used to predict the active power load. A new approach, i.e., predict the active power load based on load at last three hours and load at one day before was used with various regression models and dimensionality reduction technique was used to reduce the complexity of the model so that overfitting problem was removed. Data analytic tools were used to process the data before feeding it to the model.

Musengimana et al. [8] proposed a method that uses the DC-link voltage error deviation to generate the additional reactive current reference. The obtained compensating reactive current is added to the reactive current reference from the point of common coupling (PCC) voltage control loop. By updating the reactive current reference with respect to the DC-link voltage, the instability issues induced by the control loops coupling are overcome. Further, the proposed control strategy also benefits from easy implementation and good robustness against AC voltage control (AVC), DC bus voltage control (DVC), and phase-locked loop (PLL) bandwidth variation, as well as the grid strength variation. The performance of the proposed method is evaluated by using the small-signal stability analysis based on the derived small-signal model. The comparative evaluation of the proposed control method with the conventional control method is also performed in this work.

Devarapalli et al. [9] proposed an approach to measure the true power factor (TPF) using a four-term minimal sidelobe cosine-windowed enhanced dual-spectrum line interpolated Fast Fourier Transform (FFT). The proposed method was used to measure the TPF with a National Instruments cRIO-9082 real-time (RT) system, and four different low-power consumer electronic appliances (LPCEAs) in a smart home were considered. The RT results exhibited that the TPF uniquely identified each usage pattern of the LPCEAs and could use them to improve the TPF by suggesting an alternative usage pattern to the consumer. A positive response behavior on the part of the consumer that is in their interest can improve the power quality in a demand-side management application.

Benbouhenni et al. [10] designed the third-order sliding mode (TOSM) command to control and minimize the ripples of current, reactive power, torque, and active power of asynchronous generators (ASGs)-based variable-speed contra-rotating wind power (CRWP) systems. A TOSM technique to overcome the undulations power problem is designed for the direct field-oriented control (DFOC) technique of variable-speed CRWP systems. A DFOC control strategy with TOSM controllers is employed to enable avoidance of the power undulations and improve the response time. The system states can be ensured to improve in variation parameters. The principle of the proposed TOSM controllers is detailed in the work. Validation of the designed technology is carried out by digital simulations using MATLAB software.

Gunda et al. [11] explored an extended Kalman filter (EKF) algorithm to predict primary side currents of the transformer, using the nonlinear state-space transformer model. A residual signal is derived from the difference of measured and estimated currents on the primary side winding of the transformer. When the transformer is under normal conditions, the residual signal becomes zero based on the EKF estimation. If the transformer is in a faulty condition, large residual currents are generated. The residual signal amplitude is compared with the threshold value in order to classify the magnetic inrush current and internal fault current. In this work, the diagonal elements of the covariance matrix are utilized to analyze the severity of the fault occurrence in the transformer. The simulation results are presented to evaluate the EKF for different conditions of the transformer.

Bayati et al. [12] proposed an analytical expression of the current of a DC microgrid (MG) cluster under fault conditions. First, the analytical model of the fault current of the AC/DC converter between the grid and DC MG is presented. Second, the analytical model of a bidirectional interlink DC/DC converter of the interconnection line between two adjacent DC microgrids in a clustered DC MG is derived in detail. Then, a DC fault clearing strategy is proposed based on using a DC fault current limiter (FCL) in series with a DC circuit breaker. Finally, the performance of the proposed method is evaluated based on time-domain simulation studies on a test DC MG cluster in MATLAB/Simulink environment. The simulation results are also compared with the results of the proposed analytical model. The obtained results verify the analytical expression of the fault current and prove the effectiveness of the proposed DC fault current limiting and clearing strategy.

Dubey et al. [13] proposed a multiobjective model capable of finding a set of trade-off solutions for the joint optimization problem, considering the cost of reserve and curtailment of power from renewable sources. Managing a hybrid power system is a challenging task due to its stochastic nature mixed with the objective function and complex practical constraints associated with it. This work focuses on profit-based multiobjective scheduling with the objectives are profit maximization by selling electricity generated through various sources and minimizing emissions by conventional (thermal) sources. The generation sources include solar photovoltaic (PV), wind, and conventional plants. The optimization problem addressed has practical constraints, including power balance, generation limits, and ramp limits. The conflicting objectives: profit and emission, are aggregated using the fuzzy-min ranking method. Thermal sources are modeled as quadratic polynomial; solar PV units use probabilistic modeling, and wind power uses Weibull distribution for wind velocity variation. The PV and wind power costs use overestimation and overestimation to balance renewable power generation uncertainty. A novel metaheuristic Equilibrium Optimizer (EO) algorithm is used for computing the optimal schedule and impact of renewable energy integration on profit and emission for different optimization objectives.

Javed et al. [14] analyzed an effective scheduling of a battery energy storage system (BESS) and the effects of PV systems for a campus microgrid to minimize the energy operating costs for a prosumer microgrid with the implementation of actual load data. The suggested system utilized solar energy, a BESS, and diesel generators in several scenarios and their consequences were investigated. The optimal scheduling was implemented in MATLAB and formed as a mixed-integer linear programming (MILP) problem. The time-of-use (TOU) pricing-based demand response (DR) was investigated here as part of a financial and economic analysis, and the energy storage system (ESS) was used as a flexible DR framework that could be charged or discharged wisely at various times to meet the budget target without compromising its durability. Without the distributed generation (DG) or ESS, the utility grid supplied all the campus microgrids required energy, leading to higher operational expenses.

Pothireddy et al. [15] addressed the complication of solutions, merits, and demerits that may be encountered in today's power system and encompasses DR and its impacts in reducing the installation cost, the capital cost of DGs, and total electricity tariff. To achieve this, an objective function was formulated and an optimal sizing method has been proposed by considering the impact of DR for finding the optimal size of DGs, i.e., wind turbine, PV, and diesel generator. Further, the proposed algorithm clusters the load into interruptible loads (ILs) and non-interruptible loads (NILs) and assigns a priority to the non-essential loads with the order of scheduled times by using time of use pricing (TOU) pricing. In addition, this work suggests a limit on the amount of load shift to avoid the issues like rebound effect, increase in marginal price, and operational cost of the MG due to load recovery. The future extension of this work is studying the impact of DR programs in optimal sizing of DGs by considering the uncertainty in renewable energy sources. Finding the elasticity and cross elasticity matrix of load with respect to price changes can be performed.

Romero-Ramirez et al. [16] introduced a methodology for the detection and quantification of stationary frequency components in the electrical signals of a smart building, and fusing the spectral kurtosis (SK) with the fast Fourier transform (FFT). The use of the proposed methodology presents some advantages against the conventional approaches, for instance, the ease of its implementation since the mathematics behind this methodology are simple. This situation results in a technique that is efficient and demands a low computational burden. Moreover, it is possible to perform a time tracking of a specific stationary frequency component (SFC) without problems, such as the mode mixing introduced by time-frequency transforms, such as wavelet transform (WT) and empirical mode decomposition (EMD), this way it is possible to perform a quantification of how every SFC contributes to detriment the quality of the power grid. Although SK is a widely explored technique, its use is more extended in the identification of transient events. Additionally, the combination of SK and FFT represents a novel solution for the time tracking and quantification of SFC. The proposed methodology aims to be a helpful tool to perform an estimation of the types of loads that appear in the grid and how they impact the quality of the supply. This way it is possible to propose actions in order to mitigate the undesired effects associated with a specific type of load.

Lee et al. [17] defined the life-cycle cost for an ESS in detail based on a life assessment model and used for scheduling. The life-cycle cost is affected by four factors: temperature, average state-of-charge (SOC), depth-of-discharge (DOD), and time. In the case of the DOD stress model, the life-cycle cost is expressed as a function of the cycle depth, whose exact value can be determined based on fatigue analysis techniques such as the Rainflow counting algorithm. The optimal scheduling of the ESS is constructed considering the life-cycle cost using a tool based on reinforcement learning. Since the life assessment cannot apply the analytical technique due to the temperature characteristics and time-dependent characteristics of the ESS SOC, the reinforcement learning that derives optimal scheduling is used. The results show that the SOC curve changes with respect to weight. As the weight of life-cycle cost increases, the ESS output and charge/discharge frequency decrease.

Menniti et al. [18] illustrated the main enabling technologies, smart meter, ESS, DC Nanogrid (DCNG) and Energy Community Management (ECM) platform that must be used to support the growth of renewable energy communities (RECs). The innovative settlement of the Advanced End-User (AEU) is introduced, defined as an active and pro-active end-user equipped with the above-mentioned enabling technologies. Additionally, some use cases and the associated performance indexes have been identified and defined, respectively. The numerical results performed both, considering the operation of a single

AEU and four AEUs operating as an REC, are illustrated, and discussing when flexibility service requests are sent or not, highlighting how it is possible to both maximize the self-consumption and satisfy the flexibility service request in the case of an REC consisting of AEUs as members.

3. Closing Remarks and Future Challenges

The papers in this Special Issue reveal an exciting research area, namely the "Future Grid" that is continuing to grow. This Special Issue addresses the emerging and advanced green energy technologies for a sustainable and resilient future grid, and provides a platform to enhance interdisciplinary research and share the most recent ideas. Various high-priority areas of smart grid technologies are addressed including information and communication technology integration, transmission enhancement applications, distribution and management, advanced metering infrastructure, charging infrastructure, sharing of energy storage, demand flexibility to the grid, and trading of renewable energy. Therefore, I believe that the published papers will have practical importance for the development of the future grid. Finally, I would like to thank all our authors, reviewers, and editorial staff who have contributed for the publication of this Special Issue. I am sure all readers of this Special Issue of *Energies* will find the scientific articles interesting and beneficial to their research work.

Funding: This research work was funded by "WOOSONG UNIVERSITY's Academic Research Funding-2022".

Conflicts of Interest: The author declares no conflict of interest.

References

1. Battula, A.R.; Vuddanti, S.; Salkuti, S.R. Review of Energy Management System Approaches in Microgrids. *Energies* **2021**, *14*, 5459. [CrossRef]
2. Soliman, H.M.; Bayoumi, E.; Al-Hinai, A.; Soliman, M. Robust Decentralized Tracking Voltage Control for Islanded Microgrids by Invariant Ellipsoids. *Energies* **2020**, *13*, 5756. [CrossRef]
3. Malik, M.S.; Iftikhar, N.; Wadood, A.; Khan, M.O.; Asghar, M.U.; Khan, S.; Khurshaid, T.; Kim, K.-C.; Rehman, Z.; Rizvi, S.T.u.I. Design and Fabrication of Solar Thermal Energy Storage System Using Potash Alum as a PCM. *Energies* **2020**, *13*, 6169. [CrossRef]
4. Swain, A.; Salkuti, S.R.; Swain, K. An Optimized and Decentralized Energy Provision System for Smart Cities. *Energies* **2021**, *14*, 1451. [CrossRef]
5. Kim, H.; Yu, G.; Kim, J.; Choi, S. PV String-Level Isolated DC–DC Power Optimizer with Wide Voltage Range. *Energies* **2021**, *14*, 1889. [CrossRef]
6. Yoo, Y.; Al-Shawesh, Y.; Tchagang, A. Coordinated Control Strategy and Validation of Vehicle-to-Grid for Frequency Control. *Energies* **2021**, *14*, 2530. [CrossRef]
7. Veeramsetty, V.; Mohnot, A.; Singal, G.; Salkuti, S.R. Short Term Active Power Load Prediction on A 33/11 kV Substation Using Regression Models. *Energies* **2021**, *14*, 2981. [CrossRef]
8. Musengimana, A.; Li, H.; Zheng, X.; Yu, Y. Small-Signal Model and Stability Control for Grid-Connected PV Inverter to a Weak Grid. *Energies* **2021**, *14*, 3907. [CrossRef]
9. Devarapalli, H.P.; Dhanikonda, V.S.S.S.S.; Gunturi, S.B. Demand-Side Management for Improvement of the Power Quality in Smart Homes Using Non-Intrusive Identification of Appliance Usage Patterns with the True Power Factor. *Energies* **2021**, *14*, 4837. [CrossRef]
10. Benbouhenni, H.; Bizon, N. Third-Order Sliding Mode Applied to the Direct Field-Oriented Control of the Asynchronous Generator for Variable-Speed Contra-Rotating Wind Turbine Generation Systems. *Energies* **2021**, *14*, 5877. [CrossRef]
11. Gunda, S.K.; Dhanikonda, V.S.S.S.S. Discrimination of Transformer Inrush Currents and Internal Fault Currents Using Extended Kalman Filter Algorithm (EKF). *Energies* **2021**, *14*, 6020. [CrossRef]
12. Bayati, N.; Baghaee, H.R.; Savaghebi, M.; Hajizadeh, A.; Soltani, M.N.; Lin, Z. DC Fault Current Analyzing, Limiting, and Clearing in DC Microgrid Clusters. *Energies* **2021**, *14*, 6337. [CrossRef]
13. Dubey, S.M.; Dubey, H.M.; Pandit, M.; Salkuti, S.R. Multiobjective Scheduling of Hybrid Renewable Energy System Using Equilibrium Optimization. *Energies* **2021**, *14*, 6376. [CrossRef]
14. Javed, H.; Muqeet, H.A.; Shehzad, M.; Jamil, M.; Khan, A.A.; Guerrero, J.M. Optimal Energy Management of a Campus Microgrid Considering Financial and Economic Analysis with Demand Response Strategies. *Energies* **2021**, *14*, 8501. [CrossRef]
15. Pothireddy, K.M.R.; Vuddanti, S.; Salkuti, S.R. Impact of Demand Response on Optimal Sizing of Distributed Generation and Customer Tariff. *Energies* **2022**, *15*, 190. [CrossRef]

16. Romero-Ramirez, L.A.; Elvira-Ortiz, D.A.; Romero-Troncoso, R.d.J.; Osornio-Rios, R.A.; Zorita-Lamadrid, A.L.; Gonzalez-Gonzalez, S.L.; Morinigo-Sotelo, D. Spectral Kurtosis Based Methodology for the Identification of Stationary Load Signatures in Electrical Signals from a Sustainable Building. *Energies* **2022**, *15*, 2373. [CrossRef]
17. Lee, W.; Chae, M.; Won, D. Optimal Scheduling of Energy Storage System Considering Life-Cycle Degradation Cost Using Reinforcement Learning. *Energies* **2022**, *15*, 2795. [CrossRef]
18. Menniti, D.; Pinnarelli, A.; Sorrentino, N.; Vizza, P.; Barone, G.; Brusco, G.; Mendicino, S.; Mendicino, L.; Polizzi, G. Enabling Technologies for Energy Communities: Some Experimental Use Cases. *Energies* **2022**, *15*, 6374. [CrossRef]

Review

Review of Energy Management System Approaches in Microgrids

Amrutha Raju Battula [1], Sandeep Vuddanti [1] and Surender Reddy Salkuti [2,*]

[1] Department of Electrical Engineering, National Institute of Technology Andhra Pradesh (NIT-AP), Tadepalligudem, Andhra Pradesh 534101, India; amrutharaju.sclr@nitandhra.ac.in (A.R.B.); sandeep@nitandhra.ac.in (S.V.)

[2] Department of Railroad and Electrical Engineering, Woosong University, Daejeon 34606, Korea

* Correspondence: surender@wsu.ac.kr; Tel.: +82-10-9674-1985

Abstract: To sustain the complexity of growing demand, the conventional grid (CG) is incorporated with communication technology like advanced metering with sensors, demand response (DR), energy storage systems (ESS), and inclusion of electric vehicles (EV). In order to maintain local area energy balance and reliability, microgrids (MG) are proposed. Microgrids are low or medium voltage distribution systems with a resilient operation, that control the exchange of power between the main grid, locally distributed generators (DGs), and consumers using intelligent energy management techniques. This paper gives a brief introduction to microgrids, their operations, and further, a review of different energy management approaches. In a microgrid control strategy, an energy management system (EMS) is the key component to maintain the balance between energy resources (CG, DG, ESS, and EVs) and loads available while contributing the profit to utility. This article classifies the methodologies used for EMS based on the structure, control, and technique used. The untapped areas which have scope for investigation are also mentioned.

Keywords: renewable energy sources; microgrid; energy management system; communication technologies; microgrid standards

Citation: Battula, A.R.; Vuddanti, S.; Salkuti, S.R. Review of Energy Management System Approaches in Microgrids. *Energies* **2021**, *14*, 5459. https://doi.org/10.3390/en14175459

Academic Editor: Gianfranco Chicco

Received: 15 July 2021
Accepted: 26 August 2021
Published: 2 September 2021

Publisher's Note: MDPI stays neutral with regard to jurisdictional claims in published maps and institutional affiliations.

1. Introduction

Over the last few decades, with an increasing population, the world has gone through an exponential consumption of energy which has led to the depletion of conventional resources like coal, crude oil, and natural gas. The exploitation of these resources has a severe impact on the environment with an increase in greenhouse gases [1,2]. To mitigate these effects, a policy has been adopted by different countries to introduce non-conventional/renewable sources to support the fields of electrification and transportation. In electrification, the existing power grid uses conventional sources for generation and lacks power quality. The poor power quality of supply leads to load shedding and blackouts, thereby interrupting the day-to-day activities of the consumers. The conventional grid uses one-third of the total generation fuel to convert into electricity and, with an eight percent loss in transmission lines of the generated electricity, is used to meet the peak demand that also has a five percent probability of occurring, with reduced reliability [3]. Conventional generation does not utilize the heat produced by itself for any application. These drawbacks of the conventional grid could be compensated with penetration of renewable sources at local areas or distributed generation (DG) there by reducing the transmission losses and maximum utilization of the output including heat generated [4–6]. Integration of dispatchable energy sources like wind and PV introduces the problem of intermittent power generation as they generally depend on climatic and meteorological conditions. A hybrid energy system consisting of storage elements and renewable energy sources is used for the continuous supply of power. The future power grid needs to be intelligent to maintain a reliable supply of economical and sustainable power for consumers [7–10]. To overcome the existing challenges in the grid, a smart grid needs to be adopted which controls the complex process of power exchange and plans as well for the growing energy demand. The future grid requires the support of communication technologies and local microgrids (MG) for efficient control of the system. The integration of renewable energy resources at the load side requires a two-way flow of power and data with the capability of adapting to management applications that can leverage the technology [11]. During a fault condition, the local microgrid isolates itself from the main grid, creating a standalone/islanding mode of supply to the consumers [12,13]. This feature is known as plug and play, which allows the local generation to meet the demand by balancing the energy available. The microgrid consists of a microgrid control center (MGCC) and local controllers (LCs) to balance the energy demand. The microgrid takes the inputs from forecasted parameters (weather, generation, and market prices) to meet the uncertain load demand and also participates in the energy market. The MGCC is supported by communication technologies and equipped with processing algorithms to overcome the challenges in the generation–demand balance [14–17]. The energy management in microgrids controls the power supply of storage elements, demand response, and local controllers/local generation sources. Figure 1 shows a typical structure of a microgrid.

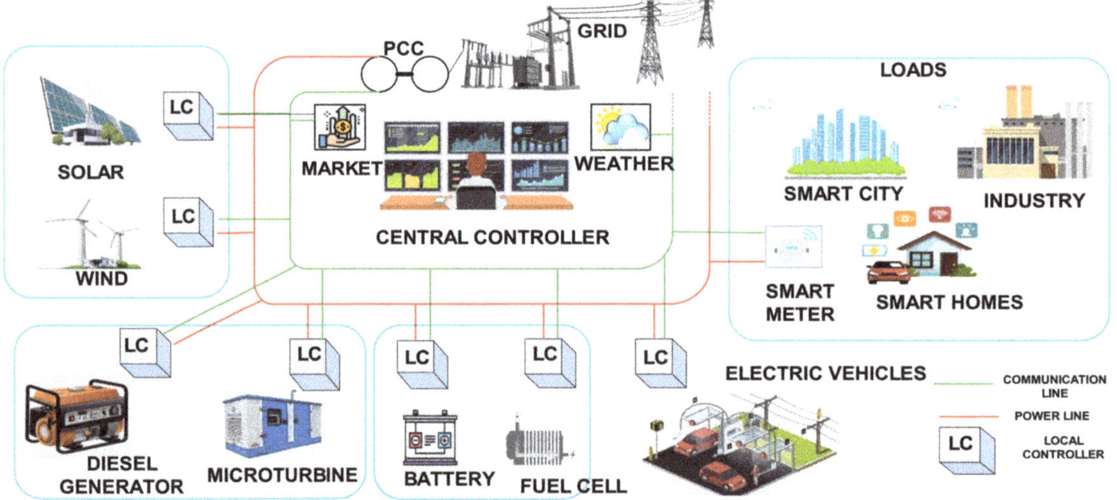

Figure 1. Structure of a typical microgrid.

The contributions of this paper are shown as below:

- This paper provides a brief introduction about the architecture of microgrids, different classifications in microgrids, components of a microgrid, communication technologies used, standards available for the implementation, and auxiliary services required.
- This paper provides a review of the recent analysis of the different energy management strategies consisting of classical, heuristic, and intelligent algorithms. The article analyzes each approach and its applications in that methodology.
- The paper addressed applications in energy management which include forecasting, demand response, data handling, and the control structure.
- This article provides insight on areas in which the scope of research and their contribution to energy management is in the nascent stage.

The energy management strategies proposed for the microgrid in the paper are structured into six sections. Section 1 is the introduction to microgrids and energy management. Section 2 provides a brief overview of microgrid elements, architecture, classification, and communication. Section 3 gives an overview of different control structures in energy management. Section 4 provides reviews on different numerical algorithms used in energy management strategies in microgrids based on the classification, control, and methods of approach. Remarks on each paper for different controls of the EMS application are given. Section 5 discusses the support infrastructure of microgrids for their efficient operation. Section 6 provides the conclusion of the paper.

2. Overview of Microgrid

2.1. Microgrid Components

A microgrid is a small or medium distribution system comprised of smart infrastructure capable of maintaining equilibrium in demand–supply while providing security, autonomy, reliability, and resilience. Sourced distributed generations (DGs) like photovoltaics (PV), wind turbines (WT), microturbine (MT), fuel cells (FC), and energy storage units (ESU) are expected to deliver electricity without interference from the main grid. This high penetration of DGs can cause challenges in the performance of power system stability in large areas. To minimize the risks, the concept of microgrids is proposed [18,19]. A microgrid is a small-scale low- or medium-level voltage distribution system consisting of distributed energy resources (DERs), intermittent storage, communication, protection, and control units that operate in coordination with each other to supply reliable electricity to end-users [20].

2.1.1. Distribution Generations (DGs)

Conventional generation (CG), such as coal-based thermal power plants, hydro power plants, wind-generation farms, and large-scale solar and nuclear power plants, are centralized to supply electricity for long distances. A decentralized generation is energy generated by the end-users by using small-scale energy resources [21,22]. Local generation when compared with the conventional power system reduces the transmission losses and the cost associated with it. The generation could be from 1 kW to a few 100 MW; the generation units are mostly used to support the peak load of the demand. Distributed generation sources consist of both renewable and non-renewable sources, i.e., wind generators, PV panels, small hydro power plants, and diesel generators [23]. Combined heat and power (CHP) is where heating is added along with electricity in the application. The sources that are being used in CHP systems are Stirling engines, internal combustion engines, and micro-turbines (MT) using biogas, hydrogen, and natural gas [24]. CHP technology stores excess allowing optimum performance, thereby attaining efficiency of more than 80%, to that of about 35% for centralized power plants [25]. Table 1 shows characteristics of distributed generation sources.

Table 1. Characteristics of distributed generation sources.

Characteristics	Solar	Wind	Micro-Hydro	Diesel	CHP
Availability	Location-Based	Location-Based	Location-Based	Anywhere	Source-Based
Output	DC	AC	AC	AC	AC
Carbon emission	Nil	Nil	Nil	High	Source-Based
Interface	Converter	Converter + IG/SG	IG/SG	Generator	Generator
Flow control	MPPT/DC Voltage	MPPT/Torque and Pitch	Controllable	Controllable	AVR and Governor

DC—Direct current, AC—Alternate current, MPPT—Maximum power point tracking, AVR—Automatic voltage regulator.

2.1.2. Energy Storage System (ESS)

Energy storage is a device that is capable of converting the electrical energy to a storable form and converting it back to electricity when it is needed. Based on the form of stored energy, there are four main categories for energy storage technologies: mechanical energy storage (MES), thermal energy storage (TES), chemical energy storage (CES), and electrical energy storage (EES). The key components for the working of MG EMS are the energy storage units, which regulate the supply–demand balance during the operation of

DGs. In [26–28], a conclusion is drawn that a system with several micro sources is modeled to support an island mode where storage systems are needed to maintain the balance of the intermittent sources. The energy storage devices that are included in microgrid systems that provide continuous power supply are batteries, flywheels, and supercapacitors [29]. In terms of the current economy, batteries are less expensive and have a high negative environmental effect compared to other storage devices. Storage in fuel cells is also another option that converts the fuel into electricity through a chemical process. These fuel cells require oxygen and hydrogen for continuous supply without discharge. A variety of fuels available for the fuel cell are propane, natural gas, anaerobic digester gas, methanol, and diesel hydrogen [30], while hydrogen has become prominent in recent years for its clean and safe operation. Table 2 shows commonly used energy storage and their characteristics.

Table 2. Different energy storage systems in microgrids.

Characteristics	Charge/Discharge Rate (MW)	Discharge Duration	Response Time	Energy Density (Wh/kg)	Power Density (W/kg)	Environmental Impact	Service (Years)	Efficiency (%)
Battery	0–40	msec–hours	msec	10–250	70–300	High	5	70–90
Flywheel	0.001–0.005	msec–1 h	msec	0.005–5	500–10,000	Low	20	75–95
Supercapacitor	0.002–0.25	msec–15 min	instantaneous	5–130	400–1500	Low	>10	90–95
Fuel Cell	0.001–50	sec-day+	m sec	800–10,000	500–1000	Moderate	>15	20–90
CES	0.1–300	Hour–day+	min	3–60	-	Low	15	40–90
SMES	0.1–10	msec–10 sec	instantaneous	0.5–5	500–2000	Low	10	>95
Pumped storage	0.1–5000	Hour–day+	Sec–min	0.5–1.5	-	Low	25	>85

2.1.3. Loads and Their Classification

Loads can be categorized as residential, commercial, industrial, and others (agriculture and public offices) from the statistical data of feeder consumption in the distribution system. Measurement-based and component-based approaches are considered for load model identification [31]. The measurement-based approach needs the measured data from the smart meters or measuring devices which derives into load model structure. The capturing of data for load characteristics needs to be composed of different environmental conditions. The data obtained from the smart devices are used to form the load model structure as static, ZIP (constant impedance-resistive components or heating, constant current-street lighting, and constant power motors), and exponential [32,33]. Then, the structure is estimated and validated with field measurements by correcting the errors using intelligent detection techniques (artificial intelligence and pattern detection). The component-based approach aggregates the load model by combining the load consumption of individual components, acquired by the information or rating of each load in the load composition. This approach needs three different datasets: (i) individual component load model, (ii) percentage of each component's load consumption, and (iii) share of the load contribution from each load class—residential, commercial, and industrial. The individual component model parameters are obtained from experiments [34–37]. Figure 2 has shown different loads classification is based on identification and control.

The above-discussed techniques and classifications are the key structures for the smart loads. Smart loads are energy-efficient sensor-based controllable load infrastructures that have real-time access to energy usage data. Smart houses control the appliances according to users' preferences, using the intelligence of the appliances to enable the consumer to use real-time energy budgeting to manage in any given day, which allows smart loads to tune the consumer's energy consumption to their daily lifestyle consumption [38,39].

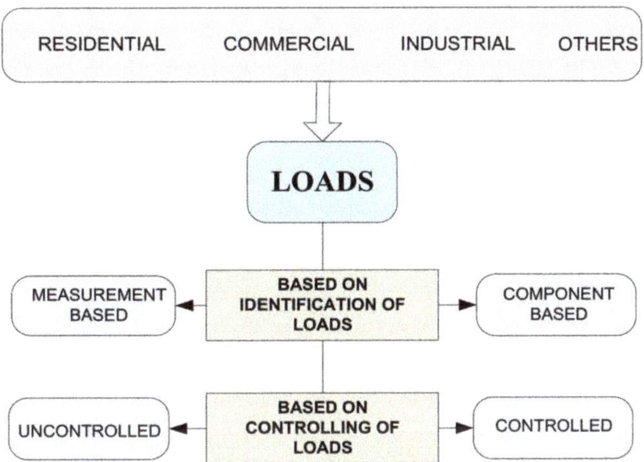

Figure 2. Loads classification is based on identification and control.

2.1.4. Integration of Electric Vehicles

Increased pollution led the world to move away from conventional fossil-powered vehicles to electric vehicles. Electric vehicles have untapped potential in both environmental and energy applications. A few of the applications of the electric vehicle are the vehicle-to-grid (V2G), vehicle-to-vehicle (V2V) supply of power [40,41]. The connection of EV connected to the grid through the charging station is shown in Figure 3.

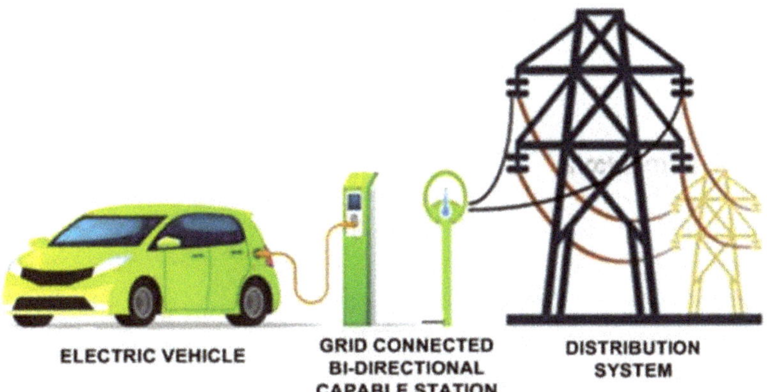

Figure 3. Electric vehicle connected to a charging station.

V2G is a process where an electric-powered vehicle supplies power to the regional local grid to meet the demand during peak demand or participate in the energy market by reducing the overall cost of bidding during peak rise in the price of power. This requires communication with the power grid to return the electricity or by controlling the charging rate which enables the EV to support the renewable energy sources from fluctuating, as they cannot be governed [42]. A few of the EVs that support the V2G are battery electric vehicles (BEV), plug-in hybrid vehicles (PHEV), and fuel cell electric vehicles (FCEV). When the electric car batteries are not in use, they can be used to provide electricity to the grid or to charge other storage devices. With an estimated increase in usage of electric vehicles in the future, it is assured to improve the storage capability to balance the demand–supply of the MG. Thus, it provides improved performance in the stability and reliability of the system.

2.2. Classifications of Microgrids

A microgrid is generally connected to the grid at the point of common coupling (PCC) through STS (static transfer switch), where voltage and frequency stability is managed by the power grid. When disturbance or failure in the grid occurs, MG maintains the system stability by isolating itself from the main power grid, forming an islanded condition. The renewable energy sources (solar, hydro, wind, and bio), which are not continuous, are connected through power electronic converters (PEC) for good power quality of output; these converters provide a resilient, reliable, continuous, and efficient power supply [43,44]. By the nature of the output obtained, MGs are classified into AC source microgrid, DC source microgrid, and (AC/DC) hybrid microgrid.

An AC microgrid is a common topology of its flexible voltage level transmission using transformers. An AC supply bus is introduced where all DERs, either with DC or AC sources, are connected using PECs to AC loads [45,46]. Almost all the loads in the power system are of AC nature; AC-MG is most sorted. Figure 4 represents a structure of an AC microgrid.

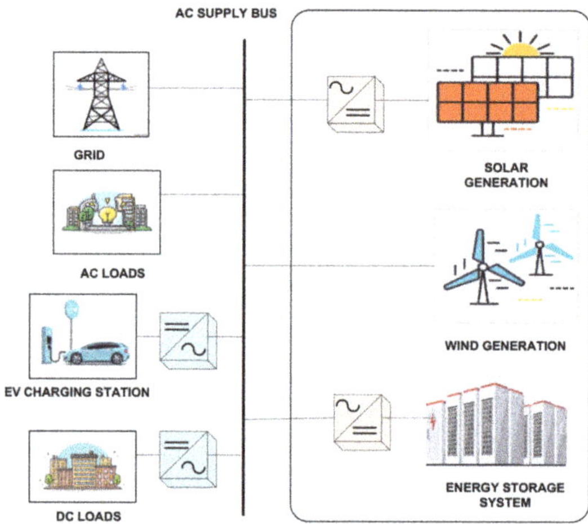

Figure 4. Structure of an AC microgrid.

In the DC-MG network, a DC bus connects both AC and DC sources from where the output is taken by the loads [47]. The concept to supply the DC supply is to reduce the number of PEC used, as the DC sources are more available compared to AC sources, which also eliminates the possibility of harmonics due to PEC, as it is not present in DC supply [48]. Increasing popularity in the usage of DC sources like mobiles, laptops, and also household items for isolated places instigated the DC-MG into existence. Figure 5 represents a typical DC microgrid structure.

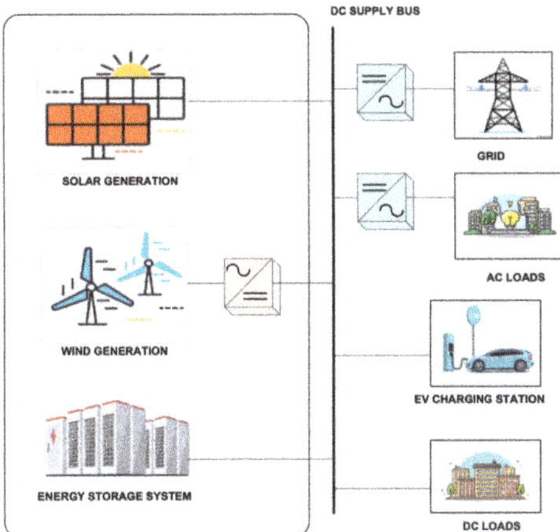

Figure 5. Structure of a DC microgrid.

An AC/DC hybrid MG is proposed to effectively introduce both AC and DC sources and consumers in a system. AC sources and DC sources are connected to their respective buses where the outputs are given to the consumers accordingly [49]. The idea of AC/DC hybrid MG is to simultaneously use the supply from both DC and AC sources and thereby reduce the overall power consumption [50]. This is possible by the PEC at both supply buses that support the bi-directional exchange of power from source end to load and vis-à-vis. Figure 6 represents a hybrid microgrid.

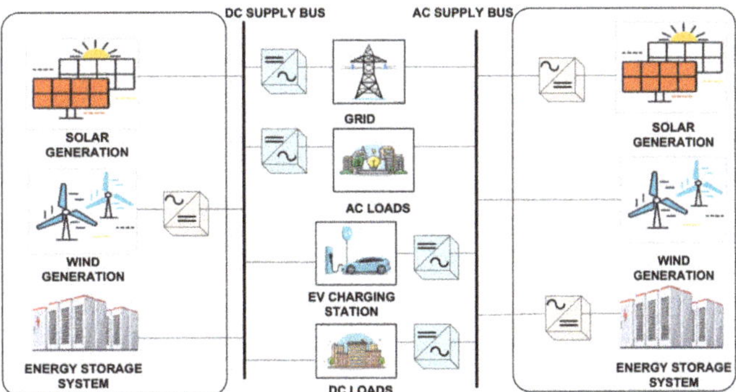

Figure 6. Structure of a hybrid microgrid.

2.3. Control Structure of a Microgrid

As a small-scale electrical distribution network, an MG has many variables and constraints to control. An energy management system plans, supervises, and manages the system's supply–demand balance while assuring dependable, cost-effective, and efficient operation [51–53]. The management of a microgrid needs to deal with different technical and economical areas, timescales, and infrastructure levels, which requires a control structure to operate the variables. One such control structure for the microgrid is the hierarchical control scheme, which is a generally accepted standardized solution [54].

The hierarchical control structure consists of three different levels operating with individual operating time, data inputs, and control equipment. The different levels in hierarchical control schemes are: (i) primary level, which supervises the control of the DER units; (ii) secondary level, which is responsible for the voltage and frequency modification of the system in coordination with the primary level; (iii) tertiary level, which is the core control of the system like demand–supply management, storage management, renewable integration, power flow control, optimization of parameters, and control strategies. The tertiary level can also be termed as the energy management system [55]. Figure 7 shows a typical hierarchal control of a MG.

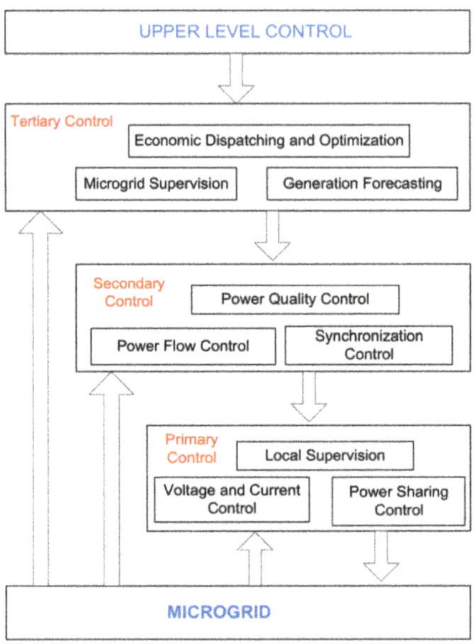

Figure 7. Hierarchal control of a microgrid.

2.4. Communication of the Microgrid

Communication is an important tool that converts the conventional power network into an intelligent system, connecting generation, transmission, distribution, and utilization systems to the central management center to maintain stability by processing the real-time data. There are several wires and wireless technologies available in the market but the selection of technologies depends on features like data rate, latency, coverage area, reliability, and consumption of power [56]. Table 3 presents various communication technologies used in microgrid. Communication equipment could increase the MG implementation cost with an increased number of communication devices like the repeater and routers for feasible and fast co-collection of data in an area. Increase data collection by the sensors and monitors in the smart homes and smart cities to compensate for the cost and the dire need to reduce the communication infrastructure while maintaining the reliable operation [57,58]. With recent trends in MG's integration and to incorporate internet of things (IoT) devices for measuring, it is better to consider wireless communication technology for its wider applications [59].

Table 3. Communication technologies in a microgrid.

Technology	Spectrum	Data Rate	Range
GSM	900–1800 MHz	14.4 Kb/s	1–10 km
GPRS	900–1800 MHz	170 Kb/s	1–10 km
3G	1.92–1.98 GHz	2 Mb/s	1–20 km
4G	2.11–2.6 GHz	100 Mb/s	1–10 km
5G	3–90 GHz	10 Gb/s	>1 km
WiMAX	2.5–5.8 GHz	75 Mb/s	10–50 km
PLC	1–30 MHz	2–3 Mb/s	1–3 km
Zigbee	800 MHz–2.4 GHz	250 Kb/s	30–50 m
Bluetooth	2.4–2.483-GHz	2.1 Mb/s	0.1–1 km

3. Energy Management System Control Structure

3.1. Structure of EMS

According to the International Electro-Technical Commission (IEC) standard application program about power systems, IEC-61,970 defines an energy management system as a "computer system comprising a software platform providing basic support services and a set of applications providing the functionality needed for the effective operation of electrical generation and transmission facilities to assure adequate security of energy supply at minimum cost" [60].

Different operations of EMS are data analytics, forecasting, optimization, and human–machine interface (HMI), and network reconfiguration for real-time interface with the EMS. Figure 8 shows the structure of the EMS of an MG.

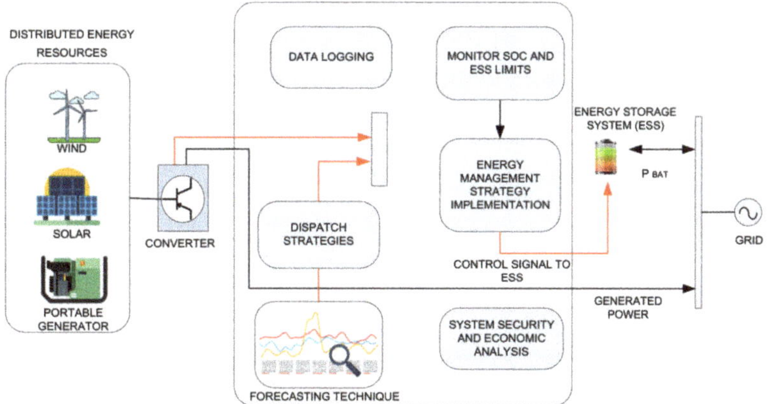

Figure 8. Structure of energy management system.

Energy management in microgrids is a complex automated system that is aimed at optimal scheduling of available resources (CG, DGs, ESS) to meet the day-to-day demand while considering the meteorological data and market price. There are three control approaches in energy management of the microgrid which are: (i) centralized, (ii) decentralized, and (iii) distributed.

The centralized control is at the core of the control in this method MGCC, which collects the information from the local controllers and analyzes it to control the system actions [61]. This process requires end-to-end communication between all local controllers to the central controller. Different EMS structures are shown in Figure 9.

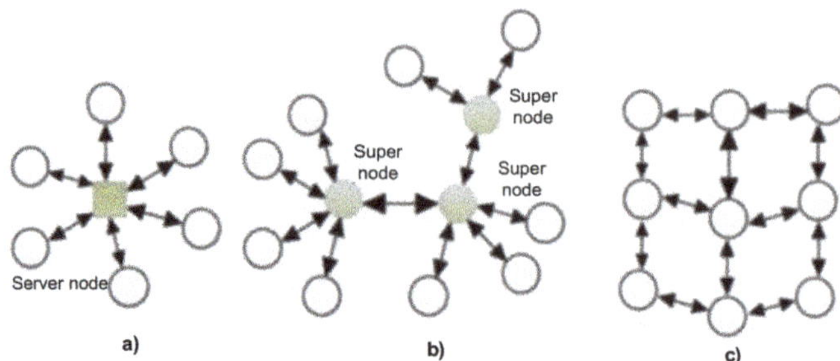

Figure 9. Types of EMS control: (**a**) centralized, (**b**) decentralized, (**c**) distributed.

With an increase in the geographical area, the system control in centralized mode becomes difficult due to the delay or lag in the communication, which leads to delay control. This process is not feasible as well as not economical; hence, we choose the decentralized mode of control. In decentralized control, each unit has its own local controller that works in an autonomous state where it receives the voltage and frequency data [62]. Here, the decentralized control does not provide the all the information to the other local controllers, but rather exchanges the global information to make the decisions of the overall system. The exchange of information is allowed in a few controllers to take action spontaneously in a state of emergency. A third approach, obtained with a combination of the above two control approaches, is the distributed control [63]. This mode of control scheme provides control to both centralized as well to decentralized property up to a certain degree of control. In this control scheme, each local controller unit uses the local information like voltage and frequency from the neighbors, which helps to obtain a global solution by the central controller while using the two-way communication link by the local controllers. Characteristics of different types of controls in the energy management system are presented in Table 4.

Table 4. Characteristics of different types of controls in the energy management system.

	Centralized	Decentralized	Distributed
Information Accessed	Microgrids pass information to the central controller	Independent control is provided with data from the other local controllers	Interoperability and data exchange between every device
Communication Information	Synchronized information from the device to the central controller	Information among local controllers is asynchronized	Communication is both locally and globally asynchronized
Function in real-time	Complex	Acceptable	Easy
Feature of Plug and play	The central controller needs to be instructed	Can be accessed by central controller	Available by the peers
Expenditure	More	Less	Less
Structure of Grid	Centrally controlled	Locally controlled	Both centrally and locally controlled
Tolerance during fault	Less tolerance capability	One router fault—tolerated N router fault—expensive	N router fault—tolerated, Possible self-healing feature
Infrastructure	Needs suggestion integrating DERs	Integration is modular and possible	No change while integration
Size (Number of nodes)	Less	IPv4-2^{12} IPv6-2^{128}	$>2^{128}$
Final Nodes	No identification	Unique identification IP	Global unique identifier
Operation Flexibility	Very less	Available	Very much needed
Bandwidth & Latencies	Low and high	Both are great	High and low
QoS	Not allowed	Allowed	Inherent
Connectivity	EPA (Physical)	TCP/IP (Physical)	TCP/IP (Virtual)
Safety measures	Less	Available	High
Individuality	No	No	Possible

3.2. Data Handling in EMS

Data handling and clustering are the prominent steps towards system management, as many intelligent measuring and sensing devices have been integrated with the MG, which generates a large amount of data per unit (hour, minutes, or seconds). The complex structure of MG system requires it to be equipped with different sensors and monitoring equipment which bring varied kinds of data, like structured data from the conventional power system, semi-structured data from the system like images (camera), unstructured data from meteorological data, network structure, and maps [64]. Figure 10 shows different data types available in EMS.

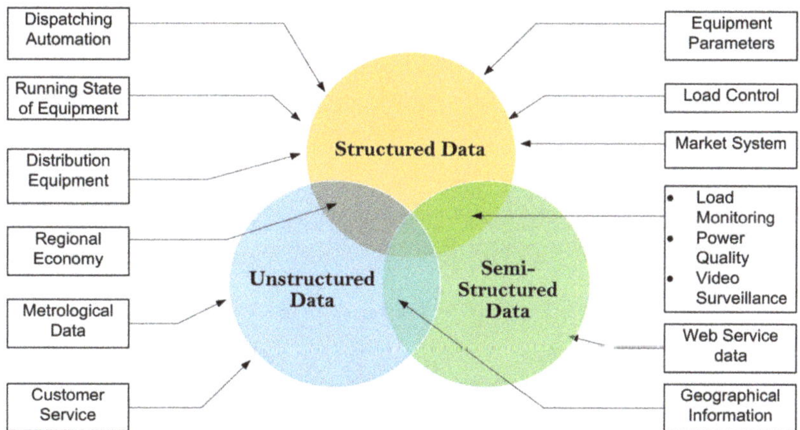

Figure 10. Data availability in a microgrid.

Usage of a wide variety of applications of communication and network has improved the speed of the data generated from the units while applications like big data are used to access the information [65]. Intelligent networks help in unfolding the unknown patterns from the data collected. Analytical software technologies like Hadoop, HBase, and Storm are used as data centers to support the vast collection of the data in a structured format by the sensors and the other measuring devices such as smart meters.

3.3. Network Reconfiguration

Network reconfiguration is an optimization problem that identifies the optimal radial topology of the distribution network based on all topologies. Network reconfiguration is generally carried out with the aim to reduce the power loss, harmonize voltage profile, and unify network loading through a multi-objective framework. The multi-objective optimal solution problem uses deterministic and stochastic methods for reconfiguration. Much work on reconfiguration is presented using the meta-heuristic method in distribution systems considering radial topologies by interchanging of tie lines [66,67].

3.4. Forecasting in EMS

EMS proceeds with data available towards analyzing different forecasting parameters like electricity price market, energy purchase, weather, demand response management, and financial planning using forecasting techniques.

Forecasting is a prominent part of energy management, which is classified in different categories concerning the period of forecast required [68]. These classifications are: (i) very short-term (seconds–$\frac{1}{2}$ h), which is used for the dynamic control of renewable energy sources according to the load requirements; (ii) short-term ($\frac{1}{2}$–6 h), which is used for energy scheduling among the sources and the storage devices; (iii) medium-term (6 h–1 day), which is used for market pricing; and (iv) long-term (1 day–1 week), which is used in load dispatch and maintenance [69]. Figure 11 shows types of forecasting techniques available in EMS of microgrid.

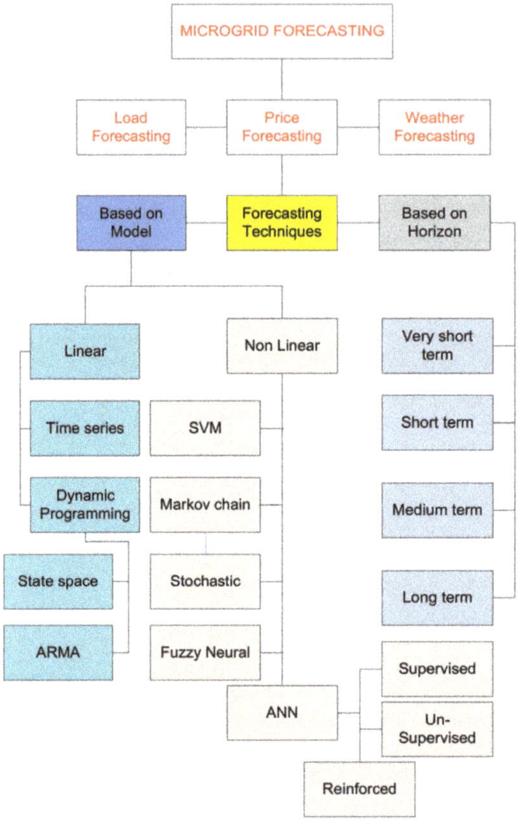

Figure 11. Forecasting techniques in EMS.

3.5. Demand Management in Microgrid

Load balance acts as a constraint between generation and demand. Load demand balance problems can be categorized in two ways: the supply-side and the demand-side [70]. Supply-side balance can be obtained by using the hierarchical control scheme for the economic scheduling for consumption by the end-users. Load control can be categorized as: (i) controllable loads, which are the loads that are managed according to the price, and (ii) shiftable loads, also known as deferrable loads, such as charging of electric vehicles, washing machines, dryers, which can provide scheduling flexibility for demand response.

The demand-side balance needs to be carefully accessed by modeling the generation in renewable energy, i.e., by forecasting for the supply to the users in the system. Demand-side control is sub-categorized into direct load control (or the demand side management) and price-based load control (or the demand response). Demand-side control is performed by the central controller by the consumer agreement to mainstream the economic agenda. In the price-based load control, the consumer is provided with options to choose their energy consumption according to the market price available. Figure 12 shows different supply and demand classification in EMS.

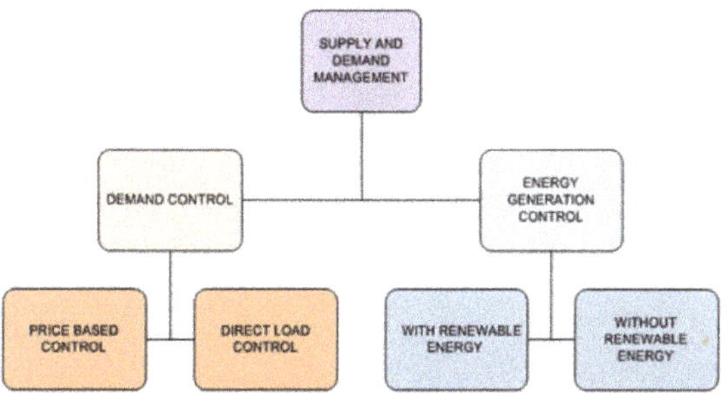

Figure 12. Supply and demand management classification.

4. Numerical Methodologies of EMS

Different EMS techniques are differentiated according to the numerical methods used for controlling the energy management system. These methods are broadly classified into three categories: (i) classical methods, (ii) metaheuristic methods, and (iii) intelligent methods.

4.1. Classical Methods

Classical methods are the mathematical programming or classical programming methods that choose certain variables to maximize or minimize a given function subject to a given set of constraints. Branch and bound are the classic components that are used for solving the classical method approach to find the optimal solution in an iterative process without integer constraints. Classical methods use both linear and nonlinear optimization models to solve the problem. The classical methods are divided into certainty- and uncertainty-constrained problems.

Under certainty linear programming (LP) are mixed integer programming (MIP) and nonlinear programming (NLP). A combination is mixed-integer non-linear (MINLP) and mixed-integer linear programming (MILP) [71–73]. Uncertainty constraints are decision theory (rule-based and deterministic-based), where the output of the model is fully determined by the parameter values and the initial values; whereas probabilistic (or stochastic) models incorporate the randomness in their approach such as dynamic programming (DP) and stochastic programming (SP) [74]. An optimization algorithm is an algorithm that uses the physical deterministic method of solving the solution without any random nature being known as deterministic. Table 5 shows a review of MG EMS by classical methods.

Table 5. A review on classical mathematical programming methods used in EMS.

Ref No.	Method	Power Sources	Ev	Dr	Grid/Island	Ems	Remarks
[75]	MILP-LP	PV, BT, FC			G/I	C	A mixed-mode of EMS is proposed with ON/OFF and continuous run mode.
[76]	MILP	PV, WT, BT		*	I	C	Cost reduced by reducing the ESS with advantageous demand response (DR) determination.
[77]	MILP	PV, BT, DE		*	G/I	C	EMS proposed to minimize the fuel cost while optimizing the diesel generators and battery sizing using a piecewise linear function.
[78]	MILP	PV, WT	*		I	C	Optimizing the day-to-day energy scheduling with DR and EVs using multiobjective constraints.

Table 5. *Cont.*

Ref No.	Method	Power Sources	Ev	Dr	Grid/Island	Ems	Remarks
[79]	MILP	PV, WT, DE, MT, FC, BT		*	G/I	C	EMS is modeled to optimize while determining the capital cost, cost of the fuel, energy cost, and penalization for emission. Energy sources and storage are considered in economical dispatch for techno-economic analysis.
[80]	MILP	PV, BT			G/I	C	A three-phase EMS model is proposed with load shedding considering outage constraints.
[81]	MINLP	PV, WT, MT, FC, BT			I	C	EMS is developed for a three-phase system to minimize the fuel, startup, and shutdown expenditure.
[82]	MINLP	PV, BT			G/I	C	Stable operation of hybrid MG with clean water supply while reducing the overall daily operating costs.
[83]	NLP	PV, FW, MT, FC, BT			G/I	C	Energy market operational cost and its profit are determined by the MG management application.
[84]	NLP	PV, FC, BT	*		G/I	C	Maximization of the cost benefiting charge–discharge scheduling of the battery considering the customers' load shifting events.
[85]	DP	DE, BT			G/I	C	EMS is modeled to optimize the operational cost of the conventional grids considering the penalty cost. Computational time is reduced using Pontryagin's Principle.
[86]	DP	WT, DE, BT			G/I	C	Minimization of the total cost of operations by scheduling the available units while predicting the wind speed by short-term forecasting and determining the real-time pricing.
[87]	Approx. DP	WT, BT			G/I	C	Optimization of the MG is proposed considering the cost function of the unit commitment and economic dispatch operations along with daily energy scheduling.
[88]	Rule based	PV, UC, MT, BT			I	C	To perform power scheduling with day-ahead forecasting for the conventional, PV generators and gas turbine are used in a deterministic power optimization.
[89]	Rule-based	PV, WT, BT			G/I	C	To configure switches operation to model different configurations considering SOC of the battery and load imbalance.
[90]	Rule based	PV, UC, BT			G	C	An energy management strategy with PV generator and SOC-based battery hierarchical structure for electricity regulation and continuous operation of the microgrid.
[91]	Deterministic based	PV, WT, MT, BT		*	G/I	C	Proposed to minimize the overall running cost of the system by reducing the industrial loads, considering TOU rate of demand response programs executed.
[92]	NP-Hard	PV, BT		*	G/I	C	Proposes polynomial–time algorithms for approximating optimal solutions and robust supplier networks of group energy communities in terms of a black start while minimizing the operational costs.

PV—Photo voltaic; WT—Wind Turbine; MT—Micro Turbine; FW—Flywheel; DE—Diesel; FC—Fuel Cell; UC—Ultra Capacitor; G—Grid; I—Islanded; C—Centralized, DC—Decentralized, DT—Distributed, *—Availability.

4.2. Metaheuristic Methods in EMS

A metaheuristic is a branch of random search and generation algorithms. These algorithms select a path through a search algorithm such as a heuristic (random) to find an optimal solution in an optimization problem with or without constraints. Metaheuristic algorithms perform computation when incomplete data or limited capacity are provided [93]; the sample set of random values are considered and explored for an optimal solution. Metaheuristic approaches use a separate search strategy to generate a random selection or assumption of the problem variables, which can be advantageous in a variety of situations.

An optimal solution can be found in the distinct search space as used in combinatorial optimization. Metaheuristic method is an iterative method that is unlikely to guarantee a global optimum solution due to its convergence properties. This can be compensated with finding the mean of the solutions; the use of Monte Carlo simulation improves the convergence of the solution. Stochastic implementation of optimization is dependent on the random variables created [94]. The metaheuristic approach works on two concepts, namely intensification and diversification. Intensification is searching a local area to find an optimal solution when we know that solution could be found in the prescribed region. The diversification process is searching the space on a global scale with no limits in the search pattern using the randomly generated variables, while randomization increases the diversity of solution when the search space exceeds the local optima. To find the global optimal or the best solution, both the intensification and diversification processes need to be in proper balance, which increases the rate of convergence in the algorithm [95–99]. A few metaheuristic algorithms are particle swarm optimization (PSO), genetic algorithm (GA), modified PSO (MOPSO), non dominated sorting genetic algorithm II (NSGA-II), enhanced velocity differential evolutionary PSO (EVDEPSO), priority PSO, multi-voxel pattern analysis (MVPA), grey wolf optimization (GWO), artificial bee colony (ABC), adaptive differential evaluation (ADE), crow search algorithm (CSA), rule-based bat optimization (BO), gravitational search algorithm (GSA), alternating direction method of multipliers (ADMM) using modified firefly algorithm (MFA), teaching–learning optimization (TLA), social spider algorithm (SSO), and whale optimization algorithm (WOA). Table 6 provides a critical review of the metaheuristic methods used in EMS.

Table 6. A review of metaheuristic methods used in EMS.

Ref No	Method	Power Sources	Ev	Dr	Grid/Island	Ems	Remarks
[100]	NSGA-II	PV, WT, BT			G/I	C	A multi-objective optimization problem is proposed to maximize the economy. Intelligent power marketing is adapted to improve the economic dispatch of the microgrid.
[101]	NSGA-II	PV, WT, BT	*		G/I	C	This paper establishes an integral objective function considering the demand response and user satisfaction constraints, which has an effect on the economy and operation of the system with the DR strategy.
[102]	PSO	PV, MT, BT, TES			G/I	C	An optimal energy planning is proposed for the recently modeled energy hub. An efficient microgrid structure is discussed along with technical and economic prospects with optimization.
[103]	CVCPSO	PV, WT, DE	*		G/I	C	Minimizing the operating costs while maximizing the utility benefit using the CVCPSO algorithm, which yielded the Pareto-optimal set for each objective, and the fuzzy-clustering technique was adopted to find the best compromise solution.

Table 6. *Cont.*

Ref No	Method	Power Sources	Ev	Dr	Grid/Island	Ems	Remarks
[104]	MPVA	PV, WT, MT, BT			G/I	C	A sports metaheuristic algorithm to minimize the overall running cost of MG while studying four different MG scenarios.
[105]	GWO	PV, WT			G/I	C	A sine cosine optimizer is used to optimally participate in the trading of energy, i.e., selling or buying the power while bringing the capital cost of the microgrid.
[106]	ABC	PV, WT, DE, BT, FC	*		G/I	C	An EMS application of the V2G economic dispatch problem is optimized in the MG while converting the multi-objective problem to a single objective using the judgment matrix methodology.
[107]	EBC	PV, WT, MT, BT			G/I	C	Different TOUs are evaluated to minimize MG operational costs and to analyze the efficiency of a typical distribution system, considering all relevant technical constraints.
[108]	ADE	DG, BT			G	C	An ADE-based optimization is proposed for the DC microgrid modeling the active power sources under real-time pricing to minimize the total operating cost.
[109]	MOPSO	PV, MT, BT, TES		*	G/I	C	EMS application is proposed to reduce the carbon dioxide emissions and payback period of the microgrid structure.
[110]	EVDEPSO	PV, BT	*	*	G/I	C	A day-ahead planning schedule is determined to improve the energy market trading while managing the resources available. Includes the electric vehicles participating in the energy market, G2V and V2G.
[111]	Rule base BO	PV, WT, MT, FC, BT		*	G	C	A bat algorithm is used to optimize the MG operation by forecasting the load power and uncertainties in RES using probabilistic methods. The weight factors are taken for tuning.
[112]	CSA	PV, FC, DE, HY		*	G/I	C	The Pareto front is considered to investigate the operating cost, solar power uncertainty, carbon emission, and the cost of the parameters. Hydrogen fuel is considered in reducing operating costs.
[113]	GSA	PV, WT, BT	*	*	G/I	C	Optimization of the overall cost considering the carbon emission and weekly generation scheduling for the small dispatchable systems.
[114]	ADMM-MFA	PV, WT, MT, FC, BT		*	G/I	C	EMS is modeled for the MG to optimize the electricity price by considering the load profile, PV irradiance, and market prices with certain constraints.
[115]	TLA	PV, WT, MT, FC	*	*	G/I	C	Hybrid MG reducing the operating cost considering thermal power recovery and hydrogen generation; V2G technology helps to convert the PEVs into active storage.
[116]	SSO	PV, WT, DE, FC		*	I	C	Optimal sizing of the renewable energy sources with conventional sources to minimize the cost of energy (COE) and power loss supply probability while analyzing the reliability.
[117]	WOA	PV, WT, DE, BT		*	I	C	EMS is proposed to optimize the load demand of the MG by minimizing the operating cost with improved reliability of the power.

PV—Photo voltaic; WT—Wind Turbine; MT—Micro Turbine; TES—Thermal energy storage; DE—Diesel; FC—Fuel Cell; HY—Hydro; C—Centralized, DC—Decentralized, DT—Distributed, *—Availability.

4.3. Intelligent Methods in EMS

4.3.1. Fuzzy Control and Neural Networks

For the computation of a large amount of numerical processing data like signals or images, fuzzy logic systems and neural networks (NN) are used. These methods are computational nonlinear algorithms with the flexibility to use a range from small software programs to large hardware systems. Through continuous decision-making by the system, learning takes place and the knowledge acquired is stored in as weights. These weights are the internal parameters of knowledge. A fuzzy logic system, when used to control a system through a set of rules considering the constraints, is known as fuzzy logic control (FLC). Applications of FLC are used to improve battery state of charge (SOC), smooth voltage profile, and grid-to-vehicle (G2V) charge transfer [118].

Neuro-fuzzy is a combination of fuzzy approach and neural network, where fuzzy inference system (FIS) is adjusted by the data provided to NN learning rules. Improved speed, accuracy, and strong learning skills along with simple execution are the advantages of this approach [119].

A neural network is an interconnection of neurons, when used in a physical system to control using different layers of connection, which are also known as an artificial neural network (ANN). These artificial NN are used for adaptive control and model predictive analysis. Applications of ANN are provided with training via a dataset. From experience or the outputs of the model, self-learning takes place. Using ANN for MG, EMS can perform complex operations such as forecasting DR and control of MG [120].

Recurrent neural network (RNN) is a classification in ANN which allows it to provide temporal dynamic behavior and the structure of RNN connects the temporal sequence through the graph between the nodes. Similar to feed forward neural networks or ANN, which process variable-length sequences using internal memory, RNN has an internal state memory to process the sequence of inputs using short-term memory (STM) or long short-term memory (LSTM) for predictions of energy and economy. A review on fuzzy and ANN-based applications in EMS are described in Table 7.

Table 7. A review on fuzzy and ANN-based applications in EMS.

Ref No.	Method	Power Sources	Ev	Dr	Grid/Island	Ems	Remarks
[121]	Fuzzy logic	PV, WT, BT		*	G/I	C	EMS for distributed generations DGs in AC MG. An adaptive neuro-fuzzy inference system (ANFIS) is developed to manage the available energy in ACMG.
[122]	Fuzzy	PV, WT, FC, BT			I	C	The system is controlled by a low complexity fuzzy system, with only 25 base rules which give better results in terms of control and energy-saving efficiency, that has been improved.
[123]	Fuzzy logic	PV, WT, DE, BT	*	*	G/I	C	Studies different fuzzy techniques for the charging/discharging of the electric vehicle while ensuring the optimal demand management from the vehicle-to-grid (V2G).
[124]	Fuzzy	PV, FC, BT		*	G/I	C	EMS is developed to manage the operating conditions with economic constraints. Operations of grid ON/OFF connections are also discussed using the fuzzy logic controller and a predictive controller.

Table 7. *Cont.*

Ref No.	Method	Power Sources	Ev	Dr	Grid/Island	Ems	Remarks
[125]	FLC	PV, BT		*	G/I	C	A fuzzy logic-based energy management system is developed to minimize the power-sharing error between renewable energy sources and demand.
[126]	Neuro-fuzzy	PV, WT, MT, FC, BT	*		G/I	C	A neuro-fuzzy Laguerre wavelet control (FRNF-Lag-WC) architecture scheme is validated for various stability, quality, and reliability factors obtained through a simulation testbed implemented.
[127]	Neuro-fuzzy	PV, FC, BT		*	I	C	A battery cycle is improved by reducing the charging/discharging period and ensuring optimal power-sharing in the microgrid.
[128]	RNN.	PV, BT		*	I	C	A control strategy is developed to maximize consumption and minimize electricity pricing by using an LSTM forecasting method for supply–demand management.
[129]	ANN	PV, WT, DE, BT		*	I	C	A real-time scheduling problem is developed for an MG with a finite horizon model using the ADP approach. The ADP approach is modeled using the RNN technique.
[130]	RNN	PV, BT		*	I	C	Discussed many algorithms for scheduling including the maximum time lap scheduling and day-ahead forecasting for a building of its energy consumption with PV installation.
[131]	ANN	PV, WT, MT, DE, BT		*	G	C	EMS application to optimize the economic dispatch and to minimize the operating cost in a hybrid microgrid using Lagrange programming neural network.

PV—Photo voltaic; WT—Wind Turbine; MT—Micro Turbine; FW—Flywheel; DE—Diesel; FC—Fuel Cell; UC—Ultra Capacitor; C—Centralized, DC—Decentralized, DT—Distributed, *—Availability.

4.3.2. Model Predictive and Multi-Agent EMS

Model predictive control (MPC) is an algorithm that regulates or controls the system based on the moving or rolling horizon approach as specified in Unnikrishnan et al. [129]. The role of MPC is to make the system less sensitive to the variables and control the physical process. MPC can be performed online with uncertainty constraints. In online methods, the current system parameter and forecasted parameters help in updating the decision variables at any instant [132,133]. The optimum solution could be obtained by updating decision variables with current system parameters with ease, and gets complex with an increase in variables. Hence, it is used in smaller systems.

In the multi-agent system (MAS), the objectives of the system are obtained by intelligent agents communicating with other nearby agents while participating to form a configuration. MAS is an online/offline approach used in MG applications as shown in [134]; this approach is utilized in the control of EMS, optimization, and managing of the energy market. In [135,136], the application of MAS is used to control the architecture of the MG while using optimization techniques for the configuration of renewable resources. Table 8 presents the review on MPC and MAS based on EMS.

Table 8. A review on MPC and MAS based on EMS.

Ref No.	Method	Power Sources	Ev	Dr	Grid/Island	Ems	Remarks
[137]	MPC	PV, FC, SC, DE, BT		*	I	DT	Energy scheduling is proposed using the MPC to optimize the dwell time of the high SoC state of the battery and to smoothen the set point deviation of the fuel cell for regenerative capability. Compared with fuzzy-based heuristic in generation and load demand.
[138]	MPC	PV, WT, BT		*	G/I	DT	MPC-based decision-making is developed by the optimization algorithm for participation in the grid electricity market with excess generation to support ancillary services of the main grid.
[139]	MPC	PV, BT		*	G/I	DT	A real-time microgrid from Athens is developed in the laboratory to study the day-ahead market and the control management of the energy profile with the energy market. User interface with the market interactions is performed for an enhanced microgrid.
[140]	Adaptive MPC	PV, DE, BT		*	I	DC	An EMS application is developed to optimize the cost function of the fuel in a diesel generator for economic dispatch using the Lagrange multiplier and lambda iteration method with battery operation constraints.
[141]	MPC	PV, BT	*	*	G/I	DT	An MPC-based control strategy is developed to sell or store the excess generated power from the solar panels while managing the overall conditions like heating, ventilation, air conditioning system, time of use pricing, and to reduce economic constraints.
[142]	MPC	PV, BT	*		G	DC	By installing an ESS at the end of the feeder, the capacity of PVs and EV connected to the bus are extended up to twice the capacity of the main power source.
[143]	MPC	PV, WT, DE, BT		*	G/I	DC	A proximate scenario is taken by the optimizer at each step, and the optimal supply of system capacity is accessed based on the scenario selected and the possible variations in the future.
[144]	MAS	PV, WT, MT, FC, BT, DE		*	I	DC	MAS-based agent optimization is developed to optimize the operation of the distribution system with DG in energy scheduling and generation. EMS is performed for the system by considering the constraints, such as generation cost and emission of carbon.
[145]	MAS	PV, BT	*	*	I	DC	A MAS-based two-stage energy management system is developed using the Kantorovich method for the energy generation scenario considering the self-healing strategy by the decentralized restoration technique and coordinated management.
[146]	MAS-CNN	PV, WT, DE, BT		*	G/I	DC	MAS-based energy management is proposed for the generation management of the PV, wind, and load. Balancing is maintained using the CNN (convolution neural network)-based load forecasting technique for the load demand.
[147]	MAS	PV, DE, BT	*	*	I	DC	This paper proposes a MAS-based intelligent energy management system to operate a hybrid microgrid in islanding mode while effectively minimizing the peak demand of the system using the V2G and LED savings.
[148]	MAS	PV, WT, FC, BT			G/I	DC	This paper proposes a communication rule for sharing the local information of the agents and getting access to the global information was based on an average consensus algorithm (ACA), and a restoration decisions strategy based on the discovered global information was developed.
[149]	MAS-RL	PV, WT, BT		*	G/I	DC	A multi-agent-based EMS is developed to manage the objectives of the system. Reinforced learning is imbibed with MAS to improve the decision-making capability by learning using the sets for the participation in the energy trade marketing.
[150]	MAS	PV, WT, BT		*	G/I	DC	Experimental results show the ability of the proposed multiagent T-Cell-based RT-EMS in maintaining the stability and smooth operation of the MG with modularity and fault tolerance features implemented through the MAS JADE platform.

PV—Photo voltaic; WT—Wind Turbine; MT—Micro Turbine; DE—Diesel; FC—Fuel Cell; SC—Supercapacitor; G—Grid; I—Islanded; C—Centralized, DC—Decentralized, DT—Distributed, *—Availability.

4.3.3. Game Theory and Deep Learning

Deep reinforcement learning (DRL) is an intelligent algorithm approach to solve complex problems like decision-making through training or learning. It is a combination of reinforced learning (RL) and deep learning (DL) where agents perform the decision-making task to a wide variety of applications. DRL is a sub-category of intelligent machine learning, which is also a part of artificial intelligence where a system learns from the actions it performs as a human learning experience [151,152]. The agent learns by a reward and penalty system on their decision policy.

Game theory (GT) brings multiple decision variables to interact using a mathematical model to analyze the environment. The objectives of the problem are achieved by introducing each strategic decision-making variable to participate in the game. Nash equilibrium is a prominent solution concept for game theory, where the actions of other players are set to constant while there is no change to the unilateral strategy by any player to change their revenue strategy. Thereby, it is possible to arrive at an optimum mutual response from all the players [153]. To find the optimal solution in the non-cooperative game theory when there is evidence that no leader–follower relationship is found, Nash equilibrium strategy is used to improve the utility parameter by making every player compete against each other. A review on game theory and deep reinforced learning in EMS has been presented in Table 9.

Table 9. A review on game theory and deep reinforced learning in EMS.

Ref No.	Method	Power Sources	Ev	Dr	Grid/Island	Ems	Remarks
[154]	DRL	WT, DE, BT		*	I	DC	An EMS is proposed for energy storage management and load shedding management with dual control policy to manage the utility of the system dual control to improve resilience. The dual controls are the energy storage and load shedding policies.
[155]	DRL	BT	*		G	DC	EMS is developed to manage fuel efficiency compared to the rule-based approach. The EMS developed makes decisions by itself from the actions of the states.
[156]	DRL	PV, WT, BT		*	I	DC	DRL-based energy management is proposed to minimize the operating cost and to improve the economic performance of the islanded microgrid by controlling the energy reserve.
[157]	DRL	PV, WT, MT, FC, BT		*	G/I	DC	An EMS is modeled with DRL and the Markov decision process (MDP) strategy to satisfy the objective function, i.e., by minimizing the overall operating cost of the MG system.
[158]	RL	WT, BT		*	G/I	C	An EMS application for the consumer-based intelligent method is developed for the consumer to explore and control the stochastic nature of the generation and load actions.
[159]	DRL	PV, WT, MT, FC, BT			G/I	DC	Paper proposes a scheduled strategy to minimize the daily operating cost of the MG using DRL architecture for addressing the problem of operating an electricity MG in a stochastic environment.
[160]	Game Theory	PV, WT			G	C	A game-theory-based EMS is modeled to minimize the utilization cost of the system using the coalition theory, the EMS is proposed to reduce the utilization cost while improving the market profit of the sellers.
[161]	Game Theory	PV, WT, BT, HYD	*	*	G/I	DC	A Nash equilibrium-based game theory EMS is modeled for controlling the power exchange and minimizing the operating cost. An optimal operation can be achieved by maximizing the preferences of the agents using the Nash equilibrium.
[162]	Game Theory	PV, BT		*	G/I	DC	An MG-based non-cooperative game theory EMS is modeled to optimally decide the electricity price for the consumers by regulating the storage capacity of the system. A mechanism for the price regulation is developed for the modeled EMS.
[163]	Game Theory	PV, BT		*	I	DC	Optimal scheduling of the energy and storage management is proposed by the continuous non-cooperative game-theory-based energy management system by considering the energy consumption scenario to reduce the overall cost.

Table 9. *Cont.*

Ref No.	Method	Power Sources	Ev	Dr	Grid/Island	Ems	Remarks
[164]	Game Theory	PV, WT, BT	*	*	I	DC	An EMS is developed by forecasting the generation of the short-term wind power plant using big data. The optimal payment period is decreased by finding the prediction error of the MG.
[165]	Game Theory	PV, WT, FC, BT	*		G/I	DT	The paper gives cooperation between the agents as a non-cooperative or a cooperative game theory approach. Nash equilibrium is used for exploring the optimum solutions of games with energy management.

PV—Photo voltaic; WT—Wind Turbine; MT—Micro Turbine; FW—Flywheel; DE—Diesel; FC—Fuel Cell; UC—Ultra Capacitor; G—Grid; I—Islanded; C—Centralized, DC—Decentralized, DT—Distributed, *—Availability.

4.4. Problem-Based Classification

The microgrid energy management strategies are discussed in previous sections, and objectives considered in the review can be further classified into problems addressed. The review methodologies that are classified based on problems addressed are shown in Table 10.

Table 10. The problem addressed in microgrid energy management.

Problems Addressed	References
Optimal storage management	[76,98,112,123]
Demand response program	[77,92,95,151,163]
On vehicle-to-grid system (V2G)	[78,108,113,118,124]
Cost minimization	[79,81,82,84,91,93,94,99,100,105,106,110,111,127,141,151,152,161,163]
Energy schedulling	[80,87,89,90,102,104,107,109,114,115,121,136,139,142,154,156,157,160]
Operating time	[85,126,150]
Reliability of operation	[116,117,143,165]
Communication and information exchange	[129,134,140]
Based on forecasting	[119,129,138,146,162,164]
Data collection and scenario generation	[125,147,155,158,159]
Based on market participation	[83,88,101,132,137,141,146,149,153]
Time response	[96,97,128,130,131]
Stability analysis	[86,120,136,145,148,150]
Generate energy with lower emissions	[103,144]

5. Microgrid Standards

Standards are the parameters or the process which ensure the product's performance levels to satisfy the safety and quality for the implementation according to utility market requirements. The standards are developed to set a standard in the market for the safety of consumers [166,167], introducing a set of verification procedures to test the performance of the quantification and their comparison with a minimum set of requirements. Standards for microgrids are set to provide configuration, topology, and laws to control the microgrid and its integration to renewable sources. Different configurations can be implemented with microgrid blocks to perform different operations. A set of testing procedures is carried out in the distributed network operator [168] (DNO) and microgrid operator with parameters to compare their control functions. These metrics or parameters are designed to test the endurance of the system. Standards that exist for the smart grid distribution network are the Institute of Electrical and Electronics Engineers (IEEE 1547) with identification code 1547, which provides guidelines for interconnecting dispatchable sources into the electric power grid; and IEEE 2030, which provides the inter-operability guide between smart grids and microgrids [169]. International Electro-Technical Commission (IEC) is another standardization for microgrids in which IEC 62,898 provides design and implementation of the microgrid. For electric vehicles in IEEE 2030.1, IEC 61851, and ISO 15118-1 give the guidelines for electric transportation and its interconnection to the power system.

IEEE 1646 and IEC 61850-7-420 provide the standards of communication in the electric network. IEEE 2413 and IEC 61,968 give the standards for connecting IoT into the system and data exchange between devices and the network, respectively. Table 11 presents the standards for microgrid and electric vehicles.

Table 11. Standards for microgrids and electric vehicles.

Standards	Description
IEEE 1547	The standard for interconnecting distributed resources with an electric power system
IEEE 1547.1	Test procedures for equipment interconnecting distributed resources
IEEE 1547.2	Application guide for IEEE 1547 for interconnecting distributed resources
IEEE 1547.3	Monitoring, information exchange, control of distributed resources
IEEE 1547.4	Design operation and integration of distributed resources
IEEE 1547.6	Interconnecting of distributed resources for distribution system secondary networks
IEEE 1547.7	Guide to conducting distribution impact studies for distributed resources interconnection
IEEE 1547.8	The practices identified in P1547.8 should lead to the development of advanced hardware and software and help streamline their implementation acceptance, resulting in higher penetration levels of DER
IEEE 2030	Guide for smart grid interoperability
IEEE 2030.1	Guide for electric power sourced transport infrastructure
IEEE 2030.2	Guide for interoperability of energy storage systems integrated with electric power infrastructure
IEEE 2030.3	The standard for test procedures of energy storage systems integrated with electric power applications
IEEE 2030.4	Guide for control and automation installations applied to the electric power infrastructure
IEEE 2030.5	The standard for smart energy profile 2.0 application protocol
IEEE 2030.6	Guide for the benefit evaluation of electric power grid customer demand response
IEEE 2030.7	The standard for the specification of microgrid controllers
IEEE 2030.8	Standard testing of microgrid controllers
IEEE 2030.9	Recommended practices for the planning and design of the microgrid
IEEE 1646	Communication requirements in substation
IEEE 2413	The standard for an architectural framework for the Internet of Things
IEC 62898-1	Guidelines for planning and design of microgrids
IEC 62898-2	Technical requirements for operation and control of microgrids
IEC 62898-3-1	Technical requirements for the protection of microgrids
IEC 62898-3-2	Technical requirements of microgrid EMS
IEC 62898-3-3	Technical requirements of self-regulation of dispatchable loads in microgrids
IEC 62257-9-2	Recommendations for renewable energy and hybrid systems for rural electrification—Part 9-2: Integrated systems—Microgrids
IEC 61850-7-420	Communication between devices in transmission, distribution, and substation automation
IEC 61968	Data exchange between devices and networks in the power distribution domain
IEC 61851-1	Electric vehicle on-board charger EMC requirements for conductive connection to AC/DC supply
IEC 61851-23	DC electric vehicle charging station
IEC 61851-24	Digital communication between a DC EV charging station and an electric vehicle for control of DC charging
ISO 15118-1	Vehicle-to-grid communication interface—Part 1: General information and use-case definition
ISO 15118-2	Network and application protocol requirements
ISO 15118-3	Physical and data link layer requirements
ISO 15118-4	Network and application protocol conformance test
ISO 15118-8	Physical layer and data link layer requirements for wireless communication

6. Auxiliary Infrastructure

In order to make a smart distribution system operable, a complex of networks and devices needs to get together for a reliable system. IoT and smart meters technologies are the primary components to make the conventional connection between the prosumer and operator into a smart interdependent system with faster and reliable communication [170].

6.1. IoT Sensors

Advancements in wireless technology with improved sensing devices using embedded processing technology have led to the Internet of Things [171], which provides efficient monitoring, measuring, and control services.

IoT connects the physical and digital components without any mediation of the operator. The connection of each network device is possible through foolproof protocols. Unique identifier (UID) is a unique identification number for each IoT device that makes it recognizable to others or the control network.

According to Gartner, the number of IoT devices in use by the year 2020 is estimated to be 20 billion. Figure 13 shows the graph of the rate of increase in IoT devices by the year. IoT devices are used in health care sectors (popular IoTs are fitness band and health monitoring devices), the industrial sector (sensing and measuring devices), security sector (cameras and positioning systems), and general devices are used in smart homes for the monitoring and control of loads. Microgrids come into this cross-industry sector: this sector specifies special devices that improve the efficiency of other network devices that include improvements in quality of monitoring and reducing the losses through effective control of failure rate in production [172]. Figure 14 shows the IoT based support to the microgrid applications.

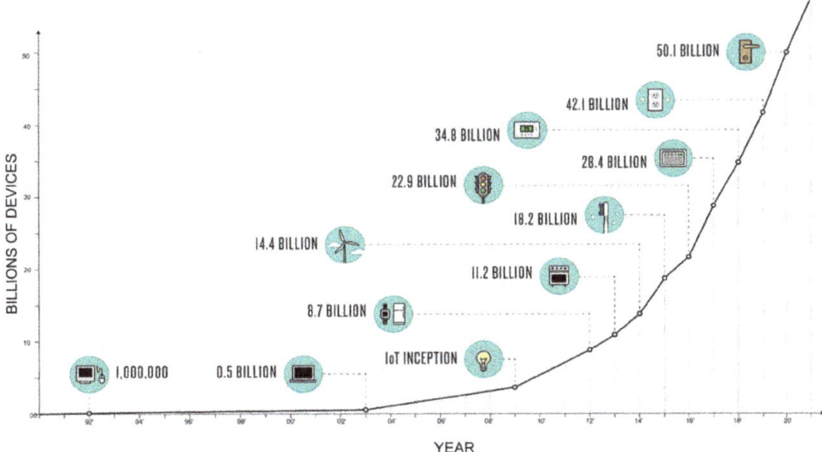

Figure 13. Internet of things (IoT) rate of increase in usage in different applications.

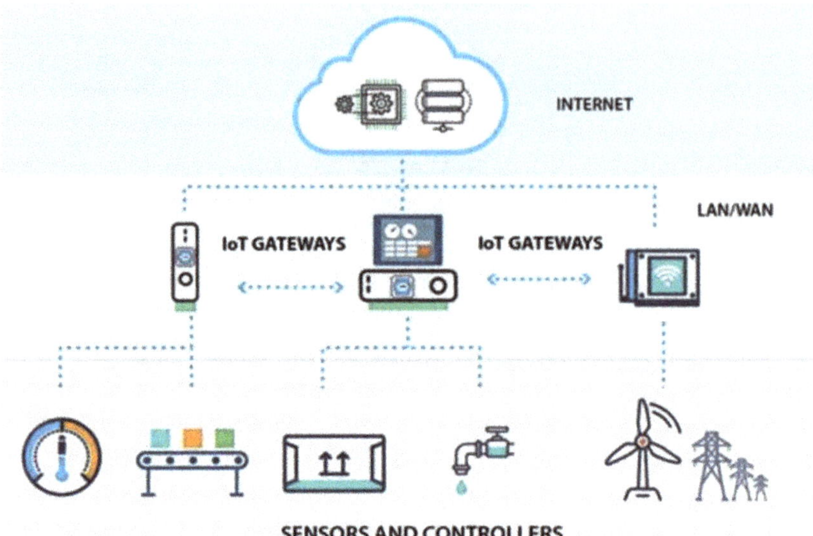

Figure 14. IoT support to the microgrid.

6.2. Smart Meters

For the last few years, disc-type meters have been replaced by electronic integrated circuit embedded meters which are used effectively by the distribution utility companies in providing authentic and electronic billing for the customers [173]. The necessity for refined flexible billing and control of billing information for two-way power flow proposes the implementation of smart meter technology. Smart meter technology provides the day-to-day of market prices of the power demand to the customer in commercial situations and industries. Previously existing automated meter reading (AMR) technology collects the energy consumption data from the customers to the utility, which is a one-way flow in power and communication. The AMR, an advanced metering infrastructure (AMI) developed in recent years, provides two-way communication and power flow between the meter and the central control system [174]. The improved functionality characteristics from the AMR to the AMI are shown in Figure 15.

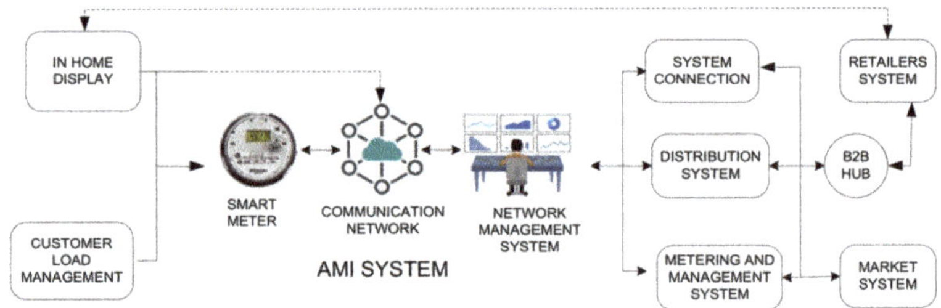

Figure 15. AMR and their functionalities in the microgrid.

In the aspect of both transmission and distribution, a smart grid is a revolutionary approach and the smart meters play a significant role as an integral part of the smart grid in communicating with the customers and data collection. Supposedly, the smart meter consists of three main components, which are communication network management, advanced metering element, and data management unit. The smart meter is equipped with a memory device that allows consumers to monitor their energy usage via a software interface, allowing it to communicate in two ways. The smart meter controls the operation distribution system switches and reclosers which provide an efficient delivery system and maintain reliability. The availability of two-way communication and the energy interface in the smart meter allows the control of distribution infrastructure by sending commands to the control center, which is also known as the distribution automation at the load end. The advantage of the smart meter is that it enables the central control to take action when tampering happens with the available rapid report sent from the smart meter as a part of collecting data [175]. This helps in reducing power theft while improving the power system security. Availability of day-to-day billing reports to the consumers helps them to manage the loads and reduce their bills through the smart meter.

The data from every meter can be collected, processed, and stored using applications like big data [176]. This makes the utility companies go towards the implementation of smart meters where two-way communication plays a prominent role.

7. Conclusions

This paper gives a detailed review of the recent analysis of the different energy management strategies proposed for the microgrid, consisting of classical, heuristic, and intelligent algorithms. Furthermore, this paper provides a brief introduction about the architecture of microgrids, different classifications in microgrids, components of a microgrid, communication technologies used, standards available for the implementation, and auxiliary services required in the microgrid. It discusses key applications in energy management, which include forecasting, demand response, data handling, and the control structure. This article also presents an insight on areas in which the scope of research and their contribution to energy management is in the nascent stage.

Optimization in cost minimization, operation control, reliability, energy scheduling, emission control, and load forecasting is the objective functions of the EMS in both the modes of microgrid operation for sustainable development. This makes the MG energy management a multi-objective optimization problem considering the economic, technical, and emission aspects as key constraints. The prime aspects that are covered in this review are on prospects, solutions, and opportunities of the objective functions of the EMS using efficient strategies. Based on the practicability, suitability, and tractability of the methods, the techniques are considered to find global solutions to the operations of the system. The microgrid energy management objectives depend on its mode of operation, whether it is centralized, decentralized, or distributed operation, several economical constraints, and the dynamic nature of dispatchable energy sources. Furthermore, few authors have considered greenhouse gas emissions as an additional objective function apart from non- renewable generators, batteries' health status, integration of active demand response, active and reactive losses along with resilience and customer management.

Many research articles have been published on the energy management of microgrids on different applications, yet the reviewed papers have been considered based on diversity of the objective functions. The areas such as customer confidentiality regulations, management of communication systems, and reliability studies on islanded mode have further scope to emphasize in future studies. Potential areas as mentioned above needed to be focused in detail along with the depth of discharge of the batteries, effect of the conventional grid on greenhouse gas emissions, and demand response integration to obtain effective and efficient operation of microgrids.

Author Contributions: Conceptualization, A.R.B. and S.V.; methodology, S.R.S.; software, A.R.B.; validation, A.R.B., S.V. and S.R.S.; formal analysis, A.R.B.; investigation, S.V.; resources, S.R.S.; data curation, A.R.B.; writing—original draft preparation, A.R.B.; writing—review and editing, A.R.B., S.V. and S.R.S.; visualization, A.R.B.; supervision, S.V. and S.R.S.; project administration, A.R.B.; funding acquisition, S.V. and S.R.S. All authors have read and agreed to the published version of the manuscript.

Funding: This research was funded by National Institute of Technology Andhra Pradesh (NIT-AP) and Woosong University's Academic Research Funding–2021.

Acknowledgments: We acknowledge National Institute of Technology Andhra Pradesh (NIT-AP), Tadepalligudem, Andhra Pradesh, India; and Woosong University, South Korea for their support in carrying out this research work.

Conflicts of Interest: The authors declare no conflict of interest.

References

1. Hu, H.; Xie, N.; Fang, D.; Zhang, X. The role of renewable energy consumption and commercial services trade in carbon dioxide reduction: Evidence from 25 developing countries. *Appl. Energy* **2018**, *211*, 1229–1244. [CrossRef]
2. Hussain, A.; Bui, V.-H.; Kim, H.-M. Microgrids as a resilience resource and strategies used by microgrids for enhancing resilience. *Appl. Energy* **2019**, *240*, 56–72. [CrossRef]
3. Farhangi, H. The path of the smart grid. *IEEE Power Energy Mag.* **2009**, *8*, 18–28. [CrossRef]
4. Qi, F.; Wen, F.; Liu, X.; Salam, A. A Residential Energy Hub Model with a Concentrating Solar Power Plant and Electric Vehicles. *Energies* **2017**, *10*, 1159. [CrossRef]
5. Pascual, J.; Sanchis, P.; Marroyo, L. Implementation and Control of a Residential Electrothermal Microgrid Based on Renewable Energies, a Hybrid Storage System and Demand Side Management. *Energies* **2014**, *7*, 210–237. [CrossRef]
6. Cai, J. Optimal Building Thermal Load Scheduling for Simultaneous Participation in Energy and Frequency Regulation Markets. *Energies* **2021**, *14*, 1593. [CrossRef]
7. Hatziargyriou, N. *Microgrids: Architectures and Control*; Wiley: Hoboken, NJ, USA, 2014.
8. Hossain, A.; Pota, H.R.; Issa, W. Overview of AC Microgrid Controls with Inverter-Interfaced Generations. *Energies* **2017**, *10*, 1300. [CrossRef]
9. Nejabatkhah, F.; Li, Y.W. Overview of Power Management Strategies of Hybrid AC/DC Microgrid. *IEEE Trans. Power Electron.* **2014**, *30*, 7072–7089. [CrossRef]
10. Patrao, I.; Figueres, E.; Garcerá, G.; González-Medina, R. Microgrid architectures for low voltage distributed generation. *Renew. Sustain. Energy Rev.* **2015**, *43*, 415–424. [CrossRef]
11. Pashajavid, E.; Shahnia, F.; Ghosh, A. A decentralized strategy to remedy the power deficiency in remote area microgrids. In Proceedings of the 50th International Universities Power Engineering Conference, Stoke on Trent, UK, 1–4 September 2015; IEEE: Piscataway, NJ, USA, 2015; pp. 1–6. [CrossRef]
12. Hussain, N.; Nasir, M.; Vasquez, J.C.; Guerrero, J.M. Recent Developments and Challenges on AC Microgrids Fault Detection and Protection Systems–A Review. *Energies* **2020**, *13*, 2149. [CrossRef]
13. Javaid, N.; Hafeez, G.; Iqbal, S.; Alrajeh, N.; Alabed, M.S.; Guizani, M. Energy Efficient Integration of Renewable Energy Sources in the Smart Grid for Demand Side Management. *IEEE Access* **2018**, *6*, 77077–77096. [CrossRef]
14. Bintoudi, A.; Zyglakis, L.; Tsolakis, A.; Gkaidatzis, P.; Tryferidis, A.; Ioannidis, D.; Tzovaras, D. OptiMEMS: An Adaptive Lightweight Optimal Microgrid Energy Management System Based on the Novel Virtual Distributed Energy Resources in Real-Life Demonstration. *Energies* **2021**, *14*, 2752. [CrossRef]
15. Takano, H.; Goto, R.; Hayashi, R.; Asano, H. Optimization Method for Operation Schedule of Microgrids Considering Uncertainty in Available Data. *Energies* **2021**, *14*, 2487. [CrossRef]
16. Hussain, A.; Bui, V.-H.; Kim, H.-M. Robust Optimization-Based Scheduling of Multi-Microgrids Considering Uncertainties. *Energies* **2016**, *9*, 278. [CrossRef]
17. Cha, H.-J.; Won, D.-J.; Kim, S.-H.; Chung, I.-Y.; Han, B.-M. Multi-Agent System-Based Microgrid Operation Strategy for Demand Response. *Energies* **2015**, *8*, 14272–14286. [CrossRef]
18. Helmi, A.M.; Carli, R.; Dotoli, M.; Ramadan, H.S. Efficient and Sustainable Reconfiguration of Distribution Networks via Metaheuristic Optimization. *IEEE Trans. Autom. Sci. Eng.* **2021**, *PP*, 1–17. [CrossRef]
19. Muhammad, M.A.; Mokhlis, H.; Naidu, K.; Amin, A.; Franco, J.F.; Othman, M. Distribution Network Planning Enhancement via Network Reconfiguration and DG Integration Using Dataset Approach and Water Cycle Algorithm. *J. Mod. Power Syst. Clean Energy* **2020**, *8*, 86–93. [CrossRef]
20. Li, F.; Xie, K.; Yang, J. Optimization and Analysis of a Hybrid Energy Storage System in a Small-Scale Standalone Microgrid for Remote Area Power Supply (RAPS). *Energies* **2015**, *8*, 4802–4826. [CrossRef]
21. Prehoda, E.; Pearce, J.M.; Schelly, C. Policies to Overcome Barriers for Renewable Energy Distributed Generation: A Case Study of Utility Structure and Regulatory Regimes in Michigan. *Energies* **2019**, *12*, 674. [CrossRef]
22. Moradi, H.; Esfahanian, M.; Abtahi, A.; Zilouchian, A. Modeling a Hybrid Microgrid Using Probabilistic Reconfiguration under System Uncertainties. *Energies* **2017**, *10*, 1430. [CrossRef]
23. Shayeghi, H.; Alilou, M. 3—Distributed generation and microgrids. In *Hybrid Renewable Energy Systems and Microgrids*; Kabalci, E., Ed.; Academic Press: Cambridge, MA, USA, 2021; pp. 73–102.
24. Lidula, N.W.A.; Rajapakse, A. Microgrids research: A review of experimental microgrids and test systems. *Renew. Sustain. Energy Rev.* **2011**, *15*, 186–202. [CrossRef]
25. Mariam, L.; Basu, M.; Conlon, M.F. A Review of Existing Microgrid Architectures. *J. Eng.* **2013**, *2013*, 937614. [CrossRef]
26. Karki, S.; Kulkarni, M.; Mann, M.D.; Salehfar, H. Efficiency Improvements through Combined Heat and Power for On-site Distributed Generation Technologies. *Cogener. Distrib. Gener. J.* **2007**, *22*, 19–34. [CrossRef]

27. Li, J.; Wei, W.; Xiang, J. A Simple Sizing Algorithm for Stand-Alone PV/Wind/Battery Hybrid Microgrids. *Energies* **2012**, *5*, 5307–5323. [CrossRef]
28. Dinh, H.T.; Kim, D. An Optimal Energy-Saving Home Energy Management Supporting User Comfort and Electricity Selling With Different Prices. *IEEE Access* **2021**, *9*, 9235–9249. [CrossRef]
29. Lasseter, R. MicroGrids. In Proceedings of the IEEE Winter Meeting Power Engineering Society, New York, NY, USA, 27–31 January 2002; IEEE: Piscataway, NJ, USA, 2003; Volume 1, pp. 305–308. [CrossRef]
30. Li, Z.; Zheng, Z.; Xu, L.; Lu, X. A review of the applications of fuel cells in microgrids: Opportunities and challenges. *BMC Energy* **2019**, *1*, 8. [CrossRef]
31. Katiraei, F.; Iravani, R.; Hatziargyriou, N.; Dimeas, A. Microgrids management. *IEEE Power Energy Mag.* **2008**, *6*, 54–65. [CrossRef]
32. Ruiz, N.; Cobelo, I.; Oyarzabal, J. A Direct Load Control Model for Virtual Power Plant Management. *IEEE Trans. Power Syst.* **2009**, *24*, 959–966. [CrossRef]
33. Graditi, G.; Di Silvestre, M.L.; Gallea, R.; Sanseverino, E.R. Heuristic-Based Shiftable Loads Optimal Management in Smart Micro-Grids. *IEEE Trans. Ind. Inform.* **2014**, *11*, 271–280. [CrossRef]
34. Asimakopoulou, G.E.; Dimeas, A.L.; Hatziargyriou, N.D. Leader-Follower Strategies for Energy Management of Multi-Microgrids. *IEEE Trans. Smart Grid* **2013**, *4*, 1909–1916. [CrossRef]
35. Fischer, D.; Härtl, A.; Wille-Haussmann, B. Model for electric load profiles with high time resolution for German households. *Energy Build.* **2015**, *92*, 170–179. [CrossRef]
36. Mohsenian-Rad, A.-H.; Leon-Garcia, A. Optimal Residential Load Control With Price Prediction in Real-Time Electricity Pricing Environments. *IEEE Trans. Smart Grid* **2010**, *1*, 120–133. [CrossRef]
37. Ângelos, E.W.S.; Saavedra, O.R.; Cortés, O.A.C.; De Souza, A.N. Detection and Identification of Abnormalities in Customer Consumptions in Power Distribution Systems. *IEEE Trans. Power Deliv.* **2011**, *26*, 2436–2442. [CrossRef]
38. Scarabaggio, P.; Grammatico, S.; Carli, R.; Dotoli, M. Distributed Demand Side Management With Stochastic Wind Power Forecasting. *IEEE Trans. Control. Syst. Technol.* **2021**, *PP*, 1–16. [CrossRef]
39. Sortomme, E.; El-Sharkawi, M.A. Optimal Power Flow for a System of Microgrids with Controllable Loads and Battery Storage. In Proceedings of the 2009 IEEE/PES Power Systems Conference and Exposition, Seattle, WA, USA, 15–18 March 2009; pp. 1–5. [CrossRef]
40. Kleiner, F.; Beermann, M.; Beers, E.; Çatay, B.; Davies, H. *HEV TCP Task 27: Final report—Electrification of Transport Logistic Vehicles (eLogV)*; Institut für Fahrzeugkonzepte: Stuttgart, Germany, 2017.
41. Savio, D.A.; Juliet, V.A.; Chokkalingam, B.; Padmanaban, S.; Holm-Nielsen, J.B.; Blaabjerg, F. Photovoltaic Integrated Hybrid Microgrid Structured Electric Vehicle Charging Station and Its Energy Management Approach. *Energies* **2019**, *12*, 168. [CrossRef]
42. Robledo, C.B.; Oldenbroek, V.; Abbruzzese, F.; van Wijk, A.J. Integrating a hydrogen fuel cell electric vehicle with vehicle-to-grid technology, photovoltaic power and a residential building. *Appl. Energy* **2018**, *215*, 615–629. [CrossRef]
43. Hatziargyriou, N.; Asano, H.; Iravani, R.; Marnay, C. Microgrids. *IEEE Power Energy Mag.* **2007**, *5*, 78–94. [CrossRef]
44. Dragičević, T.; Blaabjerg, F. Chapter 9—Power Electronics for Microgrids: Concepts and Future Trends. In *Microgrid*; Mahmoud, M.S., Ed.; Butterworth-Heinemann: Oxford, UK, 2017; pp. 263–279.
45. Vegunta, S.; Higginson, M.; Kenarangui, Y.; Li, G.; Zabel, D.; Tasdighi, M.; Shadman, A. AC Microgrid Protection System Design Challenges—A Practical Experience. *Energies* **2021**, *14*, 2016. [CrossRef]
46. Dhifli, M.; Lashab, A.; Guerrero, J.M.; Abusorrah, A.; Al-Turki, Y.A.; Cherif, A. Enhanced Intelligent Energy Management System for a Renewable Energy-based AC Microgrid. *Energies* **2020**, *13*, 3268. [CrossRef]
47. Xu, L.; Chen, D. Control and Operation of a DC Microgrid With Variable Generation and Energy Storage. *IEEE Trans. Power Deliv.* **2011**, *26*, 2513–2522. [CrossRef]
48. Lago, J.; Heldwein, M.L. Operation and Control-Oriented Modeling of a Power Converter for Current Balancing and Stability Improvement of DC Active Distribution Networks. *IEEE Trans. Power Electron.* **2011**, *26*, 877–885. [CrossRef]
49. Peyghami, S.; Mokhtari, H.; Blaabjerg, F. Autonomous Operation of a Hybrid AC/DC Microgrid With Multiple Interlinking Converters. *IEEE Trans. Smart Grid* **2017**, *9*, 6480–6488. [CrossRef]
50. Liu, X.; Wang, P.; Loh, P.C. A Hybrid AC/DC Microgrid and Its Coordination Control. *IEEE Trans. Smart Grid* **2011**, *2*, 278–286. [CrossRef]
51. Olivares, D.E.; Mehrizi-Sani, A.; Etemadi, A.H.; Canizares, C.A.; Iravani, R.; Kazerani, M.; Hajimiragha, A.H.; Gomis-Bellmunt, O.; Saeedifard, M.; Palma-Behnke, R.; et al. Trends in Microgrid Control. *IEEE Trans. Smart Grid* **2014**, *5*, 1905–1919. [CrossRef]
52. Guerrero, J.; Vasquez, J.C.; Matas, J.; de Vicuña, L.G.; Castilla, M. Hierarchical Control of Droop-Controlled AC and DC Microgrids—A General Approach Toward Standardization. *IEEE Trans. Ind. Electron.* **2010**, *58*, 158–172. [CrossRef]
53. Planas, E.; Gil-De-Muro, A.; Andreu, J.; Kortabarria, I.; de Alegría, I.M. General aspects, hierarchical controls and droop methods in microgrids: A review. *Renew. Sustain. Energy Rev.* **2013**, *17*, 147–159. [CrossRef]
54. Mehrizi-Sani, A.; Iravani, R. Potential-Function Based Control of a Microgrid in Islanded and Grid-Connected Modes. *IEEE Trans. Power Syst.* **2010**, *25*, 1883–1891. [CrossRef]
55. Su, W.; Wang, J.; Roh, J. Stochastic Energy Scheduling in Microgrids With Intermittent Renewable Energy Resources. *IEEE Trans. Smart Grid* **2013**, *5*, 1876–1883. [CrossRef]
56. Kabalci, Y. A survey on smart metering and smart grid communication. *Renew. Sustain. Energy Rev.* **2016**, *57*, 302–318. [CrossRef]

57. Bani-Ahmed, A.; Weber, L.; Nasiri, A.; Hosseini, H. Microgrid communications: State of the art and future trends. In Proceedings of the 2014 International Conference on Renewable Energy Research and Application (ICRERA), Milwaukee, WI, USA, 19–22 October 2014; pp. 780–785. [CrossRef]

58. Saleh, M.; Esa, Y.; Mohamed, A.A. Communication-Based Control for DC Microgrids. *IEEE Trans. Smart Grid* **2018**, *10*, 2180–2195. [CrossRef]

59. Saleh, M.; Esa, Y.; Mohamed, A. Hardware based testing of communication based control for DC microgrid. In Proceedings of the 6th International Conference on Renewable Energy Research and Applications, San Diego, CA, USA, 5–8 November 2017; IEEE: Piscataway, NJ, USA, 2017; pp. 902–907. [CrossRef]

60. Bose, A. Power System Stability: New Opportunities for Control. In *Stability and Control of Dynamical Systems with Applications: A Tribute to Anthony N. Michel*; Liu, D., Antsaklis, P.J., Eds.; Birkhäuser Boston: Boston, MA, USA, 2003; pp. 315–330.

61. Sun, C.; Joos, G.; Ali, S.Q.; Paquin, J.N.; Rangel, C.M.; Al Jajeh, F.; Novickij, I.; Bouffard, F. Design and Real-Time Implementation of a Centralized Microgrid Control System with Rule-Based Dispatch and Seamless Transition Function. *IEEE Trans. Ind. Appl.* **2020**, *56*, 3168–3177. [CrossRef]

62. Zhuo, W.; Savkin, A.V.; Meng, K. Decentralized Optimal Control of a Microgrid with Solar PV, BESS and Thermostatically Controlled Loads. *Energies* **2019**, *12*, 2111. [CrossRef]

63. Wang, J.; Li, K.-J.; Javid, Z.; Sun, Y. Distributed Optimal Coordinated Operation for Distribution System with the Integration of Residential Microgrids. *Appl. Sci.* **2019**, *9*, 2136. [CrossRef]

64. Liu, L.; Wang, Y.; Liu, L.; Wang, Z. Intelligent Control of Micro Grid: A Big Data-Based Control Center. *IOP Conf. Series Earth Environ. Sci.* **2018**, *108*, 52037. [CrossRef]

65. Daki, H.; El Hannani, A.; Aqqal, A.; Haidine, A.; Dahbi, A. Big Data management in smart grid: Concepts, requirements and implementation. *J. Big Data* **2017**, *4*, 13. [CrossRef]

66. Kahouli, O.; Alsaif, H.; Bouteraa, Y.; Ben Ali, N.; Chaabene, M. Power System Reconfiguration in Distribution Network for Improving Reliability Using Genetic Algorithm and Particle Swarm Optimization. *Appl. Sci.* **2021**, *11*, 3092. [CrossRef]

67. Jakus, D.; Čađenović, R.; Vasilj, J.; Sarajčev, P. Optimal Reconfiguration of Distribution Networks Using Hybrid Heuristic-Genetic Algorithm. *Energies* **2020**, *13*, 1544. [CrossRef]

68. Veeramsetty, V.; Mohnot, A.; Singal, G.; Salkuti, S. Short Term Active Power Load Prediction on A 33/11 kV Substation Using Regression Models. *Energies* **2021**, *14*, 2981. [CrossRef]

69. Ma, J.; Ma, X. A review of forecasting algorithms and energy management strategies for microgrids. *Syst. Sci. Control. Eng.* **2018**, *6*, 237–248. [CrossRef]

70. Onishi, V.C.; Antunes, C.H.; Trovão, J.P.F. Optimal Energy and Reserve Market Management in Renewable Microgrid-PEVs Parking Lot Systems: V2G, Demand Response and Sustainability Costs. *Energies* **2020**, *13*, 1884. [CrossRef]

71. Liang, H.; Zhuang, W. Stochastic Modeling and Optimization in a Microgrid: A Survey. *Energies* **2014**, *7*, 2027–2050. [CrossRef]

72. Dolara, A.; Grimaccia, F.; Magistrati, G.; Marchegiani, G. Optimization Models for Islanded Micro-Grids: A Comparative Analysis between Linear Programming and Mixed Integer Programming. *Energies* **2017**, *10*, 241. [CrossRef]

73. Stadler, M.; Pecenak, Z.; Mathiesen, P.; Fahy, K.; Kleissl, J. Performance Comparison between Two Established Microgrid Planning MILP Methodologies Tested On 13 Microgrid Projects. *Energies* **2020**, *13*, 4460. [CrossRef]

74. Karimi, H.; Jadid, S.; Karimi, H.; Jadid, S. Optimal energy management for multi-microgrid considering demand response programs: A stochastic multi-objective framework. *Energy* **2020**, *195*, 116992. [CrossRef]

75. Sukumar, S.; Mokhlis, H.; Mekhilef, S.; Naidu, K.; Karimi, M. Mix-mode energy management strategy and battery sizing for economic operation of grid-tied microgrid. *Energy* **2016**, *118*, 1322–1333. [CrossRef]

76. Amrollahi, M.H.; Bathaee, S.M.T. Techno-economic optimization of hybrid photovoltaic/wind generation together with energy storage system in a stand-alone micro-grid subjected to demand response. *Appl. Energy* **2017**, *202*, 66–77. [CrossRef]

77. Anglani, N.; Oriti, G.; Colombini, M. Optimized energy management system to reduce fuel consumption in remote military microgrids. *IEEE Trans. Ind. Appl.* **2016**, *53*, 5777–5785. [CrossRef]

78. Igualada, L.; Corchero, C.; Cruz-Zambrano, M.; Heredia, F.-J. Optimal Energy Management for a Residential Microgrid Including a Vehicle-to-Grid System. *IEEE Trans. Smart Grid* **2014**, *5*, 2163–2172. [CrossRef]

79. Murty, V.V.S.N.; Kumar, A. Multi-objective energy management in microgrids with hybrid energy sources and battery energy storage systems. *Prot. Control. Mod. Power Syst.* **2020**, *5*, 2. [CrossRef]

80. Vergara, P.; López, J.C.; da Silva, L.C.; Rider, M.J. Security-constrained optimal energy management system for three-phase residential microgrids. *Electr. Power Syst. Res.* **2017**, *146*, 371–382. [CrossRef]

81. Olivares, D.E.; Canizares, C.A.; Kazerani, M. A Centralized Energy Management System for Isolated Microgrids. *IEEE Trans. Smart Grid* **2014**, *5*, 1864–1875. [CrossRef]

82. Helal, S.A.; Najee, R.J.; Hanna, M.; Shaaban, M.F.; Osman, A.H.; Hassan, M.S. An energy management system for hybrid microgrids in remote communities. In Proceedings of the 2017 IEEE 30th Canadian Conference on Electrical and Computer Engineering (CCECE), Windsor, ON, Canada, 30 April–3 May 2017; pp. 1–4. [CrossRef]

83. Tsikalakis, A.G.; Hatziargyriou, N.D. Centralized Control for Optimizing Microgrids Operation. *IEEE Trans. Energy Convers.* **2008**, *23*, 241–248. [CrossRef]

84. Panwar, L.K.; Konda, S.R.; Verma, A.; Panigrahi, B.K.; Kumar, R. Operation window constrained strategic energy management of microgrid with electric vehicle and distributed resources. *IET Gener. Transm. Distrib.* **2017**, *11*, 615–626. [CrossRef]

85. Heymann, B.; Bonnans, J.F.; Martinon, P.; Silva, F.J.; Lanas, F.; Jiménez-Estévez, G. Continuous optimal control approaches to microgrid energy management. *Energy Syst.* **2017**, *9*, 59–77. [CrossRef]
86. Babazadeh, H.; Gao, W.; Wu, Z.; Li, Y. Optimal energy management of wind power generation system in islanded microgrid system. In Proceedings of the 2013 North American Power Symposium, Manhattan, KS, USA, 22–24 September 2013; IEEE: Piscataway, NJ, USA, 2013; pp. 1–5. [CrossRef]
87. Strelec, M.; Berka, J. Microgrid energy management based on approximate dynamic programming. In Proceedings of the IEEE PES ISGT Europe 2013, Lyngby, Denmark, 6–9 October 2013; pp. 1–5. [CrossRef]
88. Kanchev, H.; Lu, D.; Colas, F.; Lazarov, V.; Francois, B. Energy Management and Operational Planning of a Microgrid With a PV-Based Active Generator for Smart Grid Applications. *IEEE Trans. Ind. Electron.* **2011**, *58*, 4583–4592. [CrossRef]
89. Merabet, A.; Ahmed, K.T.; Ibrahim, H.; Beguenane, R.; Ghias, A.M.Y.M. Energy Management and Control System for Laboratory Scale Microgrid Based Wind-PV-Battery. *IEEE Trans. Sustain. Energy* **2016**, *8*, 145–154. [CrossRef]
90. Choudar, A.; Boukhetala, D.; Barkat, S.; Brucker, J.-M. A local energy management of a hybrid PV-storage based distributed generation for microgrids. *Energy Convers. Manag.* **2015**, *90*, 21–33. [CrossRef]
91. Nojavan, S.; Pashaei-Didani, H.; Mohammadi, A.; Ahmadi-Nezamabad, H. Chapter 10—Deterministic-based energy management of hybrid AC/DC microgrid. In *Risk-based Energy Management*; Nojavan, S., Shafieezadeh, M., Ghadimi, N., Eds.; Academic Press: Cambridge, MA, USA, 2020; pp. 177–202.
92. Shin, I. Approximation Algorithm-Based Prosumer Scheduling for Microgrids. *Energies* **2020**, *13*, 5853. [CrossRef]
93. Leonori, S.; Paschero, M.; Mascioli, F.M.F.; Rizzi, A. Optimization strategies for Microgrid energy management systems by Genetic Algorithms. *Appl. Soft Comput.* **2019**, *86*, 105903. [CrossRef]
94. Teo, T.T.; Logenthiran, T.; Woo, W.L.; Abidi, K.; John, T.; Wade, N.S.; Greenwood, D.M.; Patsios, C.; Taylor, P.C. Optimization of Fuzzy Energy-Management System for Grid-Connected Microgrid Using NSGA-II. *IEEE Trans. Cybern.* **2020**, 1–12. [CrossRef]
95. Ali, H.; Hussain, A.; Bui, V.-H.; Jeon, J.; Kim, H.-M. Welfare Maximization-Based Distributed Demand Response for Islanded Multi-Microgrid Networks Using Diffusion Strategy. *Energies* **2019**, *12*, 3701. [CrossRef]
96. Teekaraman, Y.; Kuppusamy, R.; Nikolovski, S. Solution for Voltage and Frequency Regulation in Standalone Microgrid using Hybrid Multiobjective Symbiotic Organism Search Algorithm. *Energies* **2019**, *12*, 2812. [CrossRef]
97. Jumani, T.A.; Mustafa, M.W.; Rasid, M.M.; Mirjat, N.H.; Leghari, Z.H.; Saeed, M.S. Optimal Voltage and Frequency Control of an Islanded Microgrid using Grasshopper Optimization Algorithm. *Energies* **2018**, *11*, 3191. [CrossRef]
98. Nimma, K.S.; Al-Falahi, M.D.A.; Nguyen, H.D.; Jayasinghe, S.D.G.; Mahmoud, T.S.; Negnevitsky, M. Grey Wolf Optimization-Based Optimum Energy-Management and Battery-Sizing Method for Grid-Connected Microgrids. *Energies* **2018**, *11*, 847. [CrossRef]
99. Huang, Y.; Masrur, H.; Shigenobu, R.; Hemeida, A.; Mikhaylov, A.; Senjyu, T. A Comparative Design of a Campus Microgrid Considering a Multi-Scenario and Multi-Objective Approach. *Energies* **2021**, *14*, 2853. [CrossRef]
100. Zhao, F.; Yuan, J.; Wang, N. Dynamic Economic Dispatch Model of Microgrid Containing Energy Storage Components Based on a Variant of NSGA-II Algorithm. *Energies* **2019**, *12*, 871. [CrossRef]
101. Wang, Y.; Huang, Y.; Wang, Y.; Yu, H.; Li, R.; Song, S. Energy Management for Smart Multi-Energy Complementary Micro-Grid in the Presence of Demand Response. *Energies* **2018**, *11*, 974. [CrossRef]
102. Wasilewski, J. Optimisation of multicarrier microgrid layout using selected metaheuristics. *Int. J. Electr. Power Energy Syst.* **2018**, *99*, 246–260. [CrossRef]
103. Kim, H.-J.; Kim, M.-K. Multi-Objective Based Optimal Energy Management of Grid-Connected Microgrid Considering Advanced Demand Response. *Energies* **2019**, *12*, 4142. [CrossRef]
104. Ramli, M.A.; Bouchekara, H.; Alghamdi, A.S. Efficient Energy Management in a Microgrid with Intermittent Renewable Energy and Storage Sources. *Sustainability* **2019**, *11*, 3839. [CrossRef]
105. Dey, B.; Bhattacharyya, B. Hybrid Intelligence Techniques for Unit Commitment of Microgrids. In Proceedings of the 20th International Conference on Intelligent System Application to Power Systems, New Delhi, India, 10–14 December 2019; IEEE: Piscataway, NJ, USA, 2019; pp. 1–6. [CrossRef]
106. Habib, H.U.R.; Subramaniam, U.; Waqar, A.; Farhan, B.S.; Kotb, K.M.; Wang, S. Energy Cost Optimization of Hybrid Renewables Based V2G Microgrid Considering Multi Objective Function by Using Artificial Bee Colony Optimization. *IEEE Access* **2020**, *8*, 62076–62093. [CrossRef]
107. Lin, W.-M.; Tu, C.-S.; Tsai, M.-T. Energy Management Strategy for Microgrids by Using Enhanced Bee Colony Optimization. *Energies* **2015**, *9*, 5. [CrossRef]
108. Qian, X.; Yang, Y.; Li, C.; Tan, S.-C. Operating Cost Reduction of DC Microgrids Under Real-Time Pricing Using Adaptive Differential Evolution Algorithm. *IEEE Access* **2020**, *8*, 169247–169258. [CrossRef]
109. Zhang, J.; Cho, H.; Mago, P.J.; Zhang, H.; Yang, F. Multi-Objective Particle Swarm Optimization (MOPSO) for a Distributed Energy System Integrated with Energy Storage. *J. Therm. Sci.* **2019**, *28*, 1221–1235. [CrossRef]
110. Dabhi, D.; Pandya, K. Enhanced Velocity Differential Evolutionary Particle Swarm Optimization for Optimal Scheduling of a Distributed Energy Resources with Uncertain Scenarios. *IEEE Access* **2020**, *8*, 27001–27017. [CrossRef]
111. Elgamal, M.; Korovkin, N.; Elmitwally, A.; Menaem, A.A.; Chen, Z. A Framework for Profit Maximization in a Grid-Connected Microgrid With Hybrid Resources Using a Novel Rule Base-BAT Algorithm. *IEEE Access* **2020**, *8*, 71460–71474. [CrossRef]

112. Jamshidi, M.; Askarzadeh, A. Techno-economic analysis and size optimization of an off-grid hybrid photovoltaic, fuel cell and diesel generator system. *Sustain. Cities Soc.* **2018**, *44*, 310–320. [CrossRef]
113. Swief, R.A.; El-Amary, N.H.; Kamh, M.Z. Optimal Energy Management Integrating Plug in Hybrid Vehicle Under Load and Renewable Uncertainties. *IEEE Access* **2020**, *8*, 176895–176904. [CrossRef]
114. Papari, B.; Cox, R.; Sockeel, N.; Hoang, P.H.; Ozkan, G. Supervisory Energy Management in Hybrid AC-DC Microgrids Based on a Hybrid Distributed Algorithm. In Proceedings of the 2020 Clemson University Power Systems Conference (PSC), Clemson, SC, USA, 10–13 March 2020; pp. 1–6. [CrossRef]
115. Gong, X.; Dong, F.; Mohamed, M.A.; Abdalla, O.M.; Ali, Z.M. A Secured Energy Management Architecture for Smart Hybrid Microgrids Considering PEM-Fuel Cell and Electric Vehicles. *IEEE Access* **2020**, *8*, 47807–47823. [CrossRef]
116. Fathy, A.; Kaaniche, K.; Alanazi, T.M. Recent Approach Based Social Spider Optimizer for Optimal Sizing of Hybrid PV/Wind/Battery/Diesel Integrated Microgrid in Aljouf Region. *IEEE Access* **2020**, *8*, 57630–57645. [CrossRef]
117. Diab, A.A.Z.; Sultan, H.M.; Mohamed, I.S.; Kuznetsov, O.N.; Do, T.D. Application of Different Optimization Algorithms for Optimal Sizing of PV/Wind/Diesel/Battery Storage Stand-Alone Hybrid Microgrid. *IEEE Access* **2019**, *7*, 119223–119245. [CrossRef]
118. Boglou, V.; Karavas, C.-S.; Arvanitis, K.; Karlis, A. A Fuzzy Energy Management Strategy for the Coordination of Electric Vehicle Charging in Low Voltage Distribution Grids. *Energies* **2020**, *13*, 3709. [CrossRef]
119. Hernandez, L.; Baladrón, C.; Aguiar, J.M.; Carro, B.; Sanchez-Esguevillas, A.J.; Lloret, J. Short-Term Load Forecasting for Microgrids Based on Artificial Neural Networks. *Energies* **2013**, *6*, 1385–1408. [CrossRef]
120. Sun, Q.; Sun, Q.; Qin, D. Adaptive Fuzzy Droop Control for Optimized Power Sharing in an Islanded Microgrid. *Energies* **2018**, *12*, 45. [CrossRef]
121. Bayhan, S.; Abu-Rub, H. Smart Energy Management System for Distributed Generations in AC Microgrid. In Proceedings of the 13th International Conference on Compatibility, Power Electronics and Power Engineering, Sonderborg, Denmark, 23–25 April 2019; IEEE: Piscataway, NJ, USA. [CrossRef]
122. Vasantharaj, S.; Indragandhi, V.; Subramaniyaswamy, V.; Teekaraman, Y.; Kuppusamy, R.; Nikolovski, S. Efficient Control of DC Microgrid with Hybrid PV—Fuel Cell and Energy Storage Systems. *Energies* **2021**, *14*, 3234. [CrossRef]
123. Arcos-Aviles, D.; Pascual, J.; Marroyo, L.; Sanchis, P.; Guinjoan, F. Fuzzy Logic-Based Energy Management System Design for Residential Grid-Connected Microgrids. *IEEE Trans. Smart Grid* **2016**, *9*, 530–543. [CrossRef]
124. Parmar, C.; Tiwari, S. Fuzzy logic based charging of electric vehicles for load management of microgrid. In Proceedings of the 5th International Conference on Communication and Electronic Systems ICCES, Coimbatore, India, 10–12 June 2020; pp. 1–7. [CrossRef]
125. Jafari, M.; Malekjamshidi, Z.; Zhu, J.; Khooban, M.-H. A Novel Predictive Fuzzy Logic-Based Energy Management System for Grid-Connected and Off-Grid Operation of Residential Smart Microgrids. *IEEE J. Emerg. Sel. Top. Power Electron.* **2018**, *8*, 1391–1404. [CrossRef]
126. Awais, M.; Khan, L.; Ahmad, S.; Jamil, M. Feedback-Linearization-Based Fuel-Cell Adaptive-Control Paradigm in a Microgrid Using a Wavelet-Entrenched NeuroFuzzy Framework. *Energies* **2021**, *14*, 1850. [CrossRef]
127. Zhang, R.; Samuel, R.D.J. Fuzzy Efficient Energy Smart Home Management System for Renewable Energy Resources. *Sustainability* **2020**, *12*, 3115. [CrossRef]
128. Ulutas, A.; Altas, I.H.; Onen, A.; Ustun, T.S. Neuro-Fuzzy-Based Model Predictive Energy Management for Grid Connected Microgrids. *Electronics* **2020**, *9*, 900. [CrossRef]
129. Arkhangelski, J.; Mahamadou, A.-T.; Lefebvre, G. Data forecasting for Optimized Urban Microgrid Energy Management. In Proceedings of the 2020 International Conference on Computational Intelligence for Smart Power System and Sustainable Energy (CISPSSE), Keonjhar, India, 29–31 July 2020; IEEE: Piscataway, NJ, USA, 2019; pp. 1–6. [CrossRef]
130. Dridi, A.; Moungla, H.; Afifi, H.; Badosa, J.; Ossart, F.; Kamal, A.E. Machine Learning Application to Priority Scheduling in Smart Microgrids. In Proceedings of the 2020 International Wireless Communications and Mobile Computing (IWCMC), Limassol, Cyprus, 15–19 June 2020; IEEE: Piscataway, NJ, USA, 2020; pp. 1695–1700. [CrossRef]
131. Wang, T.; He, X.; Deng, T. Neural networks for power management optimal strategy in hybrid microgrid. *Neural Comput. Appl.* **2017**, *31*, 2635–2647. [CrossRef]
132. Baez-Gonzalez, P.; Garcia-Torres, F.; Ridao, M.A.; Bordons, C. A Stochastic MPC Based Energy Management System for Simultaneous Participation in Continuous and Discrete Prosumer-to-Prosumer Energy Markets. *Energies* **2020**, *13*, 3751. [CrossRef]
133. Garcia-Torres, F.; Zafra-Cabeza, A.; Silva, C.; Grieu, S.; Darure, T.; Estanqueiro, A. Model Predictive Control for Microgrid Functionalities: Review and Future Challenges. *Energies* **2021**, *14*, 1296. [CrossRef]
134. Nguyen, T.-L.; Guillo-Sansano, E.; Syed, M.H.; Nguyen, V.-H.; Blair, S.M.; Reguera, L.; Tran, Q.-T.; Caire, R.; Burt, G.M.; Gavriluta, C.; et al. Multi-Agent System with Plug and Play Feature for Distributed Secondary Control in Microgrid—Controller and Power Hardware-in-the-Loop Implementation. *Energies* **2018**, *11*, 3253. [CrossRef]
135. Xu, S.; Sun, H.; Zhang, Z.; Guo, Q.; Zhao, B.; Bi, J.; Zhang, B. MAS-Based Decentralized Coordinated Control Strategy in a Micro-Grid with Multiple Microsources. *Energies* **2020**, *13*, 2141. [CrossRef]
136. Nassourou, M.; Blesa, J.; Puig, V. Robust Economic Model Predictive Control Based on a Zonotope and Local Feedback Controller for Energy Dispatch in Smart-Grids Considering Demand Uncertainty. *Energies* **2020**, *13*, 696. [CrossRef]

137. Nair, U.R.; Costa-Castello, R. A Model Predictive Control-Based Energy Management Scheme for Hybrid Storage System in Islanded Microgrids. *IEEE Access* **2020**, *8*, 97809–97822. [CrossRef]
138. E Silva, D.P.; Salles, J.L.F.; Fardin, J.F.; Pereira, M.M.R. Management of an island and grid-connected microgrid using hybrid economic model predictive control with weather data. *Appl. Energy* **2020**, *278*, 115581. [CrossRef]
139. Dao, L.A.; Dehghani-Pilehvarani, A.; Markou, A.; Ferrarini, L. A hierarchical distributed predictive control approach for microgrids energy management. *Sustain. Cities Soc.* **2019**, *48*, 101536. [CrossRef]
140. Baker, K.; Guo, J.; Hug, G.; Li, X. Distributed MPC for Efficient Coordination of Storage and Renewable Energy Sources Across Control Areas. *IEEE Trans. Smart Grid* **2016**, *7*, 992–1001. [CrossRef]
141. Rezaei, E.; Dagdougui, H. Optimal Real-Time Energy Management in Apartment Building Integrating Microgrid With Multizone HVAC Control. *IEEE Trans. Ind. Informatics* **2020**, *16*, 6848–6856. [CrossRef]
142. Ryu, K.-S.; Kim, D.-J.; Ko, H.; Boo, C.-J.; Kim, J.; Jin, Y.-G.; Kim, H.-C. MPC Based Energy Management System for Hosting Capacity of PVs and Customer Load with EV in Stand-Alone Microgrids. *Energies* **2021**, *14*, 4041. [CrossRef]
143. Cheng, J.; Duan, D.; Cheng, X.; Yang, L.; Cui, S. Probabilistic Microgrid Energy Management with Interval Predictions. *Energies* **2020**, *13*, 3116. [CrossRef]
144. Khan, M.W.; Wang, J.; Xiong, L. Optimal energy scheduling strategy for multi-energy generation grid using multi-agent systems. *Int. J. Electr. Power Energy Syst.* **2020**, *124*, 106400. [CrossRef]
145. Zaheri, D.M.; Khederzadeh, M. Two-stage coupling of coordinated energy management and self-healing strategies in a smart distribution network considering coupling neighbouring microgrids according to autonomous decentralized restoration technique. *Int. Trans. Electr. Energy Syst.* **2020**, *30*, e12596. [CrossRef]
146. Afrasiabi, M.; Mohammadi, M.; Rastegar, M.; Kargarian, A. Multi-agent microgrid energy management based on deep learning forecaster. *Energy* **2019**, *186*. [CrossRef]
147. Manbachi, M.; Ordonez, M. Intelligent Agent-Based Energy Management System for Islanded AC–DC Microgrids. *IEEE Trans. Ind. Inform.* **2019**, *16*, 4603–4614. [CrossRef]
148. Rokrok, E.; Shafie-Khah, M.; Siano, P.; Catalão, J.P.S. A Decentralized Multi-Agent-Based Approach for Low Voltage Microgrid Restoration. *Energies* **2017**, *10*, 1491. [CrossRef]
149. Fang, X.; Wang, J.; Yin, C.; Han, Y.; Zhao, Q. Multiagent Reinforcement Learning With Learning Automata for Microgrid Energy Management and Decision Optimization. In Proceedings of the 2020 Chinese Control And Decision Conference, Hefei, China, 22–24 August 2020; IEEE: Piscataway, NJ, USA, 2020; pp. 779–784. [CrossRef]
150. Harmouch, F.Z.; Ebrahim, A.F.; Esfahani, M.M.; Krami, N.; Hmina, N.; Mohammed, O.A. An Optimal Energy Management System for Real-Time Operation of Multiagent-Based Microgrids Using a T-Cell Algorithm. *Energies* **2019**, *12*, 3004. [CrossRef]
151. Ji, Y.; Wang, J.; Xu, J.; Li, D. Data-Driven Online Energy Scheduling of a Microgrid Based on Deep Reinforcement Learning. *Energies* **2021**, *14*, 2120. [CrossRef]
152. Jiang, K.; Wu, F.; Zong, X.; Shi, L.; Lin, K. Distributed Dynamic Economic Dispatch of an Isolated AC/DC Hybrid Microgrid Based on a Finite-Step Consensus Algorithm. *Energies* **2019**, *12*, 4637. [CrossRef]
153. Lee, J.-W.; Kim, M.-K.; Kim, H.-J. A Multi-Agent Based Optimization Model for Microgrid Operation with Hybrid Method Using Game Theory Strategy. *Energies* **2021**, *14*, 603. [CrossRef]
154. Nie, H.; Chen, Y.; Xia, Y.; Huang, S.; Liu, B. Optimizing the Post-Disaster Control of Islanded Microgrid: A Multi-Agent Deep Reinforcement Learning Approach. *IEEE Access* **2020**, *8*, 153455–153469. [CrossRef]
155. Hu, Y.; Li, W.; Xu, K.; Zahid, T.; Qin, F.; Li, C. Energy Management Strategy for a Hybrid Electric Vehicle Based on Deep Reinforcement Learning. *Appl. Sci.* **2018**, *8*, 187. [CrossRef]
156. Qazi, H.S.; Liu, N.; Wang, T. Coordinated Energy and Reserve Sharing of Isolated Microgrid Cluster using Deep Reinforcement Learning. In Proceedings of the 2020 5th Asia Conference on Power and Electrical Engineering (ACPEE), Chengdu, China, 4–7 June 2020; pp. 81–86. [CrossRef]
157. Ji, Y.; Wang, J.; Xu, J.; Fang, X.; Zhang, H. Real-Time Energy Management of a Microgrid Using Deep Reinforcement Learning. *Energies* **2019**, *12*, 2291. [CrossRef]
158. Kuznetsova, E.; Li, Y.-F.; Ruiz, C.; Zio, E.; Ault, G.; Bell, K. Reinforcement learning for microgrid energy management. *Energy* **2013**, *59*, 133–146. [CrossRef]
159. Tomin, N.; Zhukov, A.; Domyshev, A. Deep Reinforcement Learning for Energy Microgrids Management Considering Flexible Energy Sources. *EPJ Web Conf.* **2019**, *217*, 01016. [CrossRef]
160. Karimizadeh, K.; Soleymani, S.; Faghihi, F. Microgrid utilization by optimal allocation of DG units: Game theory coalition formulation strategy and uncertainty in renewable energy resources. *J. Renew. Sustain. Energy* **2019**, *11*, 025505. [CrossRef]
161. Karavas, C.-S.; Arvanitis, K.; Papadakis, G. A Game Theory Approach to Multi-Agent Decentralized Energy Management of Autonomous Polygeneration Microgrids. *Energies* **2017**, *10*, 1756. [CrossRef]
162. Naz, A.; Javaid, N.; Rasheed, M.B.; Haseeb, A.; Alhussein, M.; Aurangzeb, K. Game Theoretical Energy Management with Storage Capacity Optimization and Photo-Voltaic Cell Generated Power Forecasting in Micro Grid. *Sustainability* **2019**, *11*, 2763. [CrossRef]
163. Gao, B.; Liu, X.; Wu, C.; Tang, Y. Game-theoretic energy management with storage capacity optimization in the smart grids. *J. Mod. Power Syst. Clean Energy* **2018**, *6*, 656–667. [CrossRef]

164. Zhou, Z.; Xiong, F.; Xu, C.; Jiao, R. *Energy Management in Microgrids: A Combination of Game Theory and Big Data Based Wind Power Forecasting*; IntechOpen: London, UK, 2017.

165. Montes-Parra, M.; Garcia-Hernandez, J.; Gordillo-Sierra, J.; Jimenez-Estevez, G.; Mendoza-Araya, P. Microgrid Energy Management System Optimization using Game Theory. In Proceedings of the 2019 IEEE CHILEAN Conference on Electrical, Electronics Engineering, Information and Communication Technologies (CHILECON), Valparaiso, Chile, 13–27 November 2019; IEEE: Piscataway, NJ, USA, 2019; pp. 1–7. [CrossRef]

166. *2030.7-2017—IEEE Standard for the Specification of Microgrid Controllers*; IEEE: Piscataway, NJ, USA, 2018. [CrossRef]

167. Rebollal, D.; Carpintero-Rentería, M.; Santos-Martín, D.; Chinchilla, M. Microgrid and Distributed Energy Resources Standards and Guidelines Review: Grid Connection and Operation Technical Requirements. *Energies* **2021**, *14*, 523. [CrossRef]

168. Battula, A.R.; Vuddanti, S. Optimal reconfiguration of balanced and unbalanced distribution systems using firefly algorithm. *Int. J. Emerg. Electr. Power Syst.* **2021**. [CrossRef]

169. Joos, G.; Reilly, J.; Bower, W.; Neal, R. The Need for Standardization: The Benefits to the Core Functions of the Microgrid Control System. *IEEE Power Energy Mag.* **2017**, *15*, 32–40. [CrossRef]

170. Hirsch, A.; Parag, Y.; Guerrero, J. Microgrids: A review of technologies, key drivers, and outstanding issues. *Renew. Sustain. Energy Rev.* **2018**, *90*, 402–411. [CrossRef]

171. Rana, M.; Xiang, W.; Wang, E.; Jia, M. IoT Infrastructure and Potential Application to Smart Grid Communications. In Proceedings of the GLOBECOM 2017—2017 IEEE Global Communications Conference, Singapore, 4–8 December 2017; IEEE: Piscataway, NJ, USA, 2017; pp. 1–6. [CrossRef]

172. Wu, Y.; Wu, Y.; Guerrero, J.M.; Vasquez, J.C.; Palacios-García, E.J.; Guan, Y. IoT-enabled Microgrid for Intelligent Energy-aware Buildings: A Novel Hierarchical Self-consumption Scheme with Renewables. *Electronics* **2020**, *9*, 550. [CrossRef]

173. Le, T.N.; Chin, W.-L.; Truong, D.K.; Nguyen, T.H. Advanced Metering Infrastructure Based on Smart Meters in Smart Grid. In *Smart Metering Technology and Services—Inspirations for Energy Utilities*; IntechOpen: London, UK, 2016.

174. Palacios-Garcia, E.J.; Guan, Y.; Savaghebi, M.; Vasquez, J.C.; Guerrero, J.M.; Moreno-Munoz, A.; Ipsen, B.S. Smart metering system for microgrids. In Proceedings of the IECON 2015—41st Annual Conference of the IEEE Industrial Electronics Society, Yokohama, Japan, 9–12 November 2015; IEEE: Piscataway, NJ, USA, 2015; pp. 003289–003294. [CrossRef]

175. Pop, C.; Antal, C.; Cioara, T.; Anghel, I.; Sera, D.; Salomie, I.; Raveduto, G.; Ziu, D.; Croce, V.; Bertoncini, M. Blockchain-Based Scalable and Tamper-Evident Solution for Registering Energy Data. *Sensors* **2019**, *19*, 3033. [CrossRef]

176. Jeong, B.-C.; Shin, D.-H.; Im, J.-B.; Park, J.-Y.; Kim, Y.-J. Implementation of Optimal Two-Stage Scheduling of Energy Storage System Based on Big-Data-Driven Forecasting—An Actual Case Study in a Campus Microgrid. *Energies* **2019**, *12*, 1124. [CrossRef]

Article

Optimal Energy Management of a Campus Microgrid Considering Financial and Economic Analysis with Demand Response Strategies

Haseeb Javed [1], Hafiz Abdul Muqeet [2], Moazzam Shehzad [3], Mohsin Jamil [4,*], Ashraf Ali Khan [4] and Josep M. Guerrero [5]

[1] Department of Electrical Engineering, Muhammad Nawaz Sharif University of Engineering and Technology, Multan 60000, Pakistan; haseebjaved1996@yahoo.com
[2] Department of Electrical Engineering Technology, Punjab Tianjin University of Technology, Lahore 54770, Pakistan; abdul.muqeet@ptut.edu.pk
[3] Department of Electrical Engineering, University of Engineering and Technology (RCET), Lahore 54890, Pakistan; moazzam.shehzad@uet.edu.pk
[4] Department of Electrical and Computer Engineering, Faculty of Engineering and Applied Science, Memorial University of Newfoundland, St. John's, NL A1B 3X5, Canada; ashrafak@mun.ca
[5] The Villum Center for Research on Microgrids (CROM), AAU Energy, Aalborg University, 9220 Aalborg East, Denmark; joz@et.aau.dk
* Correspondence: mjamil@mun.ca

Citation: Javed, H.; Muqeet, H.A.; Shehzad, M.; Jamil, M.; Khan, A.A.; Guerrero, J.M. Optimal Energy Management of a Campus Microgrid Considering Financial and Economic Analysis with Demand Response Strategies. *Energies* **2021**, *14*, 8501. https://doi.org/10.3390/en14248501

Academic Editor: Surender Reddy Salkuti

Received: 4 November 2021
Accepted: 10 December 2021
Published: 16 December 2021

Publisher's Note: MDPI stays neutral with regard to jurisdictional claims in published maps and institutional affiliations.

Abstract: An energy management system (EMS) was proposed for a campus microgrid (μG) with the incorporation of renewable energy resources to reduce the operational expenses and costs. Many uncertainties have created problems for microgrids that limit the generation of photovoltaics, causing an upsurge in the energy market prices, where regulating the voltage or frequency is a challenging task among several microgrid systems, and in the present era, it is an extremely important research area. This type of difficulty may be mitigated in the distribution system by utilizing the optimal demand response (DR) planning strategy and a distributed generator (DG). The goal of this article was to present a strategy proposal for the EMS structure for a campus microgrid to reduce the operational costs while increasing the self-consumption from green DGs. For this reason, a real-time-based institutional campus was investigated here, which aimed to get all of its power from the utility grid. In the proposed scenario, solar panels and wind turbines were considered as non-dispatchable DGs, whereas a diesel generator was considered as a dispatchable DG, with the inclusion of an energy storage system (ESS) to deal with solar radiation disruptions and high utility grid running expenses. The resulting linear mathematical problem was validated and plotted in MATLAB with mixed-integer linear programming (MILP). The simulation findings demonstrated that the proposed model of the EMS reduced the grid electricity costs by 38% for the campus microgrid. The environmental effects, economic effects, and the financial comparison of installed capacity of the PV system were also investigated here, and it was discovered that installing 1000 kW and 2000 kW rooftop solar reduced the GHG generation by up to 365.34 kg CO_2/day and 700.68 kg CO_2/day, respectively. The significant economic and environmental advantages based on the current scenario encourage campus owners to invest in DGs and to implement the installation of energy storage systems with advanced concepts.

Keywords: smart grid; campus microgrid; batteries; prosumer market; energy management system; distributed generation; renewable energy resources; energy storage system

1. Introduction

Power systems have been facing a lot of issues and challenges, including greenhouse gas (GHG) emissions, complicated network overloading, and rising consumption costs. The conventional power system is not capable enough to handle these challenges and issues effectively, but the developing microgrid systems with distributed generators (DGs) integrated with automated distribution systems and energy storage technologies have the potential to alleviate such issues by applying demand-responsive solutions. A campus microgrid (μG) is made up of energy storage, onsite DGs, and a scheduled load [1].

In addition, it can operate in islanded mode and grid connection mode [2]. The advancement in the microgrids provides an efficient solution for the intelligent monitoring of the system, with an automatic recovery system, persuasive demand control, and high-tech controlling capabilities that are controlled with the help of efficient and intelligent sensors [3]. It also provides a variety of energy-saving and renewable energy integration opportunities for the microgrid for energy producers and consumers through the integrated energy management system (EMS). The energy management strategy requires secure communication between producers and consumers and utilities to operate smart control equipment [4].

The benefits described above are particularly evident for μGs with large loads. Due to the variable nature of their loads, university campuses are one of the large-load microgrid users that come under the category of mixed-load consumers. Because of the availability of onsite power generation, these buildings can export excess electricity to the power grid as a prosumer [5]. Similarly, when local DGs and energy storages are inadequate in fulfilling the overall load demand, they can import electricity from the utility grid [6]. The involvement of these μGs in power systems lowers their operating energy costs, with the focus on the benefits of the distribution system [7]. The grid operator additionally provides different price-based and incentive-based DR schemes to entice such large-scale users to actively participate in energy markets [8].

Energy management solutions are used in accordance with conventional resources to assist in the optimal dispatch to fulfill the load demand at a lower cost and to ensure their active involvement in the microgrid operations [9]. This study concentrated on the optimization of an energy management system for a campus μG with onsite DGs and a battery storage facility attached. The presented EMS concept could efficiently and effectively control the bidirectional power flow between utility networks and the μG, and appropriately schedule the battery charging or discharging patterns for the ESS to decrease the energy costs. The actual load of an existing campus commonly known as UET Taxila, as shown in Figure 1, was taken into account for a comprehensive analysis. The proposed campus μG currently has a national grid network link from a local energy market entitled the Islamabad Electric and Supply Company (IESCO) and also has an additional standby diesel generator and wind energy as an external source. In this study, the economic and environmental impact of solar PV with battery storage and electricity generation with different kinds of renewable energy resources were also addressed.

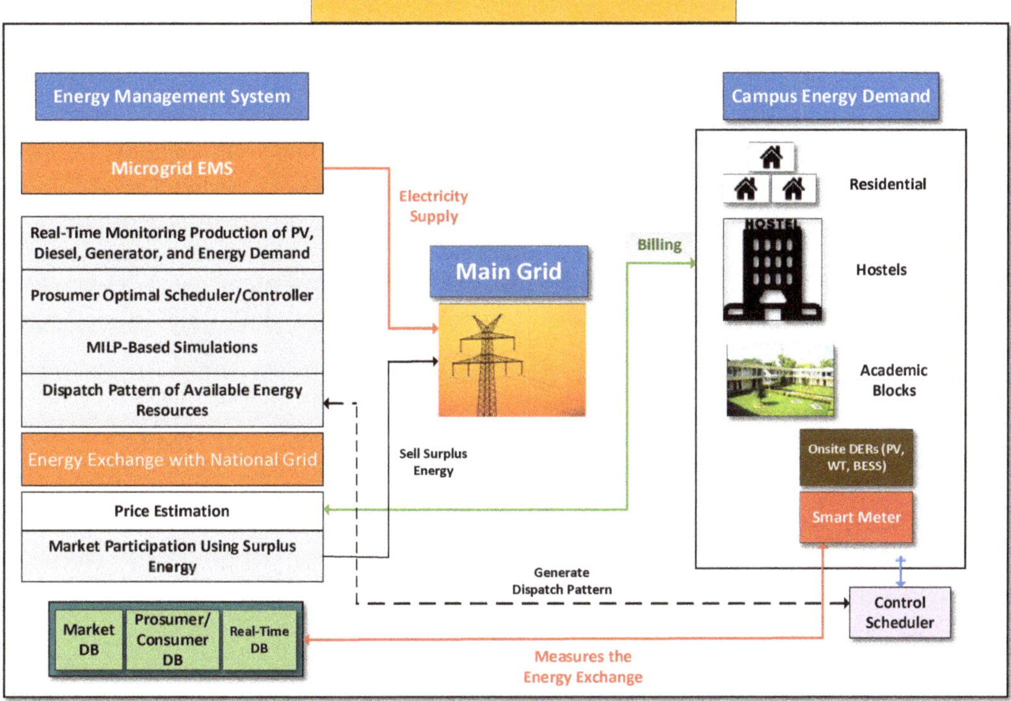

Figure 1. Proposed EMS structure.

2. Recent Research Work: A Detailed Review

A microgrid model was developed for the Malta University Campus by [10]. In this study, the design factor of such a microgrid was developed. It was analyzed under different functional controlling modes, such as peak controlling mode and continuous power mode. This design was demonstrated for the 3D Micro-Grid project [11]. The results were evaluated under different operating modes that analyzed the performance of a microgrid. It showed that the microgrid delivered a constant 50 kW of power between 8 am and 1 pm and it reduced the peak power flow between 5:35 pm and 10 pm in which its consumption was 90 kW. Another microgrid model is presented in [12] for the University of Coimbra, Portugal, which consisted of a PV plant, Li-ion batteries, uni/bi-directional charging of EV, and controllers. The main objective of this microgrid system was to achieve sustainable energy, install the RER resources, and maximize the economic benefit. The results show that it achieved less energy consumption (50 kWh/m^2) and the PV system covered 22.3% of the yearly electricity demand.

A system was proposed by [13] that comprised PV, energy storage cells, and a gas-operated small microturbine to effectively control the number of electric vehicles while making it compatible with the transformers that were also coupled with the microgrid. In the proposed system, LOL calculations were performed, which calculated the number of electric vehicle charges during the 24 h while the supporting transformer was also connected. Using this proposition, the transformer was compatible and supported 17.4 kW on average and had the charging capacity of 30 electric vehicles during a 24 h timespan. This system increased the capacity by about 33% compared with the system that was not connected with μG. The results showed that this system accomplished demand response strategies, energy management scheduling, and maintaining the level of PV through V2G (vehicle-to-grid) technology. Furthermore, advancement was needed in demand response management on the utility side where utility cost reduction was concerned.

The authors of [14] also proposed a distributed DR (demand response)-based algorithm to control the load at the peak time for the Connecticut Campus microgrid, USA. The campus microgrid consisted of a co-generating power plant. The ADMM (alternating directional multiplier method)-based strategy was implemented to analyze the energy consumption scenarios for multiple buildings on campus. The results show that it reduced the energy consumption ratio for multiple buildings (10 buildings here) and it improved the satisfaction level among customers. Fahad et al. [15] presented a cost-effective microgrid solution with the consideration of many feasible cases to optimally schedule the energy for the University AMU (Ali Garh Muslim University), India. They devised the most optimal solution for the AMU campus using HOMER software in which a wind, PV, and grid combination system was the final solution. They calculated the NPC (net present cost) as $17.3 million/year and the CO_2 emissions for the system as 35,792 kg/year.

A novel integrated design was proposed for the size of batteries in [16]. In this study, the exhaustion method was used to obtain the energy management design parameter "d", which mainly focuses on profit maximization. The effects of the SOC (state of charge) and non-operation time "T" were also calculated for the lead-acid battery. The results derived from a utility power company for the interval of 1 min from a PV plant with actual load data showed that using lithium-ion batteries maximized the profit by a 6% margin compared with the lead-acid battery, which had a negative profit margin. Franz et al. [17] initiated a microgrid project at the Siemens campus, Vienna, Austria. The microgrid campus consisted of solar PV panels (1600 m^2), a Siemens controller, Siemens building management systems, Siemens EV charging stations, and 500 kWh battery storage. The project's main objective was to facilitate researchers in the research activities in their respective areas and to optimize the microgrid with the updated energy management systems. It resulted in a peak output of 312 kW and reduced GHG emissions by almost 100 metric tons of CO_2/year.

Furthermore, Abhishek et al. [18] considered a solar PV system, a bio-gas plant diesel generator, and a BESS storage system for multiple universities based in different countries. This study demonstrated the technical aspects, architecture, and load types of different universities based in Iran, the USA, Saudi Arabia, India, China, etc. The techno-financial analysis of this proposed hybrid system was undertaken using the HOMER Pro software. The results showed that the levelized cost for the grid system was in the range of 0.18–1.39 INR/kWh in contrast to the off-grid range of 11.96–18.47 INR/kWh. Moreover, in [19], Dongshin University initiated a self-sufficient smart grid system that contained 1 MW solar PV, a CHP system, an energy storage system, and fuel cells. The main objective was to monitor the power flow information in real time, to monitor the energy consumption, and to stabilize the energy for the campus microgrid. The results showed that this combined energy management system was an optimal solution for the Dongshin Campus.

The majority of these studies were focused on the EMS of µGs and on optimal PV, ESS, and system scheduling. Some studies concentrated only on the economic viability of solar with an ESS as a campus µG, whilst it also estimated the cost savings from PV integration and a properly scheduled ESS. As demonstrated in Table 1, the economic analysis determined the LCOE while including the power interchange between utility, batteries, the ESS, photovoltaic uncertainty, and DRs. This study considered all of this research work in parallel and it provides a structural explanation of the EMS of a campus µG with an optimal sizing and the uncertainties of a photovoltaic system that was deployed in a grid exchange environment to use its real power and load data for different seasons.

Table 1. Comparison of multiple studies with various approaches.

Ref	Power Balance	DR	Grid-Connected (Bi-Directional Supply)	Generation				Optimal Strategy			GHG Emissions
				PV	Wind	DG	ESS	Optimal Scheduling of ESS	Optimal Sizing	Energy Management	
[20]	✓	✓	✓	×	✓	×	×	×	×	×	×
[21]	✓	✓	✓	×	✓	✓	✓	✓	×	×	✓
[22]	×	✓	✓	✓	✓	✓	✓	✓	×	✓	✓
[23]	×	✓	✓	✓	×	✓	✓	✓	✓	✓	✓
[24]	✓	✓	×	✓	✓	✓	✓	✓	×	×	✓
[25]	×	✓	✓	✓	✓	×	×	×	×	×	×
[26]	✓	✓	✓	✓	✓	✓	✓	✓	×	×	×
[27]	✓	×	×	×	✓	✓	✓	✓	×	×	✓
[28]	✓	✓	×	✓	✓	✓	✓	✓	✓	✓	×
[29]	×	✓	✓	✓	✓	✓	✓	✓	×	✓	×
[30]	×	✓	×	×	✓	×	✓	×	✓	✓	✓
[31]	×	✓	✓	×	✓	×	✓	✓	✓	✓	✓
[32]	×	✓	×	✓	✓	×	✓	✓	×	×	✓
[33]	✓	✓	✓	✓	✓	×	×	×	×	×	✓
[34]	✓	✓	×	×	✓	✓	✓	×	×	×	✓
[35]	✓	✓	✓	×	✓	✓	×	×	×	×	×
[36]	×	×	✓	✓	✓	✓	✓	✓	×	×	×
Proposed Model	✓	✓	✓	✓	✓	✓	✓	✓	✓	✓	✓

This study's key contributions may be summarized as follows:

(1) A smart energy management system was suggested to optimize the scheduling process of onsite DGs, ESSs, and grid energy utilizing MILP with the consideration of the TOU-based demand response to enhance the consumption from RERs and to lessen operating electricity costs and the system load during the peak consumption hours.

(2) Degrading costs of the battery are also considered with stochastic PV production that was employed in a campus prosumer μG.

(3) An economic and financial analysis was also conducted here to observe the techno-economic effects of different sizes with an environmentally friendly DG and an optimal ESS was also investigated here, which focused on a net-metering-based TOU environment.

The rest of this paper is laid out as follows. The system model and solution technique are described in Section 2. In Section 3, the detailed problem formulation is provided. Section 4 contains several case studies, as well as numerical findings. Finally, Section 5 contains the concluding thoughts, as well as some recommendations for further research.

3. Proposed Formulation of the μG System

3.1. Proposed Conceptual Model

The structural description of the proposed framework is illustrated in Figure 1, which constituted an EMS, a prosumer μG, and a utility grid. The campus prosumer μG comprised a variety of academic loads and storages facility, as well as two energy supplies (solar and a diesel engine). The prosumer, on the other hand, had a net-metering-based agreement with the electricity provider and was able to trade any excess energy back to the utility grid.

The proposed energy management structure at the prosumer building collected the load demand consumption data, weather forecast statistics, TOU pricing data, the ESS primary condition, as well as its feed-in constraints as input conditions and identified the best way to satisfy the load demand using available resources while staying within the operational and design limitations. The control scheduler used these optimal solutions to allocate the available resources. A facility for storing data of some significant parameters was indeed accessible for the suggested energy management system that will be manipulated for countless future profits. Energy trade data, TOU price data, and prosumer loading information was stored in a real-time database, marketplace database, and prosumer database. The formulation of the proposed approach is given in the next sections.

3.2. Problem Formulation

This suggested mathematical system model is presented with the linear optimization method with the aim to reduce the prosumer operational cost while considering the lifespan of the battery system.

3.3. Objective Function

The goal of the suggested model was to minimize the operating cost (J) of a microgrid that incurred costs related to the energy exchange, WT, DG, and electricity storage degradation (Equations (2)–(5)). The total costs are represented in Equation (1). As illustrated in Equations (4)–(6), the battery life is determined by a variety of parameters, including the number of cycles utilized, capital expenditures, and total system capacity, whereas the storage is expressed by η_{ch}, η_{dch}, P_t^{ch}, and $P_{(b)}$, which are separately denoted in Equation (7).

$$C_T = J = \min \sum_{t=1}^{24} (C_{it}^E + C_{it}^{DG} + C_{it}^{ES} + C_{it}^{WT} + C_{it}^{BESS}) \tag{1}$$

where

$$C_{it}^E = P_{(t)}^G \gamma_t \tag{2}$$

$$C_{it}^{DG} = \alpha T_{Gen} + \beta p_{(t)}^{DG} \tag{3}$$

$$C_{it}^{WT} = S_c \cdot P_{rated}(\$) \tag{4}$$

$$C_{it}^{ES} = \left(\frac{C_{cost}}{n \times CT \times 2} \right) \times \left(\eta_{(chrg)} p_{(t)}^{chrg} + \frac{p_{(t)}^{dchrg}}{\eta_{(dchrg)}} \right) \tag{5}$$

$$C_{it}^{BESS} = S_{BESS} \left(C_{it}^{ES} + C_m^{ESS} f_{om} \right) \tag{6}$$

$$P_{(b)} = \eta_{(chrg)} \cdot p_{(t)}^{chrg} - \frac{p_{(t)}^{dchrg}}{\eta_{(dchrg)}} \tag{7}$$

where C_{it}^E, C_{it}^{WT}, C_{it}^{ES}, and C_{it}^{DG} [37] are the costs of the energy exchange, WT, diesel generators, and battery degradation at any particular time t. IESCO's time-of-use (TOU) pricing tariff was acquired from the university. The energy trade with the utility and their unit prices are indicated with $P_{(t)}^{Grid}$ and γ_t during any hour t. C_{it}^{DG} was calculated with the help of the diesel generator nominal rated capacity (T_G = 600 kW), fuel intercept curve (α = 0.0166 L/h per kW), nominal fuels slope curve given by β = 0.277 L/h per kW, and the overall power generation from DG given by $P_{(t)}^{DG}$. S_c denotes the specific cost and P_{rated} denotes the rated power [38]. The frequent battery charging efficiency, charging power of the battery, discharging efficiency, and battery discharging powers are characterized by $\eta_{(chrg)}$, $p_{(t)}^{chrg}$, $\eta_{(dchrg)}$, and $p_{(t)}^{dchrg}$, respectively [39], while C_{cost} denotes the specific cost of energy storage indicated in Equation (5). In Equation (6), S_{BESS} denotes the size of the battery, C_m^{ESS} denotes the maintenance cost of the ESS, and f_{om} denotes the maintenance factor. The total battery power is given by $P_{(b)}$, which is indicated in Equation (7).

3.4. Load-Balancing Equality Constraint

The load-balancing constraint basically expresses the supply–demand equilibrium constraints. Equation (8) should be met and satisfied in order to achieve this equilibrium. P_t^{pv} and P_t^l [40] are the output of the solar power generation (kW) and the prosumer load, respectively.

$$P_{(t)}^G + P_{(t)}^{PV} + P_{(t)}^b + P_{(t)}^{DG} + P_{(t)}^{WT} + P_{(t)}^{BESS} = P_{(t)}^{total} \tag{8}$$

3.5. ESS Constraints

The ESS should not be overlooked in energy management, as it supports the control of the electrical load, mostly in the occurrence of a grid inability and grid failures [41]. Because the ESS is typically difficult to charge or discharge rapidly, its power limit was considered in the limitations (Equations (9)–(13)). The battery charge in the ESS relies on its earlier state $BSOC_{(t-1)}$, which was integrated into Equation (14) at any time t ($BSOC_t$). The BSOC maximum and minimum limits, denoted with $BSOC_{(minimum)}$ and $BSOC_{(maximum)}$, were included in Equation (15) to avoid the ESS overloading and complete discharge [42]. The battery's state-of-charge ($BSOC_t$) at the day's end is equivalent to the start of the battery state ($BSOC_0$) at the beginning of the day, as indicated in Equation (16).

$$\frac{BSOC_{t-1} - BSOC_{max}}{100} C^{es} \leq P^b_{(t)} \tag{9}$$

$$P^b_{(t)} \leq \frac{BSOC_{(t-1)} - BSOC_{(min)}}{100} C^{es} \tag{10}$$

$$0 \leq \eta_{(ch)} P^{chrg}_{(t)} \leq Y^{chrg}_t P^b_{(chrg,\ max)} \tag{11}$$

$$0 \leq \frac{P^{dchrg}_{(t)}}{\eta_{(dchrg)}} \leq Y^{dchrg}_t P^b_{(dchrg,max)} \tag{12}$$

$$Y^{ch}_t + Y^{dch}_t \leq 1 \forall t \tag{13}$$

$$BSOC_{(t)} = BSOC_{(t-1)} - \frac{100 \times \eta_{(dchrg)} P^{dchrg}_{(t)}}{C^{es}} - \frac{100 \times P^{dchrg}_{(t)}}{C^{es} \eta_{(dchrg)}} \tag{14}$$

$$BSOC_{(minimum)} \leq BSOC_{(t)} \leq BSOC_{(maximum)} \tag{15}$$

$$BSOC_{(t)} = BSOC_{(0)} \tag{16}$$

To properly schedule the energy usage in the EMS, the battery output power P^b_t was included in the equality constraint stated in Equation (8). ESS charging/discharging are represented by the simulated values of P^b_t. The two integer variables μ^{chrg}_t and μ^{dchrg}_t represent the ESS charging/discharging, respectively, in just about any interval "t". To simply avoid the BESS charging/discharging issue for the equivalent durations, the binary pattern characteristics that are shown in Equations (11)–(13) may not be "1" at the regular intervals. For most of these variables, a value equal to "1" expresses the activation mode. The power output gradient of its battery storage is provided below:

$$\left| P^{Battery}_{(t)} - P^{Battery}_{(t+1)} \right| \leq \Delta P^{Battery} \tag{17}$$

3.6. Limitations of the Diesel Generator and Grid

As all utility companies integrate their components depending on the load demand, users continuously sign peak periods agreements with customers. Any request that exceeds the terms of this contract will result in fines or a termination of the power supply. Diesel generators, similarly, cannot handle loads that exceed their rated capacity. Expressions are used to account for power supply constraints for the diesel generator and grid connection (Equations (18) and (19)) [43].

$$P^G_{(min)} \leq P^{Grid}_{(t)} \leq P^G_{(max)} \tag{18}$$

$$P^{DG}_{(min)} \leq P^{DG}_{(t)} \leq P^{DG}_{(max)} \tag{19}$$

3.7. Energy Exchange between the Grid and Prosumer

The power system energy (e_n^g) transacted with the grid in a day is shown below, where power import minus export from and to the grid are denoted by $p_{(t)}^g$:

$$E_n^{Grid} = \sum_{t_1}^{t_{24}} P_{(t)}^G \times h \tag{20}$$

3.8. Probabilistic PV Model

The energy generation of wind and solar energy is unpredictable and highly dependent on the environment and solar irradiance. The data from the entire year was evaluated under random situations. This analysis used a previously constructed solar irradiance model [44]. It also evaluated the variables of the probability density function (PDF). A total of 365 scenarios may be created in 24 h utilizing the Latin hypercube (LHS) universal sampling approach [45]. As previously stated, the goal was to reduce the calculation or computation load [46]. The fast-forwarding method was utilized to lessen the randomly produced scenarios to about 40 [47].

$$F_{0=} \frac{1}{\sigma\sqrt{2\pi}} (e^{-\left(\frac{1-\mu}{2\sigma^2}\right)^2} \tag{21}$$

$$P_t^{pv} = \eta_{PV} \cdot (j\alpha_{PV} \cdot I) \tag{22}$$

In Equation (21), the normal distribution function, or Gaussian function, was applied to analyze the uncertainty model for solar irradiation [48], where η_{PV}, J, I, and α_{PV} are the solar panel's efficiency (17%), total operational cost, the solar irradiance pattern (kW/m^2), and the solar panel's area (m^2), respectively, while μ and σ indicate the normal distribution mean and standard deviation, respectively. Equation (22), which is based on the solar irradiation of a specific region, indicates the output of solar PV and is given above as P_t^{pv}. Figure 2 illustrates the normal distribution with the standard deviation for the photovoltaic irradiance's predictable pattern in the region of Taxila in Pakistan. The Taxila region's latitude and longitude are 33.746° N and 72.839° E, respectively, corresponding to a daily irradiance value of 5.3 kWh/m^2 [49].

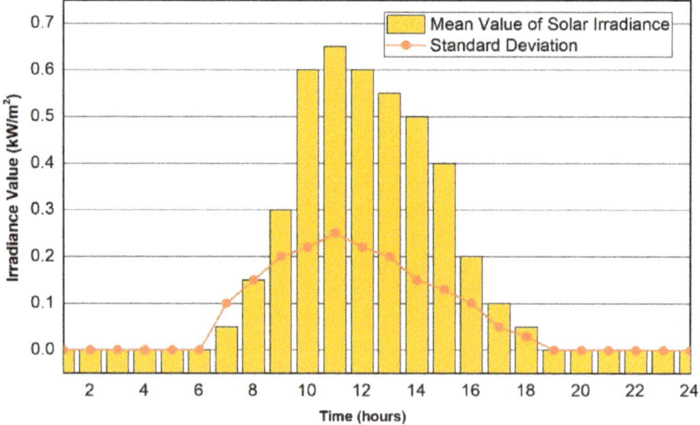

Figure 2. Mean and standard deviation curves.

3.9. Grid Energy Exchange: Wind Turbine Operation

Equation (23) expresses the wind power output $P_{(t)}^{Output}$ exchanged with the utility grid:

$$P_{(t)}^{Output} = \begin{cases} 0, & V_{(t)} < V_{ci} \\ P_{rated}^{WT} \times \left(\frac{v_w - v_{ci}}{v_r - v_{ci}} \right), & v_{ci} < v_{(t)} < v_r \\ P_{rated}^{WT} + \left(\frac{Y_w - V_r}{V_{ci} - V_r} \right) \times \left(P_{co}^{WT} - P_r^{WT} \right), & v_r < v_w < v_{co} \\ 0, & v_{co} < v_w \end{cases} \tag{23}$$

where v_{ci} is the minimal cut-in speed necessary for the *WT* to generate electricity. The maximal cut-out speed at which optimal electricity may be produced is indicated as v_{co}; if this speed is surpassed, the turbine is switched off in order to avoid damage.

3.10. Levelized Cost of Energy (LCOE)

The levelized energy cost is assessed in different scenarios when performing a fair and equitable analysis for the systems. It is defined as the ratio of the entire system cost of installation (USD) to the total energy generated (kWh). The LCOE of storage or a particular energy source is stated in USD per kilowatt-hour. It covers all the associated expenses, which include the cost of installation, operating costs, maintenance, and capital investments. It can also be defined as the lowest possible price at which electricity is generated and used during the lifetime of a particular energy source or storage equipment in order to attain the breakeven point [50]. The LCOE formula may be expressed mathematically as follows:

$$LCOE \quad \frac{\text{Lifespan Cost(USD)}}{\text{Lifetime Energy Generation (kWh)}} \tag{24}$$

3.11. Solution Methodology

Because the suggested system's model's objective function and its related constraints are generally linear models with many integer variables, MILP programming was implemented since it is excellent for solving linear programming problems. This same MILP technique is widely used globally as an optimization method for resolving various kinds of optimization problems associated with marketing and optimal scheduling [51]. Furthermore, it is compared to various metaheuristic approaches that yield inferior results, while MILP yields the most optimal results. As a result, the MILP method is widely used in EMS optimization [52]. The basic structure of a mixed-integer problem is described as follows:

$$\min_{x} \ f^t x \tag{25}$$

$$t_0 \begin{cases} B \cdot x \leq b \\ B_{eq} \cdot x = b_{eq} \\ xb \leq x \leq yb \end{cases} \tag{26}$$

In Equations (25) and (26), xb, yb, x, b, b_{eq}, and f are vectors, where B_{eq} and B are matrixes. The main flow diagram for controlling the proposed campus μG is given in Figure 3. Initially, one hour before each day's arrival, all of the input data that is required for the day is loaded; the forecasted irradiance, load patterns, temperature, ESS starting state, TOU tariff information, and its related parameters are all part of the data. The simulation of the provided optimization method was based on a regular period of each hour prior to usage. The suggested technique was emulated in MATLAB software, version R2017a, with an Intel (R) core (TM) i7-7700 @ 2.80 GHz processor with 8 GB RAM.

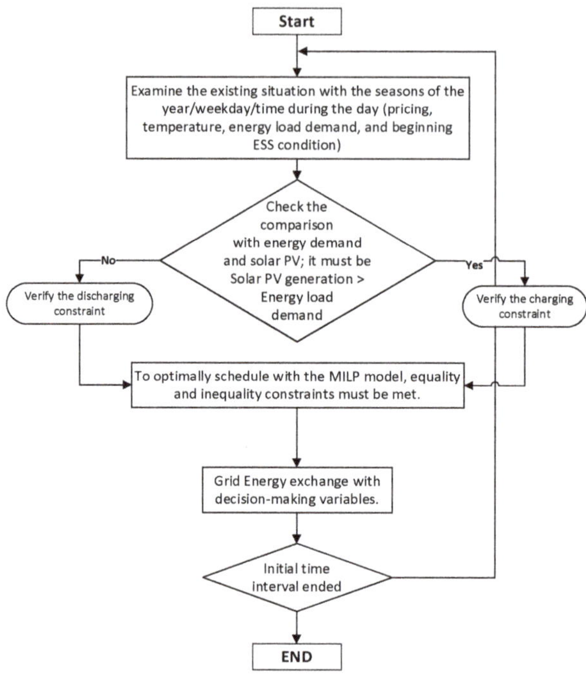

Figure 3. Proposed solution methodology flowchart.

4. Results and Discussion

The concept shown here is used for the prosumer microgrid in Section 3 of the Punjab province. There are eight hostel residences for university students, fourteen departments for different fields, and six faculties. At the moment, the university's load is fed by a 2 MW grid interconnection. The capability of the university rooftop solar panel installation is measured as being 4 MW using a concise assessment of the available space for the university rooftop.

Since NEPRA (National Electric Power Regulatory Agency, Pakistan) enables only 1 MW of energy exchange for an electricity grid, we had a capacity constraint of 4 MW for the campus due to limited resources. In our situation, as compared to the distributed generation sizing, we concentrated on the method applied in [53]. An onsite 2 MW solar photovoltaic setup was taken into consideration for extensive economic and technical analysis. Additional effects highlighted here include utilization of the existing backup generator in the case of a power grid breakdown.

Furthermore, the proposed framework was planned to have an effective net-metering infrastructure that permitted power outflow regulated up to 1 MW to offset the cost of prosumer energy use while fulfilling charging/discharging constraints. The campus load varied constantly due to the loads of the hostels, academic buildings, administrative offices, and housing colonies on campus in summer and winter, which were incorporated into the constraints, as shown in Figure 3.

According to the report of [54], Pakistan generates 5100 kWh of energy production per day through a 1 MW solar facility. As a result, in this study, we found a solution by constructing a photovoltaic system for the campus μG. Further, in our approach, a BESS system was also proposed. In this approach, Li-ion batteries were presented, with the benefits of their extended lifetime, exceptional performance, good power output, high dependability, and minimal self-discharge [55].

4.1. Case Study

In the given scenario, an optimal scheduling strategy was given for the university microgrid with different peak timings and off-peak timings year-round. Variance in the load patterns was commonly observed here in this study, and for the convenience of this study, these patterns were assumed equivalent for all seasons. In Pakistan, the maximum energy consumption periods are May–August, but the data were analyzed for the whole year [56]. Peak load statistics for all months are studied here for the economic analysis, with worst-case scenarios included. Choosing the worst-case scenario yielded the best potential result in terms of cost reduction. To maximize the benefit, the energy produced by solar can be transferred to the grid.

The actual power consumption of the institution was taken into account in regular periods and data from a nearby grid was used to assess the electricity generation expenses on a regular basis.

The load fluctuation behavior that was observed for different seasons is illustrated in Figure 4, while the load allocation patterns between academic blocks, hostels, and campus office buildings are presented in Figure 5.

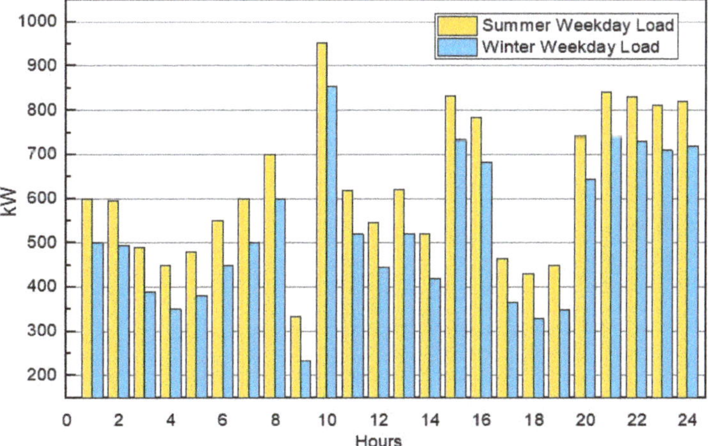

Figure 4. Campus weekdays summer and winter load behavior patterns.

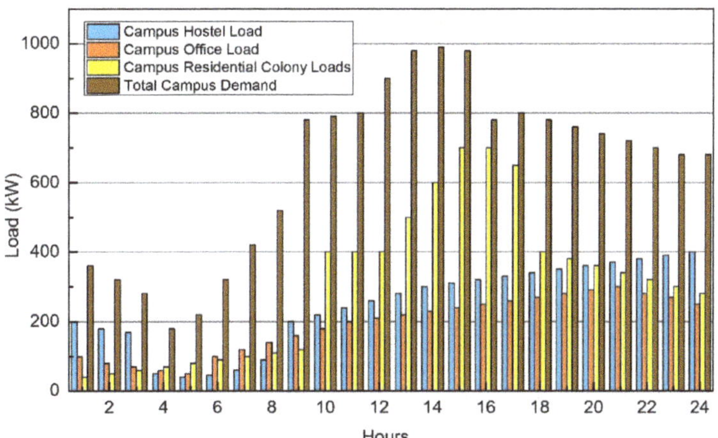

Figure 5. Average campus load demand among buildings.

The loads of the administrative and academic blocks were scheduled to be higher is when the university is on, while the peak electricity demands in the hostels and resident colonies were monitored until midnight. Table 2 represents the numerous factors that were interconnected with the system, whereas Table 3 describes the TOU agreement's power-pricing information [57]. The comprehensive solar irradiance patterns analyzed here were acquired from [58], and the data features were modeled and evaluated using the previously described probability density function (PDF) in Equation (19). The primary goal of using the PDF was to create regular irradiance patterns, while the previously generated solar irradiance pattern predicted the PV production power consumption patterns using Equation (20). Table 4 provides the case study data.

Table 2. System parameters.

Parameters	Value	Parameters	Value
P^{pv}_{rated}	2000 kW	C^{ES}	800 kWh
$P^{G}_{(t,max)}$	2000 kW	$P^{G}_{(t,min)}$	−1000 kW
$P^{b}_{(t,max)}$	800 kW	$P^{b}_{(t,min)}$	−800 kW
$BSOC^{b}_{(max)}$	90%	$BSOC^{b}_{(min)}$	10%
$BSOC_0$	50%	$P^{DG}_{(t)}$	600 kW

Table 3. Electricity prices in peak and off-peak times.

Cases	Only Grid	Solar PV	ESS	Diesel Generator	Wind	Power Load
Case 1	✓	×	×	×	×	
Case 2	✓	✓	×	×	×	Multiple
Case 3	✓	✓	✓	×	×	Load
Case 4	✓	✓	✓	✓	×	Variations
Case 5	✓	✓	✓	✓	✓	

Table 4. Different case study profiles.

Seasons/Parameters	Spring	Summer	Autumn	Winter
Months	March–April	May–August	September–October	November–February
Peak times	11:00 AM–5:00 PM	8:00 AM–6:00 PM	9:00 AM–5:00 PM	12:00 AM–4:00 PM
Unit prices in peak times (d)	0.11	0.146	0.11	0.10
Off-peak times	Rest of the day	Rest of the day	Rest of the day	Rest of the day
Unit prices in off-peak times (USD)	0.09	0.126	0.08	0.10

4.2. Different Seasons Case Study

In this case scenario, the energy exchange evaluation among the grids and energy demand were analyzed using the price-based data presented in Table 3. Several strategies were developed here to better understand the energy consumption for different seasons. These case scenarios were developed to maximize the dependability of solar PV, where the daily hours of sunlight were 8–10 h in summer and 6–8 h in winter; different case scenarios were implemented to get the most economical result by optimally scheduling the different resources to minimize the dependence on the grid and to minimize the campus grid electricity cost.

Case (1) (energy received from the grid): In the first case, the campus's energy demands were fulfilled completely by the utility. For the campus, no solar PV, ESS, wind, or DG were considered in this case. The operating costs of the electricity were determined using the time-of-use (TOU) rate, which was USD 1430.8. In this scenario, the LCOE was determined as 0.0988 USD/kWh.

The analysis indicated that the energy operational cost in the first case was exceptionally expensive, but this was utilized as a study case for comparison with other cases for evaluation regarding each season.

Case (2) (energy exchange between PV and grid): In the second case, solar PV was connected with a prosumer microgrid as shown in Figure 6, and this was interconnected to manage both the import of necessary energy from and export of surplus energy to the grid. The rooftop PV system produced 8925.7 kWh, indicating the PV's performance throughout the peak periods of the year, especially in summer. The LCOE for the rooftop PV was 0.055 USD/kWh here. Therefore, the grid power net cost for 24 h dropped by 43.6% relative to the baseline to USD 798.5.

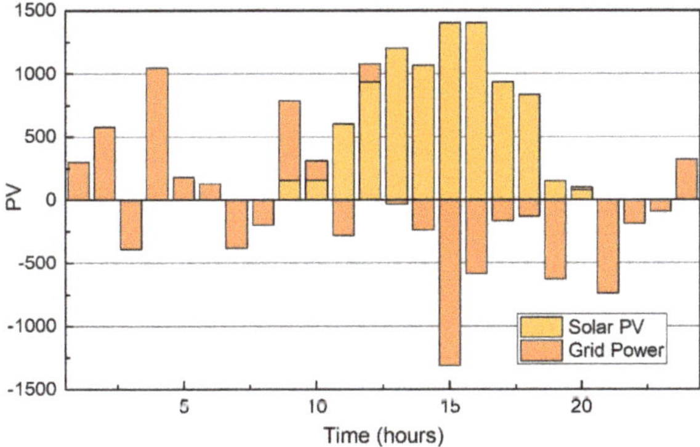

Figure 6. Case (2): Energy exchange with solar PV and the grid.

Case (3) (ESS integration with PV and the grid): In the third case, the ESS was connected with the solar and the grid. The proposed methodology was used to calculate the net electricity costs of USD 819.9 and to best plan for the behavior of battery charge–discharge patterns while taking into account all related associated costs. The LCOE was determined to be 0.056 USD/kWh using TOU-based pricing and optimal BESS scheduling, as presented in Table 5. When compared to the baseline scenario (case 1), it decreased the electricity net cost by 42.8%. Figure 7 represents the energy exchange with the electricity grid, with upper positive and lower negative values representing the energy import and export. The optimal scheduling outcome of the ESS demonstrated that the battery terminated operation at the same amount of SOC, i.e., it continued to operate at 50% precisely based on what it began the current day with. Furthermore, as shown in Figure 8, the ESS intelligently saved excess electricity during off-peak and peak hours and released it proportionately to decrease the operating cost of electricity where Figure 9 state-of-charge of a battery with unit prices with respect to time.

Table 5. Proposed system calculation using LCOE.

Different Scenarios	Imported Utility Power (kWh/Day)	Prosumer Electricity Generation (kWh/Day)	Grid Electricity Net Cost (USD/Day)	Carbon Credit (USD/Day) (A)	Electricity Net Cost without CC (USD/Day) (B)	Electricity Net Cost of CC (USD/Day) (C = B − A)	LCOE (USD/kWh)	Saving (%)
Case 1	14,472.5	-	1430.8	-	1430.8	1430.8	0.0988	-
Case 2	5546.8	8925.7	610.7	165	963.5	798.5	0.055	43.6
Case 3	5546.7	8925.7	711.5	165	984.9	819.9	0.056	42.8
Case 4	4983.2	8925.7	768.2	155	970.5	843.5	0.058	40.2
Case 5	4763.2	9295.9	546.4	145	995.9	850	0.060	38.3

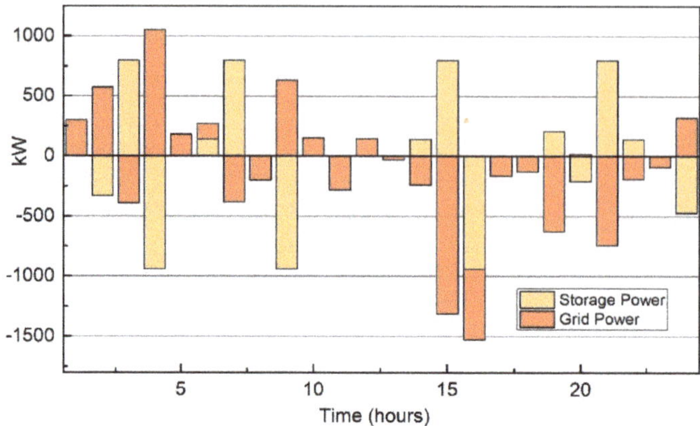

Figure 7. Case (3): Energy exchange with the ESS and grid.

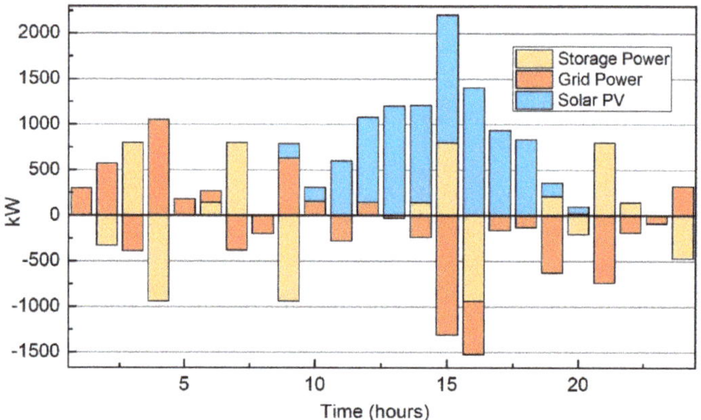

Figure 8. Case (3): Energy exchange with the ESS, grid, and PV.

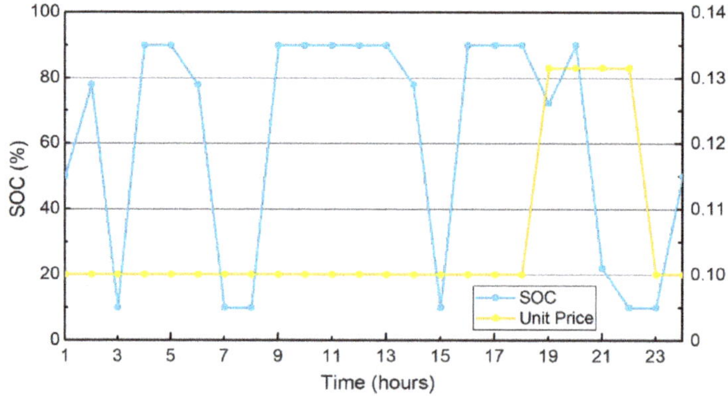

Figure 9. Case (3): State-of-charge of a battery with unit prices.

Case (4) (DG integration with the PV, ESS, and grid): In the fourth case, the campus microgrid combined a diesel generator (DGen) with rooftop solar and a BESS structure to minimize the potential peak demand of energy from the utility grid, even during the summer season (8:00 AM–6:00 PM). The grid consumed electricity up to 50 kW, which was the limit for the smart grid, while the DGen's output power consumption was restricted to 400 kW only in peak hours, as illustrated in Figure 10.

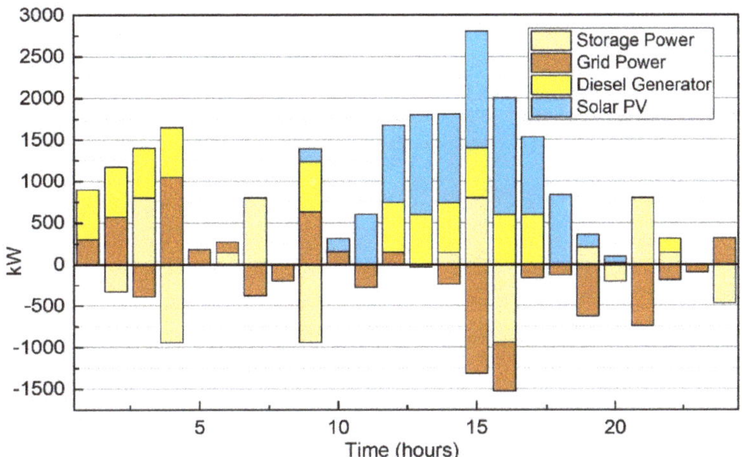

Figure 10. Optimal scheduling in case (4).

After the optimal BESS scheduling, the net cost of power was determined to be USD 843.5 each day. In this case, the measured LCOE was 0.058 USD/kWh, which when compared with the baseline (case (1)), was 40.2% lower with the 2.2 s execution time during the respective seasons, especially in summer because it experienced the lengthy peak periods.

Case (5) (proposed scheduling): The wind turbine system (100 kW) was combined with the PV, ESS, diesel generator, and grid connection in the proposed case, as shown in Figure 11. The wind turbine's power and wind speeds are often 3–5 times greater in August and September. The campus μG used a wind turbine with a power rating of 10 kW and a height of 36.6 m, a rating speed of 9 m/s, a wind cut-in velocity of 3.6 m/s, and a wind cut-out velocity of 26 m/s as the best option among the wind turbines examined. The LCOE computed for wind energy was 0.060 USD/kWh, which was 38.3% less than the baseline case (and better than the 35% found by [53]), which was calculated for a hot or windy weather condition. It was estimated that by integrating the wind turbine system with the ESS, solar PV, DG, and national grid (WAPDA), the UET Taxila university's campus energy net cost will be reduced by 3%.

4.3. Effects of the Sizing of Solar PV on Electricity Cost and Reduction in GHG Emissions with Financial Feasibility

The impact of various solar PV sizes on the obtaining cost of electricity with the utility and the decrease of carbon emissions each day were investigated. When the PV integration was doubled, greenhouse gas emissions were reduced by half, as well as the costs, as illustrated briefly in Table 6.

Figure 12 contains a bar graph that includes the solar PV incorporation in the proposed scenario and the cost consequences for the electricity obtained from the utility. We evaluated the differences in the operational costs of electricity based on the figures acquired in the prior cases. Table 7 depicts the techno-economic assessment data with various components utilized for the proposed system; this assessment included all of the suggested system's maintenance, operating, and capital expenses.

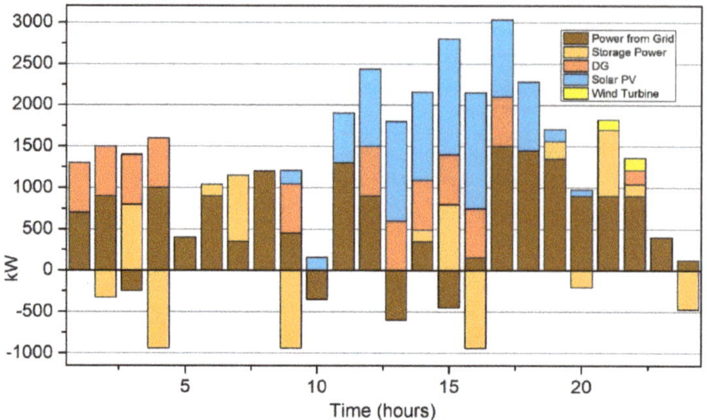

Figure 11. Grid scheduling for the proposed case (5).

Table 6. Profile of the case studies on the grid electricity cost.

Case	Solar PV Penetration Level	Electricity Imported from Utility (kWh/24 h)	Solar PV Electricity Generation (kWh/24 h)	Net Cost of Grid Electricity (USD/Day)	GHG Emissions Reduction (kg/24 h)
Summer	1000 kW	10,037.23	4462.85	1843.20	365.34
	2000 kW	5546.8	8925.7	798.5	700.68
	Pattern of Load Consumption	**Electricity Import from Grid (kWh/24 h)**	**Electricity Generated from Solar PV (kWh/24 h)**	**Grid Electricity Net Cost (USD/day)**	**LCOE (USD/kWh)**
Summer	Lowest	3545.2	8925.7	553.7	0.044
	Average	4986.3	8925.7	697.6	0.050
	Peak	5546.8	8925.7	798.5	0.055

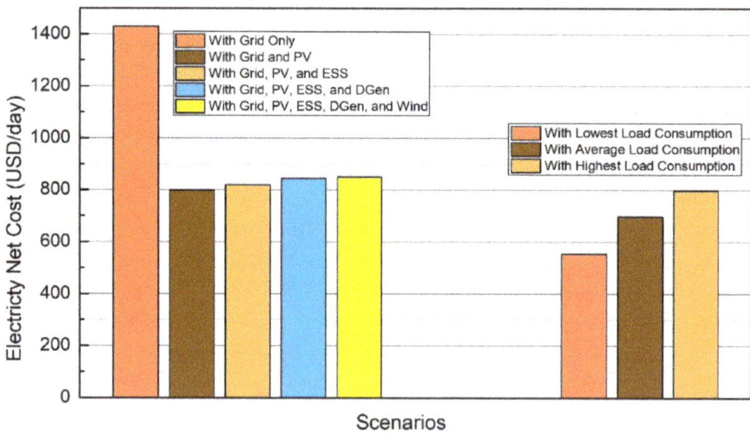

Figure 12. Cost analysis of electricity net cost (USD) in different scenarios.

Table 7. Techno-economic price comparison of different types of components connected with the proposed system.

Sr No.	Objective Components	Parameters	Values	Units
1	Solar PV	PV Rating	1	kW
		Capital Expenses for PV	933.33	USD
		Replacement Cost for PV	800.00	USD
		Maintenance and Operation Cost	13.33	USD/kW
		Derating Factors	88	%
		PV Lifetime	20	Years
2	Converter	Power Ratings	1	kW
		Converter Capital Cost	133.3	USD
		Converter Replacement Cost	106.7	USD
		Maintenance and Operation Cost	160	USD/kW
		Converter Efficiency	90	%
		Converter Lifetime	20	Years
3	BESS	Capital Cost of the Battery	133.3	USD
		Replacement Costs	56	USD
		Battery Size	2.1	kW
		Minimum State of Charge	30	%
		Maximum State of Charge	100	%
		Efficiency	95.5	%
		Battery Life	5	Years
4	WT	Wind Turbine	1	kW
		WT Capital Expenses	15,000	USD
		WT Replacement Cost	800.00	USD
		Maintenance Costs	13.33	USD/kW
		Derating Factors	88	%
		WT Lifetime	20	Years
5	DGs	Net Capital Expenses	9467	USD
		Replacement Costs	28.35	USD
		Operational Costs	2449.5	USD/kW
		Overall Efficiency	80	%
		Lifetime	25	Years
6	Grid	Supply Cost	10	USD
7	Other	Discount	6	%
		Project Lifetime	20	Years

Table 8 shows the cash flow analysis of the respective campus microgrids with the consideration of a year-wise comparison for up to 10 years. It gives a brief comparison for investments, feed-in/export tariff, electricity savings, annual cash flow, and accrued cash flow (cash balance). In Figure 13, a financial analysis was freshly proposed here in this study to observe the financial performance of the campus microgrid. The cash balance was generated with a year-wise comparison to illustrate the accrued cash flow up to almost 20 years. In Figure 14, the financial feasibility was calculated for up to almost 20 years; it shows the annual energy cost for solar PV before and after the installation.

Table 8. Cash flow (USD) analysis of the campus microgrid.

Years	Year 2021	Year 2022	Year 2023	Year 2024	Year 2025
Investments	(7020.00)	0	0	0	0
Feed-in/Export Tariff	0.00	218.55	449.84	445.39	440.98
Electricity Savings	766.29	773.88	781.54	789.28	797.09
Annual Cash Flow	(6253.71)	992.42	1231.38	1234.67	1238.07
Accrued Cash Flow (Cash Balance)	(6253.71)	(5261.29)	(4029.91)	(2795.24)	(1557.17)
Years	**Year 2026**	**Year 2027**	**Year 2028**	**Year 2029**	**Year 2030**
Investments	0	0	0	0	0
Feed-in/Export Tariff	436.61	432.29	428.01	423.77	419.58
Electricity Savings	804.98	812.95	821.00	829.13	837.34
Annual Cash Flow	1241.60	1245.24	1249.01	1252.90	1256.92
Accrued Cash Flow (Cash Balance)	(315.58)	929.67	2178.68	3431.58	4688.50

The research showed that incorporating distributed generating systems offered several advantages, including self-consumption, load demand flexibility, cost savings, and minimized GHG emissions. Therefore, due to these consequences, the proposed approach may be used to reduce the operational costs of campus energy consumption. To properly regulate the distributed generators, a control facility is required. Furthermore, unloading the grid enhances the grid efficiency by incorporating renewable sources. In other situations, capital and installation expenses will be distributed, enabling campus investors to invest further in storage installations and in DG. Grid outages (organized load shedding) are rather common in developing nations due to a variety of difficulties. When the grid is unavailable, diesel generators and the energy storage system can be utilized as backups. Scheduled high load shedding is typical throughout peak times. As a result, diesel generators are employed at peak periods in those specific situations.

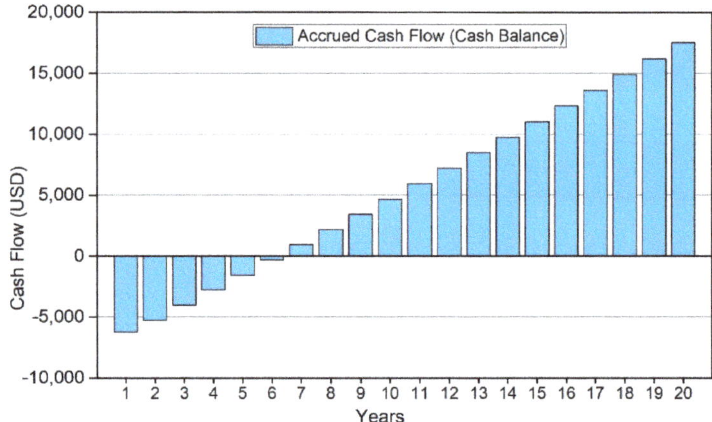

Figure 13. Cash balance comparison—year-wise comparison.

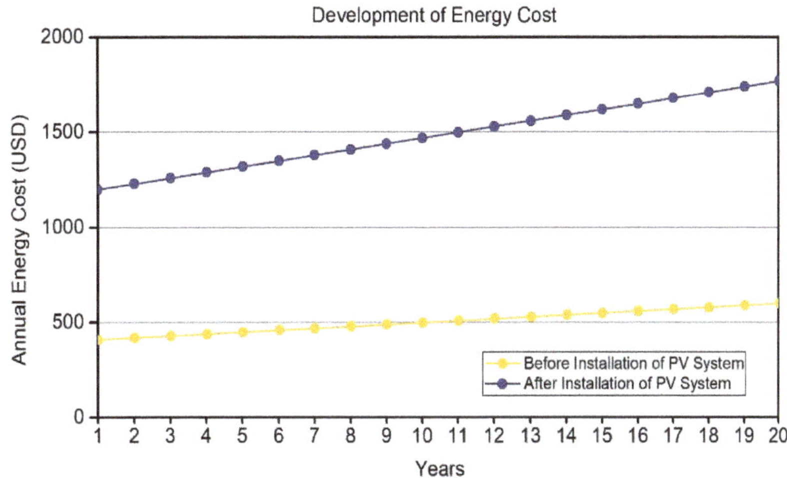

Figure 14. Financial analysis of solar PV—year-wise comparison.

The proposed model shown in Table 9 achieved the optimal results for our campus microgrid due to the utilization of demand response strategies, utilization of renewable energy resources, and optimal scheduling such that it achieved a 38.3% economic benefit to our microgrid, which is the most economical up-to-date.

Table 9. Proposed methodology comparison with the current works.

Ref.	Years	Applications	Methods	Comments	Savings
[59]	2017	IEEE-14 bus system	BBSA	Reliability, energy losses	18.26%
[60]	2018	Campus μG	MILP	ESS degradation Cost, peak demand	5.32%
[61]	2018	(IEEE-15) bus system	NA and conic technique	Financial feasibility	3.3%
[62]	2018	Residential level	MILP	Frequency regulation	7%
[63]	2019	Residential μG	LP	Grid outage	16%
[53]	2020	Campus μG	MILP	DR, ESS degradation	29%, 35%
Proposed model	2021	Campus μG	MILP	Self-consumption, ESS degradation, demand response, optimal scheduling, economic and financial analysis	38.3%

5. Conclusions

In this study, the effective scheduling of a BESS and the effects of PV systems were analyzed for a campus μG to minimize the energy operating costs for a prosumer microgrid with the implementation of actual load data. The suggested system utilized solar energy, a BESS, and diesel generators in several scenarios and their consequences were investigated. The optimal scheduling was implemented in MATLAB and formed as a MILP problem. The TOU pricing-based DR was investigated here as part of a financial and economic analysis, and the ESS was used as a flexible DR framework that could be charged or discharged wisely at various times to meet the budget target without compromising its durability.

Without the DG or ESS, the utility grid supplied all the campus μG's required energy, leading to higher operational expenses. However, when the solar PV, DGen, WT, and especially ESS were combined into the prosumer μG, the daily benefits in different seasons was an approximate 38.3% cost reduction. The environmental effects of various sizes of the installed capacity of the PV system were also investigated here, where it was discovered that installing 1000 kW rooftop solar in hot months may save between 365.34 kg CO_2/day. If 2000 kW of rooftop solar was incorporated in the network, the savings were improved by 700.68 kg CO_2/day. The cost of energy was reduced based on a variety of factors, such as energy consumption, feed-in tariff (FIT), and region. In Pakistan, the FIT has separate prices for importing and exporting energy to those in several other countries, although the cost of supplying energy to the utilities was considerably cheaper than the cost of purchasing energy from the utilities. As a consequence, by investing in on-site solar PV and ESS systems using an appropriate timetable that depends on FIT, location, and load usage, investors may expect their electricity prices to rise by 20–30%. As a result, the optimal charge–discharge method for the ESS plays an important role in the economic performance of prosumer μG with internal RER installations. In a future study, DG uncertainty will be investigated using more advanced mathematical models with many power systems and DR kinds.

Author Contributions: H.J.: writing, methodology, and original draft preparation; H.A.M. and M.S.: supervision; A.A.K.: draft revision; J.M.G. and M.J.: funding. All authors have read and agreed to the published version of the manuscript.

Funding: This research received no external funding.

Institutional Review Board Statement: Not applicable.

Informed Consent Statement: Not applicable.

Data Availability Statement: Not applicable.

Acknowledgments: The authors would like to thank the UET, Pakistan, for giving a formal atmosphere to conduct this research.

Conflicts of Interest: The authors declare no conflict of interest.

Nomenclature and Acronym

The following acronyms and nomenclature are used in this manuscript:

A	*Acronyms*
BSOC	Battery state of charge
BESS	Battery energy storage system
BBSA	Binary backtracking search algorithm
DG	Distributed generator
DERs	Distributed energy resources
DR	Demand response
ESS	Energy storage systems
DSM	Demand-side management
MILP	Mixed-integer linear programming
GHG	Greenhouse gas
FIT	Feed-in tariffs
LP	Linear programming
TOU	Time of use
RERs	Renewable energy resources
PV	Photovoltaic
WT	Wind turbine

B	**Constants and Variables**
$BSOC_{min}$	Minimum $BSOC$ level (%)
$BSOC_{max}$	Maximum $BSOC$ level (%)
$BSOC_t$	$BSOC$ value at time t
$BSOC_0$	The starting value of $BSOC$ at time 0 (%)
C_t^{es}	Cost of storage degradation (USD)
C^{ES}	Rated capacity of energy storage (kWh)
C_t^{e}	Net cost of energy (USD)
C_t^{dg}	Cost of diesel generator (USD)
C_t^{WT}	Net cost of wind energy (USD)
I	Solar irradiance
J	Overall operations cost
μG	Microgrid
E_{net}^{g}	Net energy exchange with the grid
P_t^{bat}	The output power of the battery storage system (kW)
P_t^{ch}	Charging power of the battery (kW)
P_t^{pv}	Solar PV output power (kW)
P_t^{dg}	Diesel generator output power
T	Time interval (hour)
P_t^{g}	Power taken from grid (kW)
P_{max}^{g}	Maximum power exchange limit of utility grid (kW)
P_{min}^{g}	Minimum power exchange limit of utility grid (kW)
P_t^{l}	Load demand of prosumer (kW)
T_G	Diesel generator rated capacity
Sc	Specific cost
μ_t^{ch} / μ_t^{dch}	Storage charging integers/storage discharging integers
λ_t	Electricity rate (USD/kWh)
μ	Solar irradiance mean value
σ	Solar irradiance standard deviation value
β_{pv}	Area of a solar panel
η_{pv}	The efficiency of solar panel

References

1. Hirsch, A.; Parag, Y.; Guerrero, J. Microgrids: A review of technologies, key drivers, and outstanding issues. *Renew. Sustain. Energy Rev.* **2018**, *90*, 402–411. [CrossRef]
2. Li, Y.; Yang, Z.; Li, G.; Zhao, D.; Member, S. Optimal Scheduling of an Isolated Microgrid with Battery Storage Considering Load and Renewable Generation Uncertainties. *IEEE Trans. Ind. Electron.* **2018**, *66*, 1565–1575. [CrossRef]
3. Lu, R.; Hong, S.H.; Zhang, X. A Dynamic pricing demand response algorithm for smart grid: Reinforcement learning approach. *Appl. Energy* **2018**, *220*, 220–230. [CrossRef]
4. El-Hendawi, M.; Gabbar, H.A.; El-Saady, G.; Ibrahim, E.N.A. Control and EMS of a grid-connected microgrid with economical analysis. *Energies* **2018**, *11*, 129. [CrossRef]
5. Javed, H.; Muqeet, H.A. Design, Model & Planning of Prosumer Microgrid for MNSUET Multan Campus. *Sir Syed Univ. Res. J. Eng. Technol.* **2021**, *11*, 1–7.
6. Iqbal, M.M.; Sajjad, I.A.; Nadeem Khan, M.F.; Liaqat, R.; Shah, M.A.; Muqeet, H.A. Energy Management in Smart Homes with PV Generation, Energy Storage and Home to Grid Energy Exchange. In Proceedings of the 2019 International Conference on Electrical, Communication, and Computer Engineering (ICECCE), Swat, Pakistan, 24–25 July 2019; pp. 1–7. [CrossRef]
7. Nasir, T.; Bukhari, S.S.H.; Raza, S.; Munir, H.M.; Abrar, M.; ul Muqeet, H.A.; Bhatti, K.L.; Ro, J.-S.; Masroor, R. Recent Challenges and Methodologies in Smart Grid Demand Side Management: State-of-the-Art Literature Review. *Math. Probl. Eng.* **2021**, *2021*, 5821301. [CrossRef]
8. Muqeet, H.A.; Munir, H.M.; Javed, H.; Shahzad, M.; Jamil, M.; Guerrero, J.M. An Energy Management System of Campus Microgrids: State-of-the-Art and Future Challenges. *Energies* **2021**, *14*, 6525. [CrossRef]
9. Abd, H.; Muqeet, U.L.; Member, G.S.; Ahmad, A. Optimal Scheduling for Campus Prosumer Microgrid Considering Price Based Demand Response. *IEEE Access* **2020**, *8*, 71378–71394. [CrossRef]
10. Hadjidemetriou, L.; Zacharia, L.; Kyriakides, E.; Azzopardi, B.; Azzopardi, S.; Mikalauskiene, R.; Al-Agtash, S.; Al-Hashem, M.; Tsolakis, A.; Ioannidis, D.; et al. Design factors for developing a university campus microgrid. In Proceedings of the 2018 IEEE International Energy Conference (ENERGYCON), Limassol, Cyprus, 3–7 June 2018; pp. 1–6. [CrossRef]

11. Chiou, F.; Fry, R.; Gentle, J.P.; McJunkin, T.R. 3D model of dispatchable renewable energy for smart microgrid power system. In Proceedings of the 2017 IEEE Conference on Technologies for Sustainability (SusTech), Phoenix, AZ, USA, 12–14 November 2017; pp. 1–7.
12. Moura, P.; Correia, A.; Delgado, J.; Fonseca, P.; De Almeida, A. University Campus Microgrid for Supporting Sustainable Energy Systems Operation. In Proceedings of the 2020 IEEE/IAS 56th Industrial and Commercial Power Systems Technical Conference (I&CPS), Las Vegas, NV, USA, 29 June–28 July 2020. [CrossRef]
13. Singh, R.; Kumar, K.N.; Sivaneasan, B.; So, P.; Gooi, H.; Jadhav, N.; Marnay, C. *Sustainable Campus with PEV and Microgrid*; Lawrence Berkeley National Lab (LBNL): Berkeley, CA, USA, 2012; pp. 128–139.
14. Kou, W.; Bisson, K.; Park, S.Y. A Distributed Demand Response Algorithm and its Application to Campus Microgrid. In Proceedings of the 2018 IEEE Electronic Power Grid (eGrid), Charleston, SC, USA, 12–14 November 2018; pp. 1–6. [CrossRef]
15. Iqbal, F.; Siddiqui, A.S. Optimal configuration analysis for a campus microgrid—A case study. *Prot. Control Mod. Power Syst.* **2017**, *2*, 23. [CrossRef]
16. Yang, Y.; Ye, Q.; Tung, L.J.; Greenleaf, M.; Li, H. Integrated Size and Energy Management Design of Battery Storage to Enhance Grid Integration of Large-Scale PV Power Plants. *IEEE Trans. Ind. Electron.* **2018**, *65*, 394–402. [CrossRef]
17. Mundigler, F. Microgrid Project in Vienna: Small Grid, Major Impact. Siemens Website. 2020. Available online: https://new.siemens.com/global/en/company/stories/infrastructure/2020/microgrid-project-in-vienna.html (accessed on 20 October 2021).
18. Kumar, A.; Singh, A.R.; Deng, Y.; He, X.; Kumar, P.; Bansal, R.C. Multiyear load growth based techno-financial evaluation of a microgrid for an academic institution. *IEEE Access* **2018**, *6*, 37533–37555. [CrossRef]
19. University, D. Dongshin University Smart Energy Campus—Microgrid PRJ. 2020. Available online: http://www.nuritelecom.com/service/micro-grid.html (accessed on 21 October 2021).
20. Chen, J.J.; Qi, B.X.; Rong, Z.K.; Peng, K.; Zhao, Y.L.; Zhang, X.H. Multi-energy coordinated microgrid scheduling with integrated demand response for flexibility improvement. *Energy* **2021**, *217*, 119387. [CrossRef]
21. Vahedipour-Dahraie, M.; Rashidizadeh-Kermani, H.; Anvari-Moghaddam, A.; Siano, P. Risk-averse probabilistic framework for scheduling of virtual power plants considering demand response and uncertainties. *Int. J. Electr. Power Energy Syst.* **2020**, *121*, 106126. [CrossRef]
22. Karkhaneh, J.; Allahvirdizadeh, Y.; Shayanfar, H.; Galvani, S. Risk-constrained probabilistic optimal scheduling of FCPP-CHP based energy hub considering demand-side resources. *Int. J. Hydrogen Energy* **2020**, *45*, 16751–16772. [CrossRef]
23. Ali, S.; Malik, T.N.; Raza, A. Risk-Averse Home Energy Management System. *IEEE Access* **2020**, *8*, 91779–91798. [CrossRef]
24. Bostan, A.; Nazar, M.S.; Shafie-khah, M.; Catalão, J.P.S. An integrated optimization framework for combined heat and power units, distributed generation and plug-in electric vehicles. *Energy* **2020**, *202*, 117789. [CrossRef]
25. Vahedipour-Dahraie, M.; Rashidizadeh-Kermani, H.; Anvari-Moghaddam, A.; Siano, P. Flexible stochastic scheduling of microgrids with islanding operations complemented by optimal offering strategies. *CSEE J. Power Energy Syst.* **2020**, *6*, 867–877. [CrossRef]
26. Sheikhahmadi, P.; Mafakheri, R.; Bahramara, S.; Damavandi, M.Y.; Catalão, J.P.S. Risk-based two-stage stochastic optimization problem of micro-grid operation with renewables and incentive-based demand response programs. *Energies* **2018**, *11*, 610. [CrossRef]
27. Qiu, J.; Meng, K.; Zheng, Y.; Dong, Z.Y. Optimal scheduling of distributed energy resources as a virtual power plant in a transactive energy framework. *IET Gener. Transm. Distrib.* **2017**, *11*, 3417–3427. [CrossRef]
28. Wang, Y.; Huang, Y.; Wang, Y.; Zeng, M.; Yu, H.; Li, F.; Zhang, F. Optimal scheduling of the RIES considering time-based demand response programs with energy price. *Energy* **2018**, *164*, 773–793. [CrossRef]
29. Hosseinnia, H.; Modarresi, J.; Nazarpour, D. Optimal eco-emission scheduling of distribution network operator and distributed generator owner under employing demand response program. *Energy* **2020**, *191*, 116553. [CrossRef]
30. Mansour-Saatloo, A.; Mirzaei, M.A.; Mohammadi-Ivatloo, B.; Zare, K. A Risk-Averse Hybrid Approach for Optimal Participation of Power-to-Hydrogen Technology-Based Multi-Energy Microgrid in Multi-Energy Markets. *Sustain. Cities Soc.* **2020**, *63*, 102421. [CrossRef]
31. Lekvan, A.A.; Habibifar, R.; Moradi, M.; Khoshjahan, M.; Nojavan, S.; Jermsittiparsert, K. Robust optimization of renewable-based multi-energy micro-grid integrated with flexible energy conversion and storage devices. *Sustain. Cities Soc.* **2021**, *64*, 102532. [CrossRef]
32. Khaloie, H.; Anvari-Moghaddam, A.; Hatziargyriou, N.; Contreras, J. Risk-constrained self-scheduling of a hybrid power plant considering interval-based intraday demand response exchange market prices. *J. Clean. Prod.* **2021**, *282*, 125344. [CrossRef]
33. Vahedipour-Dahraie, M.; Rashidizadeh-Kermani, H.; Anvari-Moghaddam, A.; Guerrero, J.M. Stochastic risk-constrained scheduling of renewable-powered autonomous microgrids with demand response actions: Reliability and economic implications. *IEEE Trans. Ind. Appl.* **2020**, *56*, 1882–1895. [CrossRef]
34. Vahedipour-Dahraie, M.; Rashidizadeh-Kermani, H.; Shafie-Khah, M.; Catalão, J.P.S. Risk-Averse Optimal Energy and Reserve Scheduling for Virtual Power Plants Incorporating Demand Response Programs. *IEEE Trans. Smart Grid* **2021**, *12*, 1405–1415. [CrossRef]
35. Vahedipour-Dahraie, M.; Rashidizadeh-Kermani, H.; Anvari-Moghaddam, A. Risk-Based Stochastic Scheduling of Resilient Microgrids Considering Demand Response Programs. *IEEE Syst. J.* **2020**, *15*, 971–980. [CrossRef]

36. Zhang, N.; Sun, Q.; Yang, L. A two-stage multi-objective optimal scheduling in the integrated energy system with We-Energy modeling. *Energy* **2021**, *215*, 119121. [CrossRef]
37. Hafiz Abdul Muqeet, A.A. Optimal Day-Ahead Operation of Microgrid Considering Demand Response and Battery Strategies for Prosumer Community. *Tech. J. UET Taxila* **2020**, *25*, 41–51.
38. Das, B.K.; Zaman, F. Performance analysis of a PV/Diesel hybrid system for a remote area in Bangladesh: Effects of dispatch strategies, batteries, and generator selection. *Energy* **2019**, *169*, 263–276. [CrossRef]
39. Prakash Kumar, K.; Saravanan, B. Real time optimal scheduling of generation and storage sources in intermittent microgrid to reduce grid dependency. *Indian J. Sci. Technol.* **2016**, *9*, 1–4. [CrossRef]
40. Waqar, A.; Tanveer, M.S.; Ahmad, J.; Aamir, M.; Yaqoob, M.; Anwar, F. Multi-objective analysis of a CHP plant integrated microgrid in Pakistan. *Energies* **2017**, *10*, 1625. [CrossRef]
41. Zhang, M.; Gan, M.; Li, L. Sizing and Siting of Distributed Generators and Energy Storage in a Microgrid Considering Plug-in Electric Vehicles. *Energies* **2019**, *12*, 2293. [CrossRef]
42. Meng, J.; Stroe, D.I.; Ricco, M.; Luo, G.; Teodorescu, R. A simplified model-based state-of-charge estimation approach for lithium-ion battery with dynamic linear model. *IEEE Trans. Ind. Electron.* **2019**, *66*, 7717–7727. [CrossRef]
43. Shehzad Hassan, M.A.; Chen, M.; Lin, H.; Ahmed, M.H.; Khan, M.Z.; Chughtai, G.R. Optimization modeling for dynamic price based demand response in microgrids. *J. Clean. Prod.* **2019**, *222*, 231–241. [CrossRef]
44. Raza, A.; Malik, T.N. Energy management in commercial building microgrids. *J. Renew. Sustain. Energy* **2019**, *11*, 015502. [CrossRef]
45. Cai, D.; Shi, D.; Chen, J. Probabilistic load flow computation using Copula and Latin hypercube sampling. *IET Gener. Transm. Distrib.* **2014**, *8*, 1539–1549. [CrossRef]
46. Pansota, M.S.; Javed, H.; Muqeet, H.; Khan, H.A.; Ahmed, N.; Nadeem, M.U.; Ahmed, S.U.F.; Sarfraz, A. An Optimal Scheduling and Planning of Campus Microgrid Based on Demand Response and Battery Lifetime. *Pak. J. Eng. Technol.* **2021**, *4*, 8–17.
47. Ayta, S.; Electronic, K.; Electric, F.E. Quality Lignite Coal Detection with Discrete Wavelet Transform, Discrete Fourier Transform, and ANN based on k-means Clustering Method. In Proceedings of the 2018 6th International Symposium on Digital Forensic and Security (ISDFS), Antalya, Turkey, 22–25 March 2018.
48. Aslam Chaudhry, M. Extended incomplete gamma functions with applications. *J. Math. Anal. Appl.* **2001**, *274*, 725–745. [CrossRef]
49. Tahir, Z.R.; Asim, M. Surface measured solar radiation data and solar energy resource assessment of Pakistan: A review. *Renew. Sustain. Energy Rev.* **2018**, *81*, 2839–2861. [CrossRef]
50. Lai, C.S.; McCulloch, M.D. Levelized cost of electricity for solar photovoltaic and electrical energy storage. *Appl. Energy* **2017**, *190*, 191–203. [CrossRef]
51. Bordin, C.; Anuta, H.O.; Crossland, A.; Gutierrez, I.L.; Dent, C.J.; Vigo, D. A linear programming approach for battery degradation analysis and optimization in offgrid power systems with solar energy integration. *Renew. Energy* **2017**, *101*, 417–430. [CrossRef]
52. Pickering, B.; Ikeda, S.; Choudhary, R.; Ooka, R. Comparison of Metaheuristic and Linear Programming Models for the Purpose of Optimising Building Energy Supply Operation Schedule. In Proceedings of the CLIMA 2016: 12th REHVA World Congress, Aalborg, Denmark, 22–25 May 2016; Department of Civil Engineering, Aalborg University: Aalborg, Denmark, August 2017; Volume 6.
53. Chudinҳow, D.; Haas, J.; Díaz-Ferrán, G.; Moreno-Leiva, S.; Eltrop, L. Simulating the energy yield of a bifacial photovoltaic power plant. *Sol. Energy* **2019**, *183*, 812–822. [CrossRef]
54. Reporter, S. IBA University to supply solar power to Sepco. 28 November 2019. Available online: https://nation.com.pk/26-Jan-2019/iba-university-to-supply-solar-power-to-sepco (accessed on 22 October 2021).
55. Baig, M.J.A.; Iqbal, M.T.; Jamil, M.; Khan, J. Design and implementation of an open-Source IoT and blockchain-based peer-to-peer energy trading platform using ESP32-S2, Node-Red and, MQTT protocol. *Energy Rep.* **2021**, *7*, 5733–5746. [CrossRef]
56. Jamil, M.; Hussain, B.; Abu-Sara, M.; Boltryk, R.J.; Sharkh, S.M. Microgrid power electronic converters: State of the art and future challenges. In Proceedings of the 2009 44th International Universities Power Engineering Conference (UPEC), Glasgow, UK, 1–4 September 2009.
57. Govt. Schedule of Electricity Tariffs For Islamabad Electric Supply Company (IESCO). 16 April 2019. Available online: https://www.iesco.com.pk/index.php/customer-services/tariff-guide (accessed on 25 October 2021).
58. Iqbal, M.M.; Sajjad, M.I.A.; Amin, S.; Haroon, S.S.; Liaqat, R.; Khan, M.F.N.; Waseem, M.; Shah, M.A. Optimal scheduling of residential home appliances by considering energy storage and stochastically modelled photovoltaics in a grid exchange environment using hybrid grey wolf genetic algorithm optimizer. *Appl. Sci.* **2019**, *9*, 5226. [CrossRef]
59. Abdolrasol, M.G.M.; Hannan, M.A.; Mohamed, A.; Amiruldin, U.A.U.; Abidin, I.B.Z.; Uddin, M.N. An Optimal Scheduling Controller for Virtual Power Plant and Microgrid Integration Using the Binary Backtracking Search Algorithm. *IEEE Trans. Ind. Appl.* **2018**, *54*, 2834–2844. [CrossRef]
60. Gao, H.C.; Choi, J.H.; Yun, S.Y.; Lee, H.J.; Ahn, S.J. Optimal scheduling and real-time control schemes of battery energy storage system for microgrids considering contract demand and forecast uncertainty. *Energies* **2018**, *11*, 1371. [CrossRef]
61. Zheng, Y.; Zhao, J.; Song, Y.; Luo, F.; Meng, K.; Qiu, J.; Hill, D.J. Optimal Operation of Battery Energy Storage System Considering Distribution System Uncertainty. *IEEE Trans. Sustain. Energy* **2018**, *9*, 1051–1060. [CrossRef]

62. Vahedipour-Dahraie, M.; Rashidizadeh-Kermani, H.; Anvari-Moghaddam, A.; Guerrero, J.M. Stochastic Frequency-Security Constrained Scheduling of a Microgrid Considering Price-Driven Demand Response. In Proceedings of the 2018 International Symposium on Power Electronics, Electrical Drives, Automation and Motion (SPEEDAM), Amalfi, Italy, 20–22 June 2018; pp. 716–721. [CrossRef]
63. Nasir, M.; Khan, H.A.; Khan, I.; ul Hassan, N.; Zaffar, N.A.; Mehmood, A.; Sauter, T.; Muyeen, S.M. Grid load reduction through optimized PV power utilization in intermittent grids using a low-cost hardware platform. *Energies* **2019**, *12*, 1764. [CrossRef]

Article

Optimal Scheduling of Energy Storage System Considering Life-Cycle Degradation Cost Using Reinforcement Learning

Wonpoong Lee [1], Myeongseok Chae [2] and Dongjun Won [2,*]

1   KEPCO Management Research Institute (KEMRI), Korea Electric Power Corporation (KEPCO), 55, Jeollyeok-ro, Naju 58277, Korea; poong89@gmail.com
2   Department of Electrical and Computer Engineering, Inha University, 100, Inha-ro, Michuhol-gu, Incheon 22212, Korea; eric980206@gmail.com
*   Correspondence: djwon@inha.ac.kr

Abstract: Recently, due to the ever-increasing global warming effect, the proportion of renewable energy sources in the electric power industry has increased significantly. With the increase in distributed power sources with adjustable outputs, such as energy storage systems (ESSs), it is necessary to define ESS usage standards for an adaptive power transaction plan. However, the life-cycle cost is generally defined in a quadratic formula without considering various factors. In this study, the life-cycle cost for an ESS is defined in detail based on a life assessment model and used for scheduling. The life-cycle cost is affected by four factors: temperature, average state-of-charge (SOC), depth-of-discharge (DOD), and time. In the case of the DOD stress model, the life-cycle cost is expressed as a function of the cycle depth, whose exact value can be determined based on fatigue analysis techniques such as the Rainflow counting algorithm. The optimal scheduling of the ESS is constructed considering the life-cycle cost using a tool based on reinforcement learning. Since the life assessment cannot apply the analytical technique due to the temperature characteristics and time-dependent characteristics of the ESS SOC, the reinforcement learning that derives optimal scheduling is used. The results show that the SOC curve changes with respect to weight. As the weight of life-cycle cost increases, the ESS output and charge/discharge frequency decrease.

Keywords: energy storage system; life-cycle cost; optimal scheduling; reinforcement learning

1. Introduction

Recently, consumers' perception of energy has changed due to the development and demonstration of an operating system for regional power grids characterized by VPP and MG. Under the influence of economic factors, such as decreasing installation costs of renewable energy and technological advances, consumers have become energy prosumers who can trade their own electricity through distributed power systems [1,2]. Because the power surplus can be sold to neighbors, the energy flow in the energy market has changed from one-way to two-way. In addition, the existing hierarchical market structure has transformed into a network structure.

With the adoption of distributed energy, the need to establish usage standards is increasing with an increasing use of ESSs. When conducting trading through ESSs, certain usage standards, such as the fuel cost function of the generator, must be considered. The life-cycle cost of the ESS can be considered as one of these standards. As research on the ESS life-cycle, Ref. [3] proposed the total capital cost and life-cycle cost models for ESSs. The cost function was introduced and modeled for the system, and a learning model that can accurately estimate the life-cycle cost based on various battery types was built. Reference [4] proposed an analytical optimization for the capacity and sizing of solar power and ESSs connected to the grid. Ref. [5] studied the efficiency difference between HESS and LESS in an independent microgrid. The constraint variable was set by combining

Citation: Lee, W.; Chae, M.; Won, D. Optimal Scheduling of Energy Storage System Considering Life-Cycle Degradation Cost Using Reinforcement Learning. *Energies* **2022**, *15*, 2795. https://doi.org/10.3390/en15082795

Academic Editors: Surender Reddy Salkuti and Miguel Castilla

Received: 2 March 2022
Accepted: 8 April 2022
Published: 11 April 2022

Publisher's Note: MDPI stays neutral with regard to jurisdictional claims in published maps and institutional affiliations.

the SOC with the cost function, and the stability of the system was considered in preparation for the surge currents of LIB and LAB. Ref. [6] proposed an ESS life-cycle definition using SOC and SOH models. They established an equation via correlation analysis and introduced a cycle depth variable. Ref. [7] introduced an overall cost evaluation model for ESS and used fuzzy comprehensive evaluation theory to analyze the model, considering basic facility and operating costs. Ref. [8] proposed a life assessment method for ESS in distributed energy systems and established an evaluation method classified into four scenarios. Ref. [9] introduced a P2P energy sharing scheme using ESSs. In their study, scheduling was set to maximize system profits depending on the existence of an ESS. Ref. [10] conducted a comparative study on single/hybrid ESSs to examine stable energy transfer capabilities of these systems. This model considered the charge/discharge rate function of the battery, and data analyses were performed for situations with varying supply and demand. Ref. [11] defined a cycle life model of a battery considering the SOC, DOD, average C-rate, and aging of the lithium-ion battery. A comparative analysis was performed on the battery temperature, output, and resistance values with varying parameters. Ref. [12] introduced an ESS life-cycle cost optimization method through an energy consumer scheduling scheme. The battery life was calculated using the Rainflow counting algorithm for maximizing battery life. Scheduling was configured based on unstable PV and WT output data and the composed ESS life results. Ref. [13] introduced an improvement in the prediction accuracy of lithium-ion batteries. A BP neural network was used to predict the life-cycle cost of the battery, and the weights were set using the DE algorithm. Ref. [14] analyzed the ESS life-cycle cost using various forecasting techniques, such as the RVM and CNN models. Ref. [15] conducted a study on the battery output considering the life-cycle cost of an ESS used in the grid. Their study, conducted on the stability of the system, included measuring the frequency fluctuations over time. Moreover, a scheduling scheme for an ESS used in the auxiliary service market was established. Ref. [16] presented a methodology for the optimal location, selection, and operation of battery energy storage systems (BESSs) and renewable distributed generators in medium- to low-voltage distribution systems. Ref. [17] proposed a new formulation of the battery degradation cost for the optimal scheduling of BESSs. This paper defined a one-cycle battery cost function based on the cycle life curve and an auxiliary state of charge (SOC) that tracks the actual SOC only upon discharge. Ref. [18] proposed a mixed-integer nonlinear programming (MINLP) model for the PV-battery systems which aims to minimize the life-cycle cost (LCC), and solved LCC Problem by a novel two-layer optimization, and Ref. [19] studied the multi-objective operation of BESS in AC distribution systems using a convex reformulation. Ref. [20] proposed a two-stage multi-objective optimal operation scheduling method to improve the operation efficiency and reduce the emission of a solar-power-integrated hybrid ferry with shore-to-ship (S2S) power supply, and Ref. [21] addressed the problem associated with economic dispatch of BESSs in alternating current (AC) distribution networks. Ref. [22] addressed the problem of the optimal operation of BESSs in AC grids from the point of view of multi-objective optimization. Ref. [23] proposed a distributed multi-agent consensus-based control algorithm for multiple BESSs, operating in a microgrid, for fulfilling several objectives, including: SOC trajectories tracking control, economic load dispatch, active and reactive power sharing control, and voltage and frequency regulation. Ref. [24] proposed an optimal BESS scheduling for MGs to solve the stochastic unit commitment problem, considering the uncertainties in renewables and load.

In summary, most previous studies derive their results by defining the life-cycle cost in a quadratic manner or simplifying it. This study aims to define it in detail based on a life-cycle cost assessment method and utilize it for scheduling. Because the defined life-cycle cost cannot be derived analytically and explicitly, a solution is derived using reinforcement learning techniques.

The contributions of this study are as follows:

(1)   By defining the life-cycle cost of an ESS, and deriving and utilizing it for optimal scheduling, prosumers with ESSs can make the best choice between incurring life-

cycle costs due to ESS use and profiting from transactions. In addition, because of the active adjustment of prosumers with ESSs, it is possible to reduce the line loss inside a system.

(2)   Through analysis of the trading tendency of flexible prosumers with respect to changes in ESS life-cycle cost weights, prosumers who own an ESS have the choice of participating in P2P energy trading to make profits.

2. Life Degradation Model for ESSs Based on a Life-Cycle Assessment Method

This chapter presents the design of an ESS life-cycle cost metric for prosumer participation in P2P energy trading with ESSs. ESSs can be classified according to the type of battery they use. In this study, lithium-ion batteries, which are commonly used in ESSs, are chosen, and their life-cycle cost is designed. The life-cycle cost was designed based on existing studies related to battery life assessment. The life assessment model consists of four stress models: temperature, average state-of-charge (SOC), depth-of-discharge (DOD), and time Ref. [25].

The degradation ratio of the battery life-cycle is determined by the corresponding stress models, and it can be evaluated using the corresponding degradation ratio. The degradation ratio, four stress models, and the consumption life-cycle ratio are formulated as follows:

$$f_{d,1} = [S_\delta(\delta) + S_t(t_c)]S_\sigma(\sigma)S_T(T_c) \tag{1}$$

$$S_T(T_c) = e^{k_T(T_c - T_{ref})(\frac{T_{ref}}{T_c})} \tag{2}$$

$$S_\sigma(\sigma) = e^{k_\sigma(\sigma - \sigma_{ref})} \tag{3}$$

$$S_\delta(\delta) = k_{\delta,q1}\delta^{k_{\delta,q2}} \tag{4}$$

$$S_t(t) = k_t t \tag{5}$$

$$L = 1 - \alpha_{sei}e^{-\beta_{sei}f_d} - (1 - \alpha_{sei})e^{-f_d} \tag{6}$$

where f_d is the degradation ratio, and S_T, S_σ, S_δ are the stresses for temperature, average SOC, and DOD, respectively. T_c is the battery cell temperature, T_{ref} is the reference temperature, and k_T is the temperature stress coefficient. σ is the average SOC, σ_{ref} is the reference average SOC, and k_σ is the average SOC stress coefficient. S_t is the stress for time, δ is the cycle depth, and $k_{\delta,q1}$, $k_{\delta,q2}$ are the DOD coefficients. t is time, k_t is the time stress coefficient, L is the consumed life-cycle, and α_{sei}, β_{sei} are the solid electrolyte interphase (SEI) film formation coefficients.

In the case of a DOD stress model, various models such as linear, exponential, polynomial, and power are applicable, but the power function is used according to references. Stress models for average SOC and time can be used immediately for life-cycle cost design because they are explicit. The same does not hold for DOD and temperature stress models.

First, in the case of the DOD stress model (4), which is expressed as a function of the cycle depth, the exact value of the cycle depth can be determined through a post evaluation based on fatigue analysis techniques, such as the Rainflow counting algorithm. In the case of the temperature stress model (2), additional analysis is required because a model for the internal battery temperature is required with respect to the output ESS. Therefore, additional design stages for these two models are required. The model for temperature is designed by analyzing the relationship between battery output and temperature using the thermoelectric model of the battery. Furthermore, for the DOD model in this case, an approximation that considers one charge or discharge of the battery as a half cycle is assumed.

2.1. Temperature Stress Model Formulation

A thermoelectric model is used as the temperature stress model, which is categorized into two types: an electric circuit model and a thermal model Refs. [26–29].

2.1.1. Electric Circuit Model

The electric circuit model of the battery used in the temperature stress model is shown in Figure 1 Ref. [28]. The open circuit voltage (OCV) can be expressed as a function of the SOC and internal temperature of the battery. OCV, characteristically, rises during charging and falls during discharging, and this tendency varies according to SOC. The internal resistance R_{in} can also be expressed as a function of SOC and temperature. The RC network located to the right of the internal resistance is a secondary model and represents the diffusion resistance and capacitance. R_1 and C_1 are related to the charge transfer processes occurring in the middle frequency range, whereas R_2 and C_2 are responsible for reproducing the diffusion processes. V_h is an additional voltage component caused by the hysteresis characteristics of the RC network, which refers to the fluctuations on the open voltage during charge/discharge. This component is ignored, assuming its effect is relatively small. Moreover, the corresponding model for life-cycle cost analysis does not require detailed dynamic characteristics of the battery.

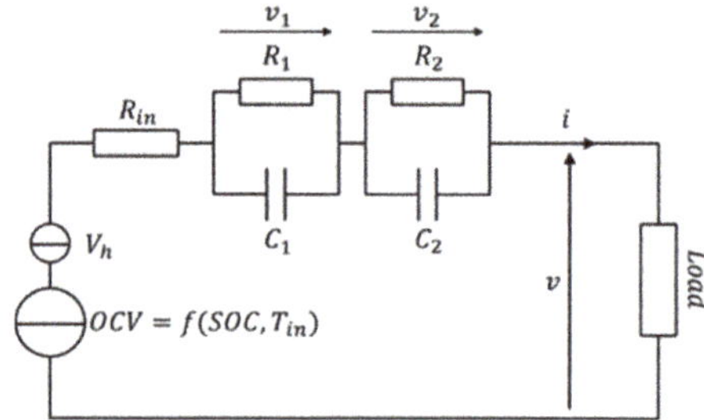

Figure 1. Battery electric circuit model.

The function for the SOC of the OCV (7) is based on the parameters listed in Table 1.

$$\text{OCV} = f(\text{SOC}) = ae^{(b \cdot \text{SOC})} + ce^{(d \cdot \text{SOC})} \tag{7}$$

Table 1. Battery OCV curve parameters.

Parameter	a	b	c	d
Value	3.263	0.02451	-0.2297	-7.666

Regarding the effect of temperature on the OCV, Equation (8) shows the correlation between the OCV bias component and temperature. It can be expressed as a polynomial with the following related parameters in Table 2:

$$\begin{aligned}
b_{\text{OCV}} &= g(\text{SOC}, T_{in}) \\
&= p00 + p10 \cdot \text{SOC} + p01 \cdot T_{in} + p20 \cdot \text{SOC}^2 + p11 \cdot \text{SOC} \cdot T_{in} + p02 \cdot T_{in}^2 + p31 \cdot \text{SOC}^3 \\
&\quad + p21 \cdot \text{SOC}^2 \cdot T_{in} + p12 \cdot \text{SOC} \cdot T_{in}^2 + p40 \cdot \text{SOC}^4 + p31 \cdot \text{SOC}^3 \cdot T_{in} + p22 \cdot \text{SOC}^2 \cdot T_{in}^2 \\
&\quad + p50 \cdot \text{SOC}^5 + p41 \cdot \text{SOC}^4 \cdot T_{in} + p32 \cdot \text{SOC}^3 \cdot T_{in}^2
\end{aligned} \tag{8}$$

Table 2. Correlation parameters between OCV bias component and temperature/SOC.

Parameter	$p00$	$p10$	$p01$	$p20$
Value	-0.001202	0.2458	8.558×10^{-5}	-1.248
Parameter	$p11$	$p02$	$p30$	$p21$
Value	-0.007113	1.552×10^{-5}	2.328	0.03044
Parameter	$p12$	$p40$	$p31$	$p22$
Value	-1.063×10^{-5}	-1.899	-0.04233	-7.069×10^{-5}
Parameter	$p50$	$p41$	$p32$	
Value	0.569	0.01919	8.263×10^{-5}	

Finally, the open circuit model is constructed as a linear sum of the OCV model for SOC in Equation (9).

$$OCV = f(SOC, T_{in}) = f(SOC) + g(SOC, T_{in}) \tag{9}$$

The SOC is updated according to the output current, whose unit can be set as % (or p.u.). The sign of the discharge current was set to positive. The discrete equation can be described as:

$$SOC(k) = SOC(k-1) - \frac{i(k-1)}{C_n}\frac{T_s}{3600} \tag{10}$$

where T_s is the sampling time (unit: second [s]), C_n is the battery capacity (unit: Ampere hour [Ah]), i is the output current, and k is a time index.

The internal resistance is also configured as a function of the internal temperature and SOC, similar to the OCV. However, the internal resistance remains constant without significant changes over the general battery SOC usage period and is dominantly affected by the internal temperature Ref. [19]. Therefore, the internal resistance is expressed as a function of the internal temperature with the battery internal resistance curve parameters in Table 3 and formulated as follows:

$$R_{in} = f_R(T_{in}) = ae^{bT_{in}} + ce^{dT_{in}} \tag{11}$$

Table 3. Battery internal resistance curve parameters.

Parameter	a	b	c	d
Value	0.0003448	−0.2954	0.01771	−0.008504

The relationship between voltage, resistance, and current in an RC network can be represented as

$$v_1(k) = a_1 v_1(k-1) + b_1 i(k-1),\ a_1 = e^{-\left(\frac{T_s}{R_1 C_1}\right)},\ b_1 = R_1(1 - a_1) \tag{12}$$

$$v_2(k) = a_2 v_2(k-1) + b_2 i(k-1),\ a_2 = e^{-\left(\frac{T_s}{R_2 C_2}\right)},\ b_2 = R_2(1 - a_2) \tag{13}$$

where T_s is the sampling time, and k is the discrete time index. The second network time constant does not change Ref. [28]; thus, what remains to be estimated are the resistance and time constant of the first RC network and the resistance of the second RC network. If R_1 and τ_1 are known, then using the time constant relational expression ($\tau = RC$) C_1 can be calculated, and C_2 can be calculated in a similar manner if R_2 and τ_2 are known. Network resistances are based on a polynomial model, whereas the first time constant is based on an exponential equation. The equations are stated below and parameters are shown in Tables 4–6.

$$\begin{aligned} R_1 = f_{R_1}(SOC, T_{in}) \\ = p00 + p10{\cdot}SOC + p01{\cdot}T_{in} + p20{\cdot}SOC^2 + p11{\cdot}SOC{\cdot}T_{in} \\ + p02{\cdot}T_{in}^2 + p21{\cdot}SOC^2{\cdot}T_{in} + p12{\cdot}SOC{\cdot}T_{in}^2 + p03{\cdot}T_{in}^3 \end{aligned} \tag{14}$$

$$\begin{aligned} R_2 = f_{R_2}(SOC, T_{in}) \\ = p00 + p10{\cdot}SOC + p01{\cdot}T_{in} + p20{\cdot}SOC^2 + p11{\cdot}SOC{\cdot}T_{in} \\ + p02{\cdot}T_{in}^2 + p30{\cdot}SOC^3 + p21{\cdot}SOC^2{\cdot}T_{in} + p12{\cdot}SOC \\ {\cdot}T_{in}^2 + p03{\cdot}T_{in}^3 + p31{\cdot}SOC^3{\cdot}T_{in} + p22{\cdot}SOC^2{\cdot}T_{in}^2 + p13 \\ {\cdot}SOC{\cdot}T_{in}^3 + p04{\cdot}T_{in}^4 \end{aligned} \tag{15}$$

$$\tau_1 = f_{\tau_1}(SOC) = ae^{b{\cdot}SOC} \tag{16}$$

Table 4. Correlation parameters between RC Network R_1 and SOC/internal temperature.

Parameter	*p00*	*p10*	*p01*	*p20*
Value	0.04375	−0.05367	−0.000974	0.01182
Parameter	*p11*	*p02*	*p21*	*p12*
Value	0.0005085	7.661×10^{-6}	0.0004819	-6.957×10^{-6}
Parameter	*p03*			
Value	1.993×10^{-8}			

Table 5. Correlation parameters between RC Network R_2 and SOC/internal temperature.

Parameter	*p00*	*p10*	*p01*	*p20*
Value	0.05018	−0.1373	−0.002979	0.1224
Parameter	*p11*	*p02*	*p30*	*p21*
Value	0.005049	0.0001103	−0.02099	−0.001888
Parameter	*p12*	*p03*	*p31*	*p22*
Value	-9.877×10^{-5}	-2.118×10^{-6}	−0.001455	5.704×10^{-5}
Parameter	*p13*	*p04*		
Value	2.744×10^{-7}	1.708×10^{-8}		

Table 6. Correlation parameter between RC Network first time constant and SOC.

Parameter	*a*	*b*
Value	53.99	−1.573

2.1.2. Lumped Thermal Model

As shown in Figure 2, the battery thermal model is affected by the temperature values at three points: the cell inside the battery shell, shell surrounding it, and environment Ref. [19].

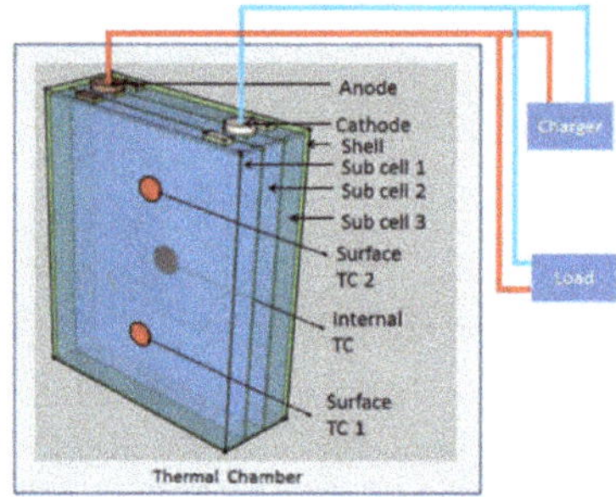

Figure 2. Configuration of battery internal system.

Therefore, the battery thermal model can be modeled in two ways: the heat generation that occurs inside the battery and heat transfer from the inside to the battery shell and from the shell to the environment. In general, heat generated by the cell is considered only as the heat generated by the internal resistance.

However, the heat generated due to the overpotential of the RC network and entropy change also need to be considered. The total heat generated by the cell is given by

$$Q = R_{in}i^2 + v_1 i + v_2 i + i \times T_{in} \frac{d\text{OCV}}{dT_{in}} \tag{17}$$

In general, the heat transfer in and out of a battery includes three mechanisms: conduction, convection, and radiation. Before modeling the heat transfer, both the battery shell temperature and internal temperature must be uniform, and the thermal characteristics must also be uniformly distributed inside the battery. Only the heat conduction between the inside and shell of the battery and between the shell and environment is considered. The heat transfer model is expressed as follows:

$$C_{q1}\frac{dT_{in}}{dt} = Q - k_1(T_{in} - T_{sh}) \tag{18}$$

$$C_{q2}\frac{dT_{sh}}{dt} = k_1(T_{in} - T_{sh}) - k_2(T_{sh} - T_{amb}) \tag{19}$$

where T_{in} is the battery internal temperature, T_{sh} is the battery shell temperature, and T_{amb} is the ambient temperature. C_{q1} and C_{q2} are the internal and shell thermal capacities of the battery, respectively, and k_1 and k_2 are the heat conduction coefficients between the battery internal and the shell, and between the battery shell and the ambience, respectively. Because Equations (18) and (19) are continuous, they are discretized as follows:

$$C_{q1}\frac{z-1}{T_s}T_{in} = Q - k_1(T_{in} - T_{sh}) \tag{20}$$

$$C_{q2}\frac{z-1}{T_s}T_{in} = k_1(T_{in} - T_{sh}) - k_2(T_{sh} - T_{amb}) \tag{21}$$

Finally, the formulae for the internal temperature and shell temperature are as given by

$$T_{in}(k+1) = \frac{1 - T_s k_1}{C_{q1}}T_{in}(k) + \frac{T_s k_1}{C_{q1}}T_{sh}(k) + \frac{T_s Q(k)}{C_{q1}} \tag{22}$$

$$T_{sh}(k+1) = \frac{T_s k_1}{C_{q2}}T_{in}(k) + \frac{1 - T_s(k_1 + k_2)}{C_{q2}}T_{sh}(k) + \frac{T_s k_2 T_{amb}}{C_{q2}} \tag{23}$$

The heat capacity coefficients and internal heat capacity are constant. The heat capacity coefficient k_2 used in this model has time-varying characteristics; and the following relation holds Ref. [28]:

$$k_2 = k_{21} + k_{22}(T_{sh} - T_{amb}) \tag{24}$$

The k_2 certainly depends on the heat dissipation condition, such as cooling wind speed and temperature. k_2 also increases with this temperature gradient $T_{sh} - T_{amb}$. To take this effect into consideration, two cases are compared here: Constants k_{21} of k_2 and time-varying k_{22} of k_2.

2.1.3. Coupled Thermoelectric Model

By combining the two previously defined thermal/electrical models into one,

$$x(k+1) = Ax(k) + B(k) \tag{25}$$

$$v(k) = OCV(k) + v_1(k) + v_2(k) - i(k)R_{in} \tag{26}$$

where,

$$x(k) = [SOC(k),\ v_1(k),\ v_2(k),\ T_{in}(k),\ T_{sh}(k)]^T$$

$$A = \begin{bmatrix} 1 & 0 & 0 & 0 & 0 \\ 0 & a_1 & 0 & 0 & 0 \\ 0 & 0 & a_2 & 0 & 0 \\ 0 & 0 & 0 & 1 - k_1\dfrac{T_s}{C_{q1}} & k_1\dfrac{T_s}{C_{q1}} \\ 0 & 0 & 0 & k_1\dfrac{T_s}{C_{q2}} & 1 - (k_1 + k_2)\dfrac{T_s}{C_{q2}} \end{bmatrix}$$

$$B(k) = \left[-i(k)\frac{T_s}{C_n},\ -b_1 i(k),\ -b_2 i(k),\ Q(k)\frac{T_s}{C_{q1}},\ k_2 T_{amb}\frac{T_s}{C_{q2}} \right]^T$$

2.2. Cycle Depth Stress Model Formulation

The cycle depth is derived after fatigue analysis using the Rainflow counting algorithm, as mentioned earlier in the study related to the life-cycle cost evaluation of the ESS. Therefore, it is impossible to determine the cycle depth before scheduling is configured. The first step in solving this problem is deriving it through dynamic programming when composing the ESS schedule. However, in dynamic programming, a cost or reward should be calculated at the transition time between states. It is necessary to define a state, which can be the SOC of the ESS. However, because SOC is a continuous variable, it cannot be determined discretely; however, the state can be defined by dividing it into a specific unit as a simplification to reduce the burden of calculation. For example, if the state is defined in units of 0.1, when the minimum SOC is 0.1 and the maximum SOC is 0.9, nine states can be defined in one time period (Stage). When the total schedule interval is T, the number of cases composed by states is 9^{T-1}. This refers to the number of cases when searching for a path from the first to the last stage in the dynamic plan. That is, the computational power required to search for an optimal point is quite large. To solve this problem, a reinforcement learning-based approach is introduced, and the cycle depth to be used in this approach is approximated. Therefore, for the cycle depth, the same half cycle was applied for all charging/discharging cycles. To check whether this approximation is appropriate, we created a random SOC candidate group and compared the difference between the complete and approximate life-cycle cost analysis results.

Figure 3 shows a flowchart depicting this process. Figure 4 shows the SOC graph where the difference between the two results is maximum and minimum when the life-cycle costs for the approximated and total cycle depths are calculated. As a result, when charging and discharging are repeatedly performed, the difference between the two life costs is small, as shown in the blue graph, whereas when charging and discharging are sequentially performed in one large cycle depth, the difference is the largest.

Figure 5 shows the approximated cycle depth for 20 candidates and the life-cycle cost for the total cycle depth, as well as the ratio between the two life-cycle costs. Although there is a difference in the ratio for each candidate group, we confirmed that even if the life-cycle cost is calculated using the approximated cycle depth, the effect could be similar to the lifetime cost calculated using the total cycle depth. When the life-cycle cost is included in the actual objective function, it may be lower than the actual expected life-cycle cost owing to the approximated cycle depth. However, this can be avoided because the life-cycle cost is used for weight and not directly converted into an actual financial cost.

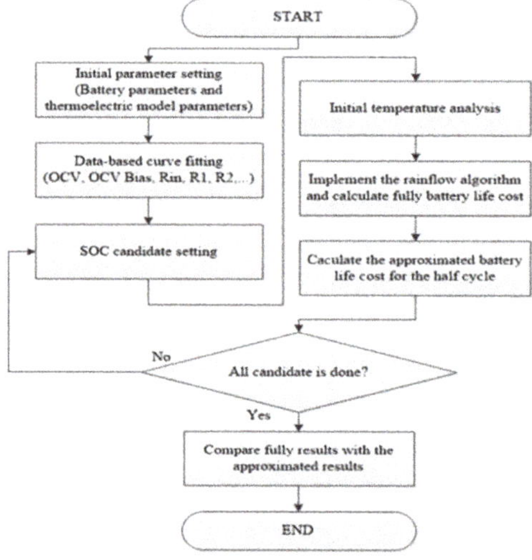

Figure 3. Life-cycle cost comparison flowchart for cycle approximation.

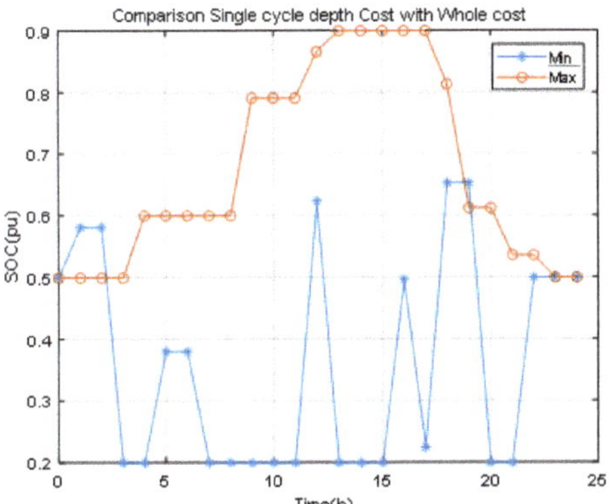

Figure 4. SOC difference (max/min) of the two lifetime costs.

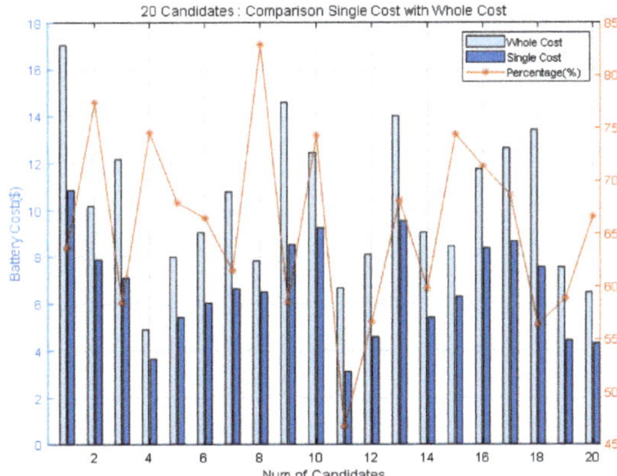

Figure 5. Comparison of two life-cycle costs regarding the SOC candidates.

2.3. ESS Life-Cycle Cost Formulation

Considering the temperature and DOD stress models, the ESS life-cycle cost can be expressed as

$$f_d = 0.5[S_\delta(\delta) + S_t(t_c)]S_\sigma(\sigma)S_T(T_c) \tag{27}$$

$$S_T(T_c) = e^{k_T(T_c - T_{ref})(\frac{T_{ref}}{T_c})} \tag{28}$$

$$T_c = f(P_{bat}) \tag{29}$$

$$S_\sigma(\sigma) = e^{k_\sigma(\sigma^t + \frac{\eta P_{bat}}{2C_{ESS}} - \sigma_{ref})} \tag{30}$$

$$S_\delta(\delta) = k_{\delta,q1}\left(\frac{\eta P_{bat}}{C_{ESS}}\right)^{k_{\delta,q2}} \tag{31}$$

$$S_t(t) = k_t t \tag{32}$$

$$L = 1 - \alpha_{sei}e^{-\beta_{sei}f_d} - (1 - \alpha_{sei})e^{-f_d} \tag{33}$$

where P_{bat} and C_{ESS} are the output and capacity of the ESS, respectively. η is the ESS charging/discharging efficiency, and σ^t is the SOC at time t. Because the consumed life-cycle L presented in Equation (33) is a cumulative expression of the battery aging, the difference in L values reflects the actual shortened lifespan. For example, if the life-cycle L_1 consumed on the first day and life-cycle L_2 consumed on the second day are determined through life assessment, the actual life-cycle consumed on the second day becomes $L_2 - L_1$. If this is applied in the dynamic programming method mentioned above, complex calculations, such as the number of cases for the path by the SOC state, must proceed. Therefore, to maintain the tendency of the life-cycle cost and lower the computational complexity, the initially consumed life-cycle value is initialized at 0. The calculation complexity can be reduced using only the degradation ratio and life-cycle cost.

Finally, the life-cycle cost of the ESS is treated as a concept of depreciation cost by multiplying the investment cost for the battery, as shown in Equation (34).

$$f_{bat}(t) = In_{bat} \times L(t) \tag{34}$$

f_{bat} is the life-cycle cost, and In_{bat} is the investment cost for the battery. The DOD stress model can be solved analytically through half-cycle approximation. However, the temperature stress model cannot be directly used for optimization problems because it is derived through dynamic characteristic analysis. Therefore, this problem is solved through a reinforcement learning approach.

3. ESS Scheduling Formulation Considering the Life-Cycle Cost

The basic optimization problem regarding a prosumer who owns an ESS is the summation of the cost of electricity purchased from the grid and life-cycle cost of the ESS, which can be expressed as follows:

$$f_{pros}^{ESS} = \sum_{t=1}^{T} [\omega_{ESS} I_{ESS} L_{ESS}^{t} + (1 - \omega_{ESS}) \pi_{grid}^{t} P_{grid}^{t}] \tag{35}$$

$$P_{grid}^{t} + P_{dch}^{t} - P_{ch}^{t} = P_{load}^{t} - P_{PV}^{t} \tag{36}$$

$$SOC^{t+1} = SOC^{t} + \frac{\left(\eta_{eff} P_{ch}^{t} - \left(\frac{P_{dch}^{t}}{\eta_{ef}}\right)\right)}{C_{ESS}} \Delta t \tag{37}$$

$$SOC^{0} = SOC_{init} \tag{38}$$

$$SOC^{t} = SOC_{end} \tag{39}$$

$$L_{ESS}^{t} = 1 - \alpha_{sei} e^{-\beta_{sei} f_{d}^{t}} - (1 - \alpha_{sei}) e^{-f_{d}^{t}} \tag{40}$$

$$f_{d}^{t} = F(S_{\delta}^{t}, S_{t}^{t}, S_{\sigma}^{t}, S_{T}^{t}) = 0.5(S_{\delta}^{t} + S_{t}^{t}) S_{\sigma}^{t} S_{T}^{t} \tag{41}$$

$$S_{\delta}^{t} = k_{\delta,q1} (\delta^{t})^{k_{\delta,q2}} \tag{42}$$

$$\delta^{t} = \frac{\eta_{eff} P_{ch}^{t} + P_{dch}^{t}/\eta_{dch}}{C_{ESS}} \tag{43}$$

$$S_{t}^{t} = k_{t} t \tag{44}$$

$$S_{\sigma}^{t} = e^{k_{\sigma}(\sigma^{t} - \sigma_{ref})} \tag{45}$$

$$\sigma^{t} = \frac{\left|SOC^{t+1} + SOC^{t}\right|}{2} = \frac{2SOC^{t} + \eta_{eff} P_{ch}^{t} + P_{dch}^{t}/\eta_{dch}}{2C_{ESS}} \tag{46}$$

$$S_{T}^{t} = e^{k_{T}(T_{c}^{t} - T_{ref})\left(\frac{T_{ref}}{T_{c}^{t}}\right)} \tag{47}$$

$$0 \leq P_{dch}^{t} \leq P_{bat}^{rated} \tag{48}$$

$$0 \leq P_{ch}^{t} \leq P_{bat}^{rated} \tag{49}$$

$$SOC^{min} \leq SOC^t \leq SOC^{max} \tag{50}$$

$$0 \leq P^t_{grid} \leq P^{max}_{grid} \tag{51}$$

In (35), weights are applied to both the ESS life-cycle cost and system purchase cost to reflect the subjective preference of the ESS operator regarding the life-cycle cost. The larger this weight is, the larger the ESS life-cycle is, which is reduced when operating the ESS, whereas the system purchase cost is considered relatively low. Regarding this problem, the state and stage for the ESS SOC are defined as shown in Figure 6, and a reward table for state transition is constructed for reinforcement learning.

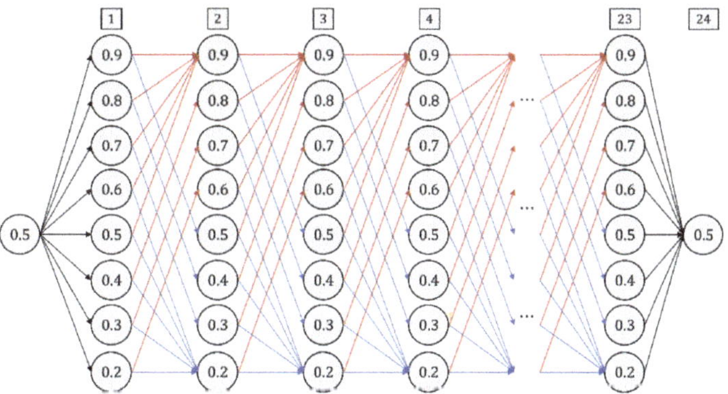

Figure 6. Conceptual diagram of status and stage definitions for SOC of ESS.

Figure 7 shows a flowchart for deriving a solution that applies reinforcement learning to the optimization problem considering the life-cycle cost of the ESS. Figures 8 and 9 show examples of internal reward tables for reinforcement learning. Because this problem is a cost minimization problem, the cost value is treated with a negative sign.

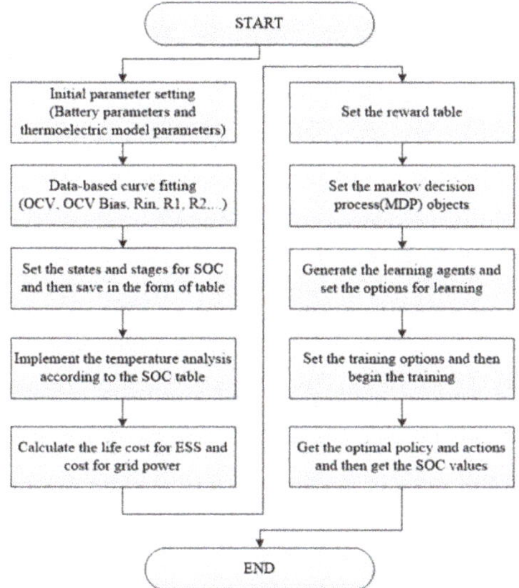

Figure 7. Flow chart of the optimization problem solution considering the life-cycle cost of ESS.

State	0.9	0.8	0.7	0.6	0.5	0.4	0.3	0.2
0.9	0	-0.1	-0.2	-0.3	-0.4	-0.5	-0.6	-0.7
0.8	0.1	0	-0.1	-0.2	-0.3	-0.4	-0.5	-0.6
0.7	0.2	0.1	0	-0.1	-0.2	-0.3	-0.4	-0.5
0.6	0.3	0.2	0.1	0	-0.1	-0.2	-0.3	-0.4
0.5	0.4	0.3	0.2	0.1	0	-0.1	-0.2	-0.3
0.4	0.5	0.4	0.3	0.2	0.1	0	-0.1	-0.2
0.3	0.6	0.5	0.4	0.3	0.2	0.1	0	-0.1
0.2	0.7	0.6	0.5	0.4	0.3	0.2	0.1	0

(a)

State	0.9	0.8	0.7	0.6	0.5	0.4	0.3	0.2
0.9	25	25.0525	25.1626	25.3227	25.5457	25.8671	0	0
0.8	25.0606	25	25.0525	25.1626	25.3227	25.5457	25.8671	0
0.7	25.2398	25.0606	25	25.0525	25.1626	25.3227	25.5457	25.8671
0.6	25.608	25.2398	25.0606	25	25.0525	25.1626	25.3227	25.5457
0.5	26.2893	25.608	25.2398	25.0606	25	25.0525	25.1626	25.3227
0.4	27.5142	26.2893	25.608	25.2398	25.0606	25	25.0525	25.1626
0.3	0	27.5142	26.2893	25.608	25.2398	25.0606	25	25.0525
0.2	0	0	27.5142	26.2893	25.608	25.2398	25.0606	25

(b)

State	0.9	0.8	0.7	0.6	0.5	0.4	0.3	0.2
0.9	0	-4.096	-8.192	-12.288	-16.384	-20.48	81.92	81.92
0.8	4.096	0	-4.096	-8.192	-12.288	-16.384	-20.48	81.92
0.7	8.192	4.096	0	-4.096	-8.192	-12.288	-16.384	-20.48
0.6	12.288	8.192	4.096	0	-4.096	-8.192	-12.288	-16.384
0.5	16.384	12.288	8.192	4.096	0	-4.096	-8.192	-12.288
0.4	20.48	16.384	12.288	8.192	4.096	0	-4.096	-8.192
0.3	81.92	20.48	16.384	12.288	8.192	4.096	0	-4.096
0.2	81.92	81.92	20.48	16.384	12.288	8.192	4.096	0

(c)

State	0.9	0.8	0.7	0.6	0.5	0.4	0.3	0.2
0.9	0	0.85	0.8	0.75	0.7	0.65	-0.7	-0.8
0.8	0.85	0	0.75	0.7	0.65	0.6	0.55	-0.8
0.7	0.8	0.75	0	0.65	0.6	0.55	0.5	0.45
0.6	0.75	0.7	0.65	0	0.55	0.5	0.45	0.4
0.5	0.7	0.65	0.6	0.55	0	0.45	0.4	0.35
0.4	0.65	0.6	0.55	0.5	0.45	0	0.35	0.3
0.3	-0.1	0.55	0.5	0.45	0.4	0.35	0	0.25
0.2	-0.1	-0.2	0.45	0.4	0.35	0.3	0.25	0

(d)

Figure 8. (**a**) Table configured for cycle depth; (**b**) Table configured for temperature; (**c**) Table configured for ESS output; (**d**) Table configured for average SOC.

```
val(:,:,1) =

   -7.7487   -8.3565   -8.9639   -9.6310  -10.2962  -10.9786 -100.0000 -100.0000
   -7.1680   -7.7487   -8.3550   -8.9791   -9.6211  -10.2792  -10.9530 -100.0000
   -6.6072   -7.1665   -7.7487   -8.3537   -8.9749   -9.6121  -10.2639  -10.9299
   -6.0675   -6.6024   -7.1652   -7.7487   -8.3525   -8.9710   -9.6040  -10.2502
   -5.5506   -6.0573   -6.5981   -7.1640   -7.7487   -8.3514   -8.9675   -9.5967
   -5.0634   -5.5328   -6.0482   -6.5942   -7.1629   -7.7487   -8.3504   -8.9644
 -100.0000   -5.0351   -5.5168   -6.0399   -6.5907   -7.1619   -7.7487   -8.3495
 -100.0000 -100.0000   -5.0096   -5.5024   -6.0325   -6.5875   -7.1610   -7.7487


val(:,:,2) =

   -7.7851   -8.3464   -8.9274   -9.5281  -10.1468  -10.7827 -100.0000 -100.0000
   -7.2508   -7.7851   -8.3450   -8.9226   -9.5181  -10.1298  -10.7571 -100.0000
   -6.7365   -7.2494   -7.7851   -8.3436   -8.9184   -9.5091  -10.1145  -10.7341
   -6.2432   -6.7317   -7.2480   -7.7851   -8.3424   -8.9145   -9.5010  -10.1008
   -5.7728   -6.2331   -6.7274   -7.2468   -7.7851   -8.3413   -8.9110   -9.4937
   -5.3320   -5.7550   -6.2239   -6.7235   -7.2457   -7.7851   -8.3404   -8.9079
 -100.0000   -5.3038   -5.7390   -6.2157   -6.7200   -7.2448   -7.7851   -8.3395
 -100.0000 -100.0000   -5.2783   -5.7246   -6.2082   -6.7168   -7.2439   -7.7851
```

Figure 9. Example of a reward table constructed on MATLAB.

All problems subject to reinforcement learning can be expressed as a Markov decision process (MDP) model, and this MDP is based on the Markov process (MP). The purpose of reinforcement learning is to solve the Bellman Equations below.

$$V_\pi(s) = \sum_{a \in \hat{A}} \pi(a|s) Q_\pi(s, a) \tag{52}$$

$$V_\pi(s) = \sum_{a \in \hat{A}} \pi(a|s) \left(R_s^a + \gamma \sum_{s' \in \hat{S}} P_{ss'}^a V_\pi(s') \right) \tag{53}$$

$$Q_\pi(s, a) = R_s^a + \gamma \sum_{s' \in \hat{S}} P_{ss'}^a \sum_{a' \in \hat{A}} \pi(a'|s') Q_\pi(s'a') \tag{54}$$

$$V_\pi^*(s) = \max(V_\pi(s)) \quad , \quad Q_\pi^*(s, a) = \max(Q_\pi(s, a)) \tag{55}$$

$$Q(s, a) = Q(s, a) + \alpha_{lr} \Delta Q \tag{56}$$

(52)–(54) are the Bellman Expectation Equations. If the optimal value of Q is found as in (55), the action state a^* can be obtained and π^* can be obtained accordingly. In (54), R_s^a, the reward of action a in state s is the sum of the negative values of the cost for energy consumption and the life-cycle cost as shown in Figure 9. $P_{ss'}^a$, the probability of transition from s to s' is set to 1 in this problem. For example, when SOC 0.5 is state s and 0.4 is s', the action is a discharge corresponding to the amount of energy for SOC 0.1 which is a difference. As a result, when the action of discharging from SOC 0.5 to 0.4 is selected, the probability of the transition becomes 1 because another state cannot exist according to this action. The discount factor γ is used to evaluate future rewards at the point in time. When determining the optimal scheduling, the γ is set to 1 in this problem because the reward is not discounted. The function approximator solves the problem by finding the value function value in the reverse order from the final state through (54) and updating the $Q(s,a)$ value as shown in (56).

4. Simulation Results

The MDP object is defined through the configured table, and the problem is solved using the reinforcement learning toolbox of MATLAB 2019a. Figure 10 shows the battery open circuit voltage fitting curve and Figure 11 shows the bias model for temperature and SOC. Figure 12 shows the battery internal resistance/temperature curve and Figures 13 and 14 show RC network R curves and surface with respect to SOC/Internal Temperature. Table 7 shows the settings for the agent and training options.

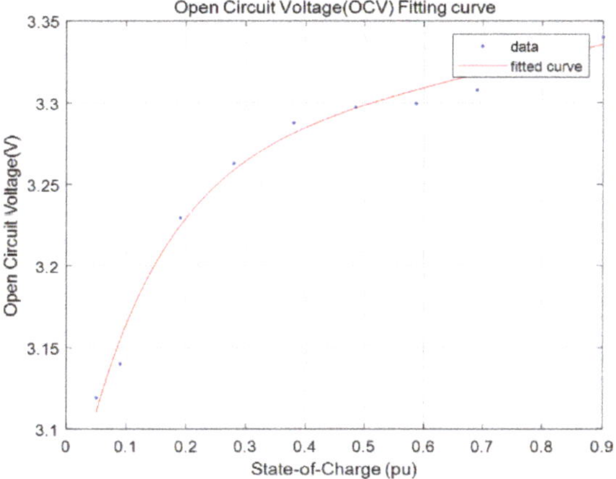

Figure 10. Battery open circuit voltage fitting curve.

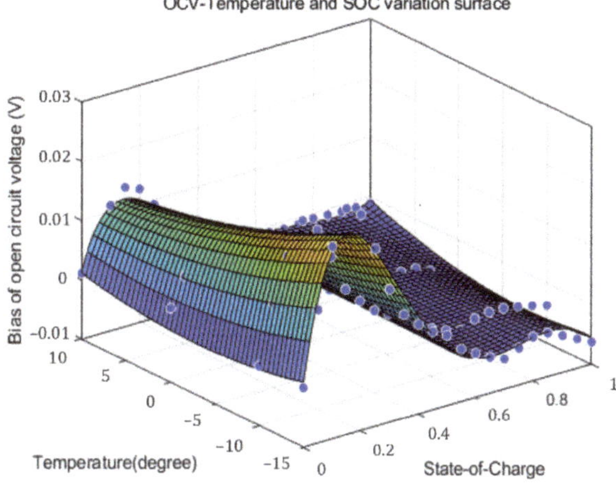

Figure 11. Bias of open circuit voltage (V).

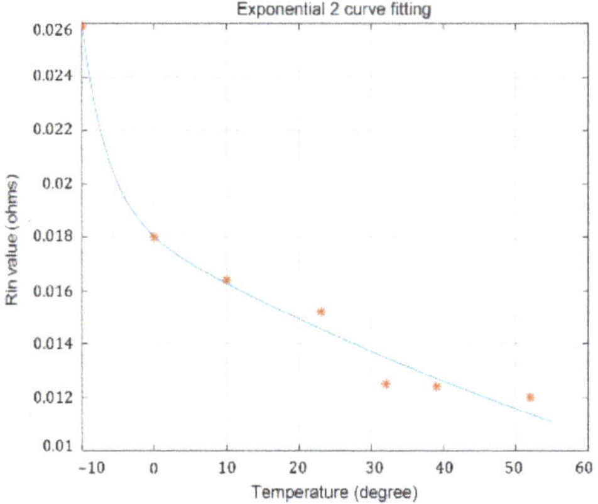

Figure 12. Battery internal resistance/temperature curve.

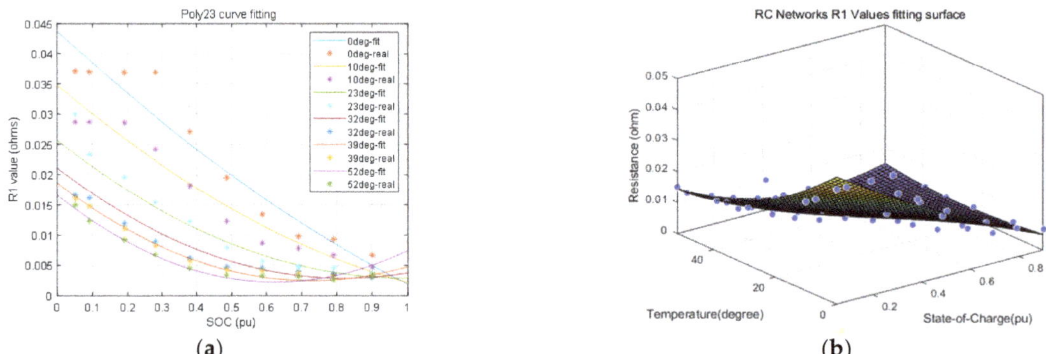

Figure 13. (**a**) RC Network R_1 curves with respect to SOC/Internal Temperature; (**b**) RC Network R_1 surface with respect to SOC/Internal Temperature.

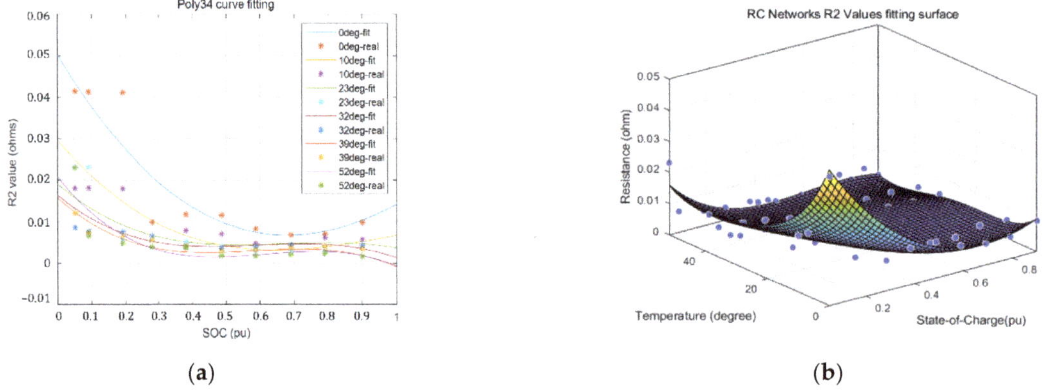

Figure 14. (**a**) RC Network R_2 curves with respect to SOC/Internal Temperature; (**b**) RC Network R_2 surface with respect to SOC/Internal Temperature.

Table 7. Setting parameters for the Q-learning agent and training options.

Q-Learning Agent Options	Learning Rate	Epsilon Greedy Exploration Probability	Epsilon Decay
Parameter value	0.1	0.9	0.01
Trading Options	Max steps per episode		Max episodes
Parameter Value	20,000		20,000

For the optimal scheduling of ESS, the power system's architecture is shown as Figure 15a and is behind the meter. Figure 15b shows the load power curve and PV output curve. In the case of the ESS used in this paper, the load and PV were modeled in the behind the meter (BTM) method. Figure 16 demonstrates the process of finding the path to the SOC through reinforcement learning.

By changing the weight for the life-cycle cost through reinforcement learning, we checked whether the effect of the life-cycle cost is reflected in the ESS SOC results in Figure 17.

The results in Figure 17 compare the optimal ESS SOC results when the life-cycle cost is not reflected and when it is reflected. In the figures, the green graphs represent the price curve. In Figure 17a, when the initial life-cycle cost is not considered, the ESS repeats a charging/discharging pattern due to the price difference and discharging during the most expensive time period to maximize profits.

However, in Figure 17b–e, when the life-cycle cost is considered, frequent charging/discharging is reduced. As the life-cycle cost weight increases, discharge is not performed even in a time period when the price is low. It was also confirmed that no charging/discharging was performed when the weight of the life-cycle cost increased by more than a certain amount in Figure 17f. This is because the investment cost value of the ESS itself dominates the difference between the system purchase cost and absolute size.

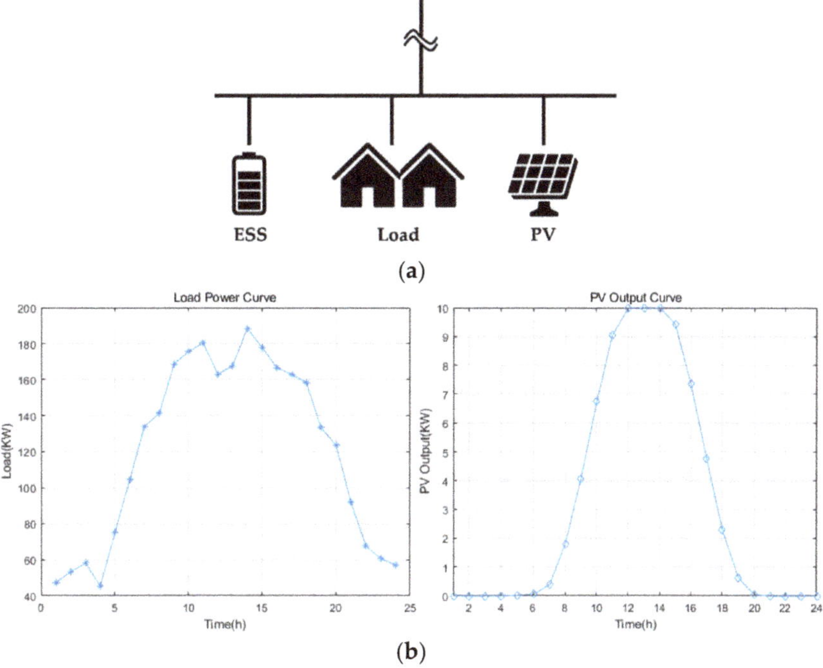

Figure 15. (**a**) The power system architecture; (**b**) The Load power curve and PV Output curve.

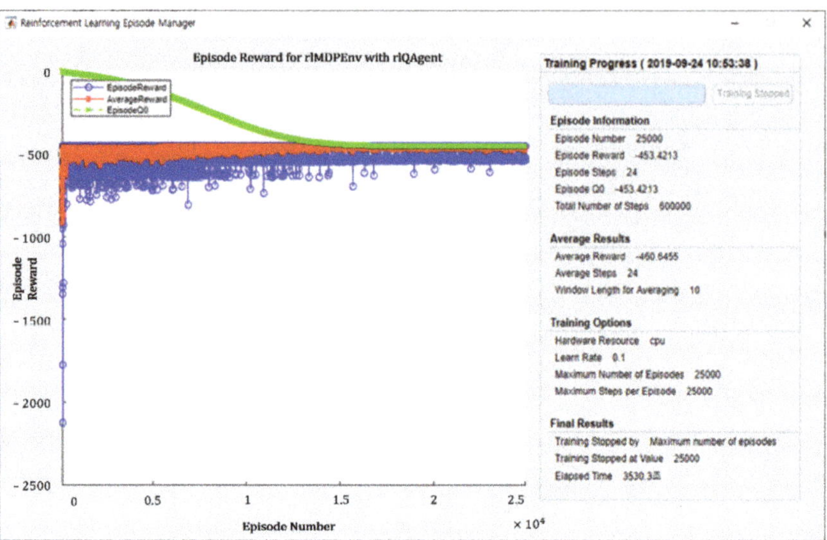

Figure 16. An example of solving an ESS optimization problem through Q-learning.

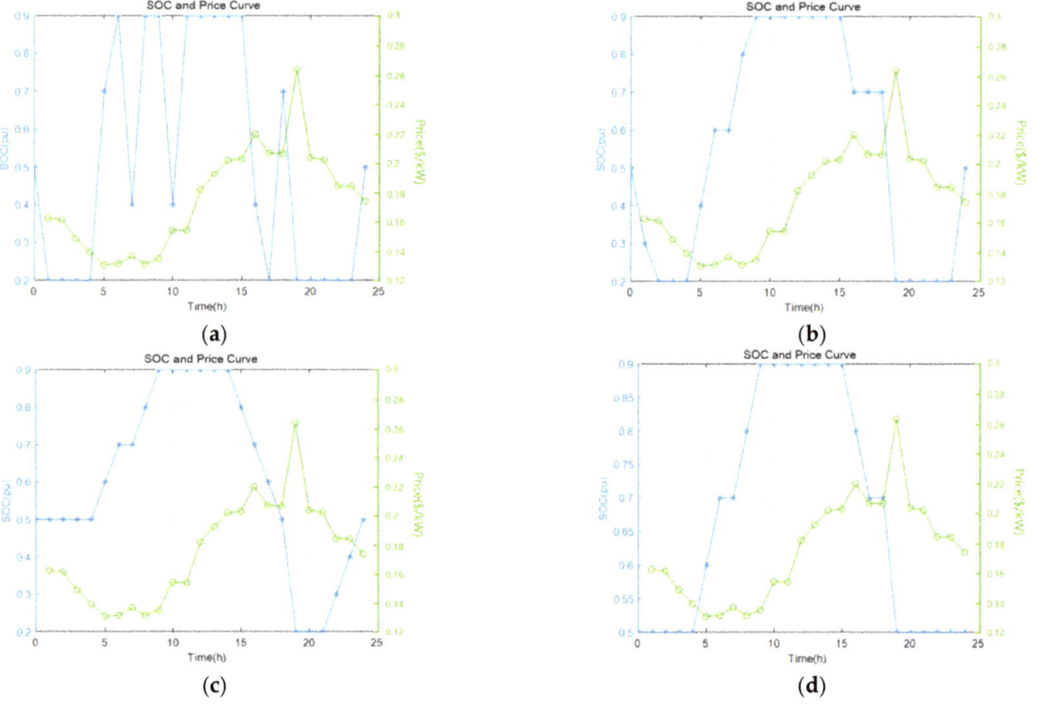

(**a**)

(**b**)

(**c**)

(**d**)

Figure 17. *Cont.*

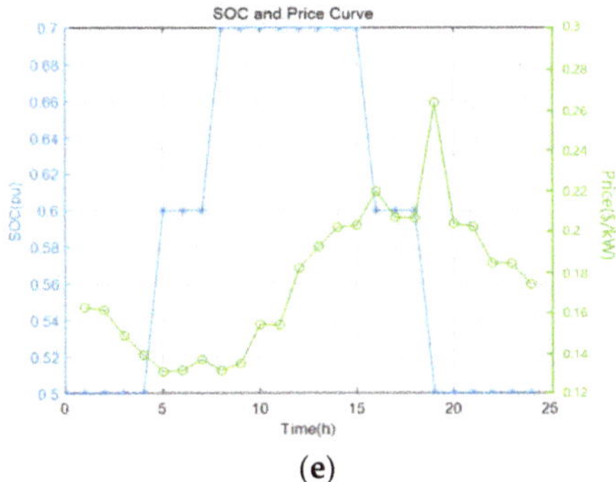

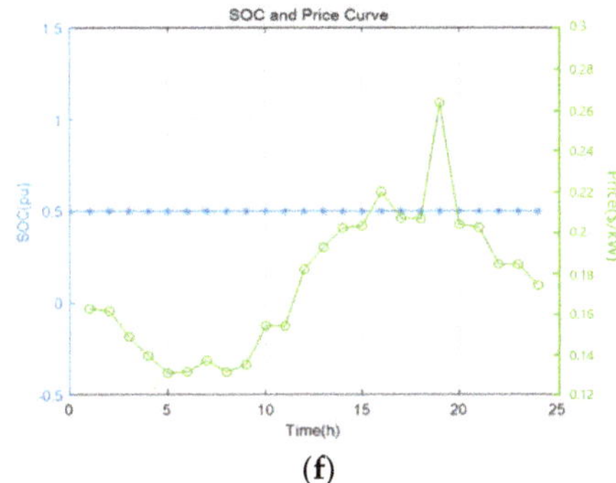

(e)   (f)

Figure 17. (**a**) SOC graph when life-cycle cost is not considered; (**b**) SOC graph when life-cycle cost weight is 0.1; (**c**) SOC graph when life-cycle cost weight is 0.4; (**d**) SOC graph when life-cycle cost weight is 0.5; (**e**) SOC graph when life-cycle cost weight is 0.6; (**f**) SOC graph when life-cycle cost weight is 0.8.

5. Conclusions

In this study, the life-cycle cost for an ESS is defined in detail based on a life assessment model and is used for scheduling. Prosumers with ESSs can make an assessment on the price of P2P energy transactions based on the defined ESS life-cycle cost. The life-cycle cost is affected by four factors: temperature, average SOC, DOD, and time. Among the four stress models, the temperature and DOD cannot be approached analytically; therefore, they are solved by approximation and reinforcement learning. The life-cycle cost of an ESS is verified through the reinforcement learning toolbox of MATLAB. Regarding the life-cycle cost, it is confirmed that the SOC result curve changes according to the weight, and as the weight of life-cycle cost increases, the ESS output and charge/discharge frequency decrease. When the initial life-cycle cost is not considered, the ESS repeats a charging/discharging pattern due to the price difference and the ESS discharges during the most expensive time period to maximize profits. However, when the life-cycle cost is considered, frequent charging/discharging is reduced. As the life-cycle cost weight increases, discharge is not performed even in a time period when the price is low. It was also confirmed that no charging/discharging was performed when the weight of the life-cycle cost increased by more than a certain amount. In the future, we shall investigate the connection between the community grid, general distribution system and a real-time P2P energy trading strategy that considers real-time uncertainty.

Author Contributions: This work W.L. devised the idea and completed the simulations M.C. prepared the manuscript. D.W. has supervised and commented on the manuscript. All authors have read and agreed to the published version of the manuscript.

Funding: This work was supported by INHA UNIVERSITY Research Grant.

Institutional Review Board Statement: Not applicable.

Informed Consent Statement: Not applicable.

Conflicts of Interest: The authors declare no conflict of interest.

Nomenclature

The following Nomenclatures are used in this manuscript:

f_d	The degradation ratio
S_T, S_σ, S_δ	The stresses for temperature, average SOC, and DOD
T_c	The battery cell temperature
T_{ref}	The reference temperature
k_T	The temperature stress coefficient
σ	The average SOC
σ_{ref}	The reference average SOC
k_σ	The average SOC stress coefficient
S_t	The stress for time
δ	The cycle depth
$k_{\delta,q1}$, $k_{\delta,q2}$	The DOD coefficients
t	Time
k_t	The time stress coefficient
L	The consumed life-cycle
α_{sei}, β_{sei}	The solid electrolyte interphase (SEI) film formation coefficients
T_s	The sampling time (unit: second [s])
C_n	The battery capacity (unit: Ampere hour [Ah])
i	The output currents
k	Time index
T_{in}	The battery internal temperature
T_{sh}	The battery shell temperature
T_{amb}	The ambient temperature
C_{q1}, C_{q2}	The internal and shell thermal capacities of the battery
k_1, k_2	The heat conduction coefficients
P_{bat}	The output of the ESS
C_{ESS}	The capacity of the ESS
η	The ESS charging/discharging efficiency
σ^t	The SOC at time t
f_{bat}	The life-cycle cost
In_{bat}	The investment cost for the battery
$V_\pi(s)$	The value of state s
$Q_\pi(s,a)$	The value of action a in state s
$\pi(a|s)$	The policy of action a in state s
R_s^a	The reward of action a in state s
$P_{ss'}^a$	The probability of transition from state s to state s' by action a
γ	The discount factors
α_{lr}	The learning rates

References

1. Jeonghwa, G. *Issue Study Report Changes in the Global Power Generation Industry Paradigm and Implications*; Korea Eximbank: Seoul, Korea, 2019; Volume 2018-04, pp. 1–51.
2. Knowledge Industry Information Institute. *Distributed Power New Technology Development Trend and Small Power Brokerage Market Promotion Status*; Knowledge Industry Information Institute: Seoul, Korea, 2020; pp. 505–593.
3. Hernandez, J.; Etemadi, A. Use of Multiple Linear Regression Techniques to Predict Energy Storage Systems' Total Capital Costs and Life Cycle Costs. In *Proceedings of the 2020 IEEE PES/IAS PowerAfrica, Nairobi, Kenya, 25–28 August 2020*; IEEE: Piscataway, NJ, USA, 2020; pp. 1–5.
4. Bonanno, G.; De Caro, S.; Sciammetta, A.; Scimone, T.; Testa, A. An Analytic Approach to Pay-Back Time Assessment of Grid-Connected PV Plants with ESS. In Proceedings of the 2015 Second International Conference on Mathematics and Computers in Sciences and in Industry (MCSI), Sliema, Malta, 17 August 2015; pp. 1–6.
5. Torkashvand, M.; Khodadadi, A.; Sanjareh, M.B.; Nazary, M.H. A Life Cycle-Cost Analysis of Li-ion and Lead-Acid BESSs and Their Actively Hybridized ESSs with Supercapacitors for Islanded Microgrid Applications. *IEEE Access* **2020**, *8*, 153215–153225. [CrossRef]
6. Yan, N.; Li, X.; Ma, S.; Zhao, H.; Zhang, B. Research on capacity configuration method of energy storage system for echelon utilization based on accelerated life test in microgrids. *CSEE J. Power Energy Syst.* **2020**, 1–11. [CrossRef]

7. Jiao, J.; Du, Y.; Yang, J.; Tian, Y.; Hu, L.; Wang, Q. Cost Management Evaluation of Power Grid Engineering: A Life Cycle Theory. In Proceedings of the 2019 IEEE Innovative Smart Grid Technologies—Asia (ISGT Asia), Chengdu, China, 21–24 May 2019; pp. 2499–2504.
8. Zhang, H.; Xue, S.; Li, X.; Li, R.; Zhang, X.; Ma, L. Evaluation Model for Life-Cycle Management Capability of Power Grid Corporation's Distribution Equipment Assets. In Proceedings of the 2020 5th Asia Conference on Power and Electrical Engineering (ACPEE), Chengdu, China, 4–7 June 2020; pp. 1961–1965.
9. Liu, J.; Zou, D. Study on the P2P Sharing Mode Operating Scheme of Consumer-side Distributed Energy Storage. In Proceedings of the 2019 IEEE 3rd Conference on Energy Internet and Energy System Integration (EI2), Changsha, China, 8–10 November 2019; Volume EI2, pp. 2097–2102.
10. Jamahori, H.F.; Rahman, H.A. Hybrid energy storage system for life cycle improvement. In Proceedings of the 2017 IEEE Conference on Energy Conversion (CENCON), Kuala Lumpur, Malaysia, 30–31 October 2017; pp. 196–200.
11. Motapon, S.N.; Lachance, E.; Dessaint, L.-A.; Al-Haddad, K. A Generic Cycle Life Model for Lithium-Ion Batteries Based on Fatigue Theory and Equivalent Cycle Counting. *IEEE Open J. Ind. Electron. Soc.* **2020**, *1*, 207–217. [CrossRef]
12. Bouakkaz, A.; Gil Mena, A.J.; Haddad, S.; Ferrari, M.L. Scheduling of Energy Consumption in Stand-alone Energy Systems Considering the Battery Life Cycle. In Proceedings of the 2020 IEEE International Conference on Environment and Electrical Engineering and 2020 IEEE Industrial and Commercial Power Systems Europe (EEEIC/I&CPS Europe), Madrid, Spain, 9–12 June 2020; pp. 1–4.
13. Yao, Z.; Lu, S.; Li, Y.; Yi, X. Cycle life prediction of lithium ion battery based on DE-BP neural network. In Proceedings of the 2019 International Conference on Sensing, Diagnostics, Prognostics, and Control (SDPC), Beijing, China, 15–17 August 2019; pp. 137–141.
14. Shen, S.; Nemani, V.; Liu, J.; Hu, C.; Wang, Z. A Hybrid Machine Learning Model for Battery Cycle Life Prediction with Early Cycle Data. In Proceedings of the 2020 IEEE Transportation Electrification Conference & Expo (ITEC), Chicago, IL, USA, 23–26 June 2020; pp. 181–184.
15. Zhang, Y.; Xu, Y.; Yang, H.; Dong, Z.Y.; Zhang, R. Optimal Whole-Life-Cycle Planning of Battery Energy Storage for Multi-Functional Services in Power Systems. *IEEE Trans. Sustain. Energy* **2019**, *11*, 2077–2086. [CrossRef]
16. Valencia, A.; Hincapie, R.A.; Gallego, R.A. Optimal location, selection, and operation of battery energy storage systems and renewable distributed generation in medium–low voltage distribution networks. *J. Energy Storage* **2021**, *34*, 102158. [CrossRef]
17. Lee, J.-O.; Kim, Y.-S. Novel battery degradation cost formulation for optimal scheduling of battery energy storage systems. *Int. J. Electr. Power Energy Syst.* **2021**, *137*, 107795. [CrossRef]
18. Wu, Y.; Liu, Z.; Liu, J.; Xiao, H.; Liu, R.; Zhang, L. Optimal battery capacity of grid-connected PV-battery systems considering battery degradation. *Renew. Energy* **2021**, *181*, 10–23. [CrossRef]
19. Gil-González, W.; Montoya, O.D.; Grisales-Noreña, L.F.; Escobar-Mejía, A. Optimal Economic–Environmental Operation of BESS in AC Distribution Systems: A Convex Multi-Objective Formulation. *Computation* **2021**, *9*, 137. [CrossRef]
20. Hein, K.; Yan, X.; Wilson, G. Multi-Objective Optimal Scheduling of a Hybrid Ferry with Shore-to-Ship Power Supply Considering Energy Storage Degradation. *Electronics* **2020**, *9*, 849. [CrossRef]
21. Montoya, O.D.; Gil-González, W.; Serra, F.M.; Hernández, J.C.; Molina-Cabrera, A. A Second-Order Cone Programming Reformulation of the Economic Dispatch Problem of BESS for Apparent Power Compensation in AC Distribution Networks. *Electronics* **2020**, *9*, 1677. [CrossRef]
22. Molina-Martin, F.; Montoya, O.; Grisales-Noreña, L.; Hernández, J.; Ramírez-Vanegas, C. Simultaneous Minimization of Energy Losses and Greenhouse Gas Emissions in AC Distribution Networks Using BESS. *Electronics* **2021**, *10*, 1002. [CrossRef]
23. Ullah, S.; Khan, L.; Badar, R.; Ullah, A.; Karam, F.W.; Khan, Z.A.; Rehman, A.U. Consensus based SoC trajectory tracking control design for economic-dispatched distributed battery energy storage system. *PLoS ONE* **2020**, *15*, e0232638. [CrossRef] [PubMed]
24. Lee, Y.-R.; Kim, H.-J.; Kim, M.-K. Optimal Operation Scheduling Considering Cycle Aging of Battery Energy Storage Systems on Stochastic Unit Commitments in Microgrids. *Energies* **2021**, *14*, 470. [CrossRef]
25. Xu, B.; Oudalov, A.; Ulbig, A.; Andersson, G.; Kirschen, D.S. Modeling of Lithium-Ion Battery Degradation for Cell Life Assessment. *IEEE Trans. Smart Grid* **2018**, *9*, 1131–1140. [CrossRef]
26. Gong, X. Modeling of Lithium-ion Battery Considering Temperature and Aging Uncertainties. Ph.D. Thesis, University of Michigan-Dearbor, Dearborn, MI, USA, 2016.
27. Liu, K.; Li, K.; Yang, Z.; Zhang, C.; Deng, J. An advanced Lithium-ion battery optimal charging strategy based on a coupled thermoelectric model. *Electrochim. Acta* **2017**, *225*, 330–344. [CrossRef]
28. Zhang, C.; Li, K.; Deng, J.; Song, S. Improved Realtime State-of-Charge Estimation of Battery Based on a Novel Thermoelectric Model. *IEEE Trans. Ind. Electron.* **2017**, *64*, 654–663. [CrossRef]
29. Zhang, C.; Li, K.; Deng, J. Real-time estimation of battery internal temperature based on a simplified thermoelectric model. *J. Power Sources* **2016**, *302*, 146–154. [CrossRef]

Article

Impact of Demand Response on Optimal Sizing of Distributed Generation and Customer Tariff

Krishna Mohan Reddy Pothireddy [1], Sandeep Vuddanti [1] and Surender Reddy Salkuti [2,*]

[1] Department of Electrical Engineering, National Institute of Technology Andhra Pradesh (NIT-AP), Tadepalligudem 534101, India; krishnamohanreddyp@gmail.com (K.M.R.P.); sandeep@nitandhra.ac.in (S.V.)

[2] Department of Railroad and Electrical Engineering, Woosong University, Daejeon 34606, Korea

* Correspondence: surender@wsu.ac.kr; Tel.: +82-10-9674-1985

Abstract: Due to the surge in load demand, the scarcity of fossil fuels, and increased concerns about global climate change, researchers have found distributed energy resources (DERs) to be alternatives to large conventional power generation. However, a drastic increase in the installation of distributed generation (DGs) increases the variability, volatility, and poor power quality issues in the microgrid (MG). To avoid prolonged outages in the distribution system, the implementation of energy management strategies (EMS) is necessary within the MG environment. The loads are allowed to participate in the energy management (EM) so as to reduce or shift their demands to non-peak hours such that the maximum peak in the system gets reduced. Therefore, this article addresses the complication of solutions, merits, and demerits that may be encountered in today's power system and encompassed with demand response (DR) and its impacts in reducing the installation cost, the capital cost of DGs, and total electricity tariff. Moreover, the paper focuses on various communication technologies, load clustering techniques, and sizing methodologies presented.

Keywords: distributed energy resources; demand response; microgrid; load clustering techniques; sizing methodologies; communication technologies

Citation: Pothireddy, K.M.R.; Vuddanti, S.; Salkuti, S.R. Impact of Demand Response on Optimal Sizing of Distributed Generation and Customer Tariff. *Energies* **2022**, *15*, 190. https://doi.org/10.3390/en15010190

Academic Editor: Terence O'Donnell

Received: 26 November 2021
Accepted: 25 December 2021
Published: 28 December 2021

Publisher's Note: MDPI stays neutral with regard to jurisdictional claims in published maps and institutional affiliations.

1. Introduction

The ever-increasing population has led to a plethora of electricity needs in the country. Existing power systems got overstressed to meet the increased load demands. Though the power generation by the conventional fossil fuel-fired generators is flexible, controllable, and dispatchable, the demerits of these sources are not economic, environmentally unfriendly, and non-sustainable [1]. Moreover, the triple bottom line [2] approach suggests the reduction in global emissions, increasing profits, and achieving maximum benefits for the people. It encourages people to develop in a sustainable manner. Its main objective is to enhance the economic, environmental, and social development of a home or community or organisation. A possible solution is to ameliorate the existing system with the DERs [3], but the output of these sources is stochastic and uncertain in nature. Another possible solution is to deploy a battery energy storage system (BESS) into the existing MG to meet the power balance condition, but it is a costly solution. A localized grouping of DGs, BESS, and scattered loads form a MG [4]. Two modes of operation of MGs exist, namely, isolated MG or off-grid modes or autonomous mode and grid-connected mode or on-grid mode. Further, the MG has three topologies, namely, alternating current MG (AC-MG), direct current MG (DC-MG), and hybrid MG. In AC-MG and DC-MG, the sources may be AC or DC but the converters convert them into one form. In the case of AC-MG the converters convert the power into AC, whereas, in case of DC-MG the converters convert all the generations to DC. This process increases the number of converter operations. Hybrid MG enhances the performance and reduces the redundancy of converters required. MG can be operated in grid-connected mode and isolated or stand-alone mode based on the system type. The higher the peak load demand on the system, the more the generation capacity to be installed is, which increases the capital cost of MG. To mitigate the present issues such as load increment, fossil fuel deficit, and environmental degradation, steps have been taken to reshape existing power systems into green and efficient systems. Therefore, an EM system [5] is essential in a MG system for achieving efficient operation. Other factors that drive the EM of MG are economic benefits [6], environmental benefits [7], energy security [8], and energy integration [9]. Developments in power systems are due to advances in technology and an increase in electricity usage. Figure 1 shows various stages of the evolved power system from a source point of view.

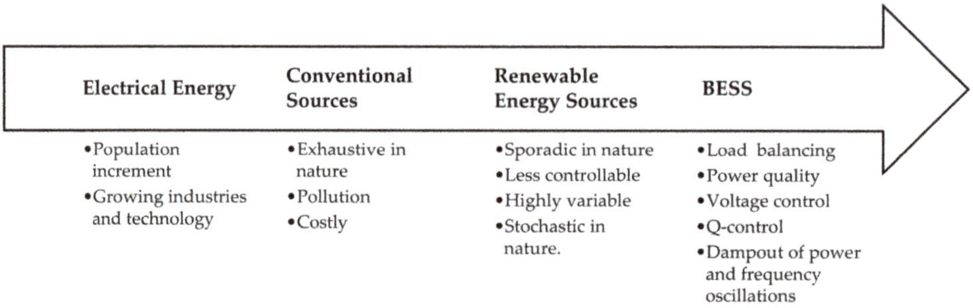

Electrical Energy	Conventional Sources	Renewable Energy Sources	BESS
•Population increment •Growing industries and technology	•Exhaustive in nature •Pollution •Costly	•Sporadic in nature •Less controllable •Highly variable •Stochastic in nature.	•Load balancing •Power quality •Voltage control •Q-control •Dampout of power and frequency oscillations

Figure 1. The evolution of power systems before DR implementation.

EM can be done in two ways: supply-side management and demand-side management (DSM). DSM focuses on DR, so as to improve the energy efficiency of the MG. DR can be stated as changes in electricity consumption by end-users based on market price fluctuation (Rs/kWh) without jeopardizing power system security. Response of the loads in accordance with the proper way benefits both the utility and the consumer. By adjusting a part of peak load to other time horizons results in reduced peak demand on the system, reduces the peak to average ratio of the load demand, and increases the load factor.

Therefore, the proposed work addresses the complication of solutions, merits, and demerits that may be encountered in today's power system and encompasses demand response (DR) and its impacts in reducing the installation cost, the capital cost of DGs, and total electricity tariff. To achieve this an objective function was formulated and an optimal sizing method has been proposed by considering the impact of DR for finding the optimal size of DGs, i.e., WT, PV, and diesel generator. Further, the proposed algorithm clusters the load into ILs and NILs and assigns a priority to the non-essential loads with the order of scheduled times by using TOU pricing. In addition, the paper suggests a limit on the amount of load shift to avoid the issues like rebound effect, increase in marginal price, and operational cost of the MG due to load recovery. Three penetration levels of demand responsive loads were considered, namely, 0%, 5%, and 10%, for studying the impact of DR programs on optimal sizing of the DGs and on consumer tariffs.

2. Literature and Contributions of the Work

Figure 2 shows a general layout of hybrid MG systems where the DC bus and AC bus are connected using a bi-directional converter. The EM achieved in grid-connected mode [10] is as follows: all the DGs, i.e., wind turbine (WT) and photo voltaic (PV) operate in maximum power point tracking (MPPT), and the dispatchable sources in the main grid supply the surplus load demand if any. During the energy surplus from MG, the BESS gets charged based on SOC. If the maximum SOC level is met, then the DGs supply the main grid and all dispatchable sources control their outputs as shown in Figure 3.

$$P_{disp} = P_{Load} - P_{Wind} - P_{PV} - P_{BESS} \tag{1}$$

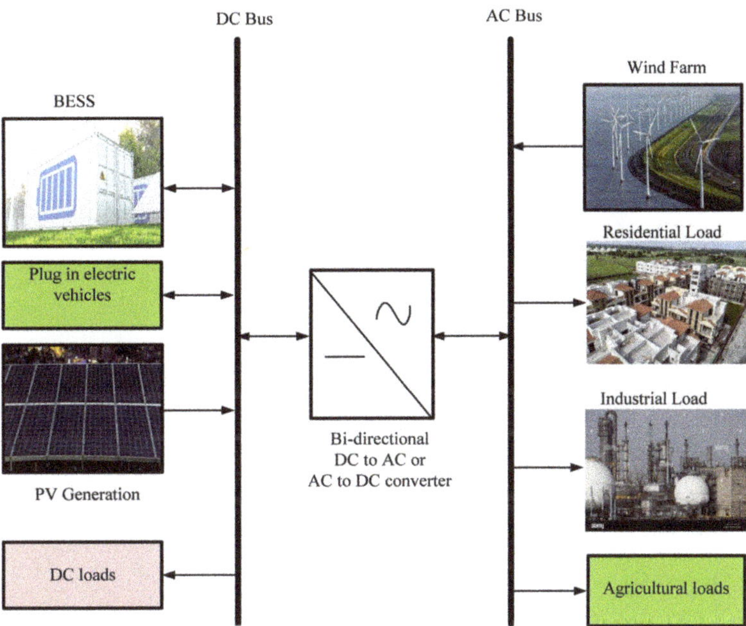

Figure 2. A general layout of hybrid MG.

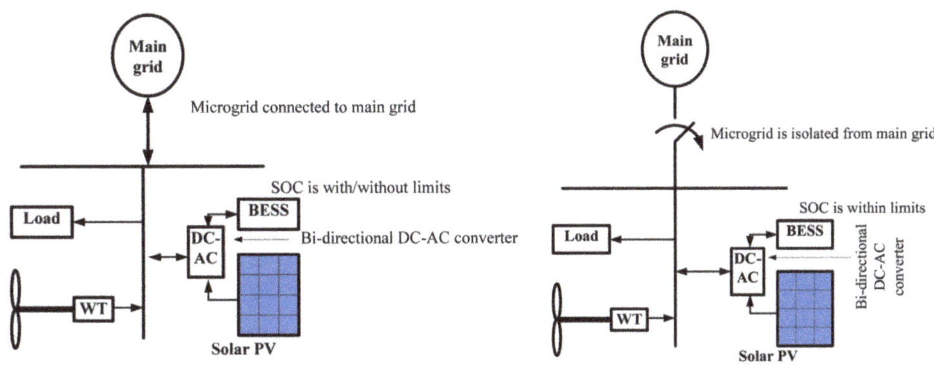

Figure 3. (**a**) Power dispatch mode in connected grid, (**b**) Un-dispatched power mode in isolated modes.

In isolated mode [11], based on the level of SOC, the DGs may or may not operate in MPPT. BESS supplies the surplus load if any. In off-MPPT, the DGs output is controlled by derating its power where PVs operate in voltage-controlled mode and WT output power operates with the pitch control mechanism. During MPPT operation, the DGs work in MPPT and the BESS charges/discharges based on the energy surplus/deficit.

$$P_{BESS} = P_{Load} - P_{Wind} - P_{PV} \tag{2}$$

The stochasticity produced by DGs and loads injects high frequency switching transients within the MG which degrades the life of the BESS, because of its low frequency and low power density capability. A super capacitor (SC) has low energy and high density;

therefore, to enhance the life of the BESS [12], the BESS operates with a SC and forms a hybrid storage energy system. Proper EM is needed to boost the efficiency of the power system and manage the ever-increasing peak demand. Deployment of new sources for meeting the shortage of electrical power is not a key solution. A promising solution to overcome the above challenges is the active participation of customers in electricity usage. In the last few decades, the power system has been in a situation where the sporadic nature is at the customer end and the generation should meet the fluctuations in the load demand.

The modern power system is quite complex, and the sporadic nature is also shifted towards the source side [13]. Therefore, it is necessary for the implementation of DR programs in the existing environment so as to manage the energy flow and control the stochasticity of loads, sources, and electricity price as shown in Figure 4.

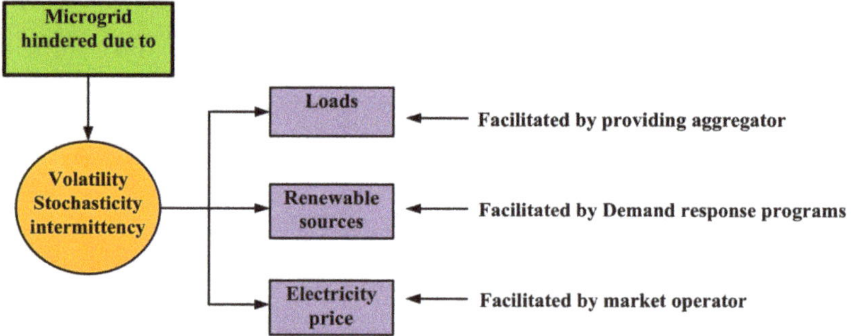

Figure 4. Stochasticity in MG.

Large integration of intermittent sources and loads into the MG needs proper EM. EM can be done at both ends, either at the source side or at the load side depending on the method of control. Optimal scheduling of DERs, optimal sizing of DERs, and optimal sizing of BESS perform supply-demand balance from the source end side, DR programs perform supply-demand balance from the end-user side. Uncertainty in RES, uncertainty in loads, and uncertainty in price require effective control. The former is mitigated by deploying fast-acting sources, the latter two addressed by performing DR programs. Power system restructuring leads to the usage of advanced metering infrastructure (AMI) which helps customers to monitor the electricity prices continuously. There can be effective scheduling of their interruptible loads (ILs) as per the charges of the real-time market by safeguarding the security of the power system. The paradigm shift from the way customers buy or sell energy has been enabled by the creation of a common marketplace or platform which establishes both energy transfer and transaction settlement on both sides. A significant change in the technologies of transmission networks is essential so as to uphold the reliability and security of the microgrid. To overcome the above challenges, there is a need to incorporate communication technologies into the distribution network such as smart metering infrastructure, supervisory control and data acquisition (SCADA), etc.

Moreover, the incorporation of EMS into the distribution network leads to the active participation of customers. The basic inputs for managing the energy in the MG are load and weather forecasting, state of charge (SOC) levels of BESS, operational, security, and reliability constraints, and the possible solutions from the EM algorithm are the schedule of DERs, load shedding/load growth, and optimal sizing of DERs. There are mainly two EMS viz., demand-side management and supply-side management and the control may be either centralized or decentralized. It is necessary to model the loads for applying the above two methods. Based on the elasticity and cross elasticity behavior of consumers, they are categorized into agricultural, industrial, residential, and commercial loads.

DR is well-defined as variations in electricity consumption by end-users based on market price fluctuation (Rs/kWh) without jeopardizing power system security [14]. DR contributes to the system reliability, security, efficiency, and economic operation of the MG. Further, DR provides a dynamic balance between supply and demand in all instances. DR is seen as one of the cornerstones of future MG for addressing the above-concerned challenges, it can be noted that resilient control can be incorporated at this stage to deal with faults [15]. Moreover, customer comfort [16] should be ensured while performing the DR operations. The load consumption pattern should match the generation profile. Loads should be diversified so that the diversity factor gets improved. DR alleviates the load profile by shifting the peak load, filling the valley point which decreases the operational cost of the MG. Various researchers are working in this field as DR techniques have the potential to solve the majority of existing power system problems. This paper focuses on the impact of DR programs on sizing problems and consumer price. Moreover, a brief review of aggregator functions, load clustering methods, and various trading models are mentioned.

Figure 5 shows the function of DSM. The collaboration between the BESS and DR programs will enhance the performance of the power system due to the uncertainty present in generation, loads, and electricity price. Decomposition algorithm with three scheduling patterns employed in [17], namely day ahead (DA) scheduling to optimize the expected operational cost of MG, an hour ahead (HA) scheduling to reduce the gap between DA, and real-time (RT) scheduling, and RT scheduling is employed to reduce the real power imbalance.

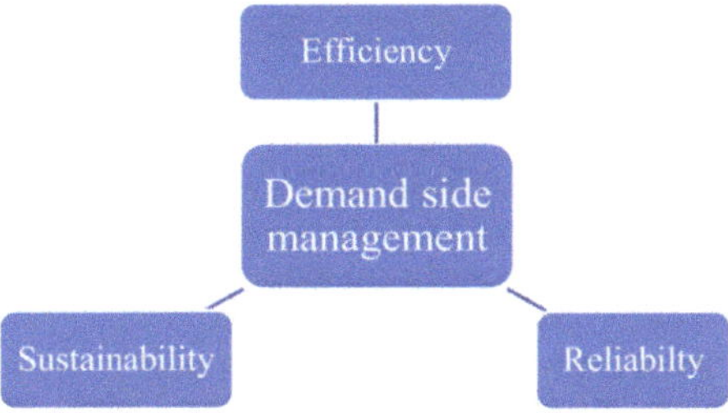

Figure 5. Functions of DSM in MG.

A review on DR programs and EMS is presented in [18]. A review on integration of DGs with the BESS, utilizing the demand side resources for increasing the flexibility of MG, and various market rules have been proposed in [19]. A review on DR strategies and EM in smart environments is presented in [20]. A detailed review on various artificial intelligence (AI)-based algorithms to forecast the energy requirement during peak hours is presented in [21], in which home energy management system (HEMS) is considered for the course of study. The forecast and EMS can be applied on any of the issues such as DGs output power, electricity price, and on loads [22–26]. A review [27] on EM in buildings based on reinforcement learning (RL) algorithms is applied for responsive sources in buildings, i.e., PV, BESS, electric vehicles (EVs), and heating, ventilating and air conditioning (HVAC) systems. Figure 6 shows various EM techniques at source side.

The deployment of RES causes a drop in the overall inertia of the modern power system. This poor inertia subsequently leads to frequency oscillations [28]. Frequency stability [29] can be analyzed by simulating a loss of a generator or a loss of the majority of a load under an aggregator or a balanced fault on the steady-state system. At present, supervision of distribution networks has become more difficult due to the penetration of intermittent renewable energy in large amounts. Volatility and uncertainty in renewable energy sources (RES) increase the burden on independent system operators (ISO) to match the generation and demand. Large-size thermal generators do not have an immediate ramp-up capability. BESS provides sufficient power balance with high speed compared to other DGs; however, it is costly. Therefore, there is a necessity to find alternate solutions. The literature and the case studies considered depict the influence of DR programs on the sizing and on consumer's tariff.

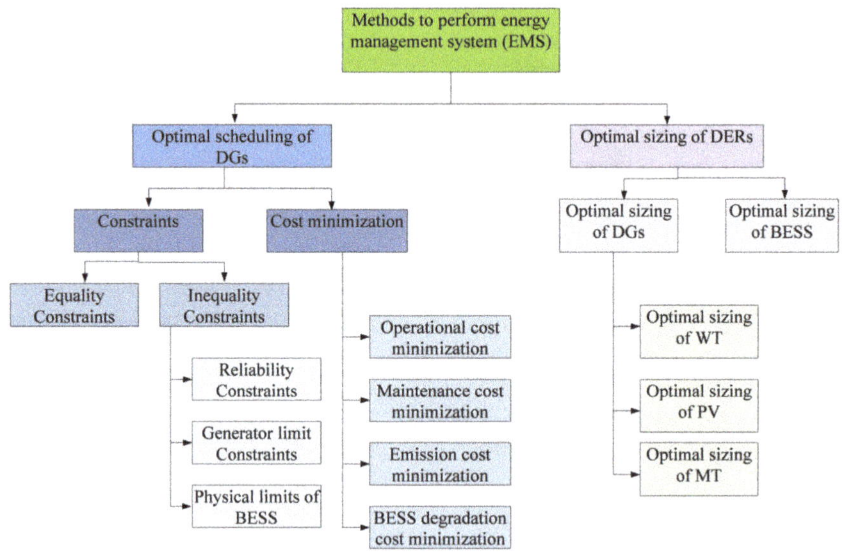

Figure 6. Long-term methods to perform EM in the MG system from source end.

In [30], a real-time hardware prototype is developed to manage the energy flow between the MG and the main grid efficiently. Three cases have been considered, namely, peak hours on the conventional grid, off-peak mode on the main grid, and isolated mode where MG is disconnected from the main grid. A peak EMS is proposed by scheduling the PV and the BESS in order to mitigate the stress developed on the main grid during peak hours and optimize the battery status based on SOC and grid status based on load profile. The failure rate on the system is a function of the number of devices connected. As the type of sources considered is of a DC nature, redundancy of equipment should be limited by considering the DC-MG to improve the reliability and reduce the conversion losses of the MG system. Therefore, the number of devices required to integrate DC sources with the AC grid gets reduced which ultimately decreases the failure rate of the system.

By performing the above two strategies as shown in Figures 6 and 7, the system operators increase the energy efficiency and improve the reliability of the MG. The area under the curve before and after valley filling should be identical to each other. Basically, the type of tariff should be simple and easy to understand by every consumer. There should be a minimum number of price updates. Depending on the number of updates and duration for each price, pricing-based tariffs are classified into three types. Table 1 shows the key differences between price-based techniques.

Table 1. Differences between various price-based tariffs [31].

S.No.	Parameter	Time of Use (TOU) Tariff [32]	Critical Peak Pricing (CPP) Tariff [33]	Real Time Pricing (RTP) [34]
1	Volatility in prices	Fixed prices during the same season.	High price during the event.	Dynamic prices.
2	Complexity	Easy to use.	Moderate and event-driven to ensure reliability. Imposed by the utility.	Complex and it needs a robust hardware setup.
3	Operation	There is no curtailment of load demand. Only shifting in time horizon takes place.	Either curtailment or shifting of load takes place.	Shifting of loads is difficult since consumers may not be able to see their incentives.
4	Frequency of imposition	Imposed on daily basis.	Not imposed on daily basis.	Imposed on hourly or minutes or seconds basis.
5	Efficiency	Highly efficient in reducing the energy cost and carbon emissions.	Less efficient in reducing the energy cost and carbon emissions.	Moderate efficient in reducing the energy cost.

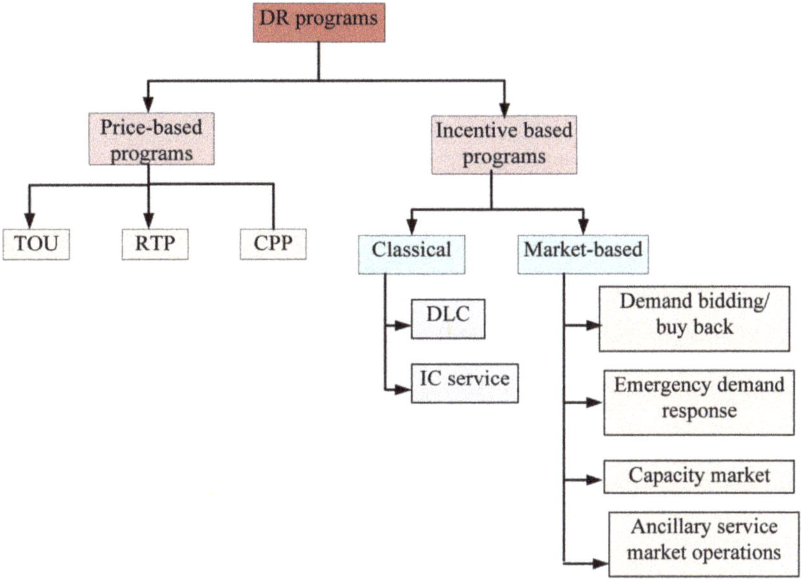

Figure 7. Methods to perform FM in the MG system from the load end.

Incentive-based tariffs are clustered into two types, namely, direct load curtailment (DLC) and indirect load curtailment. In the former one, utilities directly control the consumer appliances, therefore there is a possibility to secure threat and customer confidentiality, whereas, in the latter, customers control their loading based on the price signals displayed. Customer baseline load is used for deriving the compensations received by the consumers. However, it requires sophisticated metering infrastructure. Figure 8 shows some of the DR strategies performed by the system operator for effective management of the load whereas Figure 9 shows the benefits of applying DR strategies.

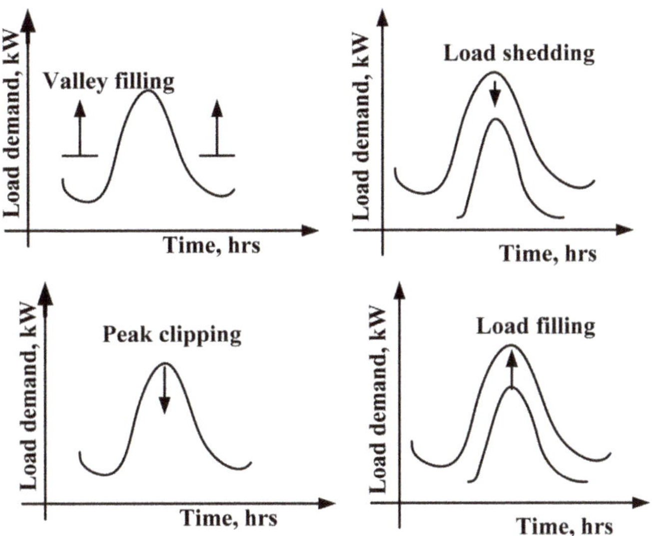

Figure 8. DR strategies.

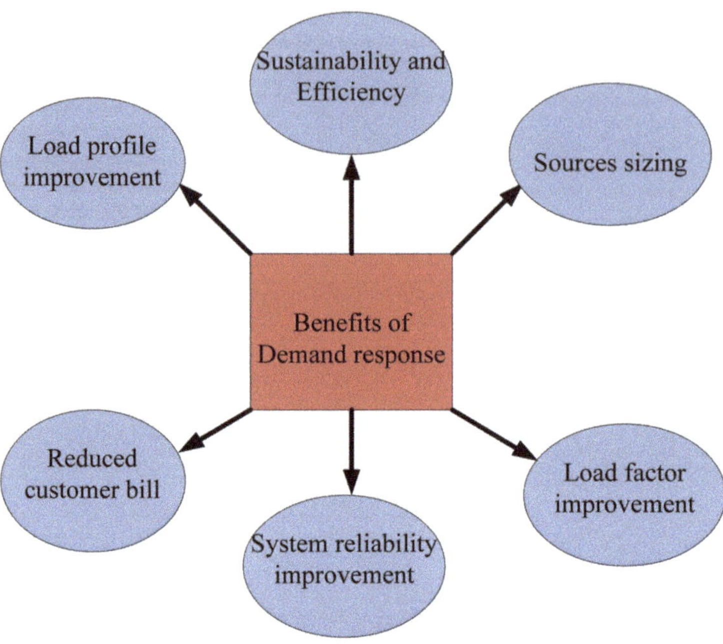

Figure 9. Merits of DR strategies.

2.1. A Brief Introduction on Optimal Scheduling in a MG

The operational cost of a MG is a function of fuel and operating crew and the amount of fuel required is a function of available load demand on the system. Therefore, proper scheduling of DERs is mandatory for reducing the operational cost of MG. Table 2 presents a brief introduction to various problem definitions for the optimal scheduling of DERs in a MG.

Table 2. Optimal scheduling of DERs in a MG.

Ref No.	Size of Distribution Network	Isolated (I) or Grid Connected (G)	Approach	Objective Function	Remarks
[35]	33 bus and 69 bus radial networks	*	Population-based incremental learning method	To minimize the power losses, improve voltage profile	Optimal power flow analysis has been done for yielding power losses and voltage profiles.
[36]	33-bus, 69-bus, and 78-bus radial networks	*	Hybridized meta-heuristic method	To optimize the active and reactive power losses	Hybridized gray wolf optimization (GWO) and particle swarm optimization (PSO) algorithms. The load power is considered as 3.715 MW and reactive power as 2.3 MVAR.
[37]	*	I	Improved multi-objective grey wolf optimization (IGWO)	To optimize the annualized cost of the system and deficiency of power supply probability	EM system is used to achieve the requirement.
[38]	IEEE 33 and 69 bus	G	Firefly algorithm	To minimize the power losses	Applied firefly algorithm for optimal sizing and siting the DGs in a radial network.

Table 2. *Cont.*

Ref No.	Size of Distribution Network	Isolated (I) or Grid Connected (G)	Approach	Objective Function	Remarks
[39]	IEEE 33 and 69 bus	G	Sine cosine algorithm (SCA) + second-order cone programming (SOCP)	To minimize the real power losses	Effective placement and sizing of DGs. Placement by using sine cosine algorithm and sizing by using the second-order cone programming.
[40]	IEEE 33 bus	G	Multi-leader PSO	Active power loss reduction	To optimally site and size DGs.
[41]	IEEE 115 bus	G	Genetic algorithm (GA) has been used	Power loss reduction, improved voltage level, and short circuit level	Proper selection and placement of DGs certainly improve the EM system.
[42]	IEEE 33 and 69 bus	*	Adaptive shuffled frog leaping algorithm (ASFLA)	Power loss minimization and voltage stability index improvement	Novel adaptive technique is used for solving the problem. The performance of the algorithm is compared with the firefly, cuckoo search, and shuffled frog leaping algorithm. Results suggest that ASFLA outperforms other algorithms.
[43]	33-bus radial distribution network	*	Modified differential evolution algorithm	To minimize DA composite economic cost, i.e., operation, economic, and transmission loss cost	The problem is modeled as a non-linear problem and scheduling has been done to optimize the fuel cost.
[44]	Loads, PV, and BESS	*	Time decision algorithm	To optimize the peak power cost	The model yields the per day load and generation capabilities further, a decision is made for BESS dispatch.
[45]	*	G	Improved particle swarm optimization (IPSO)	To achieve economic scheduling	Uncertainty in wind and solar generation and electricity prices has been considered and a two-stage stochastic model is solved to optimize the operating cost of the MG.
[46]	IEEE 33-bus system	G	Optimal power flow calculation is carried out and hierarchical distribution network integration method	To achieve optimal consumption of energy by effective economic scheduling	Effective coordination between prosumer and grid is required for the well operation of the power system. An increase in residential loads will make the power flow calculations complex. To overcome the above concern residential customers are clustered into a single residential MG.

* Not specified.

2.2. The Necessity of Optimal Sizing of DERs in a MG

For effective functioning of the power systems, there should be a dynamic balance between the supply and the demand occurring on the system. Though the peak loads on the system occur rarely, there is a need to meet the peak load demand by increasing the supply of electricity. Therefore, the size of DGs is a function of the peak load demand.

Figure 10 depicts the various costs involved while solving the optimal size problem of BESS where the trade-off point is taken between the operational, emission, and installation cost of BESS. As the size increases, the installation cost of BESS increases but the operational cost of MG decreases. The emission cost reduces to a certain point and then increases as the size of BESS increases. The underlying fact is that as the size of BESS increases the recycling cost also increases. Table 3 represents a brief study on various problem definitions for optimal sizing of DERs in a MG.

Table 3. BESS sizing for achieving EM in a MG.

Ref No.	Isolated (I) or Grid Connected (G)	Sources Considered	Objective Function	Approach	Problem Modelled as	Remarks
[47]	G	BESS alone	To improve the load factor by decreasing the peak value to valley point difference	The net energy in the BESS is considered as the initial value of the day	Mixed-integer programming	DR strategies improve the load profile which certainly improves the load factor.
[48]	G	PV and BESS	To increase the annual net profits, PV consumptive rate	Non-dominated sorting genetic algorithm 2	Multi-objective problem	For optimal sizing of BESS, the constraints considered are reliability constraints, BESS performance constraints, and user purchasing electricity cost.
[49]	G	Solar, WT, and fuel cell (FC)	To reduce electricity costs. In this study, uncertainty in the cost of electricity and load uncertainty is considered for robust sizing	Decision theory approach is applied	BESS is considered as a cluster of loads.	In this study, decision-making is based on three issues, namely, minimization of expected cost, min-max regret, and stability.
[50]	G	Solar, WT, and FC	To reduce the total harmonic distortion (THD), active power losses, and to reduce the overall cost of DERs	Size is a function of the amount of operational cost decrement	BESS supplies the deficit power and maintains the balance	Sizing of DERs in an IEEE 31 bus distribution network was performed to reduce the total cost for installing DERs.
[51]	G	PV, grid, BESS, and WT	The aim is to optimize the overall operational cost of the MG	A quasi oppositional swine influenza model model is applied	The load difference is supplied by the grid	Sizing of both rooftop PV and BESS has been done
[52]	G	PV, WT, grid, BESS	The objective is to minimize the net operational cost and installation of BESS	GWO algorithm is used	The load difference is supplied by the grid	The fuel cost is reduced so that the size of the BESS is reduced.
[53]	I	PV, WT, BESS, FC, and micro turbine (MT) are considered	The objective is to reduce the size of BESS	The problem is formulated as a mixed integer linear programming (MILP)	Grid is not present so BESS is essential	Sizing of BESS varies when the uncertainty of DERs is considered. Three cases have been considered: no BESS, BESS with no initial charge, and BESS with an initial stored charge equal to the total capacity of BESS.

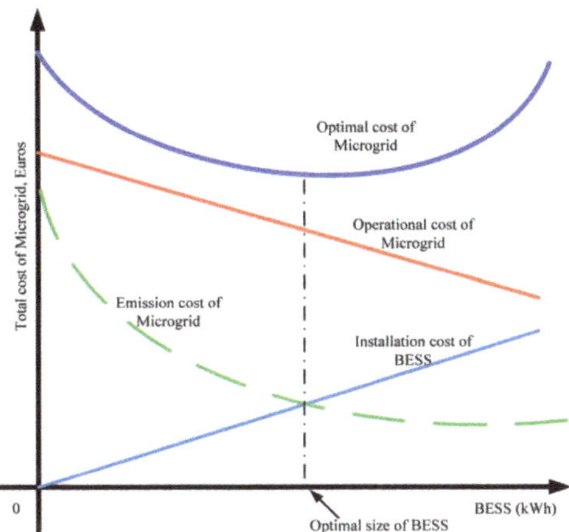

Figure 10. Trade-off points in sizing of BESS.

2.3. Functions of DR Aggregator

Customers approach a DR aggregator if they wish to participate in DR programs. Then, the DR aggregator forms a single large load by aggregating the small load demands and establishes linkage with a distributed network operator (DNO) for maximizing the profits. The aggregator acts as a mediator between loads and the ISO and the direction of data flow is bi-directional as shown in Figure 11. The below equation represents the load profiles of 'n' individual customers with 'm' load profiles submitted to the load aggregator customers, aggregates it, and changes the load pattern as per the electricity price elasticity, and submits the aggregated load data to the ISO. ISO is considered as an information hub when seen from both ends of the power system. The decision variables in DR programs are the electricity price and the incentive cost.

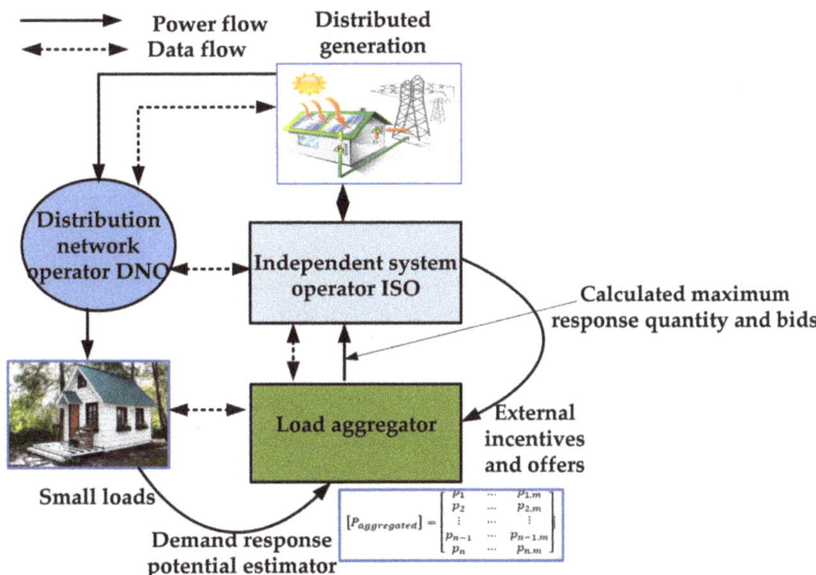

Figure 11. Function of DR aggregator.

The profit obtained by aggregators depends on the level of penetration of responsive loads. Moreover, high penetration of DERs will offer a benefit to reduce the energy price in the local community and address the undesirable line congestion issues. In [54], a framework was proposed to introduce the competition between the prosumers in which an aggregator plays as a local market operator. In [55], an aggregator model was proposed for reducing the issues in line congestion and voltage deviation, arising in response to flexible resources such as ILs, intelligent electronic devices (IEDs), and various sensors to the external price signals.

The response of loads is uncertain and the consumers participating in the DR programs should be bound to the agreed amount of curtailment or shifting of load. In [56], a method was proposed to assign priority to the reliable loads by reliability analysis. The registration period for the consumers participating in the DR is one month. During this, the aggregator derives the priority based on the consumer response in the scheduled period. Moreover, the aggregator verifies the performance of the reliable customers during the course of action by evaluating the difference between the agreed load curtailment and the actual load curtailment and changes the priority levels based on the response. Disputes in energy trading are inevitable which arises due to the multiple market players of different conflicts of interest. There should be a third party for negotiations or mediations between the prosumers and the upstream grid for suppressing the dispute. Transparency in the supply of electricity and price policies mitigates disputes.

The potential of residential consumers in DR events is increasing because the portion of load consumed by the residential loads when compared with other types of loads gets increased. In [57], the authors proposed a framework for optimal bidding strategy by considering the uncertainty in willingness to participate in DR programs of residential consumers. Table 4 represents the function of the aggregator in a MG whereas, Figure 12 indicates the functions of ISO. In general, the below Equation (3) is a function of time. It may change its shape from time to time, either in minutes, hours, or a greater number of frequent updates depending on the type of load connected, scheduled time to run, and duration of run.

$$\left[P_{aggregated}^t \right] = \begin{bmatrix} p_1^t & \cdots & p_{1,m}^t \\ p_2^t & \cdots & p_{2,m}^t \\ \vdots & \cdots & \vdots \\ p_{n-1}^t & \cdots & p_{n-1,m}^t \\ p_n^t & \cdots & p_{n,m}^t \end{bmatrix} \tag{3}$$

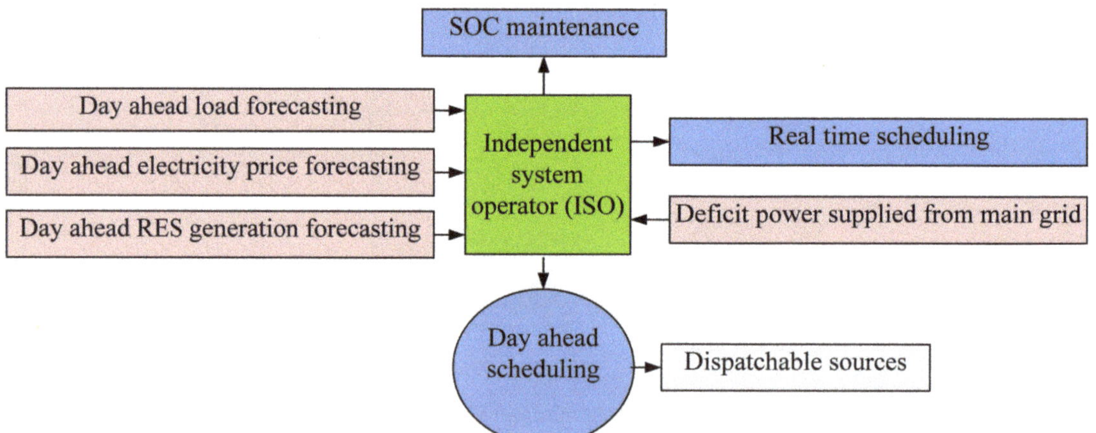

Figure 12. ISO function in load data collection, processing, and scheduling.

Table 4. Need for aggregator operations.

Ref No.	Islanded (I) (or) Grid Connected Mode (G)	Loads Are Aggregated (A) (or) No Aggregation (N)	Users	Objectives	Outcome
[58]	G	Aggregated plug-in-electric vehicle (PEV) fleet	Low volatage MG (LV-MG)	Three objective functions have been considered: Reduction of power loss, reduction of voltage deviated. and current through the lines should be optimized	The problem is modeled as a multi-objective problem in which the results are not a single value but of sequence which is called Pareto solutions therefore, it is necessary to take trade-off points which are called as Pareto optimum.
[59]	G	Clustered loads aggregated	Residential cold climate heat pump (CCHP) MG	To optimize the operating cost of the MG	A two-stage optimal dispatch problem was proposed for obtaining low operating costs.
[60]	G	EVs	Residential and commercial	Aims to reduce the running cost of the MG	A non-linear optimization problem was solved by day-ahead scheduling.
[61]	G	Residential loads within the MG were aggregated	Residential loads	To maximize the aggregator's profit	Day-ahead scheduling of DERs was considered and to mitigate the sporadic behavior a risk-constrained stochastic model was framed.
[62]	G	N	IEEE 24 bus	To achieve the economic profit, security, and stability of the MG	A multi-objective problem was solved to assess the dynamic stability.
[63]	*	A	*	To maximize the aggregator's profit by balancing the real-time deficit of power	Optimal bidding strategy of aggregator and real-time balancing at the local level.
[64]	*	A	*	Profit maximization of all the market players within the power system	Bender's decomposition is applied.
[65]	*	A	Smart distribution system	To maximize the operating cost of the agents (aggregators and prosumers) by optimally scheduling the resources and maximizing the profit of energy suppliers	Multi-follower bi-level programming is applied. The objective function is linearized and KKT is applied.
[66]	I	EVs aggregated	MG	To improve the frequency stability of the MG	In autonomous mode of operation of the MG, there is a need for the deployment of storage devices to support the frequency regulation. EVs are the active resources similar to BESS and they enhance the frequency regulation.

* Not specified.

2.4. Recap of Energy Trading Models

A novel approach for energy trading is proposed in [67], where a group of MGs has been clustered into individual MGs. In each individual cluster, the deficit power or excess power is taken from/supplied to the neighboring MGs or main grid. The price for supplying excess power is fixed in between the grid buy power and grid sell power. In [68], a fuzzy logic-based model was presented to assess the willingness of customers to participate in DR programs, and then by using the queuing method, a decision was obtained to maximize the profit of the aggregator. In [69], a method was proposed in order to directly control the heating and cooling loads as these loads have the ability to maintain the temperature within tolerable limits for subsequent hours if proper insulation is provided. A two-stage bidding strategy derived from the DA market is suggested to address the uncertainties in the electricity price.

In the majority of European Union (EU) countries, the application of DR programs is restricted to industrial applications. A method named DR-blocks of a building (DR-BOB) is proposed in [70] to apply DR programs to the BOB and aggregate many such buildings for effective bidding in electricity markets. For dynamic balance between the generation and the load, for effective monitoring of voltage levels, frequency, and phase angles at every instant and every point of MG, there is a need to incorporate fast and accurate data transferring technologies. Communication devices are ubiquitous these days in power systems because of their speed and precise data transfer. The data are bulk because more intelligent devices are connected to the distribution network. The centralized controller governs the entire MG by gathering information from all the devices. Therefore, if the size of the data is large then the time taken by the central controller to issue the governing signals gets delayed whereas, in the case of decentralized control, each agent, i.e., customer, DGs, and BESS, defines their own load schedule which employs a multi-agent system (MAS) is presented in [71]. Buildings consume 40% of the total load [72], therefore, Nikos Kampelis et al. [73] implemented a genetic algorithm (GA)-based optimization technique for EM in a building and used artificial neural networks (ANNs) prediction model for yielding DA power requirements of the customer. Time of use (TOU) pricing is used in this literature.

There is a need for a protocol [74] in order to effectively monitor, communicate, optimize, and control the information flow between the loads, distributed generators, intelligent devices, and ISO. The automation system should fetch data from the sensors, process, and be able to give necessary feedback signals to all the essential infrastructure to which it is connected. Implementation of DR-BOB by using DR-technology readiness level (DR-TRL) is proposed in [75]. The distribution system can be made smarter only when the data regarding the states of operation available at all nodes should be transferred by using a wireless communication link as shown in Figure 13. A great amount of information is generated by metering, detecting, and monitoring devices. Therefore, the MG needs dedicated and advanced communication technologies for holding, integrating, processing, and transferring the data. The function of smart meters here is to monitor, troubleshoot, and analyze the energy usage and billing for each period concerned and transfer to the grid as well as to the consumer through a mobile application for effective control. Communication infrastructures such as 5G technology are espoused seamlessly in modern power systems because of their speed, low power consumption, security, and large frequency spectrum presented in [76]. A real example for application of 5G technology in power systems is in 2019, China had carried out a pilot project in establishing protection of distribution network by using 5G technology on China southern power grid [77].

The information and communication technology (ICT) in [78] should have low latency and be able to transfer large market data for the effective operation of power systems. The below Figure 13 shows data flow in a simple MG. A low-cost and low power consumption device named Zig-bee communication in a home area network (HAN) [79–81]. Zig-bee, because of its limited range constraint, cannot be used in neighboring area networks (NAN). Wireless fidelity (Wi-Fi) can be used in HAN, NAN, and field area network (FAN). WiMAX has maximum coverage distance compared to all wireless data transfer techniques.

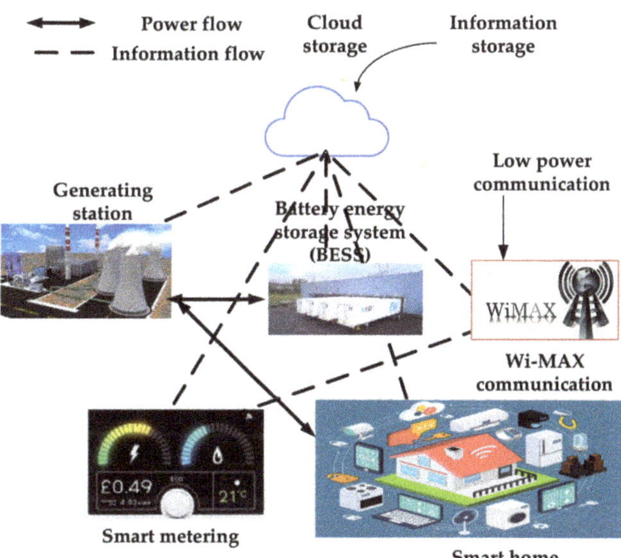

Figure 13. A detailed communication framework for effective monitoring.

In [82], a pilot project is considered for effective demand-side management and green technology improvements. Game theory is all about achieving the equilibrium point so as to maximize the profit of all the market players. Each and every agent participating in the game gets profit. The profit of each agent depends on the strategy applied by that agent and the strategy of opponents participating in the game. The market players are DSO, aggregator, and the customers. The objective function of each player is considered as the pay-off function. Consider two market players where player 1's strategy to get maximum payoff is A and player 2's strategy to get maximum payoff is B and no other strategy for player 1 will yield better outcome other than strategy A, similar for player B; then this equilibrium point is said to be NASH equilibrium. This is used in order to analyze the outcome of the tactical interaction of several choice makers. Table 5 represents a brief introduction to various trading models in a MG.

Each order (either buy or sell option) consists of TAP (time period for energy trade, amount of energy to be traded, and price of energy to be traded). After successful placement of orders, they may be modified or canceled by the same peer until the gate closure time. Once the time lapses, then the individual peer cannot alter the data; only the system operator has the right to alter the list to ensure smooth operation of the power system. A penalty should be imposed on those market players who do not meet their quoted amount of energy listed in the order.

Energy these days has become more or less a commodity. Therefore, it can be related to the stocks. The fundamental difference between them is that electricity cannot be stored, there should be a continuous balancing of supply and demand. There is a need for a flagship trading platform for placing bids and offers. The platform should encompass the following provisions: order placed, the status of orders (pending or executed), portfolio window to check the amount of energy consumed/generated with the amount to be paid/received. There should be a watchlist with all energy participants' data such as quantity, offer price, bid price, and expected time for price dip: ratings and reviews for each participant should be provided. A hierarchical decision-making trading model is proposed in [83], so as to reduce the DR contract cost at the market operator level and to reduce the incentives at the DR aggregator level so as to maximize social welfare.

Table 5. Energy trading models.

Ref No.	Size of Distribution Network	Energy Trading Model	Impact of Aggregator	Remarks
[84]	33-bus radial network	Bottom-up approach	No	An iterative algorithm was proposed for energy trade.
[85]	*	Block-chain model	Yes	The objective is to improve the reliability and user security in trading a blockchain approach.
[86]	*	Block-chain model	Yes	Developing multi-directional trading is of significance. In this paper, the authors proposed a parallel trading model.
[87]	*	Energy broker model	Yes	This method is for maximizing the energy trading between the grid and the consumer. The energy broker decides the demand and the price of electricity by using dual optimization. The problem is modeled as a convex optimization problem.

* Not specified.

2.5. A Combined Literature Survey on Various EMS

Table 6 represents various EM methods for achieving lower operational costs and reducing the installation cost and capital cost of a MG.

Table 6. A literature survey on various EMS.

Ref No.	Size of Distribution Network	Approach	Objective Function	Remarks
[88]	118-bus radial network	Unit commitment method	To minimize the total expected cost	A two-stage stochastic model for achieving effective wind power integration.
[89]	Korean electricity market	Mean-variance portfolio method	To increase expected return and to increase the profitability of the aggregator	A mean-variance portfolio method is used to avoid the risk of the DRR portfolio.
[90]	20,310 customers, 548 DGs	Resource scheduling, aggregation	To minimize the operating cost	K-means algorithm for clustering loads.
[91]	180-bus	Economic dispatch problem is taken into account	To minimize the running cost	A two-stage scheduling problem is solved to optimize the fuel cost of the test system.
[92]	218 consumers and load profile analyzed at 96-time slots.	Economic dispatch of DER	To minimize the operating cost	The allowable maximum load shift by using the DR program is less than or equal to the base load on the system. Optimization problem modeled as a linear problem. Figure 14 shows the problem associated with unjustified shifting of flexible loads.
[93]	DISCOMs	Optimal scheduling is done	To minimize the Expected cost of MG	Conditional value at risk (CVAR) index is used to analyze the uncertainty of WT. Objective function modeled as MILP. General algebraic modeling system (GAMS) and IBM ILOG CPLEX optimization studio named simply as CPLEX are used.
[94]	*	A Bayesian game model is proposed	To optimize the bidding strategy	To compensate for the power deficit/excess due to customer breach by placing auxiliary services like BESS.
[95]	20,000 heat pumps with a capacity of 1 MW.	*	Techno-economic feasibility study	Two scenarios are considered always available and always reliable.
[96]	South Korea Capacity 10 MW	Demand-side management system (DSMS)	To minimize the electricity price, energy conservation	A strategy of DSMS is proposed for calculating the customer baseline load.
[97]	*	Improved elephant herd optimization	To optimize the fuel cost of operation	A multi-objective problem has been formulated. Three cases were considered, namely scheduling to reduce operational cost, scheduling to reduce operational cost and variance, and the impact of penetrating EVs on operational cost.

* Not specified.

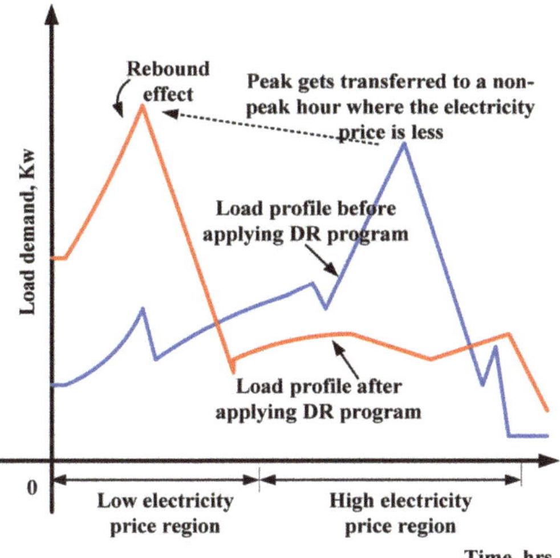

Figure 14. Impact of DR on rebound effect.

If the level of penetration of flexible or non-critical loads increases, there is a chance of occurrence of payback or rebound effect. Figure 14 shows the impact of load shifting on the rebound effect. The formation of local peaks at low electricity price zones due to the shifting of flexible loads. As the consumer's participation in DR events increases, the level of peak increases at non-peak hours. The operational cost of the MG also changes because of the load recovery. Further, the system marginal price (SMP) depends on the incremental fuel cost of the marginal generator. Hence, the SMP increases as the demand on the system increases. Therefore, unjustified shifting of non-essential load from peak duration to the off peak should be avoided to reduce the burden on the power systems.

2.6. Load Clustering and Its Significance

The first step in the application of DR programs by the ISO is load clustering. Therefore, there is a need to develop robust clustering models for avoiding the demerits of existing algorithms. There are various clustering mechanisms, among all the clustering algorithms partitional clustering is mostly used. Load clustering can be done by using optimization techniques such as ant colony optimization (ACO), bee colony optimization (BCO), and modified bee colony optimization (MBCO). There are mixed clustering methods where K-means clustering and BCO can be used for hybridization. In K-means clustering, the loads are split into K-clusters and the centroid of each cluster is found. The cluster is grouped with the nearest centroid and the process continued until the centroid gets constant. A K-means algorithm effectively works when the dataset is large. The demerit of this clustering is that there is a need to specify the number of cluster centers before starting the trial run. However, it is also to be noted that complex systems often indicate an intrinsic cluster number if an appropriate tool is chosen [98]. It further explained the necessary and sufficient conditions for cluster consensus of discrete time linear systems. Table 7 shows a brief survey on various clustering methods.

The switching signal obtained from EMS turns on or off the non-essential loads. The increment in load shift from peak to non-peak hours should be strictly controlled by the EMS to avoid the rebound effect. EMS plays a significant role in effective and efficient management of these flexible loads. Figure 15 shows the function of HEMS in a home.

Table 7. A survey on various clustering methods.

Ref. No.	Test System Considered	Clustering Method	Remarks
[99]	Residential electric water heaters	K-means clustering method is used for clustering the loads. To find the number of clusters, required Silhouette method is applied.	The goal is to reduce the impact on the availability of hot water when a DR strategy is applied. The success rate of such clustering methods is more dependent on the accuracy in load forecast.
[100]	751 residential customers	Dwelling's clustering	Based on the spot market price the decision-making algorithm shifts the load demand.
[101]	Non-residential customers	Ant colony clustering	The original electrical pattern is applied to the optimization algorithm named the ant colony clustering algorithm.
[102]	*	The spectral clustering method is applied.	By using the Euclidean function and load curve, the method performs a similarity check on all the load profiles considered. This algorithm is robust to data size.
[103]	Real-world smart meter data	Hierarchical clustering	Performs dissimilarity check.
[104]	*	Fuzzy K-means clustering	The global criterion method and Bellman–Zadeh's maximization principle are used.
[105]	Domestic customers	Fuzzy subtractive clustering is applied.	With the proposed method of load shedding and valley, the filling can be accomplished.

* Not specified.

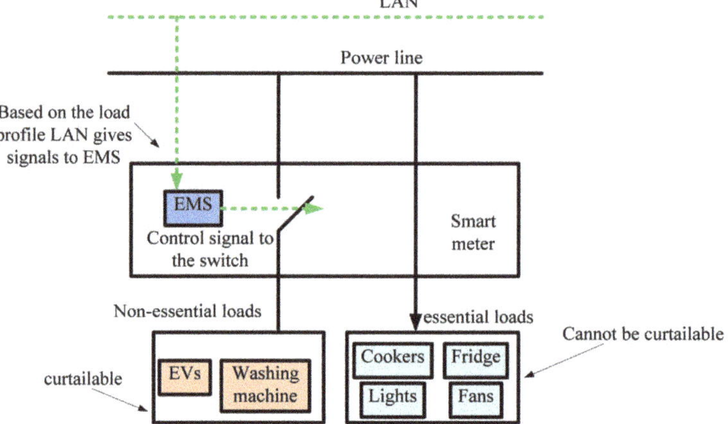

Figure 15. Home energy management system (HEMS) in a home.

2.7. Noteworthy Points

The following are the key conclusions obtained from a detailed literature survey.

- Effective load clustering is required to avoid unnecessary compromise in consumers' satisfaction and lifestyle.
- The size of DGs is proportional to the peak load demand which has to be supplied by the sources.
- The consumer's tariff is a direct function of electricity price, therefore, if there is any change in the consumers' load pattern in accordance with the electricity price, it may reduce the consumers' tariff.
- Occurrence of payback or rebound effect during low electricity price zone.

2.8. Contributions of the Work

A test system is considered and the impacts of DR programs on the supply-side and demand side analyzed. With the basic load profile and the electricity price in the area considered, the optimal size of DGs has been proposed and the electricity tariff burden on the consumer was reduced by the mentioned methods. All the loads are considered as residential loads and clustered into essential and non-essential loads without compromising the lifestyle and satisfaction of the customer. There will be no shifting of essential loads which are scheduled at a particular time. The non-essential loads may or may not be shifted based on the availability of load demand at that instant. Priority is assigned to the non-essential loads with the order of scheduled times by using TOU pricing. An objective function was formulated and an algorithm was proposed for avoiding the rebound effect during load shift. Various sizing methods have been proposed in order to optimally size the DGs with and without considering the uncertainty in PV and WT. The TOU pricing method has been applied for load shifting in DR programs in order to reduce the electricity tariff of the customer.

3. Objective Function Modeling

To reduce the above-mentioned concerns such as capital and installation cost minimization and reduction of consumers tariff, the objective function can be modeled which clusters the load as follows.

The following are the objectives to be solved.

- Minimize the size of DGs in the power system.
- Minimize the consumer's electricity bill and improve the load factor.

$$\text{Minimize, } Cost\ function = \left(F_{capital} + F_{installation} + F_{customers\ bill} \right) \tag{4}$$

Minimize total cost of installation:

$$f_1 = TC = \sum_{i=1}^{n} C_{cap} + C_{maintenance} + C_{replacement} \tag{5}$$

$$\varnothing^t = \begin{cases} \varnothing_{off}, & if\ t \in T_{off} \\ \varnothing_{mid}, & if\ t \in T_{mid} \quad \forall\ T \in 1, 2, \dots\dots 24 \\ \varnothing_{on}, & if\ t \in T_{on} \end{cases} \tag{6}$$

where C_{cap} is the capital cost, $C_{maintenance}$ is the cost involved in maintenance, $\varnothing^t$ is the electricity price at time 't', $\varnothing_{off}$, $\varnothing_{on}$, and $\varnothing_{mid}$ are the price during off-peak (T_{off}), on-peak (T_{on}), and mid-peak (T_{mid}) hours, respectively.

$$f_2 = min(TET_{customer}) = min \sum_{t=1}^{24} \sum_{i=1}^{N} P_i^t \varnothing^t \tag{7}$$

'N' indicates the number of sources available for dispatch and P_i^t is the consumed power by the consumer 'i' at time horizon 't'. The problem formulation for minimizing the consumer's tariff is represented in Equation (7). For achieving the objectives, the loads on the system are effectively clustered into essential (EL) and non-essential loads (NEL) [106] by considering all the loads as residential loads as they are sharing 25% of load demand on the system.

$$\text{Total load demand, } TL = EL + NEL \tag{8}$$

$$NEL = W_1 L_1 + W_2 L_2 + W_3 L_3 \tag{9}$$

Allocation of priorities to the non-essential loads or curtailable loads in such a way that there should not be any compromise in customer comforts and lifestyle.

L_1 Air conditioners, heating loads.

L_2 Washing machines

L_3 EV's

L_1, L_2, and L_3 are the non-essential loads and the order of priority based on importance is $L_1 > L_2 > L_3$. The higher degree of charging flexibility associated with EVs makes it less prioritized when compared to the other loads in the MG. Therefore, to represent the importance of each load according to its priority, the weights added to the loads should be of different values, where, $W_1 = 0.66$, $W_2 = 0.33$, $W_3 = 0.13$ such that all the loads should be met on the same day with some shift in time without violating the customer's satisfaction.

Subjected to the following constraints:

$$\sum_{n=1}^{N} P_{(n,t)}^{gen} = \sum_{l=1}^{L} P_{(l,t)}^{dem} - \sum_{s=1}^{S} P_{(s,t,i)}^{shift} + \sum_{s=1}^{S} P_{(s,i,t)}^{shift} \tag{10}$$

The energy recovered during peak load is shifted or re-distributed to the other time horizons where the electricity price is low. Total energy curtailed has been re-distributed to other time horizons where electricity price is low. Therefore, the total area before curtailment and after curtailment becomes identical.

The generation should meet the load demand before and after shifting. The above Equation (10) represents the equality constraint, i.e., energy balance equation. Equation (11) shows the simplified version of the energy balance equation, where the difference between generation (by all the 'N' generators) and load demand (at all the 'L' loads) on the system should be dynamically balanced throughout the considered period, i.e., 24 h. to avoid any frequency drops. The response in loads can be modeled as an ideally flexible negative generation. $\sum_{s=1}^{S} P_{(t,i)}^{shift}$ is the amount of load that has been shifted from 't' to 'i' and $\sum_{s=1}^{S} P_{(s,i,t)}^{shift}$ is the amount of load that has been shifted from 'i' to 't', $\sum_{l=1}^{L} P_{(l,t)}^{dem}$ is the amount of load demand and $\sum_{n=1}^{N} P_{(n,t)}^{gen}$ is the total generated power.

$$\sum_{n=1}^{N} P_{(n,t)}^{gen} - \sum_{l=1}^{L} P_{(l,t)}^{dem} = 0 \quad \forall \, T \in 1, 2, \ldots\ldots 24 \tag{11}$$

$$\sum_{s=1}^{S} P_{DR(s,t,i)}^{max\,shift} \leq 110\,\%(P_{base}) \quad \forall \, s \in 1, 2, \ldots\ldots S \tag{12}$$

'T' represents time in h and 'S' represents number of shifting intervals. The generator limits should not be violated while solving DR programs. The increased percentage of load shift by DR programs creates local peaks which occur due to rebound or payback phenomenon [107], at the non-peak hours. Here, P_{base} is considered an average load on the system. The maximum limit on the amount of load shift depends on the shape of load duration curves. The maximum allowable load shift by using DR programs for this considered system should be 110% of base load power, represented in Equation (12) as $\sum_{s=1}^{S} P_{DR(s,t,i)}^{max\,shift}$. This constraint is included in order to avoid payback effect or rebound effect. The operational cost of the MG also changes because of the load recovery. Further, the system marginal price (SMP) depends on the incremental fuel cost of the marginal generator. Hence, the SMP increases as the demand on the system increases.

$$P_{base} = avg \sum_{t=1}^{24} P_L \tag{13}$$

$$P_{min} \leq P_t \leq P_{max} \tag{14}$$

$$P_{size}^{Diesel} = \max \, (dispatched \; energy \; by \; that \; generator \; in \; 24 \; hours) \tag{15}$$

$$EC_{MG \; with \; DR} \leq EC_{MG \; without \; DR} \tag{16}$$

Optimal sizing of DGs without considering the impact of uncertainty is proposed. $EC_{MG\ with\ DR}$, $EC_{MG\ without\ DR}$ are the emission costs of MG with DR and without DR program, the total load demand on the system during 24 h. P_{min}, P_{max} are the minimum and maximum limits on generated power $'P'_t$ from a generator 't'. Aggregated load demand is calculated based on the following Equation (17).

$$P_{agg} = \frac{1}{T} \sum_{t=1}^{24} P_t \tag{17}$$

$$LF = \frac{P_{agg}}{P_{L,max}} \tag{18}$$

where P_{agg} is average of aggregated load demand, LF is the load factor of the MG, and $P_{L,max}$ is the peak load demand that is occurring on the system over 24 h.

4. Proposed Methodology

The solution methodology for reducing the total cost of the system is represented as follows:

- Read the load data, electricity price in each hour, generator data such as the minimum and maximum capacity, operating costs, and maintenance cost of each generator.
- Analyze the load profile and check whether the peak load is occurring at a high electricity price zone or not.
- As the total load on the system is considered as residential loads, cluster them as depicted in Equation (8), based on their priority without sacrificing the customer's satisfaction and lifestyle.
- If yes, shift a part of non-essential loads from the peak-occurring instant to the non-peak zone. Therefore, cluster the loads on the basis of an order of priority and based on electricity price, the clustered loads are allocated or dispatched at a particular time instant where the electricity price is low. Priority is added by giving large weight to the highest priority load. If the load on the system is less than the average of the aggregated load, no need to cluster. Shift a part of the peak load to those intervals where the available load is less than the average aggregated load.
- There should be an upper limit imposed on the amount of load shift on the system to overcome the rebound effect, which is represented in Equation (12).
- If yes, shift the loads to another time horizon. If no, size the sources, i.e., the sizes of WT, PV, and diesel generators that have to be installed to supply the available load demand in the area without considering the impact of uncertainty based on the peak load demand. Moreover, a portion of generator capacity allocated for any further increment on load demand is represented in Equation (19). The total capacity of generators installed is the sum of all the DGs capacity.

$$P_{size}^{DGs} = x\ \% \left(P_{size}^{MT} + P_{size}^{Diesel} + P_{size}^{Renewales} \right) \tag{19}$$

- Calculate the capital cost, installation cost, and maintenance cost of each generator by using Table 8.

$$TC = \sum_{S=1}^{n} C_{PV} S_{PV} + C_{WT} S_{WT} + C_{diesel} S_{diesel} \tag{20}$$

- C_{PV}, C_{WT}, and C_{diesel} are the total costs including capital and installation costs of PV, WT, and diesel. S_{PV}, S_{WT}, and S_{diesel} are the sizes of various sources yielded from the simulations.
- Dispatch or schedule the load on the available generators that are committed to supply and calculate the tariff of the consumer by using Equation (7).
- Calculate the load factor of the MG by using Equation (18). As the amount of load shift increases, the load curve becomes more uniform thereby improving the load factor.

Table 8. Various costs and lifetime of sources involved in installing the MG.

Source Type	Capital Cost ($/kW)	Fixed Maintenance Cost ($/kW)	Life (Years)
WT	1000	15	20
PV	1300	30.330	20
DG	800	0.012	15
MT	850	2.000	15

Figure 16 shows the proposed methodology for achieving the optimal total cost of installation and optimal consumer tariff.

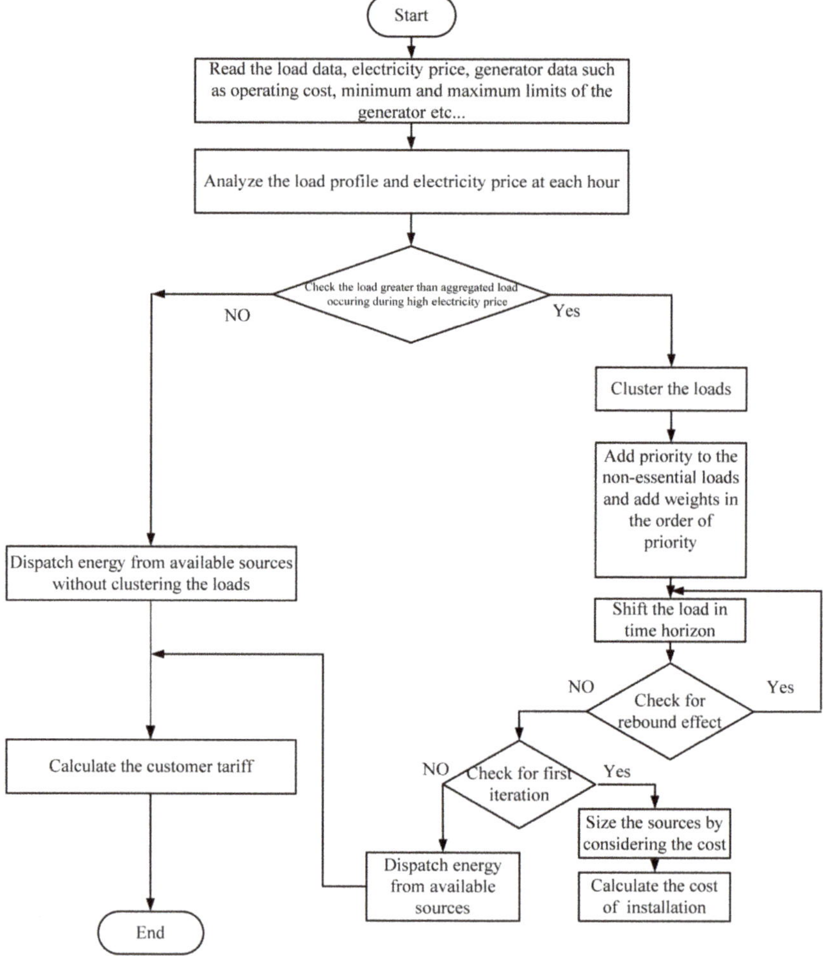

Figure 16. Methodology for analyzing the impact of load clustering on DGs sizing and tariff.

5. Results and Discussions

A test case has been considered to assess the influence of DR programs on optimal sizing of DGs and on the consumer's tariff. The IEEE-34 bus system is considered whose average load demand is 466.5 kW and the peak demand occurring on the system is 830.3 kW.

The case studies are considered simulated with a 24-h load profile and the TOU tariff is taken into consideration. In general, among all the available price-based tariffs, TOU tariff is a simple one, easy to understand, and most customers show interest in this type of tariff.

For the considered test system, the customers are charged with a time of use (TOU) tariff where the hours in a day are clustered into peak, off-peak, and moderate peak hours, and the price is fixed. The prices are fixed DA, therefore, there is no ambiguity to the customer to enter into DR programs. Figure 17 shows the variation of DA electricity price and time. The electricity price is high from hours 10:00 to 21:00 and the peak load occurring zone is also at the same time; this results in huge customer bills. One way is to curtail the loads during peak hours to control the tariff where customers' load demand is not met. Another way is to shift the load demand from the peak load time horizon to off-peak hours. Figure 18 shows the load demand for the three cases considered. Executing DR strategies will benefit not only customers but also the suppliers too. Three cases have been considered, i.e., no penetration of ILs and 5% and 10% penetration of ILs, to analyze the impact of DR programs on consumer electricity bills and load factor. The load has shifted on the time horizon and the total demand on the system per day remains the same. Figure 18 shows the different levels of penetration of loads, i.e., 0%, 5%, and 10%.

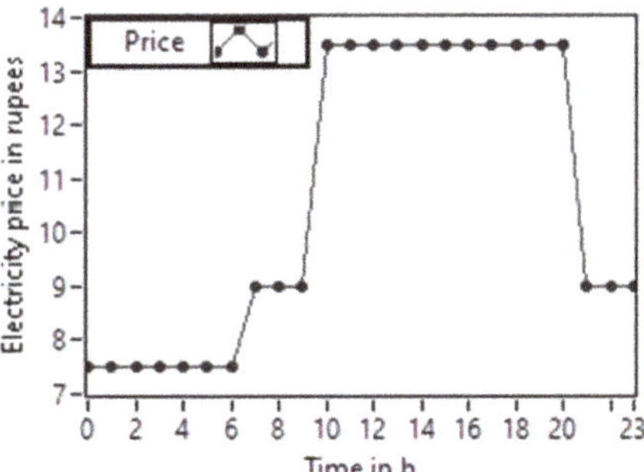

Figure 17. Time of use pricing.

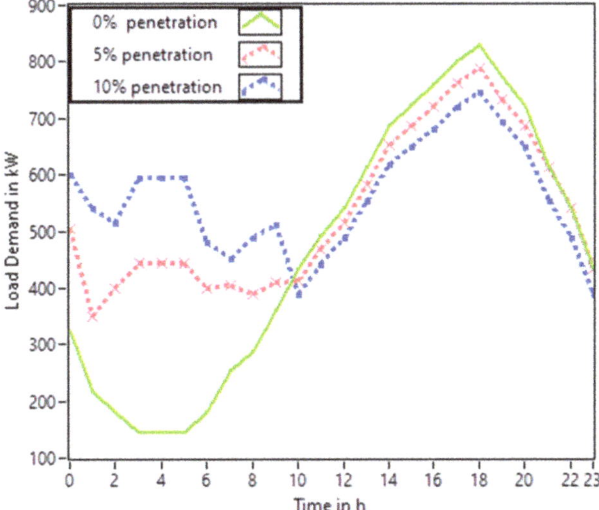

Figure 18. Load demand for three cases considered.

Depending on the objective function and the constraints considered the structure of search space changes. Therefore, to avoid this, we considered three scenarios, namely 0% penetration, 5% penetration, and 10% penetration. The algorithm is valid for all the considered scenarios even though the constraints change.

A local peak is created in the load demand curve at non-peak hour instant with a demand of 600 kW. In fact, the load demand before publishing DA tariff is 300 kW at that instant. This occurrence of local peaks at low price time horizon is considered as the rebound effect. If all the utilities supply this local peak demand, the generating companies will run in loss. Further, the emissions increase due to turning on of inefficient generators. To overcome the occurrence of the rebound effect, here, the maximum allowable load shift by using DR programs should be 110% of base load power. Hence, for the load profile considered, the average load is 466.5 kW and the maximum load that can be allowed during the non-peak hour should be less than 515 kW in the beginning hours where electricity price is minimal. Hence, ISO is responsible for maintaining the load profile within the specified limits.

5.1. Influence of DR Strategy on Optimal Sizing of DGs

The peak load on the system occurs occasionally and the generation and load demand balance should be met. The sizing of DGs will be based on the peak load that has to be supplied by the grid at any instant. Moreover, there is no need for installing new sources for supplying the occasional load. Therefore, this section investigates the impact of DR strategy in deciding the optimal capacity of DGs. It has been said that the DR program will reduce the peak demand on the system which obviously reduces the capacity of the individual generators. The lower and upper limits on the decision variables are set as [0, 500] for all the generators considered. The inputs are initialized to the algorithm; it will yield the optimal sizing of each DG, capital cost, installation cost, and total cost of each DG for the proposed size. As seen from Figure 19 and from Table 9, if the penetration of non-critical loads or flexible loads increases, then the size and cost of deploying the DGs get reduced. This is due to the fact that DR makes the load profile near flat and the peak demand on the system gets reduced, which reduces the size of DGs. Therefore, the costs involved such as installation cost and capital cost get reduced. The difference in the total cost of installing DGs in case of 0% penetration of ILs and 5% penetration of ILs is 50,675.21 $/day and of 0% penetration of ILs and 10% penetration of ILs is 93,042.89 $/day. The reduction in installation cost is large in the case of 10% penetration but the chances of getting peaks during the non-peak hours are also higher. Therefore, to nullify the occurrence of rebound effect for the considered case study, it is suggested that 5% penetration of ILs is advisable compared to 10% penetration of ILs.

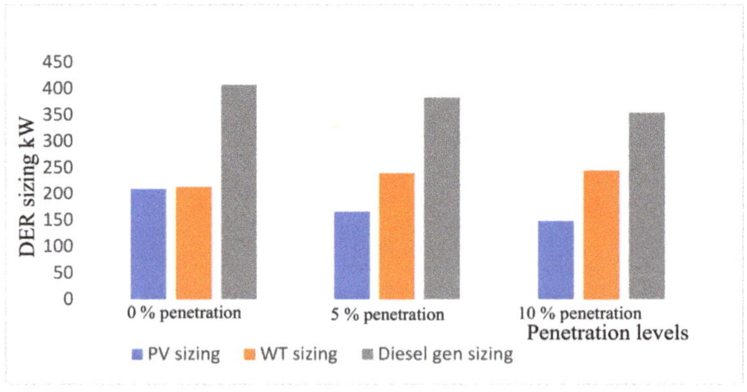

Figure 19. Sizing of sources without considering the uncertainty in RESs.

Table 9. Capital cost involved in various sources for the percentage of ILs penetration.

Percentage of ILs Penetration	Capital Cost of PV $/Day	Capital Cost of WT $/Day	Capital Cost of DG $/Day	Total Cost of MG $/Day
0%	279,374.621	216,914.635	325,225.678	821,514.900
5%	221,426.776	243,299.560	306,113.391	770,839.700
10%	196,888.840	248,550.155	283,033.045	728,472.000

5.2. Effect of DR Strategy on Consumer Electricity Bill and on Load Factor

The demonstrated work relates to the optimization of various costs, explicitly capital cost [108], installation cost, operational cost, and consumer tariff. To maintain the reliability in a heavily routed line, the ISOs basically charge more compared to the other prices. Under this scheme, the customers will get incentives for shifting [109] their loads to non-peak hours or curtailing their loads. This program is event-based, its fundamental focus is to maintain reliability in the MG. Dynamic pricing techniques alone may not fetch the feasible results as proposed in [110].

Figure 20 depicts that while increasing the level of penetration of the customer electricity bill gets reduced and the load factor gets increased. The reduction in customer bill from 0% penetration of ILs to the 5% penetration of ILs is 3709.26 Rs/day and from 0% to 10%, the reduction is 6059.025 Rs/day. Table 10 represents the reduction of electricity price and improvement in load factor for the system considered. Further, it also depicts that the reduction in peak demand from 830.300 kW to 747.270 kW from 0% penetration to 10% penetration, where the load factor is increased from 0.561957 to 0.624397.

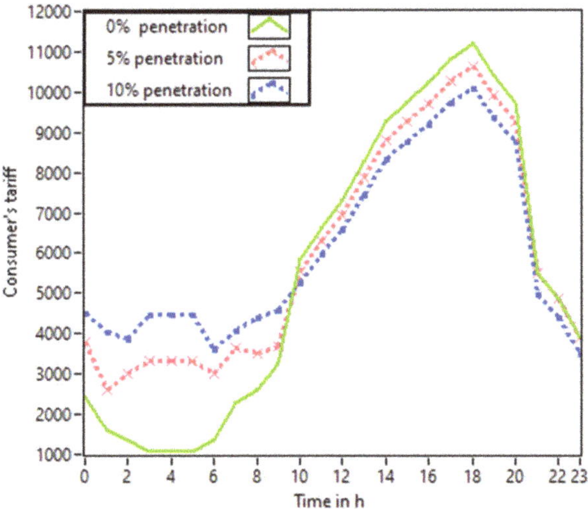

Figure 20. Electricity price of consumers for three cases considered.

Table 10. Comparison of peak demands, electricity price, and load factor with penetration of ILs.

S.No.	Penetration of ILs in %	Peak Value of Load, kW	Electricity Price, Rs/Day	Load Factor
1	0%	830.300	131,952.7	0.561957
2	5%	788.785	128,243.5	0.591540
3	10%	747.270	125,893.7	0.624397

As the level of penetration increases, the load profile becomes flat and the electricity tariff gets reduced and is clearly depicted in Figure 18 at 00:00 to 08:00 a.m. However, there is a chance of the local peak occurring in case 3 where 10% penetration of flexible or non-critical loads allowed it to penetrate the MG system. As a greater number of customers participate in the DR program, the level of peak increases. Therefore, unjustified shifting of a portion of the load from global peak to the off-peak should be avoided to reduce the burden on the power systems. Hence, there should be an upper limit on the amount of load shift. ISO should focus not only on possible reduction of consumer tariff but also ensure to overcome the rebound effect.

6. Conclusions

A detailed literature survey on various effects of DR programs for EM in the MG was done. A test system with three scenarios considered, namely no penetration of ILs (i.e., 0% load shift), 5% penetration of ILs, and 10% penetration of ILs, showed total daily load demand of 11,198.2 kW and maximum peak on the system on an hourly basis is 830.3 kW. The capacity of DERs is a function of peak load occurring on the system and the system load factor is a function of uniformity of load curve. Therefore, with the reduction in peak demand on the system, the load curve gets more uniform which reduces the size of DERs and improves the load factor of the system which further reduces the customer tariff. The reduction in peak demand for the MG from 0% penetration to 5% penetration is 41.515 kW and from 0% penetration to 10% penetration is 83.03 kW and the reduction in the cost of installation of DGs is 50,675.21 \$/day. A time of use pricing model is considered and the loads are clustered based on the prices at each interval. From the results, it is shown that the reduction in customer electricity tariff from 0% penetration to 5% penetration is 3709.26 Rs/day. The results show that, with the deployment of DR programs into the MG, there is a huge impact on the above-considered test systems.

The maximum load shift constraint is a function of shape of the load curve, therefore, for the addressed test case, the maximum shift in load is 110% of base load. This 110% is not fixed for other problems. In the future, an effective objective function in which the maximum constraint on amount of load shift can be modeled to avoid the rebound effect for any type of load curve considered. A new meta-heuristic algorithm can be developed for effective load clustering. Further, researchers can define a new set of rules for reducing the conflicts between the consumer and aggregator and between the aggregator and ISO such that there should be an imposition of penalty for violation of code of conduct. New wireless technologies can be proposed for reducing the time lapse between the IEDs and the smart meter or protection devices. One more future direction is impact of DR programs in optimal sizing of DGs by considering the uncetainty in RES. Finding the elasticity and cross elasticity matrix of load with respect to price changes can be performed.

Future researchers can further address the problems associated with load recovery, i.e., increase in production cost, increase in SMP, and increase in stress of already committed generators. In addition, there is a difference in gross load curtailed and the net load curtailed; this problem arises when the load curtailment of one customer overlaps with the load recovery period of another customer. The aggregator has to pay incentives for the gross load curtailed but the net load curtailed on the system is less. Therefore, a loss in monetary value arises. Reasearchers may develop an effective framework to avoid this situation.

Author Contributions: Conceptualization, K.M.R.P. and S.V.; methodology, S.R.S.; software, K.M.R.P.; validation, K.M.R.P., S.V. and S.R.S.; formal analysis, K.M.R.P.; investigation, S.V.; resources, S.R.S.; data curation, K.M.R.P.; writing—original draft preparation, K.M.R.P.; writing—review and editing, K.M.R.P., S.V. and S.R.S.; visualization, K.M.R.P.; supervision, S.V. and S.R.S.; project administration, K.M.R.P.; funding acquisition, S.V. and S.R.S. All authors have read and agreed to the published version of the manuscript.

Funding: This research was funded by the National Institute of Technology Andhra Pradesh (NIT-AP) and Woosong University's Academic Research Funding—2021.

Institutional Review Board Statement: Not applicable.

Informed Consent Statement: Not applicable.

Data Availability Statement: Not applicable.

Acknowledgments: We acknowledge the National Institute of Technology Andhra Pradesh (NIT-AP), Tadepalligudem, Andhra Pradesh, India and Woosong University, Korea for their support in carrying out this research work.

Conflicts of Interest: The authors declare no conflict of interest.

Abbreviations

DR program	Demand response program	AMI	Advanced metering infrastructure
MG	Microgrid	DA scheduling	Day-ahead scheduling
TOU	Time of use pricing	HA scheduling	Hour ahead scheduling
RTP	Real-time pricing	RT scheduling	Real-time scheduling
CPP	Critical peak pricing	AI	Artificial intelligence
DG	Distributed generation	SOC	State of charge
BESS	Battery energy storage system	DLC	Direct load curtailment
HEMS	Home energy management system	BEMS	Building energy management system
RES	Renewable energy system	DNO	Distribution network operator
DR	Demand response	ISO	Independent system operator
DERs	Distributed energy resources	ICT	Information and communication technology
G	Grid-connected	MAS	Multi-agent system
I	Isolated	EMS	Energy management strategies
EM	Energy management	THD	Total harmonic distortion
ILs	Interruptible loads	IEDs	Intelligent electronic devices
HVAC	Heating, ventilating, and air conditioning	RL	Reinforcement learning techniques
GAMS	General algebraic modeling system	PSO	Particle swarm optimization
GWO	Grey wolf optimization	GA	Genetic algorithm
ACO	Ant colony optimization	BCO	Bee colony optimization
MBCO	Modified bee colony optimization	DISCOMS	Distribution companies
ANN	Artificial neural networks	LV-MG	Low voltage microgrid
MILP	Mixed integer linear programming	CCHP	Cold climate heat pump
DSMS	Demand side management system	CVAR	Conditional value at risk
SCA	Sine cosine algorithm	NILs	Non-interruptible loads
HAN	Home area network	MPPT	Maximum power point tracking
PV	Photovoltaic	WT	Wind turbine
MT	Micro turbine	FC	Fuel cell
AC-MG	Alternating current microgrid	DC-MG	Direct current microgrid
SOCP	Second order cone programming	ASFLA	Adaptive shuffled frog leaping algorithm
AI	Artificial intelligence	DLC	Direct load curtailment
SCADA		DR-BOB	Demand response blocks of buildings
KKT	Karush Kuhn Tucker	SC	Super capacitor
DR-TRL	Demand response technology readiness level	MAS	Multi-agent system
NAN	Neighboring area network	FAN	Field area network
Wi-Fi	Wireless fidelity	SMP	System marginal price

References

1. Li, K.; Thompson, S.; Peng, J. Modelling and prediction of NOx emission in a coal-fired power generation plant. *Control Eng. Pract.* **2004**, *12*, 707–723. [CrossRef]
2. Ahi, P.; Searcy, C. Assessing sustainability in the supply chain: A triple bottom line approach. *Appl. Math. Model.* **2015**, *39*, 2882–2896. [CrossRef]
3. Kim, C.-Y.; Kim, C.-R.; Kim, D.-K.; Cho, S.-H. Analysis of Challenges Due to Changes in Net Load Curve in South Korea by Integrating DERs. *Electronics* **2020**, *9*, 1310. [CrossRef]

4. Mellouk, L.; Ghazi, M.; Aaroud, A.; Boulmalf, M.; Benhaddou, D.; Zine-Dine, K. Design and energy management optimization for hybrid renewable energy system- case study: Laayoune region. *Renew. Energy* **2019**, *139*, 621–634. [CrossRef]

5. Reddy, P.K.M.; Prakash, M. Optimal Dispatch of Energy Resources in an Isolated Micro-Grid with Battery Energy Storage System. In Proceedings of the 2020 4th International Conference on Intelligent Computing and Control Systems (ICICCS), Madurai, India, 13–15 May 2020; pp. 730–735. [CrossRef]

6. Aziz, A.; Tajuddin, M.; Adzman, M.; Ramli, M.; Mekhilef, S. Energy Management and Optimization of a PV/Diesel/Battery Hybrid Energy System Using a Combined Dispatch Strategy. *Sustainability* **2019**, *11*, 683. [CrossRef]

7. Shoeb, A.; Shafiullah, G. Renewable Energy Integrated Islanded Microgrid for Sustainable Irrigation—A Bangladesh Perspective. *Energies* **2018**, *11*, 1283. [CrossRef]

8. Siritoglou, P.; Oriti, G.; Van Bossuyt, D. Distributed Energy-Resource Design Method to Improve Energy Security in Critical Facilities. *Energies* **2021**, *14*, 2773. [CrossRef]

9. Sanjeevikumar, P.; Sarojini, R.K.; Palanisamy, K.; Sanjeevikumar, P. Large Scale Renewable Energy Integration: Issues and Solutions. *Energies* **2019**, *12*, 1996. [CrossRef]

10. Kumar, N.; Chopra, S.; Chand, A.; Elavarasan, R.; Shafiullah, G. Hybrid Renewable Energy Microgrid for a Residential Community: A Techno-Economic and Environmental Perspective in the Context of the SDG7. *Sustainability* **2020**, *12*, 3944. [CrossRef]

11. Vera, Y.E.G.; Dufo-López, R.; Bernal-Agustín, J.L. Energy Management in Microgrids with Renewable Energy Sources: A Literature Review. *Appl. Sci.* **2019**, *9*, 3854. [CrossRef]

12. Bai, Y.; Li, J.; He, H.; Dos Santos, R.C.; Yang, Q. Optimal Design of a Hybrid Energy Storage System in a Plug-In Hybrid Electric Vehicle for Battery Lifetime Improvement. *IEEE Access* **2020**, *8*, 142148–142158. [CrossRef]

13. Zsiborács, H.; Baranyai, N.H.; Vincze, A.; Zentkó, L.; Birkner, Z.; Máté, K.; Pintér, G. Intermittent Renewable Energy Sources: The Role of Energy Storage in the European Power System of 2040. *Electronics* **2019**, *8*, 729. [CrossRef]

14. Arteconi, A.; Polonara, F. Assessing the Demand Side Management Potential and the Energy Flexibility of Heat Pumps in Buildings. *Energies* **2018**, *11*, 1846. [CrossRef]

15. Shang, Y. Resilient Multiscale Coordination Control against Adversarial Nodes. *Energies* **2018**, *11*, 1844. [CrossRef]

16. Amer, A.; Shaban, K.; Gaouda, A.; Massoud, A. Home Energy Management System Embedded with a Multi-Objective Demand Response Optimization Model to Benefit Customers and Operators. *Energies* **2021**, *14*, 257. [CrossRef]

17. Hu, J.; Yang, G.; Ziras, C.; Kok, K. Aggregator operation in the balancing market through network-constrained transactive energy. *IEEE Trans. Power Syst.* **2019**, *34*, 4071–4080. [CrossRef]

18. Durvasulu, V.; Hansen, T.M. Benefits of a Demand Response Exchange Participating in Existing Bulk-Power Markets. *Energies* **2018**, *11*, 3361. [CrossRef]

19. Fan, S.; Ai, Q.; Piao, L. Hierarchical Energy Management of Microgrids including Storage and Demand Response. *Energies* **2018**, *11*, 1111. [CrossRef]

20. Arias, L.A.; Rivas, E.; Santamaria, F.; Hernandez, V. A Review and Analysis of Trends Related to Demand Response. *Energies* **2018**, *11*, 1617. [CrossRef]

21. Khan, A.A.; Razzaq, S.; Khan, A.; Khursheed, F.; Owais. HEMSs and enabled demand response in electricity market: An overview. *Renew. Sustain. Energy Rev.* **2015**, *42*, 773–785. [CrossRef]

22. Jang, D.; Eom, J.; Park, M.J.; Rho, J.J. Variability of electricity load patterns and its effect on demand response: A critical peak pricing experiment on Korean commercial and industrial customers. *Energy Policy* **2016**, *88*, 11–26. [CrossRef]

23. Hajibandeh, N.; Ehsan, M.; Soleymani, S.; Shafie-khah, M.; Catalão, J.P.S. The Mutual Impact of Demand Response Programs and Renewable Energies: A Survey. *Energies* **2017**, *10*, 1353. [CrossRef]

24. Silva, B.N.; Khan, M.; Han, K. Futuristic Sustainable Energy Management in Smart Environments: A Review of Peak Load Shaving and Demand Response Strategies, Challenges, and Opportunities. *Sustainability* **2020**, *12*, 5561. [CrossRef]

25. Salinas, S.; Li, M.; Li, P. Multi-Objective Optimal Energy Consumption Scheduling in Smart Grids. *IEEE Trans. Smart Grid* **2013**, *4*, 341–348. [CrossRef]

26. Adika, C.O.; Wang, L. Autonomous Appliance Scheduling for Household Energy Management. *IEEE Trans. Smart Grid* **2014**, *5*, 673–682. [CrossRef]

27. Vázquez-Canteli, J.R.; Nagy, Z. Reinforcement learning for demand response: A review of algorithms and modeling techniques. *Appl. Energy* **2019**, *235*, 1072–1089. [CrossRef]

28. Mandal, R.; Chatterjee, K. Frequency control and sensitivity analysis of an isolated microgrid incorporating fuel cell and diverse distributed energy sources. *Int. J. Hydrog. Energy* **2020**, *45*, 13009–13024. [CrossRef]

29. Golpîra, H. Bulk power system frequency stability assessment in presence of microgrids. *Electr. Power Syst. Res.* **2019**, *174*, 105863. [CrossRef]

30. Sanjeev, P.; Padhy, N.P.; Agarwal, P. Peak Energy Management Using Renewable Integrated DC Microgrid. *IEEE Trans. Smart Grid* **2017**, *9*, 4906–4917. [CrossRef]

31. Ahmed, E.M.; Rathinam, R.; Dayalan, S.; Fernandez, G.S.; Ali, Z.M.; Aleem, S.H.E.A.; Omar, A.I. A Comprehensive Analysis of Demand Response Pricing Strategies in a Smart Grid Environment Using Particle Swarm Optimization and the Strawberry Optimization Algorithm. *Mathematics* **2021**, *9*, 2338. [CrossRef]

32. Sundt, S.; Rehdanz, K.; Meyerhoff, J. Consumers' Willingness to Accept Time-of-Use Tariffs for Shifting Electricity Demand. *Energies* **2020**, *13*, 1895. [CrossRef]

33. Song, H.Y.; Lee, G.S.; Yoon, Y.T. Optimal Operation of Critical Peak Pricing for an Energy Retailer Considering Balancing Costs. *Energies* **2019**, *12*, 4658. [CrossRef]

34. Ma, T.; Wu, J.; Hao, L.; Yan, H.; Li, D. A Real-Time Pricing Scheme for Energy Management in Integrated Energy Systems: A Stackelberg Game Approach. *Energies* **2018**, *11*, 2858. [CrossRef]

35. Jadhav, A.M.; Patne, N.R.; Guerrero, J.M. A Novel Approach to Neighborhood Fair Energy Trading in a Distribution Network of Multiple Microgrid Clusters. *IEEE Trans. Ind. Electron.* **2019**, *66*, 1520–1531. [CrossRef]

36. Gazafroudi, A.S.S.; Prieto, J.; Corchado, J.M. Virtual Organization Structure for Agent-Based Local Electricity Trading. *Energies* **2019**, *12*, 1521. [CrossRef]

37. Sheikhahmadi, P.; Mafakheri, R.; Bahramara, S.; Damavandi, M.Y.; Catalão, J.P.S. Risk-Based Two-Stage Stochastic Optimization Problem of Micro-Grid Operation with Renewables and Incentive-Based Demand Response Programs. *Energies* **2018**, *11*, 610. [CrossRef]

38. Grisales-Noreña, L.F.; Montoya, D.G.; Ramos-Paja, C.A. Optimal Sizing and Location of Distributed Generators Based on PBIL and PSO Techniques. *Energies* **2018**, *11*, 1018. [CrossRef]

39. El-Salam, M.F.A.; Beshr, E.; Eteiba, M.B. A New Hybrid Technique for Minimizing Power Losses in a Distribution System by Optimal Sizing and Siting of Distributed Generators with Network Reconfiguration. *Energies* **2018**, *11*, 3351. [CrossRef]

40. Zhu, W.; Guo, J.; Zhao, G.; Zeng, B. Optimal Sizing of an Island Hybrid Microgrid Based on Improved Multi-Objective Grey Wolf Optimizer. *Processes* **2020**, *8*, 1581. [CrossRef]

41. Jalili, A.; Taheri, B. Optimal Sizing and Sitting of Distributed Generations in Power Distribution Networks Using Firefly Algorithm. *Technol. Econ. Smart Grids Sustain. Energy* **2020**, *5*, 1–14. [CrossRef]

42. Montoya, O.D.; Molina-Cabrera, A.; Chamorro, H.R.; Alvarado-Barrios, L.; Rivas-Trujillo, E. A Hybrid Approach Based on SOCP and the Discrete Version of the SCA for Optimal Placement and Sizing DGs in AC Distribution Networks. *Electronics* **2020**, *10*, 26. [CrossRef]

43. Karunarathne, E.; Pasupuleti, J.; Ekanayake, J.; Almeida, D. Optimal Placement and Sizing of DGs in Distribution Networks Using MLPSO Algorithm. *Energies* **2020**, *13*, 6185. [CrossRef]

44. Madahi, S.K.; Sarić, A. Multi-Criteria Optimal Sizing and Allocation of Renewable and Non-Renewable Distributed Generation Resources at 63 kV/20 kV Substations. *Energies* **2020**, *13*, 5364. [CrossRef]

45. Onlam, A.; Yodphet, D.; Chatthaworn, R.; Surawanitkun, C.; Siritaratiwat, A.; Khunkitti, P. Power Loss Minimization and Voltage Stability Improvement in Electrical Distribution System via Network Reconfiguration and Distributed Generation Placement Using Novel Adaptive Shuffled Frogs Leaping Algorithm. *Energies* **2019**, *12*, 553. [CrossRef]

46. Yuan, R.; Li, T.; Deng, X.; Ye, J. Optimal Day-Ahead Scheduling of a Smart Distribution Grid Considering Reactive Power Capability of Distributed Generation. *Energies* **2016**, *9*, 311. [CrossRef]

47. Galván, L.; Navarro, J.M.; Galván, E.; Carrasco, J.M.; Alcántara, A. Optimal Scheduling of Energy Storage Using a New Priority-Based Smart Grid Control Method. *Energies* **2019**, *12*, 579. [CrossRef]

48. Li, J.; Tan, Z.; Ren, Z.; Yang, J.; Yu, X. A Two-Stage Optimal Scheduling Model of Microgrid Based on Chance-Constrained Programming in Spot Markets. *Processes* **2020**, *8*, 107. [CrossRef]

49. Wang, J.; Li, K.-J.; Javid, Z.; Sun, Y. Distributed Optimal Coordinated Operation for Distribution System with the Integration of Residential Microgrids. *Appl. Sci.* **2019**, *9*, 2136. [CrossRef]

50. Lu, C.; Xu, H.; Pan, X.; Song, J. Optimal Sizing and Control of Battery Energy Storage System for Peak Load Shaving. *Energies* **2014**, *7*, 8396–8410. [CrossRef]

51. Zhou, N.; Liu, N.; Zhang, J.; Lei, J. Multi-Objective Optimal Sizing for Battery Storage of PV-Based Microgrid with Demand Response. *Energies* **2016**, *9*, 591. [CrossRef]

52. Carpinelli, G.; Di Fazio, A.R.; Khormali, S.; Mottola, F. Optimal Sizing of Battery Storage Systems for Industrial Applications when Uncertainties Exist. *Energies* **2014**, *7*, 130–149. [CrossRef]

53. HassanzadehFard, H.; Jalilian, A. Optimal sizing and location of renewable energy based DG units in distribution systems considering load growth. *Int. J. Electr. Power Energy Syst.* **2018**, *101*, 356–370. [CrossRef]

54. Sharma, S.; Bhattacharjee, S.; Bhattacharya, A. Operation cost minimization of a micro-grid using quasi-oppositional swine influenza model-based optimization with quarantine. *Ain Shams Eng. J.* **2018**, *9*, 45–63. [CrossRef]

55. Moghaddam, A.A.; Seifi, A.; Niknam, T.; Pahlavani, M.R.A. Multi-objective operation management of a renewable MG (micro-grid) with back-up micro-turbine/fuel cell/battery hybrid power source. *Energy* **2011**, *36*, 6490–6507. [CrossRef]

56. Ghaffarzadeh, N.; Zolfaghari, M.; Ardakani, F.J.; Jahanbani, A. Optimal sizing of Energy Storage System in a MG using the Mixed Integer Linear programming. *Int. J. Renew. Energy Res.* **2017**, *7*, 2004–2016.

57. Sharma, S.; Bhattacharjee, S.; Bhattacharya, A. Grey wolf optimisation for optimal sizing of battery energy storage device to minimise operation cost of MG. *IET Gener. Transm. Distrib.* **2016**, *10*, 625–637. [CrossRef]

58. Qiu, J.; Meng, K.; Zheng, Y.; Dong, Z.Y. Optimal scheduling of distributed energy resources as a virtual power plant in a transactive energy framework. *IET Gener. Transm. Distrib.* **2017**, *12*, 17–27. [CrossRef]

59. Silva, C.; Faria, P.; Vale, Z. Rating the Participation in Demand Response Programs for a More Accurate Aggregated Schedule of Consumers after Enrolment Period. *Electronics* **2020**, *9*, 349. [CrossRef]

60. Wang, N.; Xu, W.; Xu, Z.; Shao, W. Peer-to-Peer Energy Trading among Microgrids with Multidimensional Willingness. *Energies* **2018**, *11*, 3312. [CrossRef]
61. Carpinelli, G.; Caramia, P.; Mottola, F.; Proto, D. Exponential weighted method and a compromise programming method for multi-objective operation of plug-in vehicle aggregators in microgrids. *Int. J. Electr. Power Energy Syst.* **2014**, *56*, 374–384. [CrossRef]
62. Gu, W.; Lu, S.; Wu, Z.; Zhang, X.; Zhou, J.; Zhao, B.; Wang, J. Residential CCHP microgrid with load aggregator: Operation mode, pricing strategy, and optimal dispatch. *Appl. Energy* **2017**, *205*, 173–186. [CrossRef]
63. Aluisio, B.; Conserva, A.; Dicorato, M.; Forte, G.; Trovato, M. Optimal operation planning of V2G-equipped Microgrid in the presence of EV aggregator. *Electr. Power Syst. Res.* **2017**, *152*, 295–305. [CrossRef]
64. Nguyen, D.T.; Le, L.B. Risk-Constrained Profit Maximization for Microgrid Aggregators with Demand Response. *IEEE Trans. Smart Grid* **2015**, *6*, 135–146. [CrossRef]
65. Goroohi, I.; Sardou; Khodayar, M.E.; Ameli, M.T. Coordinated Operation of Natural Gas and Electricity Networks with Microgrid Aggregators. *IEEE Trans. Smart Grid* **2018**, *9*, 199–210. [CrossRef]
66. Pei, W.; Du, Y.; Deng, W.; Sheng, K.; Xiao, H.; Qu, H. Optimal Bidding Strategy and Intramarket Mechanism of Microgrid Aggregator in Real-Time Balancing Market. *IEEE Trans. Ind. Inform.* **2016**, *12*, 587–596. [CrossRef]
67. Essiet, I.O.; Sun, Y. Maximizing Demand Response Aggregator Compensation through Optimal RES Utilization: Aggregation in Johannesburg, South Africa. *Appl. Sci.* **2020**, *10*, 594. [CrossRef]
68. Gazijahani, F.S.; Salehi, J. Game Theory Based Profit Maximization Model for Microgrid Aggregators with Presence of EDRP Using Information Gap Decision Theory. *IEEE Syst. J.* **2019**, *13*, 1767–1775. [CrossRef]
69. Zhu, X.; Xia, M.; Chiang, H. Coordinated sectional droop charging control for EV aggregator enhancing frequency stability of microgrid with high penetration of renewable energy sources. *Appl. Energy* **2018**, *210*, 936–943. [CrossRef]
70. Olivella-Rosell, P.; Lloret-Gallego, P.; Munné-Collado, Í.; Villafafila-Robles, R.; Sumper, A.; Ottessen, S.Ø.; Rajasekharan, J.; Bremdal, B.A. Local Flexibility Market Design for Aggregators Providing Multiple Flexibility Services at Distribution Network Level. *Energies* **2018**, *11*, 822. [CrossRef]
71. Song, Z.; Shi, J.; Li, S.; Chen, Z.; Yang, W.; Zhang, Z. Day Ahead Bidding of a Load Aggregator Considering Residential Consumers Demand Response Uncertainty Modeling. *Appl. Sci.* **2020**, *10*, 7310. [CrossRef]
72. Chen, S.; Chen, Q.; Xu, Y. Strategic Bidding and Compensation Mechanism for a Load Aggregator with Direct Thermostat Control Capabilities. *IEEE Trans. Smart Grid* **2018**, *9*, 2327–2336. [CrossRef]
73. Sisinni, M.; Noris, F.; Smit, S.; Messervey, T.B.; Crosbie, T.; Breukers, S.; Van Summeren, L. Identification of Value Proposition and Development of Innovative Business Models for Demand Response Products and Services Enabled by the DR-BOB Solution. *Buildings* **2017**, *7*, 93. [CrossRef]
74. Karfopoulos, E.; Tena, L.; Torres, A.; Salas, P.; Jorda, J.G.; Dimeas, A.; Hatziargyriou, N. A multi-agent system providing demand response services from residential consumers. *Electr. Power Syst. Res.* **2015**, *120*, 163–176. [CrossRef]
75. UNEP. Buildings and Climate Change, Summary for Decision-Makers. 2009. Available online: https://www.uncclearn.org/wp-content/uploads/library/unep207.pdf (accessed on 17 December 2021).
76. Liu, Y.; Yang, X.; Wen, W.; Xia, M. Smarter Grid in the 5G Era: A Framework Integrating Power Internet of Things with a Cyber Physical System. *Front. Commun. Netw.* **2021**, *2*, 9590. [CrossRef]
77. Jour, D.; Zhang, Y.; Qitong, Z.W. Application Status and Prospects of 5G Technology in Distribution Automation Systems. *Wirel. Commun. Mob. Comput.* **2021**, *2021*, 5553159. [CrossRef]
78. Kampelis, N.; Tsekeri, E.; Kolokotsa, D.; Kalaitzakis, K.; Isidori, D.; Cristalli, C. Development of Demand Response Energy Management Optimization at Building and District Levels Using Genetic Algorithm and Artificial Neural Network Modelling Power Predictions. *Energies* **2018**, *11*, 3012. [CrossRef]
79. Crosbie, T.; Broderick, J.; Short, M.; Charlesworth, R.; Dawood, M. Demand Response Technology Readiness Levels for Energy Management in Blocks of Buildings. *Buildings* **2018**, *8*, 13. [CrossRef]
80. Hui, H.; Ding, Y.; Shi, Q.; Li, F.; Song, Y.; Yan, J. 5G network-based Internet of Things for demand response in smart grid: A survey on application potential. *Appl. Energy* **2020**, *257*, 113972. [CrossRef]
81. Cruz, C.; Palomar, E.; Bravo, I.; Gardel, A. Cooperative Demand Response Framework for a Smart Community Targeting Renewables: Testbed Implementation and Performance Evaluation. *Energies* **2020**, *13*, 2910. [CrossRef]
82. Shakeri, M.; Pasupuleti, J.; Amin, N.; Rokonuzzaman; Low, F.W.; Yaw, C.T.; Asim, N.; Samsudin, N.A.; Tiong, S.K.; Hen, C.K.; et al. An Overview of the Building Energy Management System Considering the Demand Response Programs, Smart Strategies and Smart Grid. *Energies* **2020**, *13*, 3299. [CrossRef]
83. Basu, K.; Hawarah, L.; Arghira, N.; Joumaa, H.; Ploix, S. A prediction system for home appliance usage. *Energy Build.* **2013**, *67*, 668–679. [CrossRef]
84. Bera, S.; Gupta, P.; Misra, S. D2S: Dynamic Demand Scheduling in Smart Grid Using Optimal Portfolio Selection Strategy. *IEEE Trans. Smart Grid* **2015**, *6*, 1434–1442. [CrossRef]
85. Ko, H.; Praca, I. Design of a Secure Energy Trading Model Based on a Blockchain. *Sustainability* **2021**, *13*, 1634. [CrossRef]
86. Mohammad, N.; Mishra, Y. The Role of Demand Response Aggregators and the Effect of GenCos Strategic Bidding on the Flexibility of Demand. *Energies* **2018**, *11*, 3296. [CrossRef]

87. Yoo, Y.-S.; Jeon, S.H.; Newaz, S.H.S.; Lee, I.-W.; Choi, J.K. Energy Trading among Power Grid and Renewable Energy Sources: A Dynamic Pricing and Demand Scheme for Profit Maximization. *Sensors* **2021**, *21*, 5819. [CrossRef]
88. Ko, W.; Vettikalladi, H.; Song, S.-H.; Choi, H.-J. Implementation of a Demand-Side Management Solution for South Korea's Demand Response Program. *Appl. Sci.* **2020**, *10*, 1751. [CrossRef]
89. Wang, B.; Liu, X.; Zhu, F.; Hu, X.; Ji, W.; Yang, S.; Wang, K.; Feng, S. Unit Commitment Model Considering Flexible Scheduling of Demand Response for High Wind Integration. *Energies* **2015**, *8*, 13688–13709. [CrossRef]
90. Chae, J.; Joo, S.-K. Demand Response Resource Allocation Method Using Mean-Variance Portfolio Theory for Load Aggregators in the Korean Demand Response Market. *Energies* **2017**, *10*, 879. [CrossRef]
91. Faria, P.; Spínola, J.; Vale, Z. Distributed Energy Resources Scheduling and Aggregation in the Context of Demand Response Programs. *Energies* **2018**, *11*, 1987. [CrossRef]
92. Faria, P.; Spínola, J.; Vale, Z. Reschedule of Distributed Energy Resources by an Aggregator for Market Participation. *Energies* **2018**, *11*, 713. [CrossRef]
93. Faria, P.; Vale, Z. A Demand Response Approach to Scheduling Constrained Load Shifting. *Energies* **2019**, *12*, 1752. [CrossRef]
94. Haghifam, S.; Dadashi, M.; Zare, K.; Seyedi, H. Optimal operation of smart distribution networks in the presence of demand response aggregators and microgrid owners: A multi follower Bi-Level approach. *Sustain. Cities Soc.* **2020**, *55*, 102033. [CrossRef]
95. Huq, M.Z.; Islam, S. Home Area Network technology assessment for demand response in smart grid environment. In Proceedings of the 2010 20th Australasian Universities Power Engineering Conference, Christchurch, New Zealand, 5–6 December 2010; pp. 1–6.
96. Samadi, P.; Mohsenian-Rad, H.; Wong, V.W.S.; Schober, R. Tackling the Load Uncertainty Challenges for Energy Consumption Scheduling in Smart Grid. *IEEE Trans. Smart Grid* **2013**, *4*, 1007–1016. [CrossRef]
97. Liu, X.; Gao, B.; Li, Y. Bayesian Game-Theoretic Bidding Optimization for Aggregators Considering the Breach of Demand Response Resource. *Appl. Sci.* **2019**, *9*, 576. [CrossRef]
98. Posma, J.; Lampropoulos, I.; Schram, W.; Van Sark, W. Provision of Ancillary Services from an Aggregated Portfolio of Residential Heat Pumps on the Dutch Frequency Containment Reserve Market. *Appl. Sci.* **2019**, *9*, 590. [CrossRef]
99. Ponds, K.T.; Arefi, A.; Sayigh, A.; Ledwich, G. Aggregator of Demand Response for Renewable Integration and Customer Engagement: Strengths, Weaknesses, Opportunities, and Threats. *Energies* **2018**, *11*, 2391. [CrossRef]
100. Shang, Y. A combinatorial necessary and sufficient condition for cluster consensus. *Neurocomputing* **2016**, *216*, 611–616. [CrossRef]
101. Alvarez, M.A.Z.; Agbossou, K.; Cardenas, A.; Kelouwani, S.; Boulon, L. Demand Response Strategy Applied to Residential Electric Water Heaters Using Dynamic Programming and K-Means Clustering. *IEEE Trans. Sustain. Energy* **2019**, *11*, 524–533. [CrossRef]
102. Mancini, F.; Romano, S.; Lo Basso, G.; Cimaglia, J.; de Santoli, L. How the Italian Residential Sector Could Contribute to Load Flexibility in Demand Response Activities: A Methodology for Residential Clustering and Developing a Flexibility Strategy. *Energies* **2020**, *13*, 3359. [CrossRef]
103. Chicco, G.; Ionel, O.; Porumb, R. Electrical Load Pattern Grouping Based on Centroid Model with Ant Colony Clustering. *IEEE Trans. Power Syst.* **2013**, *28*, 1706–1715. [CrossRef]
104. Fu-Lin, M.; Hong-yang, L. Power load classification based on spectral clustering of dual-scale. In Proceedings of the 2014 IEEE International Conference on Control Science and Systems Engineering, Yantai, China, 29–30 December 2014; pp. 162–166. [CrossRef]
105. Alonso, A.M.; Nogales, F.J.; Ruiz, C. Hierarchical Clustering for Smart Meter Electricity Loads Based on Quantile Autocovariances. *IEEE Trans. Smart Grid* **2020**, *11*, 4522–4530. [CrossRef]
106. Binh, P.T.T.; Ha, N.H.; Tuan, T.C.; Khoa, L.D. Determination of representative load curve based on Fuzzy K-Means. In Proceedings of the 2010 4th International Power Engineering and Optimization Conference (PEOCO), Shah Alam, Malaysia, 23–24 June 2010; pp. 281–286. [CrossRef]
107. Pereira, R.; Fagundes, A.; Melício, R.; Mendes, V.M.F.; Figueiredo, J.; Martins, J.; Quadrado, J.C. A fuzzy clustering approach to a demand response model. *Int. J. Electr. Power Energy Syst.* **2016**, *81*, 184–192. [CrossRef]
108. Schwarz, P.; Mohajeryami, S.; Cecchi, V. Building a Better Baseline for Residential Demand Response Programs: Mitigating the Effects of Customer Heterogeneity and Random Variations. *Electronics* **2020**, *9*, 570. [CrossRef]
109. Agnetis, A.; de Pascale, G.; Detti, P.; Vicino, A. Load Scheduling for Household Energy Consumption Optimization. *IEEE Trans. Smart Grid* **2013**, *4*, 2364–2373. [CrossRef]
110. Siano, P. Demand response and smart grids—A survey. *Renew. Sustain. Energy Rev.* **2014**, *30*, 461–478. [CrossRef]

Article

Multiobjective Scheduling of Hybrid Renewable Energy System Using Equilibrium Optimization

Salil Madhav Dubey [1], Hari Mohan Dubey [2], Manjaree Pandit [1] and Surender Reddy Salkuti [3,*]

[1] Department of Electrical Engineering, Madhav Institute of Technology & Science, Gwalior 474005, India; salil.dubey3107@gmail.com (S.M.D.); manjaree_p@hotmail.com (M.P.)
[2] Department of Electrical Engineering, Birsa Institute of Technology, Dhanbad 828123, India; hmdubey.ee@bitsindri.ac.in
[3] Department of Railroad and Electrical Engineering, Woosong University, Daejeon 34606, Korea
* Correspondence: surender@wsu.ac.kr; Tel.: +82-10-9674-1985

Abstract: Due to increasing concern over global warming, the penetration of renewable energy in power systems is increasing day by day. Gencos that traditionally focused only on maximizing their profit in the competitive market are now also focusing on operation with the minimum pollution level. The paper proposes a multiobjective model capable of finding a set of trade-off solutions for the joint optimization problem, considering the cost of reserve and curtailment of power from renewable sources. Managing a hybrid power system is a challenging task due to its stochastic nature mixed with the objective function and complex practical constraints associated with it. A novel metaheuristic Equilibrium Optimizer (EO) algorithm incepted in the year 2020 utilizes the concept of control volume and mass balance for finding equilibrium state is proposed here for computing the optimal schedule and impact of renewable energy integration on profit and emission for different optimization objectives. In this paper, EO has shown dominant performance over well-established metaheuristic algorithms such as particle swarm optimizer (PSO) and artificial bee colony (ABC). In addition, EO produces well-distributed Pareto-optimal solutions and the fuzzy min-ranking is used as a decision maker to acquire the best compromise solution.

Keywords: multi-objective; renewable energy; profit-based scheduling; Equilibrium Optimizer

Citation: Dubey, S.M.; Dubey, H.M.; Pandit, M.; Salkuti, S.R. Multiobjective Scheduling of Hybrid Renewable Energy System Using Equilibrium Optimization. *Energies* **2021**, *14*, 6376. https://doi.org/10.3390/en14196376

Academic Editor: José Matas

Received: 25 August 2021
Accepted: 30 September 2021
Published: 5 October 2021

Publisher's Note: MDPI stays neutral with regard to jurisdictional claims in published maps and institutional affiliations.

1. Introduction

The electricity demand is increasing day by day due to the growth and evolution of industrial establishments and changing lifestyles. A substantial part of the demand is still being met by thermal power generation, which depends on fossil fuels such as coal, natural gas and petroleum, which are considered the main sources of harmful emissions and air pollution. The burning of fossil fuels releases harmful gases into the atmosphere. Globally, the power generation sector contributes more than 30% of carbon dioxide emissions to the atmosphere [1]. These pollutant gases affect not only humans but are also responsible for the destruction of other lifeforms. Due to growing concern over environmental considerations, there is a demand for sufficient and secured electricity at the lowest price along with a minimum level of pollution to stabilize the environment. It is possible by multi-objective optimization that considers power generation cost and emission both for minimization. Economic emission dispatch (EED) is a key optimization problem of the power system. The objective is to schedule the committed generator optimally, so that generation cost and emission are minimized simultaneously while satisfying all operational constraints associated with it [2]. Various solution approaches have been reported for the EED problem, broadly categorized into classical, metaheuristic and hybrid approaches [3]. As a practical EED problem, is a highly nonlinear, complex constrained optimization problem, and it is not easy to find the optimal solution to such a problem by classical methods. The metaheuristic approach can overcome the difficulties, but its computational time is more. Metaheuristic and hybrid approach includes evolutionary algorithm (EA) [2], genetic algorithm (GA) [4], non-dominating sorting genetic algorithm (NSGA) [5–7], particle swarm optimization (PSO) [8–10], harmony search (HS) [11,12], differential evolution (DE) [13], hybrid bat algorithm (HBA) [14], kernel search optimization (KSO) [15], time-varying acceleration coefficient particle swarm optimization (TVAC-PSO) with exchange market algorithm (EMA) [16], interior search optimization (ISO) [17], grass shopper optimization (GSO) [18], sine cosine algorithm (SCA) [19], hybrid bacterial foraging with Nelder-mead [20] and many more. A detailed review metaheuristics approach for the solution of the EED problem can be found in references [21,22].

Creating new efficient power plants or identifying and expanding existing ones that produce low emission is time-consuming and require significant capital investment to fulfill the ever-increasing power demand. Another way is to integrate renewable energy resources (RER) such as wind and solar power into the existing grid. However, the integration of RER in the existing power grid creates several operational issues due to the unpredictable nature of wind speed and solar irradiation. Therefore, fluctuation in the wind and solar power generation needs more consideration. Further load demand is a random variable, and we cannot predict it accurately. The system operator can anticipate the uncertainty associated with wind power generation/solar power generation/load demand using the forecast. Generally, the probability distribution function is used to modeling the uncertainty related to RER integration. Consequently, integration of RER complicates the formulation of the EED problem significantly [21].

Simultaneous minimization of cost, emission and loss for the wind-thermal system with complex operational constraints such as valve point loading (VPL) effect, ramp rate limits (RRL), prohibited operating zones (POZ) and spinning reserve (SR) are available in reference [23]. Here, the time-varying fuzzy selection mechanism is used to rank the conflicting objective. Wind-based combined economic emission dispatch (WCEED) has been investigated in reference [24] to acquire Pareto optimal solution. Here piecewise linear approximation method is used to model wind power. The multi-area dynamic economic emission dispatch (MADEED) of the complex system comprises a cascaded hydro system, uncertain wind power and thermal generator system investigated in reference [25]. Here Weibull pdf is used for wind power calculation.

CEED of a grid comprising of wind and PV generation systems is presented in [26]. Here, a linear relationship between day-ahead forecasted output power by wind and PV system with operation and maintenance cost is used to model the objective function. Optimal generation scheduling of a hybrid generation system comprised of thermal, wind and solar has been investigated in reference [27]. Here Weibull pdf and bimodal distribution are separately used to handle the uncertainty associated with wind speed and solar irradiation. The scenario-tree technique was used to model the uncertainties associated with solar radiation, wind speed and load demand [28]. This approach is found to be effective while handling uncertainties but requires heavy calculation. An optimal generation scheduling strategy with total contributions of wind farms, solar parks and thermal plants for economic benefit and environmental impact is presented in reference [29]. Here uncertainty associated with wind, solar power and few coal units is described by fuzzy numbers.

The techno-economic analysis under distinct scenarios has been investigated with different combinations of renewable energy resources. Results show that integrating multiple RER and its appropriate scheduling helps minimize cost and emission [30]. A detailed review of various computation methods for planning a hybrid renewable system is presented in reference [31]. Inspired by distinct work carried out by various researchers, in this paper, a novel optimization method incepted in 2020, Equilibrium Optimizer (EO) [32], inspired by dynamic and equilibrium states of physics, is used to solve the complex multi-objective problem with and without integration of RES. The schematic diagram of the proposed model is shown in Figure 1.

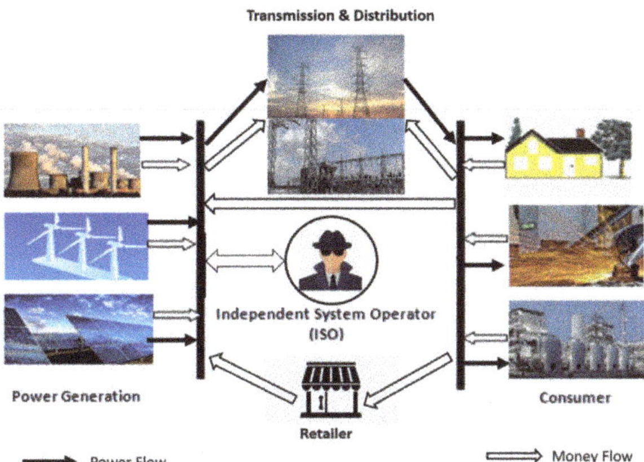

Figure 1. The grid with RES under deregulated environment.

The main contribution in the paper is as follows: (i) Equilibrium Optimizer (EO) is implemented to solve the multi-objective problem in a dynamic environment, (ii) impact of RER integration was analyzed in terms of reduction in operational cost and emission (iii) fuzzy-min ranking method is used to aggregate two conflicting objectives and finally, (iv) profit achieved by the reduction in operational cost due to RER integration were analyzed.

The remaining paper organization is as follows: problem formulation is given in Section 2, the optimization method is described in Section 3, results and discussion are presented in Section 4 and the conclusions are summarized in Section 5.

2. Formulation of the Profit-Based Multi-Objective Scheduling Problem

2.1. Objective Function I: Profit Maximization

The difference between revenue (RV) collected from the sale of electricity and the total cost of electrical power generation (TC) is taken as profit [33]. The objective is to maximize the profit defined as (1):

$$Max(Profit) = Max\,(RV - TC) \tag{1}$$

where,

$$RV = \sum_{t=1}^{24} \left(\sum_{j=1}^{NT} Gth_{jt} + \sum_{k=1}^{NS} Gs_{kt} + \sum_{l=1}^{NW} Gw_{lt} \right).SP_t \tag{2}$$

$$TC = \sum_{t=1}^{24} \left(\sum_{j=1}^{NT} C(Gth)_{jt} + \sum_{k=1}^{NS} C(Gs)_{kt} + \sum_{l=1}^{NW} C(Gw)_{lt} \right) \tag{3}$$

The first part of (3) is the generation cost of thermal units defined in (4) as:

$$C(Gth)_{jt} = \left(a_j \times Gth_{jt}^2 + b_j \times Gth_{jt} + c_j \right) \tag{4}$$

where Gth_{jt} is generated power by j^{th} thermal unit at t^{th} time interval, a_j, b_j, c_j are fuel coefficients of jth thermal unit, SP_t is the selling price at t^{th} time interval, NT, NS and NW, are the number of thermal, solar and wind units present in the hybrid system [34]. Total 24-time steps representing 24 h in a day are considered.

The second part represents the generation cost of the solar PV system. Solar power output depends on incident solar radiation (R_t) and the difference of ambient (θ_{amb}) and reference temperature (θ_{ref}). Therefore, it increases the uncertainty in computing the availability of the solar power output. Thus, underestimation and overestimation cost is added here to balance the variation due to uncertainty.

The solar power output of kth plant at time t, (Gs_{kt}) is given as [34]:

$$Gs_{kt} = Pr\left\{ \left(1 + \left(\theta_{amb} - \theta_{ref} \right)\xi \right) \times (R_t/1000) \right\} \tag{5}$$

Total solar cost function at any time t is calculated as [34]:

$$\sum_{k=1}^{NS} C(Gs_{kt}) = \sum_{k=1}^{NS} (DC_s \times Gs_{kt}) + \sum_{k=1}^{NS} k_p(Gsav_{kt} - Gs_{kt}) + \sum_{k=1}^{NS} k_r(Gs_{kt} - Gsav_{kt}) \qquad (6)$$

where,

$$k_p \times (Gsav_{kt} - Gs_{kt}) = k_p \times \int_{Gs_{kt}}^{P_r} (s - Gs_{kt}) f_s(s) ds \qquad (7)$$

$$k_r \times (Gs_{kt} - Gsav_{kt}) = k_r \times \int_{0}^{Gs_{kt}} (Gs_{kt} - s) f_s(s) ds \qquad (8)$$

k_p, k_r represents the penalty cost factor for overestimation and underestimation, Gs_{kt}, $Gsav_{kt}$ are the generated and available power for k^{th} solar plant at t^{th} time, respectively, and $f_S(s)$ is probability density function (pdf) of solar power.

The third part of (3) represents the cost due to wind power integration. Wind power depends on wind velocity, and at any given location, it is observed to follow the Weibull probability distribution [35], as shown in Figure 2. The random wind speed variable can be computed using the Weibull pdf as:

$$f_v(v) = \left(\frac{k}{c}\right)\left(\frac{v}{c}\right)^{k-1} e^{-(vc)^k} \quad 0 < v < \infty \qquad (9)$$

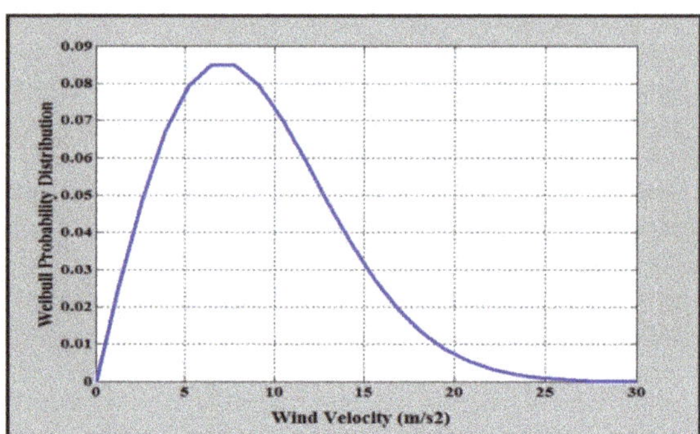

Figure 2. The Weibull probability distribution for wind velocity.

Wind power (Gw) at different velocities is calculated as:

$$Gw = \begin{cases} 0 & for \quad 0 \leq v < v_{ci} \\ Wr\left(\frac{v - v_{ci}}{v_r - v_{ci}}\right) & for \quad v_{ci} \leq v < v_r \\ Wr & for \quad v_r \leq v < v_{co} \\ 0 & for \quad v > v_{co} \end{cases} \qquad (10)$$

where, v, v_{ci}, v_{co} and v_r, are the wind velocity at any instant, cut-in speed, cut-out speed and rated wind turbine speed in m/s, respectively. Wr is the rated power of wind turbine in MW [36]. The wind speed cost has three terms; the first term is a fixed cost term while the second and third terms arise due to the uncertainty involved with wind power generation.

After including the over and under estimation costs, the total cost of lth wind plant at time t can be mathematically expressed as:

$$\sum_{l=1}^{NW} C(Gw_{lt}) = \sum_{l=1}^{NW} (DC_w \times Gw_{lt}) + \sum_{l=1}^{NW} k_p(Gwav_{lt} - Gw_{lt}) + \sum_{l=1}^{NW} k_r(Gw_{lt} - Gwav_{lt}) \qquad (11)$$

The second and third terms in (11) represent the penalty cost due to underestimation and reserve cost due to overestimation and are expressed as (12) and (13), respectively.

$$k_p \times (Gwav_{lt} - Gw_{lt}) = k_p \times \int_{Gw_{lt}}^{W_r} (w - Gw_{lt}) f_W(w) \tag{12}$$

$$k_r \times (Gw_{lt} - Gwav_{lt}) = k_r \times \int_{0}^{Gw_{lt}} (Gw_{lt} - w) f_W(w) dw \tag{13}$$

2.2. Objective Function II: Emission Minimization

The total emission (*TE*) emitted from the thermal power plant [37] can be symbolically represented as:

$$\text{Min}(TE) = Min\left\{ \sum_{j=1}^{NT} E(Gth) \right\} \tag{14}$$

where,

$$E(Gth)_j = \left(\alpha_j \times Gth_j^2 + \beta_j \times Gth_j + \gamma_j \right) \tag{15}$$

Here $\alpha_j, \beta_j, \gamma_j$ are the emission coefficients of jth thermal unit.

2.3. Objective Function III: Simultaneous Optimization of Profit and Emission

The multi-objective problem of optimization of profit and emission can be formulated as:

$$\text{Min}\left\{ \left(\tfrac{1}{Profit} \right), TE \right\} = \frac{u}{\sum_{t=1}^{24}\left(\sum_{j=1}^{NT} Gth_{jt} + \sum_{k=1}^{NS} Gs_{kt} + \sum_{l=1}^{NW} Gw_{lt} \right).SP_t - \sum_{t=1}^{24}\left(\sum_{j=1}^{NT} C(Gth)_{jt} + \sum_{k=1}^{NS} C(Gs)_{kt} + \sum_{l=1}^{NW} C(Gw)_{lt} \right)} + (1-u).ppf. \sum_{j=1}^{NT} E(Gth) \tag{16}$$

where ppf is the price penalty factor which is assumed to be 1 for simplicity and u is the weight factor. Maximization of profit is an objective function I and minimization of total emission is objective function II. Profit and cost of generation are two cornerstones of the electrical market. A decrease in the cost of generation will increase profit. This reciprocal relation between cost and profit is modelled as objective function III or simultaneous optimization of profit and emission.

It is subjected to the following operational constraints:

2.3.1. Real Power Balance: Power Demand at Any Instant (t) Must Be Equal to the Sum of the Power Output of Associated Generation Units

$$PD(t) = \sum_{j=1}^{NT} Gth_j + \sum_{k=1}^{2NS} Gs_k + \sum_{l=1}^{NW} Gw_l \tag{17}$$

2.3.2. Generation Limit: The Power Produced by Each Thermal, Wind and the Solar Unit Must Always Be between Their Respective Specified Bounds, as Given by

$$Gth_j^{min} \leq Gth_j \leq Gth_j^{max} \tag{18}$$

$$Gs_k^{min} \leq Gs_k \leq Gs_k^{max} \tag{19}$$

$$Gw_l^{min} \leq Gw_l \leq Gw_l^{max} \tag{20}$$

2.3.3. Ramp-Rate Limit Constraints: The Thermal Units Have Limited Ramping (Up as Well as Down) Capacity, and Therefore the Output of a Unit between two Consecutive Time-Intervals Must Obey the Inequality Constraint Given by

$$Gth_{jt} - Gth_{j(t-1)} \leq UR_j \tag{21}$$

$$Gth_{j(t-1)} - Gth_{jt} \leq DR_j \tag{22}$$

Here, UR_j, DR_j are the up and down rates of the jth thermal unit.

Combining the ramp rate limits of generating units given by (21) and (22) with the thermal generation limits provided by (18), the modified binding constraints can be written as:

$$Max\left(Gth_j^{min}, Gth_{j(t-1)} - DR_j\right) \leq Gth_{jt} \leq Min\left(Gth_j^{min}, Gth_{j(t-1)} + UR_j\right) \tag{23}$$

2.4. Ranking Approach

The fuzzy-min ranking method is used to aggregate two conflicting objectives: profit and emission [38]. Linear membership function, $\mu_{i,r}$ (r^{th} is the objective function for the ith solution) is described for each objective function in (24) as [22].

$$\mu_{i,r} = \begin{cases} 1 & if \quad F_{i,r} \leq F_r^{min} \\ \frac{F_r^{max} - F_{i,r}}{F_r^{max} - F_r^{min}} & if \quad F_r^{min} \leq F_{i,r} \leq F_r^{max} \\ 0 & if \quad F_{i,r} \geq F_r^{max} \end{cases} \tag{24}$$

For i^{th} solution with n number of objectives, the rank is computed as:

$$fuzzy\_rank_i = min(\mu_{i,r}) \quad for \quad r = 1, 2 \ldots. n \tag{25}$$

The solution with the maximum value of $fuzzy\_rank$ for $\forall r$ is considered as the best compromise solution.

3. Equilibrium Optimization

Equilibrium Optimization is a physics-based algorithm that follows the concept of dynamic mass balance is given control space. The first-order differential equation, which relates the mass-generated in a dynamic system with mass entering and mass leaving the system, can be expressed as:

$$V\frac{d\mathbb{C}}{dt} = Q\mathbb{C}_{eq} - Q\mathbb{C} + \mathcal{G} \tag{26}$$

where, $\mathbb{C}$ is the concentration in volume (V), $V\frac{d\mathbb{C}}{dt}$ is the rate of change in mass, Q is the flow rate, $\mathbb{C}_{eq}$ is the concentration at equilibrium state and $\mathcal{G}$ represents the rate of generation of mass. The equilibrium state is supposed to be achieved whenever $V\frac{d\mathbb{C}}{dt}$ reaches zero. The derivative $\frac{d\mathbb{C}}{dt}$, can be solved as a function of $\left(\frac{Q}{V}\right)$. The ratio $\frac{Q}{V} = \mu$ is called the turnover rate.

Equation (26) can be rearranged and written as:

$$\frac{d\mathbb{C}}{\mu\mathbb{C}_{eq} - \mu\mathbb{C} + \frac{\mathcal{G}}{V}} = dt \tag{27}$$

and,

$$\int_{\mathbb{C}_o}^{\mathbb{C}}\left(\frac{d\mathbb{C}}{\mu\mathbb{C}_{eq} - \mu\mathbb{C} + \frac{\mathcal{G}}{V}}\right) = \int_{t_o}^{t} dt \tag{28}$$

The final concentration update equation after rearranging and integrating becomes [32]:

$$\mathbb{C} = \mathbb{C}_{eq} + (\mathbb{C}_o - \mathbb{C}_{eq})\mathcal{F} + \frac{\mathcal{G}}{\mu V}(1 - \mathcal{F}) \tag{29}$$

where,

$$\mathcal{F} = exp\{-\mu(t - t_o)\} \tag{30}$$

Equation (29) provides the search mechanism for finding an optimal solution during the optimization process of EO. Here $\mathbb{C}_{eq}$, is a solution that is selected randomly from a pool consisting of 3 to 5 best solutions collected after solving the problem for different conditions. The second term $(\mathbb{C}_o - \mathbb{C}_{eq})$ is the difference in the position of a solution and the randomly selected equilibrium state. This term provides direct search and persuades particles to conduct a global search and explore the solution space extensively and effectively. The third term $\left\{\frac{G}{\mu V}(1-\mathcal{F})\right\}$ is the term associated with the generation rate and turnover rate. This term improves/updates the solution through exploitation; hence the steps are small, resulting in small changes to fine-tune the solution; however, sometimes, it allows exploration. EO follows five steps during the optimization process described as follows:

3.1. Initialization

The initialization procedure in EO is similar to other population-based metaheuristics. The initial population is created by randomly generating concentrations within the minimum and maximum limits for each dimension of the vector. The ith population vector can be constructed as follows:

$$\mathbb{C}_i^{initial} = \mathbb{C}_{min} + rand(\mathbb{C}_{max} - \mathbb{C}_{min}) \tag{31}$$

Here $\mathbb{C}_{max}$ and $\mathbb{C}_{min}$, represents vectors representing maximum and minimum concentrations of the different dimensions of the solution vector. The generated particles (solutions) are evaluated, and their fitness value is determined. Then the equilibrium pool is set up by using 3 to 5 best solutions.

3.2. Equilibrium Pool and Candidates

The equilibrium state is the global best solution of the problem, which is obtained after convergence. The equilibrium pool is created by storing the best solutions of runs conducted under different conditions. The arithmetic mean of these best solutions is also stored in the pool as shown:

$$\vec{\mathbb{C}}_{eq,pool} = \left\{\vec{\mathbb{C}}_{eq,1}, \vec{\mathbb{C}}_{eq,2}, \vec{\mathbb{C}}_{eq,3}, \vec{\mathbb{C}}_{eq,4}, \vec{\mathbb{C}}_{eq,ave}\right\} \tag{32}$$

For updating the position of a particle using (29), one of these best solutions from (32) is randomly selected. The probability of selection is uniform for all equilibrium concentrations of the pool. Suppose there are 5 candidate solutions as shown above. In that case, the new solution will be generated by exploration if any of the first four equilibrium states/solutions in the pool are selected for the position update mechanism. On the other hand, if the fifth candidate is chosen for position update, then exploitation is carried out to generate a new solution/state.

3.3. Exponential Term

The third term in the concentration update Equation (29) is the exponential term $(\mathcal{F})$. This term is designed to provide an effective balance between exploration and exploitation in the EO algorithm.

$$\vec{\mathcal{F}} = exp\left(-\vec{\mu}(t - t_0)\right) \tag{33}$$

The turnover rate (μ) is a random number ranging from 0 to 1. Time is represented by t, which decreases with iteration, as expressed below.

$$t = \left(1 - \frac{Iter}{Max\_iter}\right)^{a_2 \times \frac{Iter}{Max\_iter}} \tag{34}$$

$$\vec{t_0} = \frac{1}{\vec{\mu}} ln\left\{-a_1 sign\left(\vec{r} - 0.5\right)\left[1 - e^{-\vec{\mu}t}\right]\right\} + t\} \tag{35}$$

Substituting of (35) in (33) gives the final expression for the exponential term $(\mathcal{F})$ presented in Equation (36). The plot of $\mathcal{F}$ for four different combinations of a_1 and a_2 is shown in Figure 3. The exponential term $(\mathcal{F})$ variation with iteration can be seen to decrease (in both directions) and finally converge to zero for all four combination cases. The nature of variation indicates the effectiveness of this term in conducting exploration and exploitation.

$$\vec{\mathcal{F}} = a_1 sign\left(\vec{r} - 0.5\right)\left[e^{-\vec{\mu}t} - 1\right] \tag{36}$$

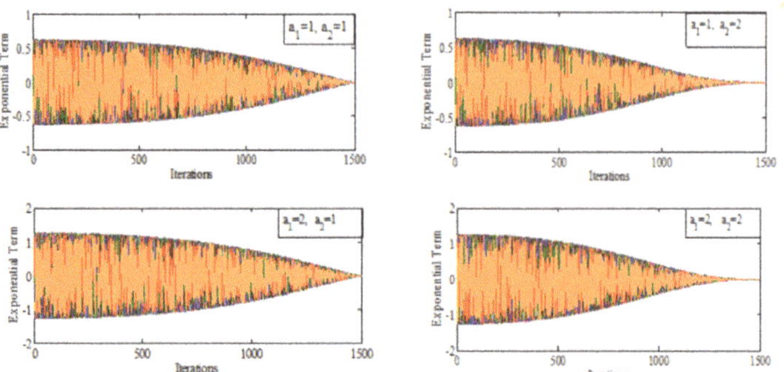

Figure 3. Variation of the exponential term for different combinations of a$_1$ and a$_2$.

In EO exploration and exploitation are controlled by the constants a_1 and a_2, respectively. Other than these two, the term $sign\left(\overrightarrow{r} - 0.5\right)$ affects the direction of exploration and exploitation during the search.

3.4. Generation Rate

This term constitutes the third term of the concentration update equation given by (29). The generation rate ensures the convergence of the algorithm to the optimal global solution $(\overrightarrow{\mathcal{G}})$ which facilitates smooth convergence by tuning the solutions using small updates. By modeling the generation rate using an exponential decay term of the first order and assuming the decay constant to be equal to the turnover rate [32], the generation rate can be expressed to be decreasing from an initial value $(\mathcal{G}_0)$ as:

$$\overrightarrow{\mathcal{G}} = \overrightarrow{\mathcal{G}_0}e^{-\overrightarrow{\mu}(t-t_0)} = \overrightarrow{\mathcal{G}_0}\overrightarrow{\mathcal{F}} \tag{37}$$

$$\overrightarrow{\mathcal{G}_0} = \overrightarrow{\lambda}\left(\overrightarrow{\mathbb{C}_{eq}} - \overrightarrow{\mu}\overrightarrow{\mathbb{C}}\right) \tag{38}$$

The generation rate control parameter (λ) decides what role will be played by the generation rate term in updating the particle position in (29). This parameter is designed to controls the exploitation and exploration of the particle as follows:

$$\overrightarrow{\lambda} = \begin{cases} 0.5r_1 & r_2 > \rho \\ 0 & r_1 < \rho \end{cases} \tag{39}$$

The probability of using the generation rate term by the particle while updating its concentration using (29) is also determined by the generation probability expressed by (ρ).

Here r_1 and r_2 are uniformly distributed random numbers in $[0, 1]$. If the first condition in (39) is true, then the generation rate parameter will be small, and the update step size will be small, causing exploitation. But if the second condition is true, then the particle is updated without any contribution from the generation rate term as (λ) and $(\mathcal{G})$ both become zero. Experiments have shown that when (ρ) is set at 0.5, the search undergoes balanced exploitation and exploration. As the generation probability ρ is increased beyond 0.5, exploration increases, and as ρ is decreased below 0.5, exploitation is observed to increase.

3.5. Particle's Memory Saving

In random operator-based optimization algorithms, some kind of memory mechanism must be used to avoid losing a better solution during the process. EO also has a procedure somewhat similar to the *pbest* in PSO, where the best position and corresponding fitness of each particle are stored and updated whenever there is an improvement in subsequent iterations. The flow chart of EO is shown in Figure 4.

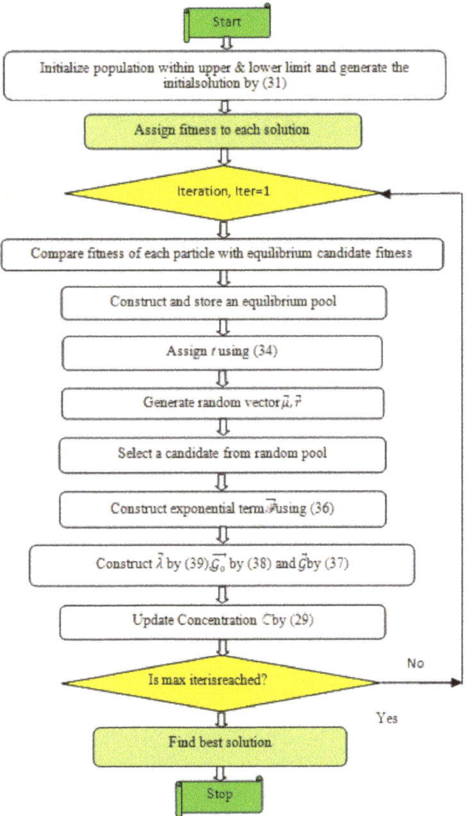

Figure 4. Flow chart of Equilibrium Optimization.

3.6. Steps for Implementation of EO for Profit Based Generation Scheduling

Step 1: Define population size, the maximum number of iterations and the number of runs for the algorithm.

Step 2: Initialize the population within lower and upper concentration limits following (31) and check the constraints (18)–(20) and (23).

Step 3: Select the equilibrium candidates from the initial population assign a_1, a_2, ρ and evaluate the values of each equilibrium candidate.

Step 4: Compare the fitness of equilibrium candidates with that of each particle in the updated population. Replace the equilibrium candidate and corresponding fitness value with that of the particle from the population, if fitness is better.

Step 5: Continue step 4 in loops equal to the number of particles in the population. At the end of the loop, one will acquire the Equilibrium pool $\vec{\mathbb{C}}_{eq,pool}$ as in (32)

Step 6: Accomplish memory saving

Step 7: Construct 't' as per (34)

Step 8: Run a loop equal to the number of particles in the population and randomly choose one candidate from the equilibrium pool (vector).

Step 9: Construct $\vec{\mu}$, r_1, $\vec{\mathcal{F}}$, $\vec{\mathcal{G}}$, $\vec{\mathcal{G}_0}$, $\vec{\lambda}$ using (36)–(39)

Step 10: Update concentrations using (29) till the number of iterations is less than the maximum number of iterations. After the loop is finished, the final concentrations are the power output for the thermal generators for the given time period of 24 h.

Step 11: Apply a Fuzzy selection mechanism to find out the best compromise solution

Step 12: Store the best compromise solution.

4. Results and Discussion

The performance of the EO algorithm is tested on standard test cases under dynamic constraints [38–40]. Impact on operational cost and emission due to RER integration are also investigated here. The objective function is written in MATLAB R2013a environment and executed on Intel core i7 processor with 2 GB RAM and 3.40 GHz speed.

4.1. Description of Test Cases

Test Case 1 has six thermal power generating unit system. Its selling cost, minimum and maximum power limits and cost/emission coefficients are listed in Table A1 [40], along with power demand and hourly selling prices on a particular day.

Test Case 2 is a modified test case created by adding two solar PV units to test case 1. The data of solar plants are listed in Table A2. Its data related to radiation and corresponding temperature are shown in Figure 5.

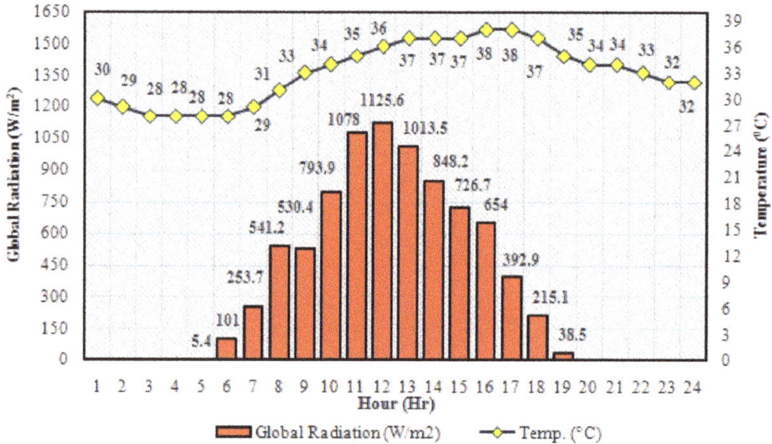

Figure 5. Solar PV data of Temperature (°C) and Radiation (W/m^2).

Test Case 3 is also a modified test case obtained by the addition of two wind generators to test case 1. The data for wind generators are listed in Table A2.

Test Case 4 is a hybrid thermal-wind-solar PV system that integrates two wind generators and two solar PV systems with thermal units of Test case 3, Test case 2 and Test case 1, respectively.

4.2. Effect of Number of Particles

To analyze optimal particle size (NP), experiments are conducted on Test Case 1 with different values of NP. Its effect on optimal generation cost is plotted in Figure 6. Here, it is observed that with an NP of 200, the mean operation cost is the lowest. By increasing NP beyond it, no significant change was observed; however, computational time increases. Hence, a particle size of 200 is considered for further analysis of the problem. The performance of EO is also validated by the comparison of results obtained by simulation of two well-established algorithms: particle swarm optimization (PSO) and artificial bee colony (ABC) algorithm, keeping the same population size. The statistical results in terms of operational cost are tabulated in Table 1 over 30 repeated runs. The cost convergence curve of the three algorithms is compared in Figure 7. The above two results show the superiority of EO over the other two in terms of better search capability and fast convergence.

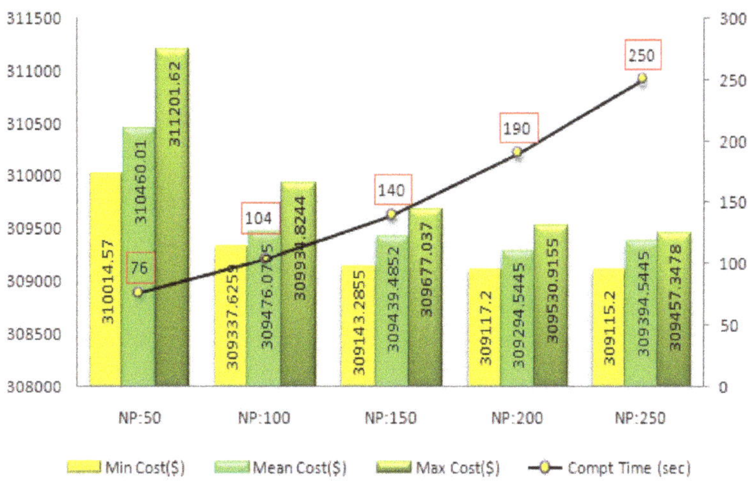

Figure 6. Effect of Number of particles (NP).

Table 1. Comparison of EO with ABC and PSO for Test Case 1.

Method	Parameters	Min Cost ($)	Mean Cost ($)	Max Cost ($)	SD	CPU Time/Iter. (s)
PSO	$c_1 = c_2 = 2.1$ $w_{min} = 0.4$ $w_{max} = 0.9$	309,125.58	309,133.97	309,170.86	5.3818	0.0215
ABC	Limit=100	309,126.34	309,154.07	309,164.77	10.05	0.0203
EO	a_1=2, a_2=1, ρ=0.5	309,117.20	309,125.54	309,139.91	0.9103	0.0188

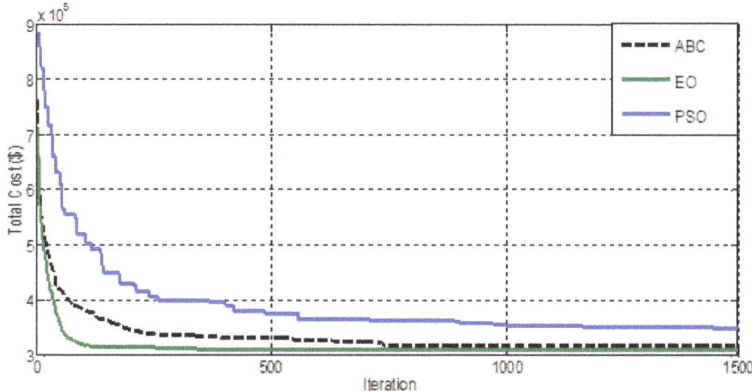

Figure 7. Cost Convergence curve of ABC, PSO and EO for test case 1.

4.3. Effect of Control Parameters on the Performance of EO

EO has three control parameters:

- constants a_1 and a_2 which control the exploration and exploitation, and
- generation probability (ρ) decides whether exploration or exploitation of search space will occur.

To analyze the impact of the above three control parameters, various tests are conducted on Test Case 4 with the variation of parameters in the prescribed range, and obtained results are tabulated in Tables 2 and 3. In Table 2, the value of ρ is kept at 0.5, and results are computed for different combinations of constants a_1 and a_2. Here, it is observed that best results are obtained when $a_1 = 2$ and $a_2 = 1$. Further ρ is varied from 0.1 to 0.9, keeping $a_1 = 2$ and $a_2 = 1$ fixed and obtained simulation results are summarised in Table 3. It is evident that the best result, with the highest profit and lowest emission content, is obtained with $\rho = 0.5$ where the exploration and exploitation have an equal chance of occurrence. Therefore, these combinations of control parameters are considered for further analysis.

Table 2. Best Compromise Solution with the variation of parameters a_1, a_2 for Test Case 4 ($\rho = 0.5$).

Parameters		Total Cost ($)	Profit ($)	Emission (Kg)
a_1	a_2			
1	1	301,550.07	337,807.18	24,412.00
1	2	309,167.46	330,189.73	28,484.83
2	**1**	**297,031.57**	**342,325.68**	**24,763.26**
2	2	297,620.53	341,736.72	26,759.37

Table 3. Best Compromise Solution with variation of generation probability 'ρ' for Test Case 4 ($a_1 = 2$ and $a_2 = 1$).

	ρ	Total Cost ($)	Profit ($)	Emission (Kg)
	0.1	298,680.22	340,677.03	28,639.24
	0.2	298,215.49	341,141.76	27,678.89
	0.3	298,261.84	341,095.41	27,217.14
	0.4	298,310.80	341,046.45	28,180.66
$a_1 = 2$, $a_2 = 1$	0.5	297,031.57	342,325.68	24,763.26
	0.6	297,845.79	341,511.46	27,399.93
	0.7	297,553.80	341,803.45	26,343.26
	0.8	297,624.74	341,732.51	26,728.33
	0.9	298,367.13	340,990.12	26,233.64

4.4. Effect of RER Integration on Profit Maximization

The simulation results for different test cases under the scenario of cost minimization/profit maximization given by (1) are shown in Figure 8. Comparing test case 1 and test case 2 shows a reduction in power generation cost by 3883.3 $ ($\approx$1.26%), and hence the profit is increased by 3883.3 $ ($\approx$1.18%). While comparison of test case 1 and test case 3, it is observed that the reduction in power generation cost is found to be 13,461.09 $ ($\approx$4.35%), and hence the profit is increased by 13,461 $ ($\approx$4.08%). Similarly, a comparison of test case 1 and the hybrid test case 4, shows that the power generation cost is reduced by 14,285.21 $ ($\approx$4.62%) and the profit increased by 14,285.18 $ ($\approx$4.32%). Hence, it is clear that the higher the integration of RER, the higher the profit even after the inclusion of cost due to uncertainty.

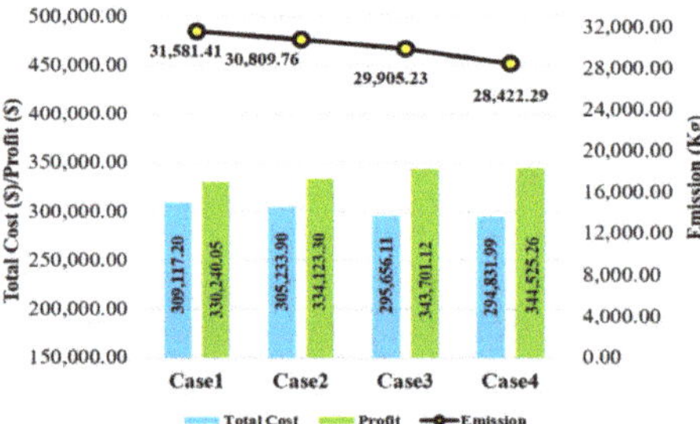

Figure 8. Comparison of cost, profit and emission for profit maximization.

The optimum generation schedule for test cases 1 and 4 are shown in Figures 9 and 10, where operational constraints are fully satisfied. Profit and emission for these two cases are compared and listed in Table 4.

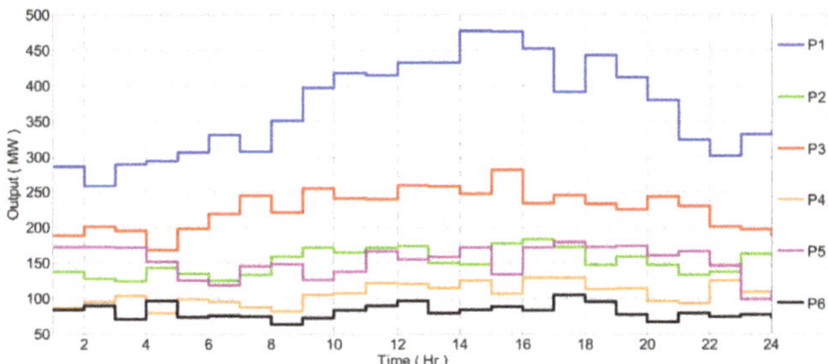

Figure 9. Optimal generation schedule under the scenario of Cost minimization (Test Case 1).

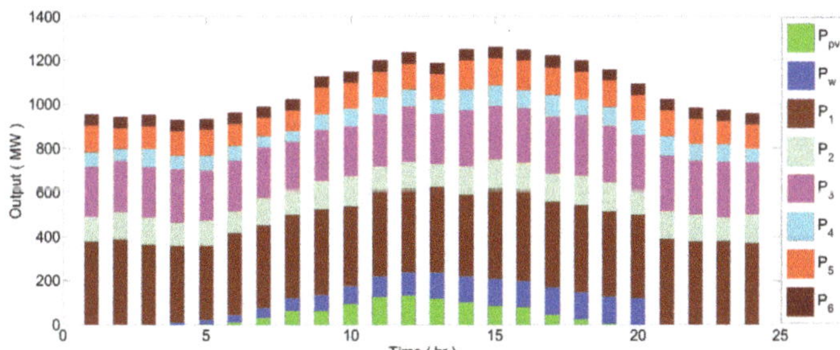

Figure 10. Optimal generation schedule under the scenario of Cost minimization (Test Case 4).

Table 4. Cost, Selling price, Profit and Emission under the scenario of profit maximization.

	ThC ($)	WC ($)	PVC ($)	TC ($)	SP ($)	Profit ($)	Emission (Kg)
Test Case 1	309,117.20	–	–	309,117.20	639,357.25	330,240.05	31,581.41
Test Case 4	276,906.41	6290.20	11,635.38	294,831.99	639,357.25	344,525.26	28,422.29

4.5. Effect of RER Integration on Emission Minimization

For minimization of emission (14), all the test cases under consideration are carefully analyzed to find the impact of the integration of (i) solar, (ii) wind and (iii) both solar and wind resources. The objective is to determine the optimal schedule for all four test cases, which will produce minimum emission content. The results for each test case are presented in Figure 11.

Comparing test case 1 and case 2, the reduction in emission content is 2255.94 kg ($\approx$8.92%) due to the solar share of 975.71 MW (4%). While Comparing test case 1 and test case 3, the reduction in emission content is 3216.74 kg ($\approx$12.72%) due to 1463.08 MW (5%) of wind share. Similarly, while comparing test case 1 and test case 4, emission reduction is 5524.63 kg ($\approx$21.84%).

Comparing all the cases, it is observed that there is a significant reduction in pollution by integration of RER.

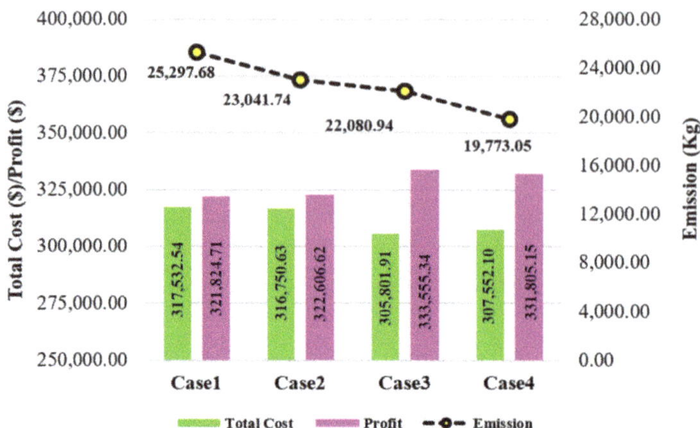

Figure 11. Comparison of cost, profit and emission under the scenario of emission minimization.

4.6. Effect of RER Integration for the Multiobjective Case

For the simultaneous optimization of profit and emission both, formulated in (16) with the help of the fuzzy min approach (24), the best compromise solution is obtained. Pareto front for all test cases obtained by EO is plotted and compared in Figure 12 and the top 10 optimal solutions and their fuzzy min rank are tabulated in Table 5. The best-compromised solution with the highest fuzzy rank is indicated for each test case.

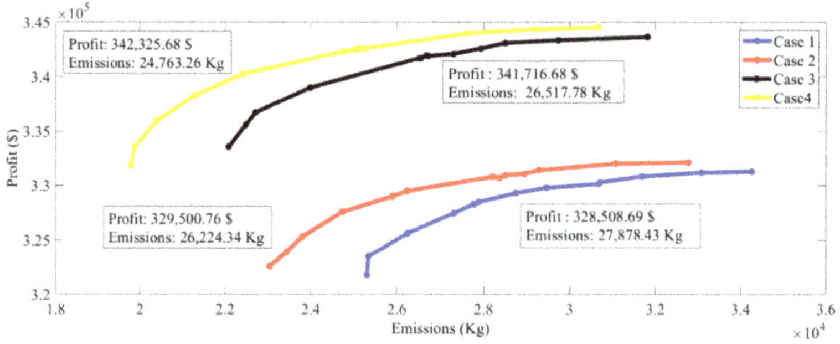

Figure 12. Pareto-front of all the non-dominated solutions obtained for different cases.

Table 5. Top 10 optimal solutions and their fuzzy min rank.

Test Case 1					Test Case 2				
Profit	Emission	μ_1	μ_2	min (μ)	Profit	Emission	μ_1	μ_2	min (μ)
328,508.692	27,878.429	0.705	0.712	0.705	329,500.761	26,224.344	0.725	0.672	0.672
328,312.919	27,759.219	0.684	0.725	0.684	329,502.218	26,226.754	0.725	0.672	0.672
328,312.663	27,759.160	0.684	0.725	0.684	329,503.106	26,227.526	0.725	0.672	0.672
328,311.440	27,758.630	0.684	0.725	0.684	327,594.083	24,734.770	0.524	0.825	0.524
329,335.245	28,738.401	0.792	0.616	0.616	327,593.259	24,733.423	0.524	0.825	0.524
329,336.053	28,739.487	0.792	0.616	0.616	327,591.770	24,732.427	0.524	0.826	0.524
329,336.009	28,739.710	0.792	0.616	0.616	330,805.950	28,204.049	0.862	0.468	0.468
327,479.749	27,313.518	0.596	0.775	0.596	330,805.924	28,204.710	0.862	0.468	0.468
327,478.885	27,312.756	0.596	0.775	0.596	330,805.931	28,205.116	0.862	0.468	0.468
327,478.002	27,312.382	0.596	0.775	0.596	330,689.142	28,348.341	0.849	0.454	0.454
Test Case 3					Test Case 4				
Profit	Emission	μ_1	μ_2	min (μ)	Profit	Emission	μ_1	μ_2	min (μ)
341,716.684	26,517.776	0.760	0.567	0.567	342,325.682	24,763.258	0.827	0.542	0.542
341,716.956	26,518.660	0.760	0.567	0.567	342,325.842	24,763.330	0.827	0.542	0.542
341,717.139	26,518.991	0.760	0.567	0.567	342,477.714	25,040.173	0.839	0.516	0.516
341,708.322	26,549.079	0.759	0.564	0.564	342,478.170	25,040.791	0.839	0.516	0.516
341,708.646	26,549.144	0.759	0.564	0.564	342,532.703	25,090.462	0.843	0.512	0.512
341,708.329	26,549.322	0.759	0.564	0.564	342,533.058	25,090.864	0.843	0.512	0.512
341,960.680	26,683.972	0.791	0.550	0.550	338,289.653	21,279.033	0.510	0.862	0.510
341,960.976	26,684.062	0.791	0.549	0.549	338,288.413	21,278.272	0.510	0.862	0.510
341,961.230	26,684.330	0.791	0.549	0.549	342,585.154	25,130.773	0.847	0.508	0.508
341,924.268	26,723.629	0.786	0.545	0.545	342,584.919	25,130.786	0.847	0.508	0.508

The optimal power generation schedule for test case 1 and hybrid thermal-wind-solar PV system, i.e., test case 4 with cost, selling price, profit and corresponding emission are listed in Tables 6 and 7. In addition, a comparison of cost, profit and emission for all four cases is presented in Figure 13. Here it is seen that the cost is reduced by 992.07 $ ($\approx$0.32%) after integration of two solar units in test case 2. The cost was reduced by 13,207.97$ ($\approx$4.25%) in test case 3 when two wind power units were added and by 13,816.99 $ ($\approx$4.44%) for test case 4 when two solar and wind units, respectively, were integrated with the existing thermal system. The profit is found to increase by 992.07 $ ($\approx$0.30%), 13,208$ ($\approx$4.02%), 13,816.66 $ ($\approx$4.21%), respectively, for test cases 2, 3 and 4.

The emission content is observed to reduce by 1654.09 kg ($\approx$5.93%) in case 2, by 1360.65 kg ($\approx$4.88%) in case 3, and by 3115.17 kg ($\approx$11.17%) in case 4 with respect to test case 1 where only thermal units are present.

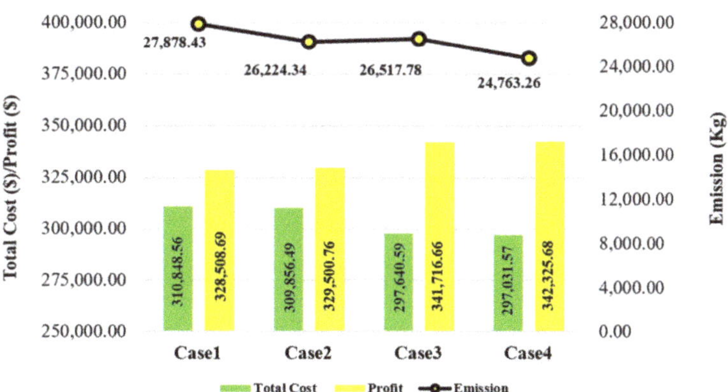

Figure 13. Comparison of cost, profit and emission under the scenario of emission minimization.

Table 6. Optimal generation schedule, cost, selling price and profit for Test Case 1 (Best Compromise Solution).

Hour	P1 (MW)	P2 (MW)	P3 (MW)	P4 (MW)	P5 (MW)	P6 (MW)	TC ($)	SP ($)	Profit ($)	Emission (Kg)
1	267.18	116.94	189.30	137.60	125.15	118.82	11,431.4097	21,630.75	10,199.34	884.64
2	290.12	120.26	180.63	108.07	135.22	107.69	11,208.0562	20,724.00	9515.94	900.50
3	273.65	147.61	201.12	89.01	122.96	118.66	11,357.6823	21,537.80	10,180.12	920.34
4	258.12	147.41	159.08	113.96	146.97	104.46	11,145.9213	20,553.00	9407.08	831.10
5	282.57	118.86	190.11	87.82	156.66	98.97	11,113.801	21,505.00	10,391.20	894.89
6	268.65	134.94	188.96	131.20	133.12	106.13	11,506.8602	22,293.45	10,786.59	903.10
7	282.25	110.86	200.46	116.44	161.67	117.31	11,814.661	24,329.40	12,514.74	943.67
8	318.19	126.99	194.56	113.38	162.78	107.09	12,181.5529	25,830.73	13,649.18	1047.15
9	324.15	164.39	226.84	118.69	198.86	93.06	13,471.5765	27,936.05	14,464.47	1239.81
10	344.50	155.59	211.73	122.96	196.45	118.77	13,786.5861	30,475.00	16,688.41	1265.30
11	350.92	183.98	240.31	139.40	167.58	118.81	14,416.1753	32,607.14	18,190.97	1395.27
12	386.77	178.95	231.68	148.91	169.39	119.30	14,838.8277	37,420.49	22,581.67	1501.13
13	358.79	159.28	228.62	149.05	174.61	119.65	14,275.8021	35,342.99	21,067.19	1368.31
14	379.97	165.36	249.43	146.40	190.98	118.86	15,041.8218	35,778.61	20,736.78	1522.54
15	387.37	178.49	247.30	148.43	182.26	119.15	15,195.1235	31,890.75	16,695.63	1558.50
16	364.96	197.62	245.53	132.81	189.51	119.58	15,050.0093	31,000.00	15,949.99	1499.40
17	358.21	169.93	256.18	129.27	189.03	118.39	14,658.0997	28,754.55	14,096.45	1450.01
18	360.92	172.60	225.78	148.92	174.26	119.51	14,434.0764	27,465.70	13,031.62	1390.41
19	322.15	154.47	229.83	133.87	199.14	119.54	13,920.6668	26,714.95	12,794.28	1255.61
20	335.55	147.30	209.25	108.47	172.87	118.56	13,035.613	25,389.00	12,353.38	1174.63
21	296.40	167.11	206.24	121.43	142.45	89.38	12,187.7462	24,040.49	11,852.75	1055.36
22	275.73	153.09	199.98	95.28	164.90	95.01	11,732.3309	22,887.84	11,155.51	959.52
23	267.59	160.44	225.76	94.03	156.50	70.68	11,596.8066	21,937.49	10,340.69	996.03
24	295.98	99.08	185.82	115.74	144.14	119.23	11,447.352	21,312.00	9864.65	921.21
Total	**7650.70**	**3631.54**	**5124.51**	**2951.15**	**3957.45**	**2656.65**	**310,848.56**	**639,357.25**	**328,508.69**	**27,878.43**

Table 7. Optimal generation schedule, cost, selling price and profit for Test Case 4 (Best Compromise Solution).

A. Optimal Generation Schedule for Test Case 4 (Best Compromise Solution)									
Hour	P1 (MW)	P2 (MW)	P3 (MW)	P4 (MW)	P5 (MW)	P6 (MW)	TS (MW)	Emission (Kg)	WS (MW)	PV Share (MW)
1	326.45	101.35	186.46	98.17	150.86	91.71	955.00	991.51	0.00	0.00
2	303.95	107.68	198.92	101.68	128.74	101.03	942.00	947.51	0.00	0.00
3	281.69	119.22	232.89	81.07	143.44	94.68	953.00	975.98	0.00	0.00
4	258.20	135.39	186.99	102.72	149.44	88.95	921.69	849.56	8.31	0.00
5	268.76	145.56	193.86	86.33	139.38	80.16	914.06	882.95	20.31	2.63
6	292.74	121.33	177.03	93.73	132.47	101.55	918.84	885.64	32.31	13.85
7	291.77	131.62	183.65	96.06	121.16	90.69	914.93	900.13	44.31	30.76
8	298.97	120.68	193.18	90.61	119.46	80.31	903.21	916.65	56.31	63.48
9	324.98	125.01	202.09	118.72	136.55	88.13	995.48	1056.59	68.31	62.22
10	327.58	135.25	204.20	106.08	142.04	61.42	976.57	1070.75	80.31	93.12
11	352.28	112.17	217.72	103.03	122.32	74.72	982.24	1134.22	92.31	120.00
12	347.07	122.85	186.34	108.57	144.80	89.03	998.66	1086.44	104.31	120.00
13	314.33	127.82	176.08	116.53	143.55	76.50	954.81	975.73	116.31	118.88
14	335.61	146.32	213.07	112.85	143.72	79.95	1031.51	1143.19	120.00	99.49
15	322.65	155.12	241.44	96.25	142.49	99.82	1057.76	1180.47	120.00	85.24
16	334.44	157.54	226.78	108.63	139.08	86.81	1053.29	1185.25	120.00	76.71
17	342.95	162.21	186.79	125.49	159.30	78.16	1054.91	1161.91	120.00	56.09
18	351.52	163.81	180.67	101.87	156.07	102.82	1056.77	1161.48	120.00	26.23
19	306.51	160.14	194.76	120.31	150.81	101.95	1034.48	1057.68	120.00	6.52
20	331.46	140.31	196.15	92.82	129.68	81.58	972.00	1055.65	120.00	0.00
21	351.75	122.82	222.79	98.26	135.87	91.50	1023.00	1168.49	0.00	0.00
22	327.44	136.26	204.65	92.46	144.33	78.86	984.00	1066.00	0.00	0.00
23	302.88	129.99	188.70	97.67	170.58	85.18	975.00	980.05	0.00	0.00
24	265.65	140.16	211.95	95.83	154.01	92.41	960.00	929.40	0.00	0.00
Total	**7561.66**	**3220.61**	**4807.17**	**2445.70**	**3400.14**	**2097.93**	**23,533.20**	**24,763.26**	**1463.08**	**975.72**

Table 7. *Cont.*

B. Cost, Selling Price and Profit for Test Case 4 (Best Compromise Solution)					
Hour	Th Cost ($)	WC ($)	PV Cost ($)	TC ($)	SP ($)	Profit ($)
1	11,313.10	26.59	0.00	11,339.69	21,630.75	10,291.06
2	11,163.59	26.59	0.00	11,190.18	20,724.00	9533.82
3	11,300.92	26.59	0.00	11,327.51	21,537.80	10,210.30
4	10,982.15	30.71	0.00	11,012.86	20,553.00	9540.14
5	10,848.76	55.33	7.55	10,911.64	21,504.92	10,593.28
6	10,907.79	102.07	141.28	11,151.14	22,293.51	11,142.37
7	10,840.90	157.43	354.88	11,353.21	24,329.42	12,976.21
8	10,666.11	214.29	757.03	11,637.43	25,830.68	14,193.25
9	11,798.47	271.29	741.92	12,811.68	27,936.16	15,124.48
10	11,535.66	328.29	1110.51	12,974.46	30,474.88	17,500.42
11	11,585.74	385.29	1507.91	13,478.94	32,607.17	19,128.22
12	11,834.05	442.29	1574.49	13,850.83	37,420.41	23,569.58
13	11,321.05	499.29	1417.69	13,238.03	35,342.89	22,104.87
14	12,224.89	516.83	1186.46	13,928.18	35,778.49	21,850.31
15	12,566.34	516.83	1016.51	14,099.68	31,890.71	17,791.02
16	12,496.84	516.83	914.82	13,928.49	30,999.90	17,071.41
17	12,556.04	516.83	549.59	13,622.46	28,754.62	15,132.16
18	12,588.57	516.83	300.88	13,406.28	27,465.67	14,059.40
19	12,338.22	516.83	53.85	12,908.90	26,715.04	13,806.14
20	11,488.50	516.83	0.00	12,005.33	25,389.00	13,383.67
21	12,101.88	26.59	0.00	12,128.47	24,040.50	11,912.03
22	11,634.79	26.59	0.00	11,661.38	22,887.84	11,226.46
23	11,583.47	26.59	0.00	11,610.06	21,937.50	10,327.44
24	11,428.16	26.59	0.00	11,454.75	21,312.00	9857.25
Total	279,105.98	6290.22	11,635.37	297,031.57	639,357.25	342,325.68

4.7. Analysis and Discussion

The EO algorithm is employed to analyze the optimal generation schedule for a hybrid thermal-solar PV-wind system under the deregulated environment with the objective is to maximize the profit of the operator and to minimize emission content for the given power demand and tariff. The cost due to uncertainty of RER in meeting the load demand is also included in the model. The effect of integrating solar and wind power units is studied under (i) profit maximization, (ii) emission minimization and (iii) profit-emission optimization.

According to the results mentioned in Sections 4.4–4.6, it is observed that profit increases when more and more renewable units are added to the thermal system. On the other hand, emission content becomes reduced with the addition of more renewable units. For the multiobjective profit-emission optimization case, the improvement in both profit and emission can be seen to lie between the conditions (i) and (ii).

5. Conclusions

In this paper Equilibrium Optimization (EO) is applied for the solution of the optimal generation schedule problem of a hybrid thermal-solar-wind test system such that the profit is maximized and the pollution content becomes reduced. The practical constraints of non-convexity, non-linearity associated with the thermal unit, probabilistic terms due to wind and solar system are included in the cost function and analyzed under dynamic conditions. The performance of EO is also compared and validated with well-known PSO and ABC algorithms.

The simulation results indicate that:

- The EO is not significantly dependent on algorithm-specific control parameters. The results are found to vary in a very narrow band with variations in control parameters.
- As the population size increases, EO gives more promising results. However, an increase in population size leads to increased computational time too.
- EO has a unique embedded mechanism for exploration and exploitation which leads to the global best solution.
- The increase in profit and decrease in emission are computed for the integration of solar, wind and wind-solar units in the existing thermal power generation system.
- It is verified that the higher the integration of RER, the greater is the profit, even after including the uncertainty costs of the renewable energy in the model.
- EO is found to produce well-distributed Pareto-optimal solutions for the multiobjective problem. For all the tested cases it is observed that EO is capable of dealing with complex operational constraints under the dynamic environment in an efficient manner.
- The proposed work is beneficial for designing hybrid renewable power systems with optimal capacities for given conditions to achieve desired profit and to reduce emission.

Author Contributions: Conceptualization, S.M.D. and H.M.D.; methodology, M.P.; software, S.R.S.; validation, S.M.D., H.M.D. and M.P.; formal analysis, S.R.S.; investigation, S.M.D.; resources, H.M.D.; data curation, M.P.; writing—original draft preparation, S.M.D. and H.M.D.; writing—review and editing, H.M.D., M.P. and S.R.S.; visualization, S.M.D.; supervision, M.P. and S.R.S.; project administration, S.M.D. and H.M.D.; funding acquisition, M.P. and S.R.S. All authors have read and agreed to the published version of the manuscript.

Funding: Woosong University's Academic Research Funding—2021.

Institutional Review Board Statement: Not applicable.

Informed Consent Statement: Not applicable.

Data Availability Statement: Not applicable.

Acknowledgments: The authors acknowledge the support provided by Woosong University's Academic Research Funding—2021.

Conflicts of Interest: The authors declare no conflict of interest.

Appendix A

Table A1. Cost and emission coefficients and generation limits of thermal units, power demand and respective market selling price.

Unit	a_i($/MW^{2}h)	b_i($/MWh)	c_i($/h)	Pmin (MW)	Pmax (MW)	α_i(Kg/MW^{2}h)	β_i(Kg/MWh)	γ_i(Kg/h)	UR (MW/h)	DR (MW/h)
1	0.007	7	240	100	500	0.00419	0.32767	13.8593	80	120
2	0.0095	10	200	50	200	0.00419	0.32767	13.8593	50	90
3	0.009	8	220	80	300	0.00683	−0.54551	40.2669	65	100
4	0.009	11	200	50	150	0.00683	−0.54551	40.2669	50	90
5	0.008	10.5	220	50	200	0.00461	−0.51116	42.8955	50	90
6	0.0075	12	190	50	120	0.00461	−0.51116	42.8955	50	90

Hour	1	2	3	4	5	6	7	8	9	10	11	12
PD (MW)	955	942	953	930	935	963	989	1023	1126	1150	1201	1235
SP ($/MW)	22.65	22	22.6	22.1	23	23.15	24.6	25.25	24.81	26.5	27.15	30.3

Hour	13	14	15	16	17	18	19	20	21	22	23	24
PD (MW)	1190	1251	1263	1250	1221	1202	1159	1092	1023	984	975	960
SP ($/MW)	29.7	28.6	25.25	24.8	23.55	22.85	23.05	23.25	23.5	23.26	22.5	22.2

Table A2. Data for Solar PV units and wind farm.

Type of System	No. of Units	Rated Power (MW/Unit)	DC ($/MWh)	kp	kr	k	c	Vci (m/s^2)	Vr (m/s^2)	Vco (m/s^2)
Solar PV	2	60	12	1.5	3	-	-	-	-	-
Wind	2	60	1.75			2	10	3	16	25

References

1. Reddy, S.S.; Bijwe, P.R.; Abhyankar, A.R. Multiobjective market-clearing of electrical energy, spinning reserves and emission for wind-thermal power system. *Int. J. Electr. Power Energy Syst.* **2013**, *53*, 782–794. [CrossRef]
2. Abido, M.A. Multiobjective evolutionary algorithms for electric power dispatch problem. *IEEE Trans. Evol. Comput.* **2006**, *10*, 315–329. [CrossRef]
3. Delshad, M.M.; Rahim, N.A. Multiobjective backtracking search algorithm for economic emission dispatch problem. *Appl. Soft Comput.* **2016**, *40*, 479–494. [CrossRef]
4. Liang, Y.C.; Juarez, J.R.C. A normalization method for solving the combined economic and emission dispatch problem with meta-heuristic algorithms. *Int. J. Electr. Power Energy Syst.* **2014**, *54*, 163–186. [CrossRef]
5. Basu, M. Fuel constrained economic emission dispatch using nondominated sorting genetic algorithm-II. *Energy* **2014**, *78*, 649–664. [CrossRef]
6. Kuk, J.N.; Gonçalves, R.A.; Pavelski, L.M.; Venske, S.M.G.S.; Almeida, C.P.; Pozo, A.T.R. An empirical analysis of constraint handling on evolutionary multiobjective algorithms for the Environmental/Economic Load Dispatch problem. *Expert Syst. Applt.* **2021**, *165*, 113774. [CrossRef]
7. Muthuswamy, R.; Krishnan, M.; Subramanian, K.; Subramanian, B. Environmental and economic power dispatch of thermal generators using modified NSGA-II algorithm. *Int. Trans. Electr. Energy Syst.* **2014**, *25*, 1552–1569. [CrossRef]
8. Wang, L.; Singh, C. Environmental/economic power dispatch using a fuzzified multi-objective particle swarm optimization algorithm. *Electr. Power Syst. Res.* **2007**, *77*, 1654–1664. [CrossRef]
9. Agrawal, S.; Panigrahi, B.K.; Tiwari, M.K. Multiobjective particle swarm algorithm with fuzzy clustering for electrical power dispatch. *IEEE Trans. Evol. Comput.* **2008**, *12*, 529–541. [CrossRef]
10. Goudarzi, A.; Li, Y.; Xiang, J. A hybrid non-linear time-varying double-weighted particle swarm optimization for solving non-convex combined environmental economic dispatch problem. *Appl. Soft Comput.* **2020**, *86*, 105894. [CrossRef]
11. Jeddi, B.; Vahidinasab, V. A modified harmony search method for environ-mental/economic load dispatch of real-world power systems. *Energy Convers. Manag.* **2014**, *78*, 661–675. [CrossRef]
12. Rezaie, H.; Kazemi-Rahbar, M.H.; Vahidi, B.; Rastegar, H. Solution of combined economic and emission dispatch problem using a novel chaotic improved harmony search algorithm. *J. Comput. Des. Eng.* **2019**, *6*, 447–467. [CrossRef]
13. Pandit, M.; Srivastava, L.; Sharma, M. Environmental economic dispatch in multi-area power system employing improved differential evolution with fuzzy selection. *Appl. Soft Comp.* **2015**, *28*, 498–510. [CrossRef]
14. Liang, H.; Liu, Y.; Li, F.; Shen, Y. A multiobjective hybrid bat algorithm for combined economic/emission dispatch. *Int. J. Electr. Power Energy Syst.* **2018**, *101*, 103–115. [CrossRef]
15. Dong, R.; Wang, S. New Optimization Algorithm Inspired by Kernel Tricks for the Economic Emission Dispatch Problem with Valve Point. *IEEE Access* **2020**, *8*, 16584–16594. [CrossRef]
16. Nourianfar, H.; Abdi, H. Solving the multiobjective economic emission dispatch problems using Fast Non-Dominated Sorting TVAC-PSO combined with EMA. *Appl. Soft Comp.* **2019**, *85*, 105770. [CrossRef]
17. Karthik, N.; Parvathy, A.K. Multiobjective economic emission dispatch using interior search algorithm. *Int. Trans. Electr. Energy Syst.* **2018**, *29*, e2683. [CrossRef]
18. Karthikeyan, R. Combined economic emission dispatch using grasshopper optimization algorithm. *Mater. Today Proc.* **2020**, *33*, 3378–3382. [CrossRef]
19. Guesmi, T.; Farah, A.; Marouani, I.; Alshammari, B.; Abdallah, H.H. Chaotic sine–cosine algorithm for chance-constrained economic emission dispatch problem including wind energy. *IET Renew. Power Gener.* **2020**, *14*, 1808–1821. [CrossRef]
20. Hooshmand, R.A.; Parastegari, M.; Morshed, M.J. Emission, reserve, and economic load dispatch problem with non-smooth and non-convex cost functions using the hybrid bacterial foraging-Nelder–Mead algorithm. *Appl. Energy* **2012**, *89*, 443–453. [CrossRef]
21. Qu, B.Y.; Zhu, Y.S.; Jiao, Y.C.; Wu, M.Y.; Suganthan, P.N.; Liang, J.J. A survey on multiobjective evolutionary algorithms for the solution of the environmental/economic dispatch problems. *Swarm Evol. Comp.* **2018**, *38*, 1–11. [CrossRef]
22. Dubey, H.M.; Pandit, M.; Panigrahi, B.K. An overview and comparative analysis of recent bio-inspired optimization techniques for wind integrated multiobjective power dispatch. *Swarm Evol. Comp.* **2018**, *38*, 12–34. [CrossRef]

23. Dubey, H.M.; Pandit, M.; Panigrahi, B.K. Hybrid flower pollination algorithm with time-varying fuzzy selection mechanism for wind integrated multiobjective dynamic economic dispatch. *Renew Energy* **2015**, *83*, 188–202. [CrossRef]
24. Hazra, S.; Roy, P.K. Quasi-oppositional chemical reaction optimization for combined economic emission dispatch in power system considering wind power uncertainties. *Renew. Energy Focus* **2019**, *31*, 45–62. [CrossRef]
25. Basu, M. Multi-area dynamic economic emission dispatch of hydro-wind-thermal power system. *Renew. Energy Focus* **2019**, *28*, 11–35. [CrossRef]
26. Dey, B.; Roy, S.K.; Bhattacharyya, B. Solving multiobjective economic emission dispatch of a renewable integrated microgrid using latest bio-inspired algorithms. *Int. J. Eng. Sci. Tech.* **2019**, *22*, 55–66.
27. Reddy, S. Optimal scheduling of thermal-wind-solar power system with storage. *Renew. Energy* **2017**, *101*, 1357–1368. [CrossRef]
28. Zubo, R.H.A.; Mokryani, G.; Abd-Alhameed, R. Optimal operation of distribution networks with high penetration of wind and solar power within a joint active and reactive distribution market environment. *Appl. Energy* **2018**, *220*, 713–722. [CrossRef]
29. Xu, J.; Wang, F.; Lv, C.; Huang, Q.; Xie, H. Economic-environmental equilibrium based optimal scheduling strategy towards wind-solar-thermal power generation system under limited resources. *Appl. Energy* **2018**, *231*, 355–371. [CrossRef]
30. Panda, A.; Mishra, U.; Tseng, M.L.; Ali, M.H. Hybrid power systems with emission minimization: Multiobjective optimal operation. *J. Clean. Prod.* **2020**, *268*, 121418. [CrossRef]
31. Swain, A.; Salkuti, S.R.; Swain, K. An Optimized and Decentralized Energy Provision System for Smart Cities. *Energies* **2021**, *14*, 1451. [CrossRef]
32. Emad, D.; El-Hameed, M.A.; Yousef, M.T.; El-Fergany, A.A. Computational Methods for Optimal Planning of Hybrid Renewable Microgrids: A Comprehensive Review and Challenges. *Arch. Comput. Methods Eng.* **2020**, *27*, 1297–1319. [CrossRef]
33. Park, H. A Stochastic Planning Model for Battery Energy Storage Systems Coupled with Utility-Scale Solar Photovoltaics. *Energies* **2021**, *14*, 1244. [CrossRef]
34. Faramarzi, A.; Heidarinejad, M.; Stephens, B.; Mirjalili, S. Equilibrium optimizer: A novel optimization algorithm. *Knowl.-Based Syst.* **2020**, *191*, 105190. [CrossRef]
35. Battula, A.R.; Vuddanti, S.; Salkuti, S.R. Review of Energy Management System Approaches in Microgrids. *Energies* **2021**, *14*, 5459. [CrossRef]
36. Lotfi, H.; Dadpour, A.; Samadi, M. Solving economic dispatch in competitive power market using improved particle swarm optimization algorithm. In Proceedings of the Conference on Electrical Power Distribution Networks Conference (EPDC2017), Semnan, Iran, 19–20 April 2017; pp. 188–195.
37. Joshi, P.M.; Verma, H.K. An improved TLBO based economic dispatch of power generation through distributed energy resources considering environmental constraints. *Sustain. Energy Grids Netw.* **2019**, *18*, 1–18. [CrossRef]
38. Hetzer, J.; Yu, D.C.; Bhattarai, K. An Economic Dispatch Model Incorporating Wind Power. *IEEE Trans. Energy Conver.* **2008**, *23*, 603–611. [CrossRef]
39. Menezes, R.F.A.; Soriano, G.D.; de Aquino, R.R.B. Locational Marginal Pricing and Daily Operation Scheduling of a Hydro-Thermal-Wind-Photovoltaic Power System Using BESS to Reduce Wind Power Curtailment. *Energies* **2021**, *14*, 1441. [CrossRef]
40. Khan, N.A.; Awan, A.B.; Mahmood, A.; Razzaq, S.; Zafar, A.; Sidhu, G.A. Combined emission economic dispatch of power system including solar photovoltaic generation. *Energy Conver. Manag.* **2015**, *92*, 82–91. [CrossRef]

Article

An Optimized and Decentralized Energy Provision System for Smart Cities

Ayusee Swain [1], Surender Reddy Salkuti [1,*] and Kaliprasanna Swain [2]

[1] Department of Railroad and Electrical Engineering, Woosong University, Daejeon 34606, Korea; ayusee.1998@gmail.com
[2] Department of Electronics and Communication Engineering, Gandhi Institute of Technological Advancement, Bhubaneswar 752054, India; kaliprasanna_ece@gita.edu.in
* Correspondence: surender@wsu.ac.kr; Tel.: +82-10-9674-1985

Abstract: Energy efficiency and data security of smart grids are one of the major concerns in the context of implementing modern approaches in smart cities. For the intelligent management of energy systems, wireless sensor networks and advanced metering infrastructures have played an essential role in the transformation of traditional cities into smart communities. In this paper, a smart city energy model is proposed in which prosumer communities were built by interconnecting energy self-sufficient households to generate, consume and share clean energy on a decentralized trading platform by integrating blockchain technology with a smart microgrid. The efficiency and stability of the grid network were improved by using several wireless sensor nodes that manage a massive amount of data in the network. However, long communication distances between sensor nodes and the base station can greatly consume the energy of sensors and decrease the network lifespan. Therefore, bio-inspired algorithm approaches were proposed to improve routing by obtaining the shortest path for traversing the entire network and increasing the system performance in terms of the efficient selection of cluster heads, reduced energy consumption, and extended network lifetime. This was carried out by studying the properties and mechanisms of biological systems and applying them in the communication systems in order to obtain the best results for a specific problem. In this comprehensive model, particle swarm optimization and a genetic algorithm are used to search for the optimal solution in any problem space in less processing time.

Keywords: microgrid; bio-inspired algorithms; wireless sensor network; genetic algorithm; particle swarm optimization; advanced metering infrastructure; blockchain; Ethereum

Citation: Swain, A.; Salkuti, S.R.; Swain, K. An Optimized and Decentralized Energy Provision System for Smart Cities. *Energies* **2021**, *14*, 1451. https://doi.org/10.3390/en14051451

Academic Editors: Gabriele Grandi and Antonio Jesus Torralba Silgado

Received: 22 January 2021
Accepted: 3 March 2021
Published: 7 March 2021

Publisher's Note: MDPI stays neutral with regard to jurisdictional claims in published maps and institutional affiliations.

1. Introduction

A smart city is an idealistic city where the quality of life and quality of services for citizens is significantly improved by promoting innovative solutions. By integrating information and communication technology (ICT) and the Internet of Things (IoT), the efficiency and effectiveness of city functionality is impressively improved. Over the years, urban life within smart cities had undergone many challenges and transformations so as to contribute to a better quality of life. In recent developments of smart cities, the quality of power has been improved significantly by developing smart grids and by the emergence of smart communities, despite the high level of energy consumption in smart cities.

Smart grids are intelligent and flexible power grids that use two-way communication technologies which enable safe, efficient and sustainable energy consumption, unlike electro-mechanically controlled conventional electric grid. Information and communication technology is used for aggregating and analyzing large amounts of data generated by various sources, such as sensor networks, wearable devices, and IoT devices deployed across the city [1]. Within this smart network, consumers can actively participate in renewable energy production and consumption at individual and community levels and finally, store the surplus energy for later use. Due to its robustness, self-healing capabilities, and cost-effective energy generation, distribution, and consumption, the usage of smart grid technology is highly encouraged in smart city transformation.

One of the key applications of smart grids is microgrids that significantly contribute to the level of reliability and resiliency of the smart grid. These are small, localized grids that can operate either as standalone systems or grid-connected ones [2]. During power disturbances or blackouts, the microgrid disconnects from the main utility network and operates autonomously in islanded mode. It consists of distributed energy sources (solar panels, wind turbines, microturbines, etc.), energy storage systems, and the cluster of loads, all of which are connected through a bidirectional, and efficient communication network. It can greatly reduce the capacity costs and energy losses during transmission and distribution, thus increasing the efficiency of the energy delivery system [3]. It can provide energy independence to individual communities who aim to manage their own power generation and distribution as well as fulfill the energy demands in remote areas.

The significant boost in the energy demand across the world has led to the formation of a new type of electricity consumers known as "prosumers" [4], who can both produce and consume electricity from renewable sources and share it within the energy community or supply the surplus energy production back to the grid if required. This is facilitated by the installation of smart microgrids, which is characterized by renewable energy integration, advanced metering infrastructure (AMI), bi-directional communication technology, distribution automation, and the monitoring and control of the entire power grid system. In order to improve the efficiency and stability of the grid network and achieve the intelligent management of smart microgrids, a large number of monitoring devices such as wireless sensors are used in the communication network [5]. A wireless sensor network (WSN) is a self-configured network of spatially dispersed sensor nodes that are used to monitor various environmental conditions such as pressure, temperature, humidity, motion, etc. and collect their data in real-time to process and forward them to the base station or sink node that acts as an interface between sensor nodes and end-users.

Sensor nodes are deployed in large numbers for a wide range of applications in the fields of IoT, smart grids, healthcare, military, surveillance, industrial sector, agriculture, transportation, and logistics, etc. However, there are still several issues and challenges associated with WSNs such as limited power, storage, and computational capabilities, localization of nodes, routing, poor maintenance, network security, and communication costs. Due to limited power supply in nodes and long communication distances, network lifetime maximization has been a critical issue in WSNs, for which certain optimization techniques are required in order to reduce data redundancy and energy consumption and increase network lifespan.

With the handling of a massive amount of data in WSNs, the privacy and security of the systems have become a major concern, since communications take place over an open channel, that is, the internet. Additionally, the systems cannot be protected beyond a certain range, since the sensor nodes operate on the limited power supply and computational resources. A large number of security solutions have been designed for the existing infrastructure, which is normally based on centralized control. There are many issues associated with centralized control, such as the involvement of intermediaries and third parties, leading to an increase in operational and transactional costs and lack of security, which makes the system vulnerable to unauthorized data tampering. Additionally, due to the rapid digitalization of the energy market, it is essential to keep track of energy transaction records in a secure and decentralized way. This can be achieved by introducing a blockchain-based energy monitoring system in which blockchain technology can be integrated with an advanced metering infrastructure (AMI), which is generally equipped with various software and hardware components such as WSNs that collect energy usage data from the smart meter and transmit them to the base station in real time. Blockchain systems can provide real-time digitally maintained energy records and make information available instantly and securely to authorized users [6].

This paper presents a smart city energy model in which smart microgrid technology is incorporated with blockchain technology to improve the energy distribution capability between citizen houses in a smart community. Since WSNs form an essential component of an AMI, this paper presents a technique using two population-based bio-inspired algorithms, i.e., particle swarm optimization (PSO) and genetic algorithm (GA), to obtain an energy-efficient strategy for cluster head selection and achieve the optimal route connecting the sensor nodes to minimize total communication distance for transmission of data in less processing time. This helps in preventing data loss and maximizing energy efficiency and lifetime of the system. The strategies used in these algorithms are simple, robust, and adaptive since they imitate the nature of biological systems to solve real-world problems.

For the secure transmission of information in the network, blockchain technology is used to maintain the security and privacy of the systems [7,8]. In this paper, this technology is mainly used in the context of peer-to-peer energy trading in microgrids, which allows prosumers to sell excess energy directly to other households in the community in a transparent and cost-effective way. A blockchain-based solution is implemented, which enables local consumers to purchase electricity directly with cryptocurrency without any third-party intervention and maintain energy transaction records in a decentralized manner in blockchain.

The remainder of this paper is organized as follows. Section 2 is divided into three sub-sections. Section 2.1 presents an overview of the literature related to wireless sensor network optimization and blockchain-based systems. The papers have been referenced as well as their authors and works are highlighted in this section. Section 2.2 provides the background study necessary to understand the concepts involved. Section 2.3 presents the proposed methodology, working, as well as its possible implementation. Section 3 contains the results and Section 4 contains the analysis of the implementation. Section 5 concludes the paper.

2. Materials and Methods

2.1. Literature Review

As the theme of the article is based on blockchain-based energy systems and optimization techniques in smart grid using bio-inspired algorithms, the entire literature review is based on two broad divisions: bio-inspired algorithms and blockchain technology for the smart energy system. Being an important building block of the smart grid technology, the wireless sensor network (WSN) has always attracted significant interest in both industries as well as research areas in recent years. In this context, much recent research has been conducted to optimize WSNs in order to improve the performance of the smart energy grid. Various applications and issues associated with the WSN in smart grids are discoursed in [9]. Several algorithms have been implemented with respect to the localization of nodes, the formation of clusters and selection of cluster heads, optimal routing, network lifetime, energy constraints, data compression and aggregation, network security, and self-organization, which are the current focus areas of research in the field of WSNs. The quality of service (QoS) protocols, energy efficiency, bandwidth utilization, and secured routing in WSN are the major areas of concern discussed in [10,11].

Several metaheuristic algorithms used for optimizing various aspects in WSNs to provide sufficiently good solutions in acceptable time constraints are presented in [12]. A firefly-based approach to perform the clustering of nodes in the WSN and the optimization of packet delivery ratio and network lifetime is proposed in [13]. In [14], the ant colony optimization (ACO) algorithm is combined with the harmonic search algorithm (HSA) to find the optimal cluster head with minimum routing path and achieve a faster transmission of packets without losing data accuracy, which results in a higher network throughput and reduced energy consumption. A novel nature-inspired algorithm named the salp swarm algorithm (SSA) is presented in [15] for accurate node localization in WSNs and is compared with four other optimization algorithms, namely the firefly algorithm, butterfly optimization algorithm, particle swarm optimization, and grey wolf optimizer based on its performance and simulation results [16]. An energy-efficient routing technique based on the artificial cee colony (ABC) algorithm that mimics the foraging behavior and waggle dance of bees is proposed in [17].

Different routing protocols in WSN are discussed and analyzed in terms of performance, limitations, and security issues in [18]. A survey based on the multipath routing techniques is presented along with its fundamental challenges and future guidelines in [19]. It also emphasizes the relevant pros and cons associated with the routing protocols which can be used as a road map for further research. An improved genetic algorithm technique is presented in [20], which optimizes the performance of mobile agents in finding the shortest route for traversing the sensor network. A hybrid GA is implemented in heterogeneous WSNs, which integrates greedy initialization and bidirectional mutation operation in order to achieve full coverage of the monitoring area and prolong the network lifetime [21]. In research conducted in [22], genetic algorithm-based clustering and routing are proposed in which both clustering and routing methods are combined into a single chromosome to construct a fitness function that improves the energy efficiency in the network. In this work, the best load balancing is achieved, which results in the lowest average energy being consumed by the cluster heads. In [23], a GA-based classification method is used to categorize the DOA (direction-of-arrival) estimates for multiple sources' position of each sensor node in a three-dimensional space. It is seen that this method could lower the computational burden to a greater extent as compared to other conventional methods without reducing the accuracy of the estimation. In [24], a multi-objective GA is proposed to optimize the problem of node deployment in WSNs based on topology, environment, the specifications of various applications and network designers' preferences.

A precise technique based on the GA-NSGAII (non-dominated sorting genetic algorithm) algorithm is designed in [25] to develop a framework for a smart cyber physical energy system which is used to manage the energy sources for electric vehicles. The cloud computing is also implemented to gather the sensing information through a Wi-Fi system from the various sensors which are deployed in the EV (electric vehicles) to realize the health monitoring system of EVs. In [26], the PSO is integrated with a multi-hop routing protocol to obtain uneven dynamic clustering in a WSN that makes the cluster distribution change dynamically upon the failure of some nodes. This improves the energy efficiency of multi-hop transmission between the base station and cluster head nodes and increases the scalability for various sizes of the network. In [27], a PSO approach is implemented to produce energy-aware clusters by the optimal selection of cluster head nodes, which is based on residual energy, least average distance from member nodes and head count of possible head nodes. The performance of this algorithm is compared with other protocols such as LEACH-C (low energy adaptive clustering hierarchy centralized) and PSO-C (particle swarm optimization centralized). In [28], a PSO technique is proposed for improving the network lifespan by forming clusters and selecting the cluster head, and the results are compared with previously proposed LEACH algorithm. Traditional clustering algorithms such as the LEACH algorithm are presented in [29], in which uniform clustering results in high communication costs in single-hop inter-cluster routing, whereas in the EEUC (energy efficient unequal clustering) algorithm [30], non-uniform clustering results in a lack of the node's residual energy consideration during the cluster head selection phase. This leads to the premature death of cluster head nodes present far away as well as near to the sink node because of the high communication cost and insufficient energy to forward data, respectively, thus affecting the overall network lifecycle. Taking the drawbacks of LEACH and EEUC algorithms into account, the PSO algorithm is presented, which considers the node's residual energy, the number of neighbors, and distance from the base station to select the cluster head node in an energy-efficient manner that can improve the quality of network monitoring and extend the network lifespan. In this paper, we mainly focus on the two nature-inspired metaheuristic algorithms, i.e., particle swarm optimization and genetic algorithm, to solve the optimization problems in WSNs for smart grid applications. GA-based route optimization is presented to obtain the shortest path connecting the cluster head nodes to the base station for the cost-effective transmission of data in less processing time. This significantly reduces the energy consumption by sensor nodes and enhances the performance and lifetime of the network.

Data security is also achieved in this work along with the enrichment of node lifetime. With the handling of the massive amount of data in WSNs, the privacy and security of energy systems also need to be achieved, since communications take place over the energy internet. Moving up to blockchain technology, different sectors such as power, energy, electrical network, and smart grid are the key areas. The literature from [31–38] focuses on the application of blockchain technology to power sectors. A vision with analysis is offered in [31] for a networked microgrid by utilizing blockchain technology for the optimum use of a power distribution system. This also suggests an additional smart contract for secured energy operation amongst local grids and network microgrids. A virtual power plant model based on blockchain technology is proposed in [32] for a distributed energy resource to reduce the cost and a secure grid connection and similarly in [33], to improve energy efficiency and power quality in a power grid blockchain expertise is implemented. The work [34] represents the optimal utilization of a power distribution energy system using blockchain technology where multiple control, operation, planning are mainly addressed, and [35] proposes the blockchain application in the electric power system area with the latest technology and innovative applications. An overview and classification of blockchain technology for the power industry is discussed with some notable projects in [36]. The appositeness and outlook of blockchain technology in the energy sector are discussed in [37] by considering the socio-technical aspects related to trading, information storage, energy flow, and service in the energy sector. In [38], the utilization of blockchain technology is discussed to form the energy communities where energy is exchanged among the prosumers.

Regarding the electrical network in a smart city, the notion of a grid consisting of many hybrid micro grids for a smart city is presented in [39] where many advantageous characteristics such as better power quality, consistency, safety, low cost etc. over a general power grid are illustrated. Pertaining to the application of blockchain towards the power grid, reference [40] proposes blockchain-based hierarchical bidding with a transaction structure for local electricity networks with the use of the Bayesian theorem and the Nash equilibrium for the accurate probability of cost and optimum quotation, respectively, and the double auction mechanism along with blockchain technology is combinedly realized in [41] for a decentralized microgrid. The blockchain (Hyperledger) with IoT devices (Raspberry Pi) are combinedly realized in [42] to generate the bill with smart metering for an electrical network. A state-of-art regarding the security issues of a blockchain-based smart city is discussed in [43] where different issues related to medical facilities, conveyance, smart grid, logistics network are focused.

Pertaining to blockchain technology in the field of the microgrid, this is discussed in [44–49]. A centralized blockchain-controlled system with advanced metering infrastructures is discussed in [44] with a bandwidth requirement comparison for a microgrid infrastructure. It is found that an about 10 times higher demand in bandwidth is found in the blockchain-based solution as compared with an advanced metering type structure. A holistic exploration is proposed in [45] about the different challenges faced in the practical implementation of blockchain-based peer-to-peer (P2P) microgrid networks. The article [46] presents a solution of double energy spending in a microgrid by applying hybrid blockchain technology by sharing a secure environment for energy exchange between the consumer and prosumer. A smart transaction is performed in [47] using the blockchain technology in the smart grid and a smart contract. In [48], a microgrid design problem is addressed by blockchain technology for a real time-based demand response program. A fuzzy optimization approach is proposed for the localization and determination of renewable generation units with dynamic pricing. This study reveals a profitable growth of 1.68% and increased consumer satisfaction of 2.61% in the area of Vietnam. The estimation of power and price with an optimum power loss for a microgrid using blockchain technology with some differential evolution algorithms is aimed in [49]. In this, the blockchain is mainly involved in reactive power pricing which is acquired from service providers.

2.2. Background Study

2.2.1. Particle Swarm Optimization

The particle swarm optimization (PSO) algorithm is a metaheuristic population-based algorithm that was proposed by James Kennedy and Russell Eberhart in 1995. It helps in obtaining the optimal solution of an optimization problem by simulating the movement and foraging of a bird flock or school of fish [11]. In this algorithm, a population of candidate solutions, named particles, explores the search landscape by initiating with a random direction and following a search strategy that results in finding the best solution. Every particle stochastically traverses a certain distance towards the personal best location (pbest) and global best location (gbest) to reach a new location (shown in Figure 1), where it updates its velocity vector and position vector for the next iteration, only if the pbest and gbest values of the current iteration are better than those of the previous iteration.

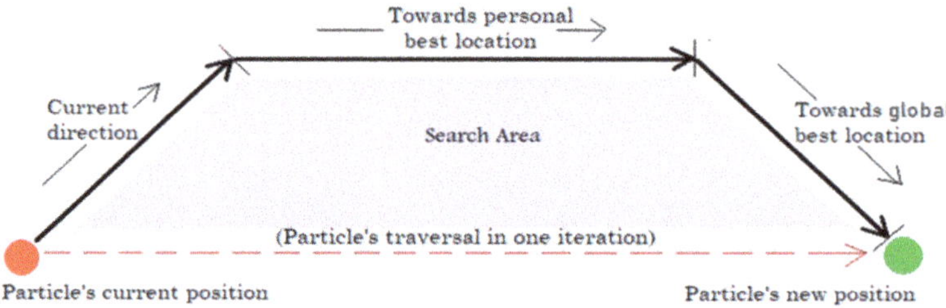

Figure 1. Particle swarm optimization (PSO) search strategy.

The particle's position and velocity are updated by using the following equation:

$$\overrightarrow{s_i^{t+1}} = \vec{s} + \overrightarrow{v_i^{t+1}} \tag{1}$$

$$\overrightarrow{v_i^{t+1}} = \omega \vec{v}_i^t + k_1 r_1 \left(\overrightarrow{pbest_i^t} - \vec{s}_i^t \right) + k_2 r_2 \left(\overrightarrow{gbest^t} - \vec{s}_i^t \right) \tag{2}$$

where t is the iteration number, $\vec{v}_i^t$ is velocity of ith particle and $\vec{s}_i^t$ is its position vector in tth iteration. Here, $\omega \vec{v}_i^t$ is the inertia, $k_1 r_1 \left(\overrightarrow{pbest_i^t} - \vec{s}_i^t \right)$ is the cognitive component and $k_2 r_2 \left(\overrightarrow{gbest^t} - \vec{s}_i^t \right)$ is the social component, where r_1 and r_2 are random components lying in the range [0, 1] and k_1 and k_2 are acceleration coefficients. Tuning the parameters such as k_1, k_2 and ω will affect the movement of particles and the exploration and exploitation of the search space. As the swarm keeps track of the best solution in the problem space and searches around them only, the time complexity of performing a search operation is greatly reduced and the probability of finding the global optimum increases, as compared to other deterministic algorithms.

2.2.2. Genetic Algorithm

The genetic algorithm (GA) is one of the first evolutionary algorithms, proposed by John Henry Holland in 1992, which mimics the Darwinian theory of evolution [16]. It was inspired by the natural selection or survival of the fittest in the biological evolution of living beings, in which genes having desirable traits are selected by nature and passed to the next generation through inheritance. In the genetic algorithm, a chromosome represents a candidate solution, and a set of candidate solutions represents a population, as shown in Figure 2. The GA is applied to an optimization problem by using the genetic representation of candidate solutions known as genes, and a fitness function or objective function. Each gene is represented by an array of bits or bit string and evaluated by the fitness function to obtain a unique value that determines the fitness of a given combination of bits with respect to the problem in consideration.

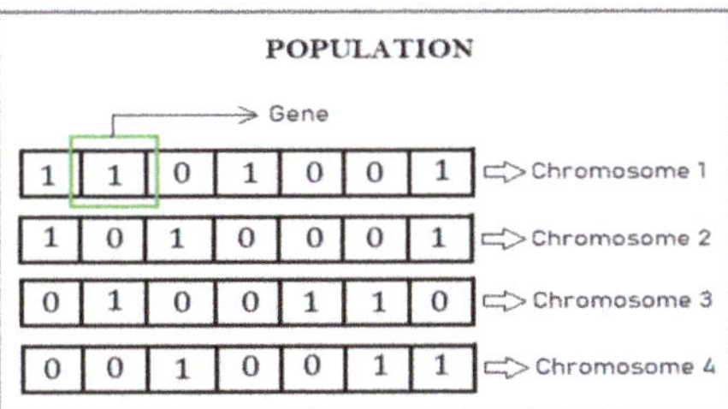

Figure 2. Representation of a population of chromosomes.

The initial population is generated with random solutions and subjected to changes in subsequent generations by applying four genetic operators to every individual solution with a certain probability, i.e., selection, crossover, mutation, and elitism. The natural selection is simulated by a roulette wheel mechanism which is implemented by first creating a sorted array of normalized fitness values of all individuals, calculating their cumulative sums, and then generating a random number in the interval [0, 1] and looping through the array of cumulative sums to find the first value greater than the random number. Each cumulative sum in the array represents the probability of the corresponding individual getting selected. Thus, individuals having cumulative sums nearly equal to 1 have greater chances of being selected since they occupy larger parts in the simulated roulette wheel.

Crossover or recombination is used to combine the genetic information of two parents by swapping their genes before and after a point which is chosen randomly in each generation, to generate a new population with a fixed number of chromosomes. The rate of crossover is determined by using the probability of crossover (P_c).

Mutations are small random changes in the genetic sequence of a chromosome that result in the addition of new features to the population and increasing its genetic variation. It also helps in preventing the GA from converging to local optima. Mutated genes that result in less fit solutions are eliminated by selection and crossover in each generation. A predefined parameter named the probability of mutation (P_m) is used to determine the rate of mutation by generating a random number in the range [0, 1] and comparing it with P_m. A gene is mutated only if P_m is greater than the random number generated.

The fourth operator, i.e., elitism, is used to preserve a small portion of the best individuals in a population, known as elites, and transfer them unchanged to the next generation in order to prevent the loss of good traits due to crossover and mutation operators. The portion of elite individuals is determined based on the elitism ratio (E_r), which can then be utilized in the process of selection, recombination, and mutation in the next generation for improving the quality of solutions. This is how the evolutionary mechanisms in the genetic algorithm are used to maximize the fitness of the population through each generation.

2.2.3. Blockchain Technology

Blockchain is one of the most cutting-edge technologies, which is a highly secure, immutable and decentralized database that stores transactional data in an encrypted manner in the form of blocks. It is a distributed ledger system consisting of a chain of blocks that keeps growing every time a transaction is carried out by users and validated by network engineers. These blocks are linked to each other through cryptographic hashing [7], as shown in Figure 3. The basic advantage of blockchain is that it is a decentralized system that operates as a layer on top of the internet, which reduces the expenses of buying servers, eliminates the intermediaries and third parties and ensures autonomy, data integrity and security to various systems [8].

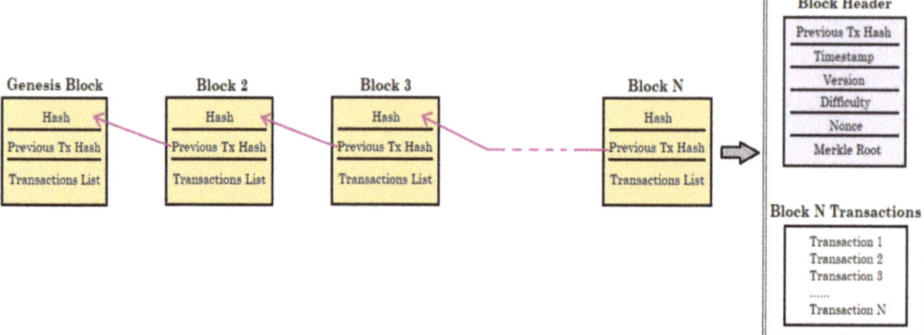

Figure 3. Structure of a block.

The encryption in the blockchain is implemented using a 256-bit hashing algorithm, for example, SHA-256 in the case of Bitcoin and ETHash in the case of Ethereum cryptocurrency, for providing a high level of security. Since each block contains the unique transaction hash of the previous block (Figure 3), tampering with the data of one block will completely change the transaction hash of that block as well as all subsequent blocks, which will require immense computing power to break all the encryptions, making it extremely difficult for attackers to manipulate the information. In order to initiate transactions as well as decrypt the secured information by the users, every blockchain user is provided with unique public and private keys which form a digital signature of the user. Blockchain also provides security at the node level. A node is a participating computer in a peer-to-peer (P2P) network in which a ledger or blockchain is stored. In a distributed ledger system, each node is provided with a copy of the ledger in order to achieve transparency and consensus among other nodes within the network at the time of mining and adding a new block to the ledger. Mining is the process of verifying and validating the transactions by solving the proof of work (PoW) consensus algorithm before adding them to the blockchain. In the case of data tampering in one node, the ledger in that node will automatically become invalid and get discarded by other nodes, due to the presence of a copy of the ledger in every node that will create an inconsistency with the tampered node.

One of the popular and leading blockchain-based platforms is Ethereum, which is an open-source, public blockchain, developed by Vitalik Buterin in 2015, that features smart contract functionality. The smart contract is an immutable computer program written mainly in Solidity programming language in case of Ethereum, which cannot be changed once deployed on the blockchain and is executed only upon meeting a certain predefined set of instructions or conditions written in the program [40]. It enables users to build and deploy decentralized applications (dApps) on top of the Ethereum blockchain, which are controlled by the logic written into the smart contract. Hence, every process in the application is automated and the system verifies and validates all the transactions without any third-party intervention.

The decentralized digital currency of Ethereum is 'Ether' (ETH) which is currently worth INR 123,743.91. It is used to carry out all the transactions on a P2P network and to deploy smart contracts and run dApps. In order to perform a transaction or execute certain tasks, a smaller unit of the Ethereum token, known as "Gas" (1 Gwei = 10^{-9} ETH), is used as a fuel to drive the process of adding a block to the public ledger so as to compensate for the computational power spent by the miners to verify and validate the transactions. This is how the consensus mechanism and hence a trustless computation works in blockchain technology.

2.3. Proposed Work

In order to promote renewable energy adoption in households and optimize their local energy resources efficiently, a smart energy community model is proposed in which a coalition of neighboring prosumers is formed based on an agreed sharing mechanism which is implemented as a smart contract on blockchain. In this model, blockchain technology is integrated with the smart microgrid for clean and decentralized energy trading between potential prosumers and consumers within the community, as shown in Figure 4. It mainly consists of energy-independent households in an interconnected energy network and decentralized storage systems that are coupled with distributed renewable energy generations of each prosumer. For the intelligent management of the microgrid, it is incorporated with a wireless sensor network and advanced metering infrastructure which monitor the generation, transmission, and consumption of electricity in real-time and send energy transaction information to the blockchain via smart contract.

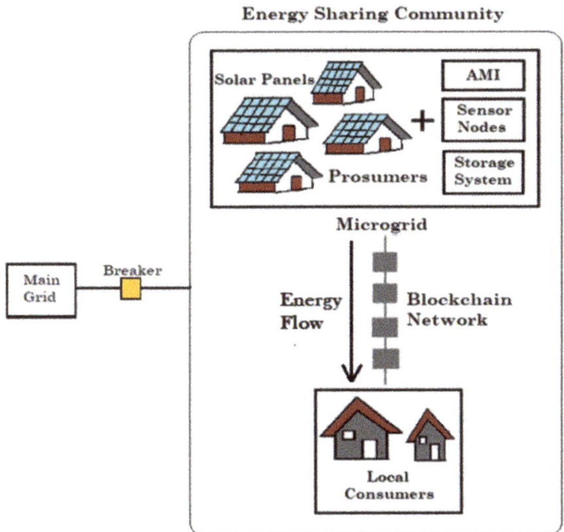

Figure 4. A smart energy community model. AMI: advanced metering infrastructure.

In this model, the microgrid acts as a small energy market in which prosumers can change their roles to sellers or buyers, depending on their net energy profiles in different time periods. For example, if the load demand of prosumers is not met by their own power generation, they can buy electricity from other prosumers in the network who have surplus energy production. The excess energy can also be exported to the utility grid for which prosumers are incentivized in the form of a feed-in tariff for generating electricity from renewable sources. The rate at which energy is shared among the households is determined by the amount of energy produced by PV prosumers on a daily basis. The energy pricing will be based on the availability of generation sources, demand, economic cost and pricing regulations, considering the power consumption flexibility of prosumers. The smart contract regulates the energy sharing activities by taking into account the energy prices agreed among all the prosumers and managing their energy supply in the community. This peer-to-peer energy trading mechanism results in sharing energy economically among the households and maximizing the consumption of power from distributed energy systems, while reducing the impact on utility grid.

Apart from managing the P2P energy sharing, it is required to handle the complexities in energy systems and uncertainty of renewable energy by monitoring and reporting the instantaneous energy flows in real time. In the proposed system, a blockchain-based advanced metering infrastructure is considered, which consists of various sensor nodes that collect energy usage data from the smart meters installed in all the houses and transmit it to the blockchain database system in the decentralized data and control center (DCC) by means of a two-way communication network in real-time, as shown in Figure 5. The energy information and transactions are securely stored in the blockchain ledger to track the electricity usage for analysis and management of energy systems without the aid of a central authority. It also enables secured access to authorized users, in this case prosumers and consumers, participating in the blockchain network.

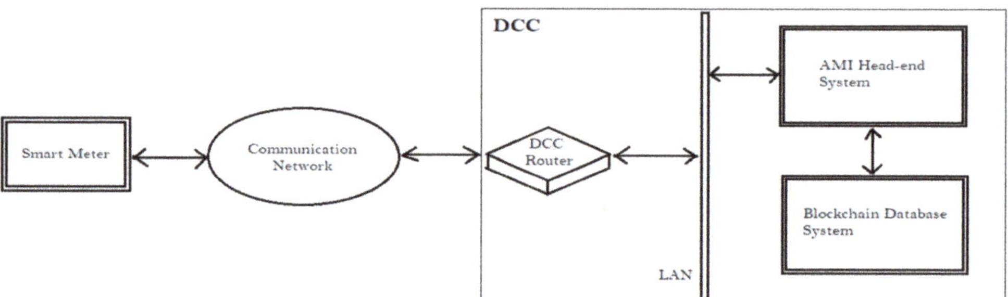

Figure 5. Blockchain-based AMI. DCC: data and control center.

In order to regulate the peer-to-peer energy trading in our smart city model, a wireless sensor network is extensively used for monitoring the energy systems and disseminating the data in a timely manner to the base station. Since energy and lifetime are two major constraints in developing a sensor network, an optimization technique is proposed for the optimal selection of cluster heads and energy-efficient routing in the network that extends node lifetime and improves the performance of the network.

Initially, the sensor nodes are deployed in random locations within an area and each node is equipped with a finite energy supply. The base station is located outside the sensing area and it is grid powered, for which it has an unlimited supply of energy. In our proposed methodology, it is assumed that the WSN is a homogeneous network, the sensor nodes, and base station are stationary after deployment and all the nodes, except the sink node, have the same amount of initial energy and are left unattended once deployed. These nodes interact with each other to aggregate data and transmit it to the target nodes at regular intervals. Due to limited energy constraints, these nodes are grouped into clusters, where the most resourceful node in each cluster is selected as the cluster head (CH) for energy-efficient communication within the network, as shown in Figure 6.

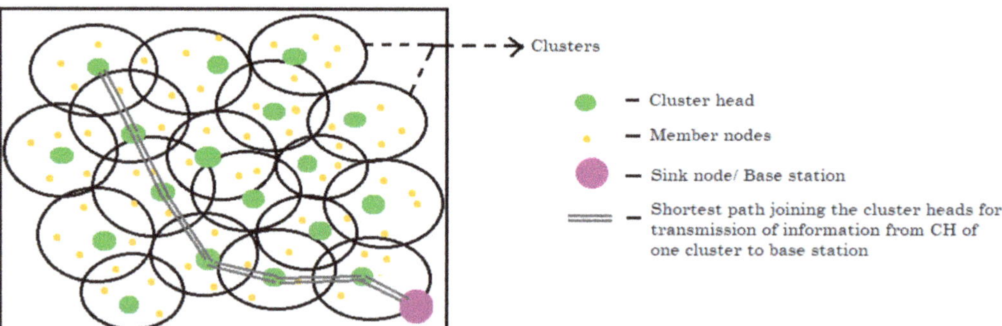

Figure 6. Cluster formation and optimal routing. CH: cluster head.

Each node in the network can operate as both sensors as well as a cluster head. A cluster head is selected based on the node's residual energy, number of neighbor nodes, and the Euclidean distance to the base station. The member nodes in a cluster transmit their data to the CH which then processes the information and forwards it to the base station directly or through multiple CHs by using the shortest path from the CH to the base station. This is carried out to minimize the cost of communication and the draining of energy during data transmission. In the proposed approach, an energy-efficient selection of the CH in a cluster of nodes is carried out by using the PSO algorithm, and the optimal route connecting the CHs to the base station is achieved by using the genetic algorithm.

The PSO algorithm is implemented by creating a cost function or objective function for the efficient selection of a cluster head. An ideal cluster head must have the highest residual energy, least average distance from the neighbor nodes, and minimum distance from the base station. So, the structure of the cost function is as follows:

$$\text{Cost function, } f = \alpha * d_m + \beta * E_r + \gamma * d_b \tag{3}$$

where, d_m is the average Euclidean distance of a node from all other member nodes in a cluster, i.e., $d_m = \sqrt{(xi - xj)^2 + (yi - yj)^2}$, d_b is the distance from the base station and E_r is the total residual energy of all alive member nodes divided by the residual energy of node in consideration. The residual energy of a node will be maximum if the factor [sum of residual energy of all memner nodes/residual energy of current node] is minimum. α, β and γ are the constants having values equal to 0.38, 0.38 and 0.18 respectively. So, all the three factors sum-up to the structure of the cost function. The aim is to minimize the cost function that will give the optimal location of the cluster head which has the maximum residual energy and minimum distance from the member nodes and base station. The steps involved in the PSO algorithm are as follows:

1. Initialization of a set of particles and their position, velocity, and residual energy;
2. Initialization of PSO parameters such as swarm size, number of iterations, inertia weight, personal acceleration, and social acceleration coefficients;
3. Calculation of the cost function of each particle;
4. Finding the personal best and global best location;
5. Updating particles' position, velocity and energy lost in each iteration;
6. Updating the particle with the least value of cost function as the cluster head in every iteration;
7. Repeating steps 3 to 6 till all the nodes become dead.

After nominating the cluster heads, data packets are needed to be sent from the cluster head node to the base station directly or through a series of other cluster heads in the network. Therefore, the optimal route is needed to be determined to reduce the cost of communication of data packets across the nodes and balance the energy and data load in the network. For this, we have used the genetic algorithm to find the shortest route that connects a set of sensor nodes from the source node to the sink node and visits every node exactly once.

In order to implement the genetic algorithm, a set of points representing sensor nodes (CH) are randomly placed on a 2-D space. The Euclidean distance between consecutive nodes serves as the fitness function for the algorithm. A population of random solutions is generated in which each candidate solution or chromosome represents a route. The steps involved in the genetic algorithm to find the shortest route connecting all the points are as follows:

1. Initialization of population;
2. Calculation of fitness value of each chromosome;
3. Selection of best chromosomes as parents for the next generation using the roulette wheel mechanism;
4. Crossover of parent chromosomes;
5. Mutation of chromosomes;

6. Evaluating fitness of each chromosome in the new population;
7. Repeating steps 3 to 6 until the shortest path is established.

With the optimization of wireless sensor networks, we will be able to monitor the energy consumption in urban households in order to regulate decentralized energy trading within the community. To implement P2P energy trading in our model, prosumers are equipped with residential PV panels and energy storage systems to produce and store clean energy and supply the excess energy to local consumers by using a P2P energy trading platform. Therefore, a blockchain-based web application powered by smart contracts is developed, which enables houses to sell and purchase clean energy without any third-party intervention. It enables authorized users' easy access to the Ethereum blockchain network and provides each node with a copy of the ledger. Thus, secured communication between energy suppliers and buyers is achieved through the AMI and distributed ledger feature of the blockchain under a decentralized keyless signature scheme.

The blockchain web application is a prototype for decentralized trading systems that allows local consumers to purchase energy directly from prosumers with Ether cryptocurrency on the web app. A model of this web app is shown in Figure 7.

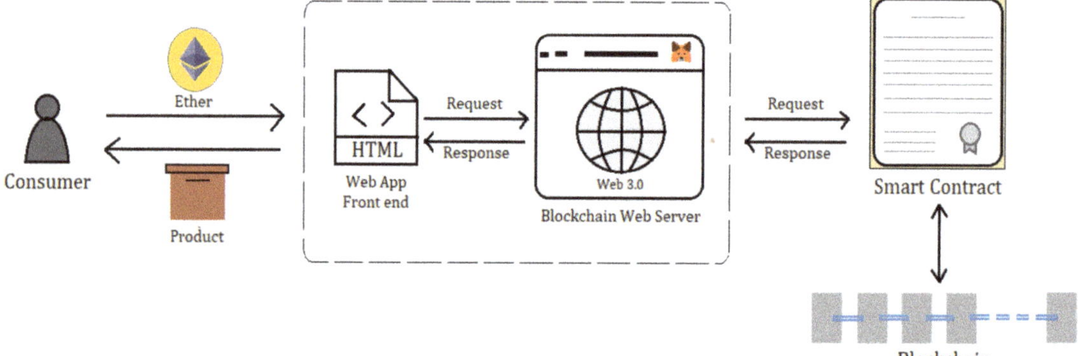

Figure 7. Blockchain web app model.

The front end of the web app is built using the HTML5 and React library of JavaScript that creates an interactive user interface for the web app. The web3.js library is used to develop a website that can interact with the Ethereum blockchain. A smart contract, which contains a predefined set of rules and agreement between buyer and seller embedded into lines of code, is created using Solidity programming language and deployed on Ethereum. In this web app prototype, we have used "Ganache" that creates and runs a virtual Ethereum node on our PC and allows us to develop, deploy and test decentralized applications (or dApps) without spending any real money. The Ganache user interface contains a list of user accounts, each credited with a balance of 100 ETH, along with unique public and private key for each account, as shown in Figure 8. These accounts are considered as the Ethereum accounts of prosumers and consumers, which can be used for transactions in the web app.

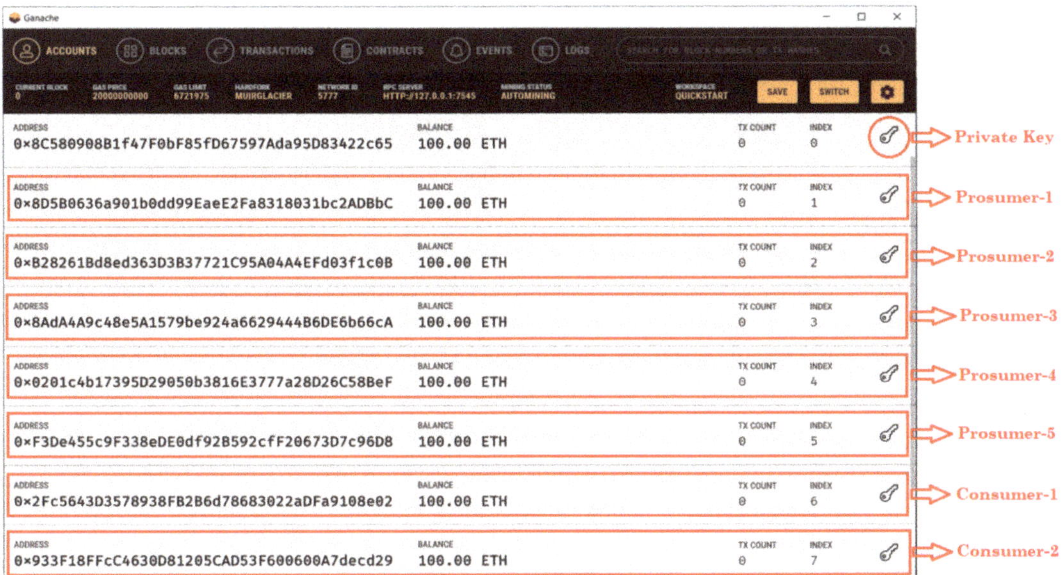

Figure 8. Ganache console.

In order to access our dApp and interact with the Ganache blockchain from the web app, we have used the "MetaMask" browser extension that serves as an Ethereum wallet and works as a bridge between regular web browsers (such as Chrome, Firefox, etc.) and Ethereum blockchain. To connect MetaMask to our virtual Ethereum blockchain, all the Ganache accounts are imported into the MataMask wallet using the private key of each user, and the default Main Ethereum Network of MetaMask is changed to Localhost 7545.

3. Results

This section describes the experimental setup and various parameters of the proposed protocol and also demonstrates their results. The proposed method of WSN optimization is implemented in the MATLAB 2020b environment using a Windows 10-based PC with 2.71 GHz, Intel Core i5 processor, and 8 GB RAM. In order to implement the PSO algorithm for the optimal selection of a cluster head, the following parameter settings are considered;

- Maximum number of iteration = 30,
- Swarm size = 10,
- Inertia coefficient = 0.9,
- Personal acceleration coefficient = 2,
- Social acceleration coefficient = 2,
- Initial energy of a node = 45 units,
- Location of the base station on the graph = (40, 40).

The inertia coefficient is varied from 0.9 to 0.3 over 30 iterations. The particles explore the search landscape, optimizing the objective function for finding the best location of cluster head and in every iteration, the best particle and its corresponding cost function value are obtained, as shown in Figure 9. The node nearest to the optimal location of the best particle is selected as the cluster head, which has the highest residual energy, maximum number of neighbor nodes, and the smallest distance from the base station.

Iteration # 1: Least Cost = 14.842: Particle =9	Iteration # 16: Least Cost = 3.8421: Particle =3
Iteration # 2: Least Cost = 14.1673: Particle =10	Iteration # 17: Least Cost = 3.6896: Particle =8
Iteration # 3: Least Cost = 13.6036: Particle =10	Iteration # 18: Least Cost = 3.3815: Particle =4
Iteration # 4: Least Cost = 12.9836: Particle =8	Iteration # 19: Least Cost = 3.3423: Particle =10
Iteration # 5: Least Cost = 12.0149: Particle =8	Iteration # 20: Least Cost = 3.1299: Particle =5
Iteration # 6: Least Cost = 11.5106: Particle =8	Iteration # 21: Least Cost = 2.9925: Particle =9
Iteration # 7: Least Cost = 10.5528: Particle =3	Iteration # 22: Least Cost = 2.6684: Particle =7
Iteration # 8: Least Cost = 9.5757: Particle =10	Iteration # 23: Least Cost = 1.5109: Particle =6
Iteration # 9: Least Cost = 8.3691: Particle =4	Iteration # 24: Least Cost = 1.0903: Particle =2
Iteration # 10: Least Cost = 7.5795: Particle =9	Iteration # 25: Least Cost = 0.65428: Particle =1
Iteration # 11: Least Cost = 6.7352: Particle =5	Iteration # 26: Least Cost = 0.65428: Particle =1
Iteration # 12: Least Cost = 5.9557: Particle =7	Iteration # 27: Least Cost = 0.65428: Particle =1
Iteration # 13: Least Cost = 5.0905: Particle =6	Iteration # 28: Least Cost = 0.65428: Particle =1
Iteration # 14: Least Cost = 4.7966: Particle =1	Iteration # 29: Least Cost = 0.65428: Particle =1
Iteration # 15: Least Cost = 3.9131: Particle =2	Iteration # 30: Least Cost = 0.65428: Particle =1

Figure 9. Location of the CH and its corresponding cost function value.

Due to energy dissipation by both cluster head and non-cluster head nodes caused by the transmission of data, the total energy consumed by the alive nodes gradually decreases in every round, as shown in Figure 10.

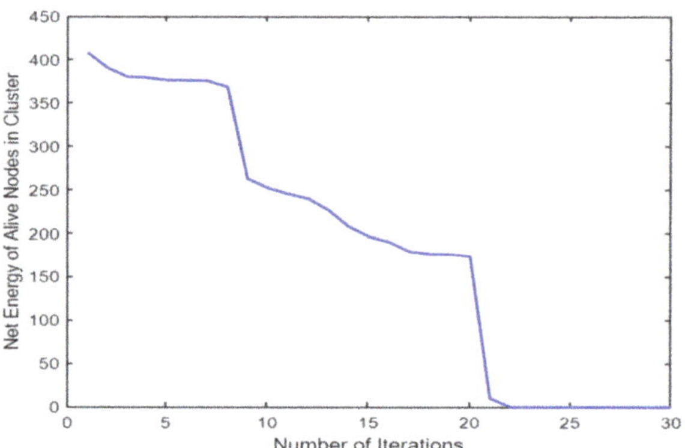

Figure 10. Decrease in total residual energy of alive nodes with the iterations as per the PSO algorithm.

It is evident from the above results that with a swarm size of 10 particles and 30 iterations, all the nodes die, or the total residual energy becomes zero on iteration number 25. In contrast to the EEUC algorithm which does not consider the node's residual energy during the cluster head selection phase, the proposed PSO technique gives better results, since EEUC's non-uniform clustering technique leads to the premature death of cluster head nodes present far away as well as near to the sink node due to the high communication cost and insufficient energy to forward data, which affects the overall network lifecycle. The proposed PSO algorithm selects a cluster head node in an energy-efficient manner that can improve the quality of network monitoring and extend the network lifespan.

After selecting the cluster head nodes using the PSO algorithm approach, an optimal path is established connecting all the CH nodes to the base station for faster transmission of data from sending node to receiving node. This is carried out by using the genetic algorithm to find the shortest route in the network. To implement this algorithm, initially, a group of 20 sensor nodes is randomly placed on a 50 × 50 sq. unit area, assuming each node except the last node is a cluster head, as shown in Figure 11.

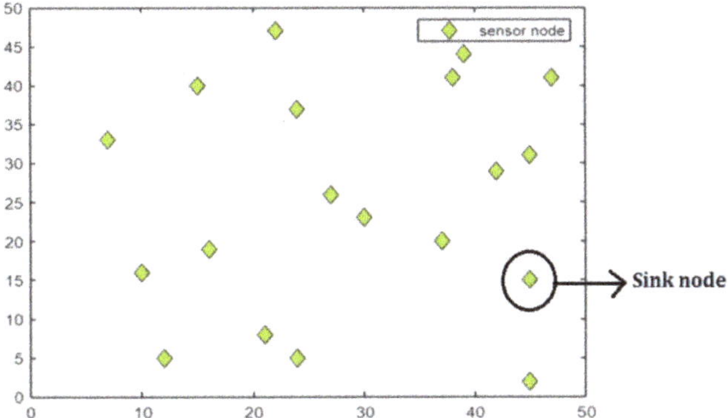

Figure 11. Position of sensor nodes.

The following parameter settings are considered for the proposed algorithm;

- Size of population = 75,
- Number of generations = 100,
- Mutation rate = 0.04,
- Crossover rate = 0.78.

The genetic algorithm improves the individual solutions in every generation by using selection, crossover, and mutation operators and creates a sequence of new populations. The algorithm thus establishes the shortest route in the network, as shown in Figure 12.

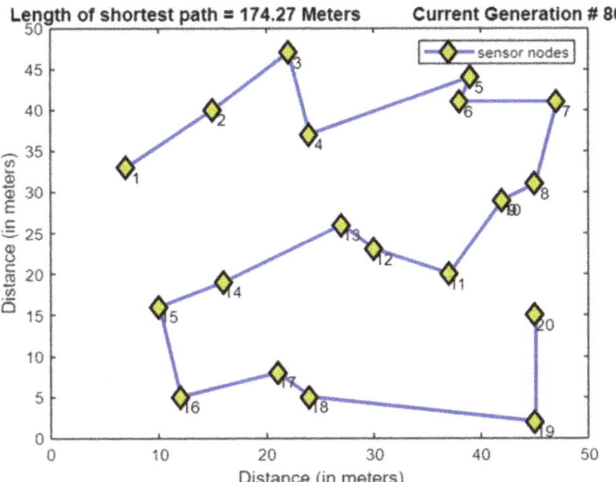

Figure 12. Shortest path obtained by the genetic algorithm (GA).

The GA determined the optimal path of length 174.27 m in only 86 generations. The performance of this algorithm can be visualized from the path length versus generations graph shown in Figure 13.

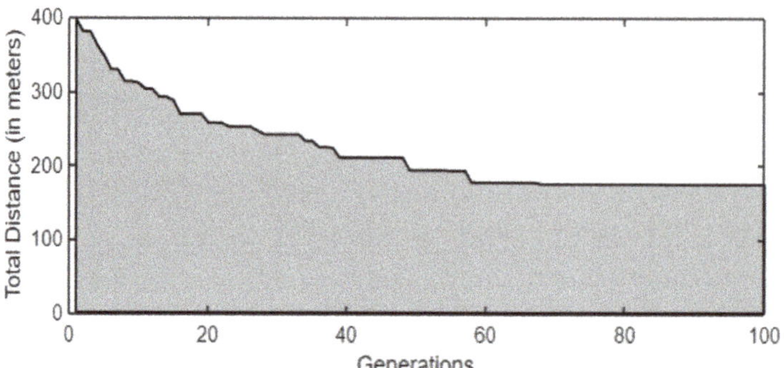

Figure 13. Distance versus generation graph.

After successfully optimizing the wireless sensor networks, P2P energy trading can be carried out between households by implementing a real-time decentralized application using the blockchain-based web app. The user interface of our web application is shown in Figure 14. In this application, it is assumed that the average energy produced by a single PV prosumer on a good day of 5 h of direct sunlight is 38 kW-h. As per the survey by the US Energy Information Administration (EIA) in 2019, the cost of 1 kW-h of electricity is USD 0.14 (i.e., 0.0000938 ETH) for households and their average power consumption in a day is 29 kW-h. Based on this information and our assumption, the energy pricings and supply are produced accordingly on this web app.

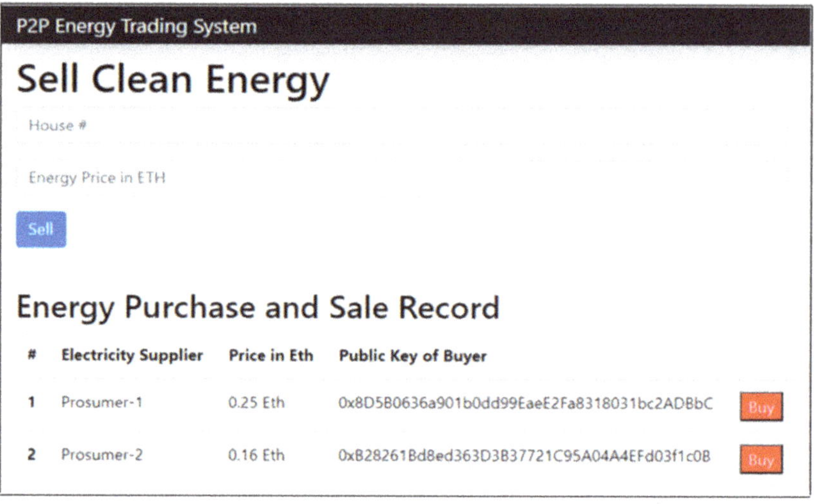

Figure 14. Web App UI in MetaMask browser.

Using this web application, prosumers can sell clean energy by adding the price of electricity supplied by them in the given field and selling it on the web app by clicking on the "sell" button. A block will be created containing the digital signature of the seller and the energy information in the Ganache blockchain and consequently a record containing the electricity price in Ether and the public key of the owner will be added in the web app. A "buy" button will be added across each record for other users to buy electricity directly from the web app. As soon as a consumer purchases energy from a prosumer, the public key of the seller will change into the consumer's public key, which indicates a transfer of ownership of energy from prosumer to consumer. The "buy" button is the payment link that will redirect to the MetaMask wallet when clicked by respective consumers due to contract interaction, where they have to sign their transactions in order to confirm payment of Ethers. Thus, blocks with encrypted transaction details will be created and added to Ganache blockchain and the red "buy" buttons will automatically change to green "sold" marks, as shown in Figure 15, which signifies the successful transaction of Ethers from the consumer's account to the seller's account. The smart contract is written in a way that restricts sellers from buying their own energy; it will throw a transaction error in MetaMask which means the block is discarded and the transaction is not successful.

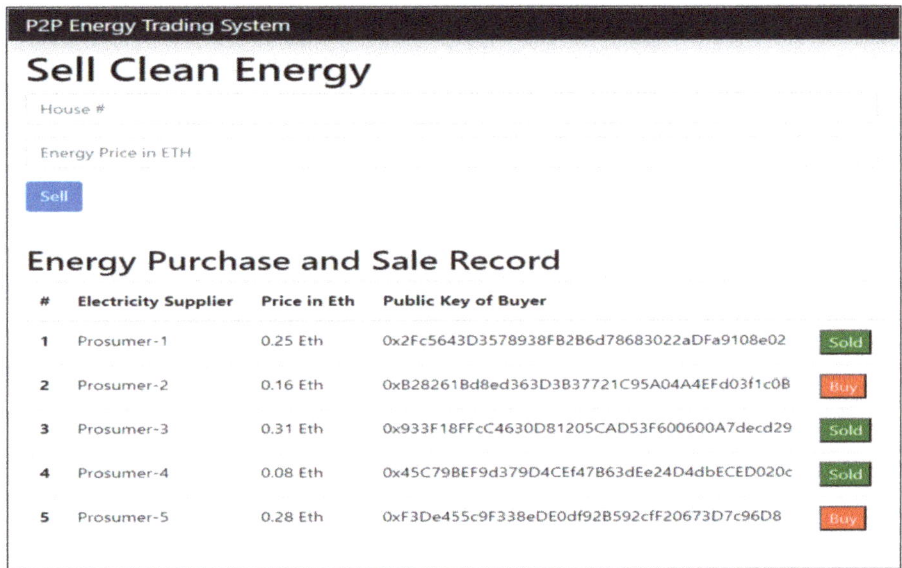

Figure 15. Energy transaction record in the web app.

We can view a list of transactions in the form of blocks wrapped with a transaction hash in the Ganache blockchain, as shown in Figure 16. With this method, we can securely store the history of transactions, which is beneficial for audit.

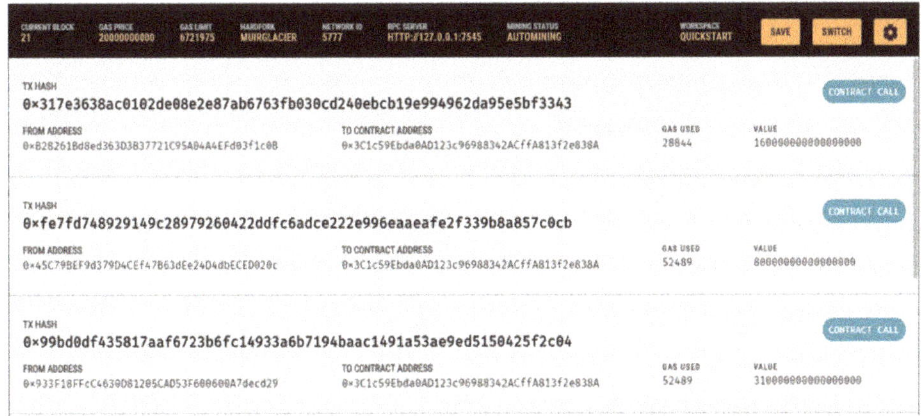

Figure 16. Blocks encrypted with 256-bit transaction hash.

We can also implement the proposed model on a public blockchain network such as the "Kovan test network" where the households can pay real Ethers for grid energy transactions directly through the web app if they are connected to a public blockchain network.

4. Discussion

It is evident from the above work on WSN optimization that the framework of both the GA and the PSO is similar, in that they initiate with a population of random solutions, explore and exploit the search space with specific parameters and estimate the global optimum for a given optimization problem. The control parameters of each algorithm depend on the nature of a problem and thus, selecting their appropriate values can have a great impact on the performance of the algorithms. In our proposed methodology, the population size in the GA should not be very large since it can slow down the algorithm, increasing the computational time significantly. The diversity of the population should be maintained in order to avoid premature convergence. For this, the mutation rate should be kept small to search locally around the promising regions found by the crossover operator without affecting the convergence of GA. Similarly, choosing appropriate values for the initial and final inertia weight in the PSO can tune its global and local searchability. Additionally, smaller values of acceleration constants c_1 and c_2 may result in the termination of the algorithm before finding a good solution and larger values can increase the acceleration, resulting in the abrupt movement of particles over the search space. So, the summation of c_1 and c_2 should be maintained less than or equal to 4 for the smooth trajectories of particles. Therefore, the parameters in these stochastic algorithms should be set carefully. Even though the genetic algorithm is easy to implement and uses less computational resources, it might not give the best results in the case of complex problems. In the case of PSO, the algorithm is simple, efficient, and robust, however, it is computationally expensive as it consumes a lot of memory, which can limit its usage in high-speed applications. Based on these limitations, future work can be carried out by developing more efficient computational intelligence strategies that can improve data aggregation and transmission and prolong network lifetime in heterogeneous sensor networks.

From the proposed work on blockchain technology in smart grid applications, it is seen that the use of decentralized energy trading platform ensures data security and immutability and offers tamper-proof transactions. The use of smart contracts in this system makes end-to-end power delivery efficient and simpler by making executed trades automatically validated and recorded in the blockchain, thereby significantly reducing delays and cost of the settlement. It empowers consumers with transparency and flexibility over transactions by eliminating intermediaries and third parties and enables prosumers to actively participate in the energy market. However, the optimal usage of blockchain technology is prevented due to its nascent stage of evolution and experimental stage of blockchain-based projects. Hence, future work can be carried out by increasing the scalability in the blockchain, introducing incentive mechanisms for using blockchain-based applications, encouraging decentralized power generation, and making this technology commercially viable.

5. Conclusions

The main goal of a smart city is to improve the quality of life of urban residents by making better utilization of resources while reducing energy usage and operational costs. In this work, an optimized and decentralized energy model for smart cities is presented that deals with the optimization of wireless sensor networks in the smart grid and facilitates peer-to-peer energy sharing among prosumers in a decentralized energy trading system. A bio-inspired approach for energy-efficient routing in a wireless sensor network is also realized by using particle swarm optimization and the genetic algorithm, which can greatly minimize the energy consumption by sensor nodes and extend network lifetime. This also helps in disseminating the data in a timely manner to the base station by finding the shortest path connecting the cluster head nodes in the network for data transmission. The privacy of a massive amount of data and security of communication networks is achieved by integrating blockchain technology with the smart grid. With the aid of smart contracts, energy transactions are maintained securely on the blockchain in a consensus manner. The proposed model for a peer-to-peer energy trading system not only promotes the sustainable use of energy but also encourages the use of renewable energy sources in microgrids.

Author Contributions: Conceptualization, A.S. and K.S.; methodology, S.R.S.; software, A.S.; validation, A.S., K.S. and S.R.S.; formal analysis, A.S.; investigation, K.S.; resources, A.S.; data curation, K.S.; writing—original draft preparation, K.S.; writing—review and editing, A.S.; visualization, A.S.; supervision, S.R.S.; project administration, A.S.; funding acquisition, S.R.S. All authors have read and agreed to the published version of the manuscript.

Funding: This research was funded by WOOSONG UNIVERSITY's Academic Research Funding (2020–2021).

Institutional Review Board Statement: Not applicable.

Informed Consent Statement: Not applicable.

Data Availability Statement: Not applicable.

Acknowledgments: We acknowledge Woosong University, South Korea for their support in carrying out this research work.

Conflicts of Interest: The authors declare no conflict of interest.

References

1. Gungor, V.C.; Sahin, D.; Kocak, T.; Ergüt, S.; Buccella, C. Smart Grid Technologies: Communication Technologies and Standards. *IEEE Trans. Ind. Inform.* **2011**, *7*, 529–539. [CrossRef]
2. Warsi, N.A.; Siddiqui, A.S.; Kirmani, S.; Sarwar, M.D. Impact Assessment of Microgrid in Smart Cities: Indian Perspective. *Technol. Econ. Smart Grids Sustain. Energy* **2019**, *4*, 1–16. [CrossRef]
3. Hirsch, A.; Parag, Y.; Guerrero, J. Microgrids: A review of technologies, key drivers, and outstanding issues. *Renew. Sustain. Energy Rev.* **2018**, *90*, 402–411. [CrossRef]
4. Rathnayaka, V.M.; Potdar, A.J.D.; Kuruppu, S.J. An innovative approach to manage prosumers in smart grid. In Proceedings of the 2011 World Congress on Sustainable Technologies (WCST), London, UK, 7–10 November 2011.
5. Sharma, P.; Pandove, G. A Review Article on Wireless Sensor Network in Smart Grid. *Int. J. Adv. Res. Comput. Sci.* **2017**, *8*, 1–5.
6. Mylrea, M.; Gourisetti, N.G. Blockchain for smart grid resilience: Exchanging distributed energy at speed, scale and security. In Proceedings of the Conference Resilience Week (RWS), Wilmington, DE, USA, 18–22 September 2017.
7. Nakamoto, S. Bitcoin: A Peer-to-Peer Electronic Cash System. Available online: www.bitcoin.org (accessed on 31 March 2009).
8. Brilliantova, V.; Thurner, T.W. Blockchain and the future of energy. *Technol. Soc.* **2019**, *57*, 38–45. [CrossRef]
9. Yide, L. Wireless Sensor Network Applications in Smart Grid: Recent Trends and Challenges. *Int. J. Distrib. Sens. Netw.* **2012**, *8*, 492819. [CrossRef]
10. Kumar, J.; Rishiwal, V.; Ansari, M.I. Quality of Service in Wireless Sensor Networks: Imperatives and Challenges. *Int. J. Sens. Wirel. Commun. Control* **2019**, *9*, 419–431. [CrossRef]
11. Shi, Y. Particle swarm optimization: Developments, applications and resources. In Proceedings of the Congress on Evolutionary Computation, Seoul, South Korea, 27–30 May 2001.
12. Gouda, O.; Nassif, A.B.; Talib, M.A.; Nasir, Q. A Systematic Literature Review on Metaheuristic Optimization Techniques in WSNs. *Int. J. Math. Comput. Simul.* **2020**, *14*, 187–192.
13. Manshahia, M.S.; Dave, M.; Singh, S.B. Firefly algorithm based clustering technique for Wireless Sensor Networks. In Proceedings of the International Conference on Wireless Communications, Signal Processing and Networking (WiSPNET), Chennai, India, 23–25 March 2016.
14. Poonguzhali, P.K.; Ananthamoorthy, N.P. Improved energy efficient WSN using ACO based HSA for optimal cluster head selection. *Peer-to-Peer Netw. Appl.* **2020**, *13*, 1102–1108. [CrossRef]
15. Kanoosh, H.M.; Houssein, E.H.; Selim, M.M. Salp Swarm Algorithm for Node Localization in Wireless Sensor Networks. *J. Comput. Netw. Commun.* **2019**, *4*, 1–12. [CrossRef]
16. Haldurai, L.; Madhubala, T.; Rajalakshmi, R. A Study on Genetic Algorithm and its Applications. *Int. J. Comput. Sci. Eng.* **2016**, *4*, 139–143.
17. Fahim, H.; Javaid, N.; Khan, Z.A.; Qasim, U.; Javed, S.; Hayat, A.; Iqbal, Z.; Rehman, G. Bio-inspired Routing in Wireless Sensor Networks. In Proceedings of the 9th International Conference on Innovative Mobile and Internet Services in Ubiquitous Computing, Blumenau, Brazil, 8–10 July 2015.

18. Majumdar, A.; Sarkar, D. Various types of routing protocols in Wireless Sensor Network with vulnerabilities: A review. In Proceedings of the International Conference and Workshop on Computing and Communication (IEMCON), Vancouver, BC, Canada, 15–17 October 2015.
19. Radi, M.; Dezfouli, B.; Bakar, K.A.; Lee, M. Multipath Routing in Wireless Sensor Networks: Survey and Research Challenges. *Sensors* **2012**, *12*, 650–685. [CrossRef]
20. Wang, J.; Guan, T.; He, H.; Chen, Y. Research of Improved GA in WSN Mobile Agent Routing Algorithm. In Proceedings of the International Conference on Computational Intelligence and Software Engineering, Wuhan, China, 11–13 December 2009.
21. Li, J.; Luo, Z.; Xiao, J. A Hybrid Genetic Algorithm with Bidirectional Mutation for Maximizing Lifetime of Heterogeneous Wireless Sensor Networks. *IEEE Access* **2020**, *8*, 72261–72274. [CrossRef]
22. Tianshu, W.; Gongxuan, Z.; Xichen, Y.; Ahmadreza, V. Genetic algorithm for energy-efficient clustering and routing in wireless sensor networks. *J. Syst. Softw.* **2018**, *146*, 196–214.
23. Zhang, Y.; Wu, Y.I. Multiple Sources Localization by the WSN using the Direction-of-Arrivals Classified by the Genetic Algorithm. *IEEE Access* **2019**, *7*, 173626–173635. [CrossRef]
24. Bouzid, S.E.; Seresstou, Y.; Raoof, K.; Omri, M.N.; Mbarki, M.; Dridi, C. MOONGA: Multi-Objective Optimization of Wireless Network Approach Based on Genetic Algorithm. *IEEE Access* **2020**, *8*, 105793–105814. [CrossRef]
25. Tehrani, K. A smart cyber physical multi-source energy system for an electric vehicle prototype. *J. Syst. Archit.* **2020**, *111*, 101804. [CrossRef]
26. Ruan, D.; Huang, J. A PSO-Based Uneven Dynamic Clustering Multi-Hop Routing Protocol for Wireless Sensor Networks. *Sensors* **2019**, *19*, 1835. [CrossRef]
27. Singh, B.; Lobiyal, D.K. *Energy-Aware Cluster Head Selection Using Particle Swarm Optimization and Analysis of Packet Retransmissions in WSN*; Elsevier: Amsterdam, The Netherlands, 2012; Volume 4, pp. 171–176.
28. Yadav, A.; Kumar, S.; Singh, V. Network Life Time Analysis of WSNs Using Particle Swarm Optimization. *Int. Conf. Comput. Intell. Data Sci.* **2018**, *132*, 805–815. [CrossRef]
29. Al-Baz, A.; El-Sayed, A. A new algorithm for cluster head selection in LEACH protocol for wireless sensor networks. *Int. J. Commun. Syst.* **2018**, *31*, 1–13. [CrossRef]
30. Arjunan, S.; Sujatha, P. A survey on unequal clustering protocols in Wireless Sensor Networks. *J. King Saud Univ. Comput. Inf. Sci.* **2019**, *31*, 304–317. [CrossRef]
31. Li, Z.; Bahramirad, S.; Paaso, A.; Yan, M.; Shahidehpour, M. Blockchain for decentralized transactive energy management system in networked micro grids. *Electr. J.* **2019**, *32*, 58–72. [CrossRef]
32. Shao, W.; Xu, W.; Xu, Z.; Liu, B.; Zou, H. A Grid Connection Mechanism of Large-scale Distributed Energy Resources based on Blockchain. In Proceedings of the Chinese Control Conference (CCC), Guangzhou, China, 27–30 July 2019.
33. Zamfirescu, A.; Şuhan, C.; Golovanov, N. Blockchain Technology Application in Improving of Energy Efficiency and Power Quality. In Proceedings of the 54th International Universities Power Engineering Conference (UPEC), Bucharest, Romania, 3–6 September 2019.
34. Adeyemi, A.; Yan, M.; Shahidehpour, M.; Botero, C.; Guerra, A.V.; Gurung, N.; Zhang, L.; Paaso, A. Blockchain technology applications in power distribution systems. *Electr. J.* **2020**, *33*, 106817. [CrossRef]
35. Silvestre, M.L.D.; Gallo, P.; Guerrero, J.M.; Musca, R.; Sanseverino, E.R.; Sciumè, G.; Vásquez, J.C.; Zizzo, G. Blockchain for power systems: Current trends and future applications. *Renew. Sustain. Energy Rev.* **2020**, *119*, 109585. [CrossRef]
36. Jabbarpour, M.R.; Joozdani, M.R.; Farshi, S.S. Blockchain Applications in Power Industry. In Proceedings of the 28th Iranian Conference on Electrical Engineering (ICEE), Tabriz, Iran, 4–6 August 2020.
37. Teufel, B.; Sentic, A.; Barmet, M. Blockchain energy: Blockchain in future energy systems. *J. Electron. Sci. Technol.* **2019**, *17*, 100011. [CrossRef]
38. Petri, I.; Barati, M.; Rezgui, Y.; Rana, O.F. Blockchain for energy sharing and trading in distributed prosumer communities. *Comput. Ind.* **2020**, *123*, 103282. [CrossRef]
39. Kazerani, M.; Tehrani, K. Grid of Hybrid AC/DC Microgrids: A New Paradigm for Smart City of Tomorrow. In Proceedings of the 15th International Conference of System of Systems Engineering (SoSE), Budapest, Hungary, 2–4 June 2020.
40. Yu, Y.; Guo, Y.; Min, W.; Zeng, F. Trusted Transactions in Micro-Grid Based on Blockchain. *Energies* **2019**, *12*, 1952. [CrossRef]
41. Wang, J.; Wang, Q.; Zhou, N.; Chi, Y. A Novel Electricity Transaction Mode of Microgrids Based on Blockchain and Continuous Double Auction. *Energies* **2017**, *10*, 1971. [CrossRef]
42. Gür, A.O.; Öksüzer, S.; Karaarslan, E. lockchain Based Metering and Billing System Proposal with Privacy Protection for the Electric Network. In Proceedings of the 2019 7th International Istanbul Smart Grids and Cities Congress and Fair (ICSG), Istanbul, Turkey, 25–26 April 2019.
43. Bhushan, B.; Khamparia, A.; Sagayam, K.M.; Sharma, S.K.; Ahad, M.A.; Debnathh, N.C. Blockchain for smart cities: A review of architectures, integration trends and future research directions. *Sustain. Cities Soc.* **2020**, *61*, 102360. [CrossRef]
44. Meeuw, A.; Schopfer, S.; Wortmann, F. Experimental bandwidth benchmarking for P2P markets in blockchain managed micro grids. *Energy Procedia* **2019**, *159*, 370–375. [CrossRef]
45. Ahl, A.; Yarime, M.; Tanaka, K.; Sagaw, D. Review of blockchain-based distributed energy: Implications for institutional development. *Renew. Sustain. Energy Rev.* **2019**, *107*, 200–211. [CrossRef]

46. Jeon, J.M.; Hong, C.S. A Study on Utilization of Hybrid Blockchain for Energy Sharing in Micro-Grid. In Proceedings of the 20th Asia-Pacific Network Operations and Management Symposium (APNOMS), Matsue, Japan, 18–20 September 2019.
47. Agung, A.; Agung, G.; Handayani, R. Blockchain for smart grid. *J. King Saud Univ. Comput. Inf. Sci.* **2020**. [CrossRef]
48. Tsao, Y.C.; VanThanh, V.; Wu, Q. Sustainable micro grid design considering blockchain technology for real-time price-based demand response programs. *Int. J. Electr. Power Energy Syst.* **2021**, *125*, 106418. [CrossRef]
49. Danalakshmi, D.; Gopi, R.; Hariharasudan, A.; Otola, I.; Bilan, Y. Reactive Power Optimization and Price Management in Micro grid Enabled with Blockchain. *Energies* **2020**, *13*, 6179.

energies

Article

Coordinated Control Strategy and Validation of Vehicle-to-Grid for Frequency Control

Yeong Yoo [1,*], Yousef Al-Shawesh [2] and Alain Tchagang [3]

[1] Energy, Mining and Environment Research Centre, National Research Council Canada, 1200 Montréal Road, M-12, Ottawa, ON K1A 0R6, Canada
[2] Faculty of Engineering and Applied Science, Ontario Tech University, 2000 Simcoe Street N., Oshawa, ON L1G 0C5, Canada; yousef.alshawesh@ontariotechu.net
[3] Digital Technologies Research Centre, National Research Council Canada, 1200 Montréal Road, M-50, Ottawa, ON K1A 0R6, Canada; alain.tchagang@nrc-cnrc.gc.ca
* Correspondence: yeong.yoo@nrc-cnrc.gc.ca; Tel.: +1-613-993-5331

Abstract: The increased penetration of renewable energy sources (RES) and electric vehicles (EVs) is resulting in significant challenges to the stability, reliability, and resiliency of the electrical grid due to the intermittency nature of RES and uncertainty of charging demands of EVs. There is a potential for significant economic returns to use vehicle-to-grid (V2G) technology for peak load reduction and frequency control. To verify the effectiveness of the V2G-based frequency control in a microgrid, modeling and simulations of single- and multi-vehicle-based primary and secondary frequency controls were conducted to utilize the integrated components at the Canadian Centre for Housing Technology (CCHT)-V2G testing facility by using MATLAB/Simulink. A single-vehicle-based model was validated by comparing empirical testing and simulations of primary and secondary frequency controls. The validated conceptual model was then applied for dynamic phasor simulations of multi-vehicle-based frequency control with a proposed coordinated control algorithm for improving frequency stability and facilitating renewables integration with V2G-capable EVs in a microgrid. This proposed model includes a decentralized coordinated control of the state of charge (SOC) and charging schedule for five aggregated EVs with different departure times and SOC management profiles preferred by EV drivers. The simulation results showed that the fleet of 5 EVs in V2B/V2G could effectively reduce frequency deviation in a microgrid.

Keywords: coordinated control; vehicle-to-grid; primary frequency control; secondary frequency control; state of charge; decentralized; Simulink model; microgrid

Citation: Yoo, Y.; Al-Shawesh, Y.; Tchagang, A. Coordinated Control Strategy and Validation of Vehicle-to-Grid for Frequency Control. *Energies* **2021**, *14*, 2530. https://doi.org/10.3390/en14092530

Academic Editor: Manuela Sechilariu

Received: 9 March 2021
Accepted: 23 April 2021
Published: 28 April 2021

Publisher's Note: MDPI stays neutral with regard to jurisdictional claims in published maps and institutional affiliations.

1. Introduction

The increased penetration of renewable energy sources (RES) and electric vehicles (EVs) has led to significant challenges to the stability, reliability, and resiliency of the electrical grid due to the intermittency nature of RES and uncertainty of charging demands of EVs. It is expected that 54% of new car sales and 33% of the global car fleet will be electric by 2040 [1]. Such a large-scale integration of EVs into power grids poses risks to the grid safety and power quality, especially when EVs are uncontrollably charged at peak demand. Vehicle grid integration (VGI) can play an important role in resolving several issues. For example, a mobile distributed energy storage system can be resolved by the operation of the power grid while minimizing the electricity infrastructure upgrading cost. The energy stored in electric vehicles (EVs) would be made available to homes, buildings, and the electrical grid to actively manage energy consumption and costs, known as vehicle grid integration (VGI) or vehicle-to-X (V2X) (where X = home (H), building (B), or grid (G)) technologies. Vehicle-to-home/grid (V2H/V2G) or vehicle-to-building/grid (V2B/V2G) has the potential to provide a storage capacity for the benefit of owners of EVs and homes/buildings, aggregators, and utilities by reducing EV cost and home/building energy cost, and enhancing grid reliability and resilience [2].

Previous studies [3–6] have shown that there is potential for significant economic returns to use vehicle-to-grid (V2G) technology for peak load reduction and frequency control. The daily electricity bill of an industrial building or a commercial unit is the summation of energy charge and peak demand charge. The peak demand charge is based on the maximum power consumption, and it is calculated from a running average of power consumption over 15 or 30 min. Peak demand charges can significantly increase the electricity bill. Smoothing these peak demands represents one of the best approaches to reducing electricity costs. Each EV has a mobile battery energy storage system (MBESS), which can be connected to the electricity grid via a fast bidirectional charger installed in locations, such as commercial buildings parking lots. The MBESS can discharge energy to the grid when it is connected, and electrical demand is high, and charge at other times to smooth the building's energy consumption profiles [7].

Since frequency is one of the most important indexes of power system stability, the MBESS in EV can also be used for frequency regulation as a tool employed by power grid operators. They can be used for maintaining the balance between supply and demand of active power in cases when the system frequency has deviated from its nominal value (50 or 60 Hz) [8]. When supply exceeds demand, the grid frequency increases; when demand exceeds supply, the grid frequency decreases [9]. When over-frequency or under-frequency events happen, the recovery of the system frequency to its nominal value involves three phases, which are collectively known as "frequency control". These include the primary frequency control (PFC), secondary frequency control (SFC), and tertiary frequency control (TFC), and these three phases are triggered within the first few seconds, within tens of seconds, and within a few minutes, respectively after the events [8].

V2G-capable EV chargers can provide PFC because of their fast response and seamless bidirectional power flow capabilities. This means they cannot only draw energy from the grid to charge EV batteries, but they could also provide power back to a high voltage grid or an islanded low voltage microgrid as energy becomes available and the V2G controller is enabled. Since the electric power from a single EV is too weak to provide frequency control, a small or large fleet of EVs should be aggregated to act as a virtual power plant [10,11]. SFC or load frequency control (LFC), which is also a part of the automatic control system and the PFC, can also be provided by V2G-capable EVs. PFC is to bring a deviated frequency back to an acceptable value in the short-term, leaving a frequency error due to a proportional droop control, but SFC or LFC is to compensate for the remaining frequency error after providing the PFC [10].

For V2X applications, the charging and discharging of EV batteries can be achieved through bidirectional AC onboard or DC off-board chargers. The impact of bidirectional charging on the electrical grid should be investigated in detail to identify the technical issues of grid connection and interactions of EVs, determine its interoperability with the grid, and develop optimized control algorithms and strategies that can be validated by simulation.

Recently, various vehicle-to-grid (V2G) control strategies have been proposed in the literature [12–15] for both PFC and SFC. In [12], a simple frequency control droop loop with a dead-band function was adopted to adjust the active power set-points of a smart EV charging interface that responds locally to frequency changes. In [13], an adaptive decentralized V2G control strategy was proposed to simultaneously control the EV charging schedule and system frequency. The droop coefficient was adjusted according to the frequency deviation and the state of charge (SOC) of EV batteries. In [14], an aggregate model of EV fleets with a participation factor was proposed to evaluate the dynamic response in PFC, incorporate EV's technical constraints flexibly, and reduce computational complexity based on an average technique. The participation factor was proposed to identify the participation of each EV in PFC according to EV's operating modes of disconnected, charging, or idle. In [15], a state-of-the-art V2G control strategy, consisting of frequency droop control and scheduled charging, was proposed to simultaneously suppress frequency fluctuation and satisfy the charging demands of EV drivers in frequency regulation. However, none of the control strategies simultaneously reflected more detailed preferences of EV drivers and potential prevention of unnecessary EV battery degradation in the practical V2G operation, including the departure time, maximum allowable C-rate for charging and discharging of EV batteries, maximum usable EV battery energy capacity, initial and end SOCs, and upper and lower SOC limits.

Therefore, it is required to develop an intelligent optimal control algorithm for coordinated control of SOC and charging schedule of aggregated EVs with different departure times, allowable ranges of C-rate and EV battery capacity, and SOC management profiles. In addition, most of the simulation validation of these V2G control studies was based on a frequency-domain two-area power system with a transient response stability analysis. Its frequency simulation was characterized only by a sudden change in load or circuit condition. There is a lack of dynamic analysis, which is dependent on the oscillation characteristics of generating units under continuous small disturbances that occur due to random fluctuations in loads and generation levels, to examine the system stability, particularly of the grid frequency on a timescale of several hours.

This paper presents a demonstration and validation of PFC and SFC with a V2G-capable Nissan leaf EV using a commercially available supervisory control and data acquisition (SCADA) and MATLAB/Simulink, based on droop characteristics designed for PFC and the dispatch of area control error (ACE) signals for SFC at a V2G testing facility of the National Research Council Canada (NRC). The simulation models derived, implemented, and described in this paper are based on time-domain dynamic analysis using the phasor modeling technique. Then, model-based simulations of PFC and SFC with an intelligent optimal control algorithm, including a coordinated control of the SOC and charging schedule for five aggregated EVs with different departure times and SOC management profiles, are presented to validate the effectiveness of frequency control using the integrated fleet of V2G-capable EVs and to verify its relevant technical issues.

2. Methodology

2.1. Vehicle-to-Grid (V2G) Testing Facility

The framework of the V2G experimental operation for this study was conducted using the V2G testing facility located at the Canadian Centre for Housing Technology (CCHT) within the Montreal Road Campus in Ottawa, ON, Canada. Power system components, such as a 10 kW bidirectional off-board EV DC fast charger (DCFC), a V2G-capable Nissan Leaf EV, a 15 kVA grid simulator, two 4.5 kW AC/DC electronic loads, and a 5 kW PV simulator combined with a 5 kW grid-tied micro-inverter, were integrated at the CCHT and controlled and monitored by SCADA. A commercially available supervisory control software, InduSoft Web Studio, was utilized for operating these components through Modbus, Ethernet, RS-232, and GPIB communication protocols. Schneider Electric's Wonderware InduSoft Web Studio® is a powerful collection of automation tools that provide all the automation building blocks to develop human–machine interfaces (HMIs), SCADA systems, and embedded instrumentation solutions [16]. However, the utilized SCADA has a limited time interval for communication and data collection in the range of one to a few seconds.

To perform the real time-based simulation and validation of the dynamic performance and steady-state operation of the integrated system components under the real electrical grid environments, a real-time simulator connected to the grid simulator as an amplifier and the integrated system components were installed, as shown in Figure 1. This integration enabled performing software-in-the-loop (SIL), hardware-in-the-loop (HIL), and power hardware-in-the-loop (PHIL) simulations.

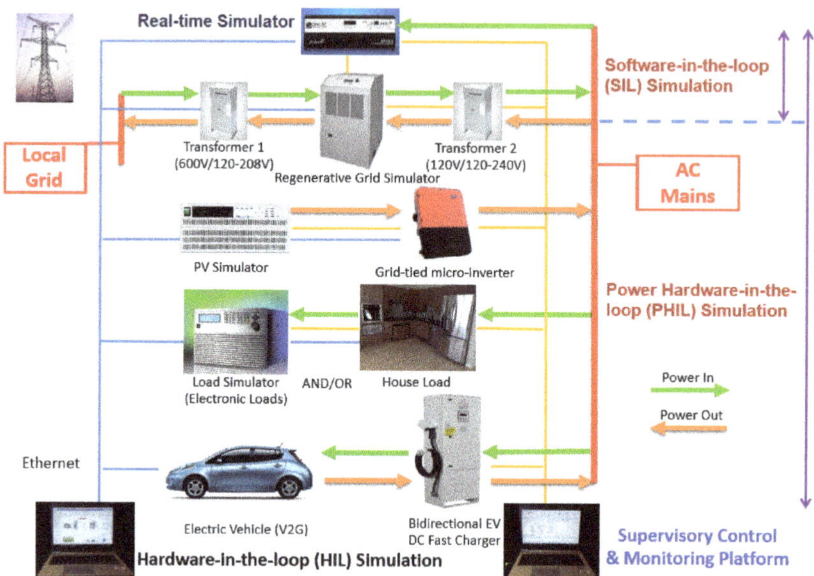

Figure 1. An integrated validation system based on power hardware-in-the-loop (PHIL) simulation for Vehicle-to-Grid (V2G) and microgrid at the Canadian Centre for Housing Technology (CCHT)-V2G testing facility.

2.2. Primary Frequency Control (PFC)

The PFC is provided by fast-reacting conventional generating units by increasing or decreasing their power production depending on detecting under-frequency or over-frequency on the grid. The PFC of EVs is a decentralized V2G control that directly responds to the system frequency deviation [17]. This PFC test requires using a grid simulator (MX-15, Ametek, Berwyn, PA, USA), a power analyzer (WT332E, Yokogawa, Tokyo, Japan), DCFC, EV, and SCADA, as well as droop characteristics to control the bidirectional power flow based on the deviations from the nominal frequency of 60 Hz. The local grid frequency was measured by the power analyzer connected to the output of the grid simulator. The frequency data detected were transferred to SCADA through Ethernet for controlling the bidirectional power flow of DCFC.

This PFC test was successfully conducted by utilizing real grid frequency data collected from the input grid at the NRC campus on 15 May 2018, at 08:30 for 100 min, as shown in Figure 2a. The frequency data collected from the NRC campus was rounded as input values to the grid simulator for PFC simulation testing, as shown in Figure 2b. Figure 3 below shows a histogram of the frequency data plotted with the Gaussian distribution, a normal distribution as a type of continuous probability distribution. The normal distribution function has a median value of 60.005 Hz and a standard deviation of 0.015297 Hz.

The grid simulator generated the rounded frequency data in AC mains, as shown in Figure 2b and its output frequency data. They were collected by the power analyzer connected to the grid simulator. The power analyzer transferred the output frequency data to a supervisory computer through Ethernet for controlling the bidirectional power flow of DCFC based on a droop characteristic designed for grid simulation purposes at the CCHT-V2G testing facility, as shown in Figure 4. The typical droop setting (R) for the existing conventional generator fleet is less than 3–5%, with a dead-band of ±16.67 mHz [18]. In this study, a specific droop setting with 0.083% (R = (0.05 Hz/60 Hz) × 100%), a dead-band of ±10 mHz, and a constant droop coefficient K_p of 200 kW/Hz were designed and used for testing and simulating the primary frequency control of the integrated V2G validation system because the range of frequency deviation from 60 Hz at the input grid of the NRC campus was less than ±50 mHz, as shown in Figure 2a.

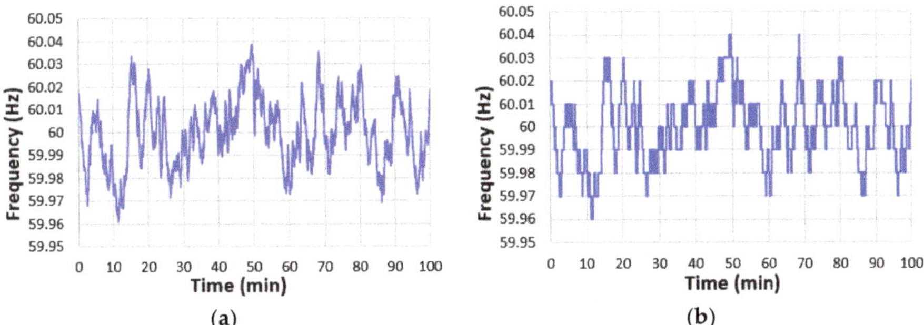

Figure 2. Frequency data (**a**) collected from the input grid at the NRC campus at 08:30 for 100 min on 15 May 2018, and (**b**) rounded as for the inputs to the grid simulator for primary frequency control (PFC) simulation testing.

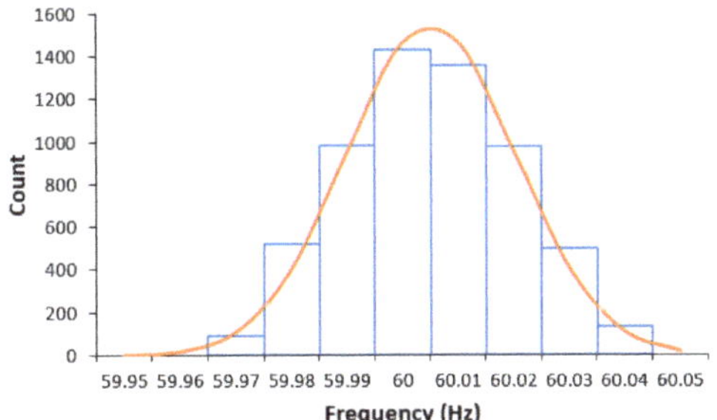

Figure 3. Histogram of frequency data collected from the input grid at the NRC campus at 08:30 for 100 min on 15 May 2018.

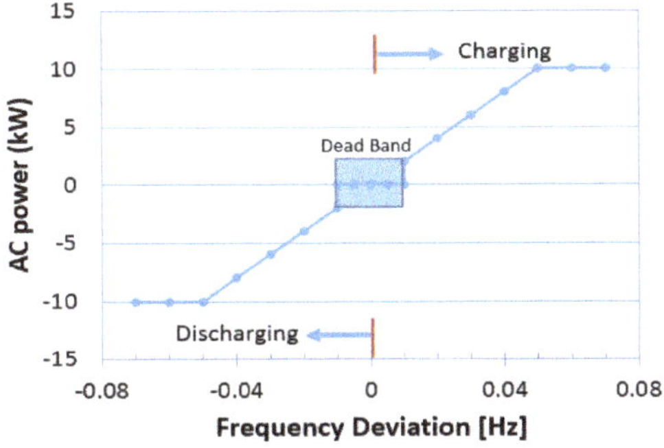

Figure 4. Droop-control characteristic designed for grid simulation on primary frequency control using V2G at the NRC-CCHT.

2.3. Secondary Frequency Control (SFC)

The SFC, also known as the secondary frequency regulation (SFR) or load frequency control (LFC), is also a part of the automatic control system together with the PFC. The purpose of the PFC is to bring the frequency back to acceptable values in the short term, leaving a frequency error due to the fully proportional control law. On the other hand, the SFC is employed to compensate for the remaining frequency error after the primary control has acted and to ensure the same frequency levels between interconnected systems [10]. The PFC is a deviating type of regulation as it can only moderate the change in the grid frequency and cannot make the frequency return to the nominal value. However, the SFC is a non-deviating adjustment and can further reduce the requirements of the controllable power supply capacity of a microgrid and reduce the dynamic frequency deviation [19]. The SFR of EVs is a dispatched-based V2G control for which the charging/discharging power of EVs is determined based on the regulation signal from the control centre [17].

In this study, a 2 h average standard deviation signal as a part of the frequency regulation duty-cycle signal in a Sandia National Lab's report (SAND2013-7315P [20]) was chosen and used for testing and simulating the SFC at the CCHT-V2G facility, as shown in Figure 5. The duty cycle was determined by analyzing the PJM balancing signal between 1 April 2011 and 31 March 2012 [21].

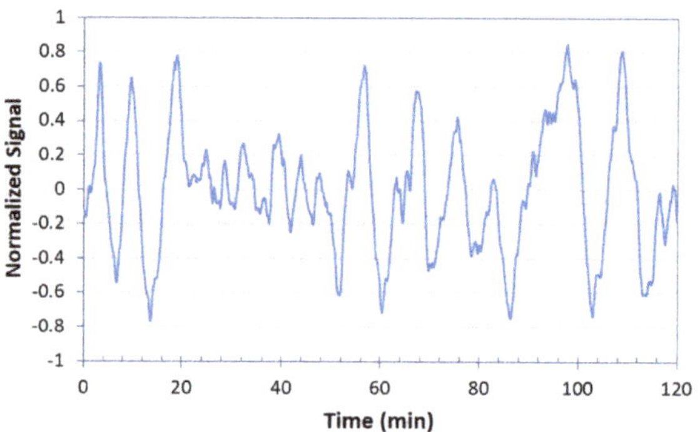

Figure 5. Normalized 2 h average standard deviation signal as a part of the 24 h frequency regulation duty cycle determined by Sandia National Laboratories [20].

3. Modeling, Simulation, and Validation

To verify the effectiveness of the proposed coordinated control strategy for V2G-based frequency control in a microgrid or V2B/V2G application, modeling and simulations of single- and multi-vehicle-based primary and secondary frequency controls are conducted to utilize the integrated validation system components at the CCHT-V2G testing facility in the MATLAB/Simulink software environment. A single-vehicle-based model with a simplified V2G control method is presented for PFC and SFC in V2H/V2G application. This time-domain phasor model is validated by comparing the empirical testing and simulation results. After validating the conceptual models of the integrated system components, Simulink-based dynamic phasor simulations of multi-vehicle-based frequency control with the proposed coordinated control algorithm are performed to validate the improvement of the frequency stability and to facilitate renewables integration with V2G in a microgrid or V2B application as a virtual power plant (VPP).

3.1. Single-Vehicle-based Vehicle-to-Grid (V2G)

As shown in Figure 6, a simplified single-vehicle V2H/V2G model based on the integrated EV and power system components at NRC's CCHT-V2G testing facility was designed to conduct PFC and SFC dynamic simulations with and without solar PV and house loads on MATLAB/Simulink. The comparison of power outputs resulting from the empirical testing and simulation was conducted to validate the effectiveness of the simulation model. The single vehicle-based model consists of four sub-models, including control for power quality and bidirectional power flow, PV renewables, DCFC/EV, and house loads.

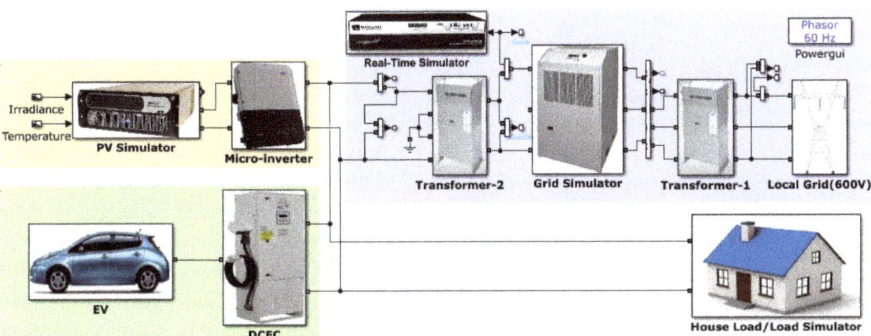

Figure 6. Simulink model of single-vehicle-based vehicle-to-home/grid (V2H/V2G).

3.1.1. Simulated Profiles of Photovoltaic (PV) Renewables and House Loads

Both rooftop solar PV panels and PV simulators were installed and operated at NRC's CCHT-V2G testing facility. For this simulation, daily PV predictive profiles were generated by using machine learning based on PV datasets collected from the existing PV panels on the roof of the CCHT InfoCentre facility in 2011–2012. A maximum power point tracker (MPPT) connected to the PV panels comprising of 14 thin-film amorphous silicon laminated modules can produce the maximum power of 2 kW.

15 May 2012, a sunny day, was selected as a sample day for this predictive PV profile, as shown in Figure 7. For this prediction, a neural network (NN) with one hidden layer and 6 neurons was utilized. The neural network has 6 inputs, including temperature, irradiance, PV power at time $t - 1$, day of the month, hour, and minute. Its output is PV power at time t. The PV historic datasets were manually divided into two subsets: 93% of the data for the training set and the remaining 7% for the testing set. Any missing historic data on the day was included in the testing/prediction set. Bayesian regularization algorithm was employed to predict a PV profile to be generated on the sunny day selected as the sample. The highest peak of PV power of 2.3 kW shown at around 15:00 in the measured profile seems to be an error in the monitoring device. ML-based prediction smoothed the erroneous peak, as shown in Figure 7b.

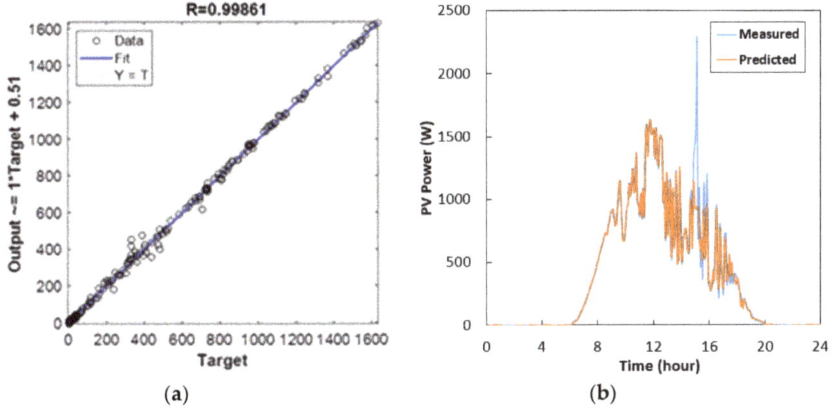

(a)   (b)

Figure 7. (**a**) Simulation result of a neural network with one hidden layer and 6 neurons and (**b**) machine learning (ML)—based prediction of photovoltaic (PV) power generation profile for 24 h on 15 May 2012, at the CCHT facility in Ottawa, ON, Canada.

Figure 8 shows the simulated load profile based on average weekday load profiles at the CCHT facility. The energy consumption rate is 5 kWh per day, and the maximum allowable load is 1 kW at the facility. However, to observe clear bidirectional power flow from the grid in the single EV model-based simulation, the house load level was intentionally increased to 10 kWh energy consumption per day and 2 kW maximum load. In the real scenarios, V2H/V2G includes various additional usage of bidirectional power flow from EV batteries, such as PV smoothing, peak shaving, and emergency back-up; however, this simulation work mainly focuses on frequency control during the day time.

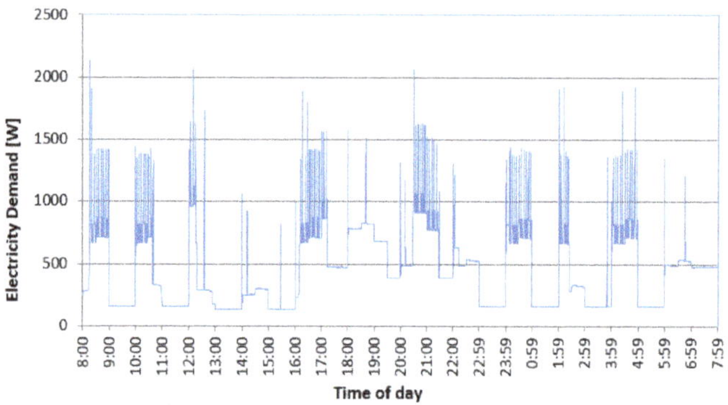

Figure 8. Simulated house load profile (24 h).

3.1.2. Primary and Secondary Frequency Control

Figure 9 shows a combined primary and secondary frequency control droop loop adopted to adjust the active power set-point of the EV battery storage for single- and multi-vehicle-based V2G simulation in this work [10,13]. The PFC is based on an open-loop where the dead-band block is added to the frequency deviation signal to reduce the charging/discharging operations of EV batteries. When the system frequency deviation is out of the predefined dead-band, the power is exchanged between the EV and the power grid to suppress the frequency fluctuation. Then, the resulting signal is multiplied by the PFC droop constant K_P followed by a saturation block with upper and lower values to limit the bidirectional power according to the designed droop characteristics described in Section 2.2. Constant charging power is also added to the PFC loop to achieve the scheduled charging to reach the SOC set-point desired by the EV driver during the plug-in period. Once frequency deviation is out of the dead-band, both scheduled charging power and PFC droop control are applied for V2G simulation. On the other hand, the SFC uses normalized standard deviation signals without any droop control characteristic designed for grid simulation, as described in Section 2.3.

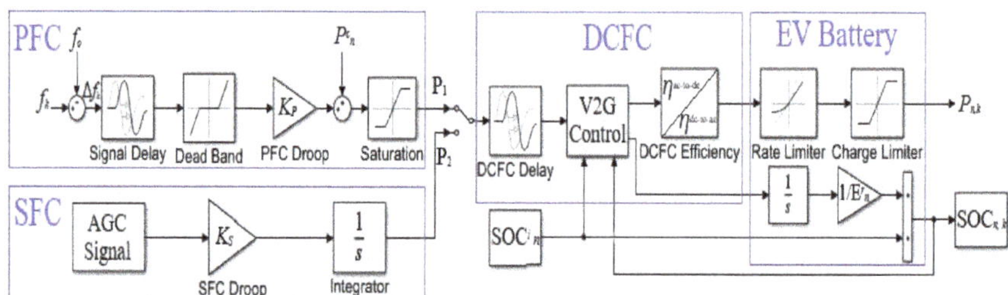

Figure 9. Combined primary and secondary frequency control droop loop and electric vehicle (EV) battery model.

The power conversion efficiencies of an off-board DCFC between AC and DC are practically varied as a function of power levels, but in this work, 92% and 95% were applied as for AC-to-DC and DC-to-AC power conversion efficiencies, respectively, for the simplified simulation. In addition, 99% and 100% were used for the energy conversion efficiencies of the Li-ion EV batteries for charging and discharging steps, respectively, in this simulation.

The EV frequency control consists of four blocks, including primary frequency control, secondary frequency control, DCFC, and EV batteries. The droop control in the PFC and SFC blocks was utilized for estimating DCFC target power values converted from grid frequency deviation in PFC and normalized automatic gain control (AGC) signals in SFC. Time delay blocks are used in PFC and DCFC models to simulate the latency in transferring the frequency and power signals. The EV battery model also includes rate and charge limiter blocks to restrict the maximum allowable rate and amount of active bidirectional power flow.

The V2G control of the DCFC and EV battery blocks includes the control of bidirectional power flow and the adjustment of SOC, as shown in Equation (1):

$$SOC_{n,k} = SOC^i_n + (1/E^r_n) \times \Delta En \tag{1}$$

3.2. Multi-Vehicle-Based V2G

3.2.1. Models

Figure 10 shows a multi-vehicle V2B/V2G model developed based on the intelligent optimal control algorithm, including a coordinated control of the state of charge (SOC) and charging schedule for five aggregated EVs with different departure times and SOC management profiles. It was assumed that a commercial building would have a daily electricity consumption of 50 kWh and a maximum load of 20 kW. 5 EVs with an EV battery energy capacity in the range of 24~40 kWh are individually connected to bidirectional off-board DCFCs with a power rating of 10~60 kW. It is expected that bidirectional onboard chargers can be introduced in the EV markets soon. For this multi-vehicle-based V2B/V2G simulation on the above-mentioned commercial building, it was also assumed that the maximum renewable power of 20 kW would be produced from PV panels on the roof of the building or a PV tracker in the parking lot. The multi-vehicle-based model consists of four sub-models, including control for power quality and bidirectional power flow, PV renewables, EVs/DCFCs, and building loads.

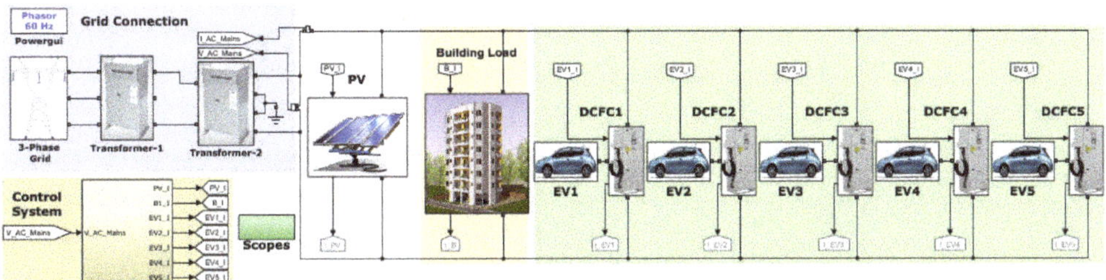

Figure 10. Simulink model of multi-vehicle-based vehicle-to-building/grid (V2B/V2G).

3.2.2. Coordinated Control Strategy and Algorithm

One of the most challenging aspects of the control of power flow from integrated system components in V2B/V2G is to reflect the preferences of EV drivers, such as the departure time, maximum allowable C-rate for charging and discharging of EV batteries, maximum usable EV battery energy capacity, and initial and end SOCs during V2B/V2G operations. Therefore, it is required to develop an intelligent optimal control algorithm for coordinated control of SOC and charging schedule of aggregated EVs with different departure time and SOC management profiles.

Table 1 shows detailed V2B/V2G operational conditions and schedule of 5EVs based on the assumed preferences of EV drivers. Each EV driver has his or her own preferred conditions. The aggregator or building energy system operator may have their own guidance on V2B/V2G operations; however, this simulation study utilizes EV driver's preferences. The maximum C-rate options depend on the power rating of DCFC available in the building or V2G service zone. The specified operational conditions include 7 input parameters, such as the initial SOC, end SOC, ΔSOC, departure time, EV battery energy capacity, maximum C-rate allowable, and maximum usable charging/discharging battery energy capacity. To prevent unnecessary battery degradation during V2B/V2G operations, the upper and lower limits of SOC with 90% and 30%, respectively, were applied for this simulation regardless of input parameters.

Table 1. Vehicle-to-building/grid (V2B/V2G) operational conditions and schedule of 5 electric vehicles (EVs) based on the preferences of EV drivers.

No	Conditions	EV-1	EV-2	EV-3	EV-4	EV-5	Remarks
1	Initial SOC (%)	40	80	50	90	70	- When EV is connected to DCFC
2	End SOC (%)	60	80	80	70	60	- Specified by EV driver for departure
3	ΔSOC	20	0	30	-20	-10	- End SOC (SOCe) minus Initial SOC (SOCi)
4	Departure time	15:00	16:00	18:00	17:00	16:00	- Specified by EV driver
5	EV battery energy capacity (kWh)	24	30	40	24	30	- Depending on EV model-year
6	Max. C-rate allowable	0.5	2.0	1.0	0.5	0.8	- Permitted by EV driver - Options 1/2/3/4: 0.5/0.8/1.0/2.0C
7	Max. usable charging/discharging battery energy capacity (kWh)	10	15	15	10	15	- Specified by EV driver for PFC/SFC

Figure 11 shows a flowchart of the coordinated control algorithm proposed for multi-vehicle-based V2B/V2G operations. Initially, monitoring devices installed in the microgrid are checking if the coordinated control strategy mode is activated to perform V2G. If V2G is deactivated or there is any fault occurring in the system, then this disturbance must be eliminated first by the existing resources in the microgrid without any frequency regulation, which may unbalance the microgrid. However, if V2G mode is active, then this control algorithm estimates the energy capacity required for the PFC and SFC before participating in ancillary services and automatically receives the power adjustment instructions from the decentralized controller based on the feedback information of various parameters, such as the initial SOC, end SOC, ΔSOC, departure time, EV battery energy capacity, maximum C-rate allowable, maximum usable charging/discharging battery energy capacity, and the plug-in-state of EVs. Then this control algorithm determines the current SOC of EV batteries and predicts an EV charging schedule based on the informed various parameters. This process enables selecting an EV operational mode between frequency control and EV charging/discharging for departure. Frequency measurement in the grid is continuously performed to determine frequency deviations during the EV's plug-in period.

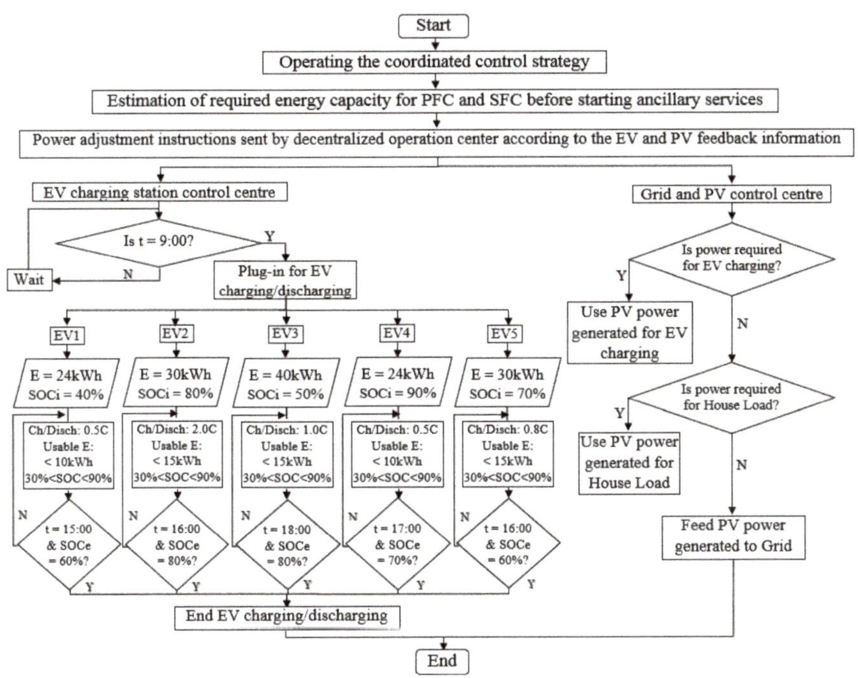

Figure 11. Coordinated control algorithm for frequency control.

The electrical energy production from PV renewables, the energy consumption for building loads, and bidirectional power flow for frequency control as a function of time can be estimated at the initial step based on predicted and simulated profiles of PV, load, and primary/secondary frequency control. In this simulation, it was assumed that the EV operational time for frequency control would be maximized to benefit the building energy system operator. The latest time to start charging EVs using PV renewables and/or grid power for preparing EV departure can be determined by energy capacities estimated for PV/loads/frequency control and operational conditions preferred by EV drivers. The maximum C-rate allowable for charging each EV using PV and/or grid power depends on the specifications of DCFC connected and the EV driver's preferences.

4. Results and Discussion

4.1. Experimental Frequency Control Testing and Simulation on Single-Vehicle-Based Frequency Control

Figure 12a shows frequencies produced by the grid simulator in AC mains at the CCHT and monitored by the power analyzer. The output frequencies measured by the power analyzer are transferred to the supervisory computer through Ethernet for controlling the bidirectional power flow of DCFC/EV based on the droop characteristic designed for grid simulation purposes at the CCHT-V2G testing facility, as shown in Figure 4. In addition, input frequencies rounded for the grid simulator are identical to output frequencies generated, as shown in Figures 2b and 12a.

Figure 12b shows the droop control-based AC target power for DCFC/EV and AC output actual power responding from DCFC/EV for primary frequency control at the CCHT-V2G facility. The positive and negative values in Figure 12b indicate power flow for charging and discharging, respectively. There is a very small discrepancy between the target power and actual response under the charging power of 5 kW, whereas the discrepancy increases over 5 kW; therefore, it seems that target values in the utilized DCFC are slightly less accurate for charging power control when over the range of 5 kW than those for discharging power control.

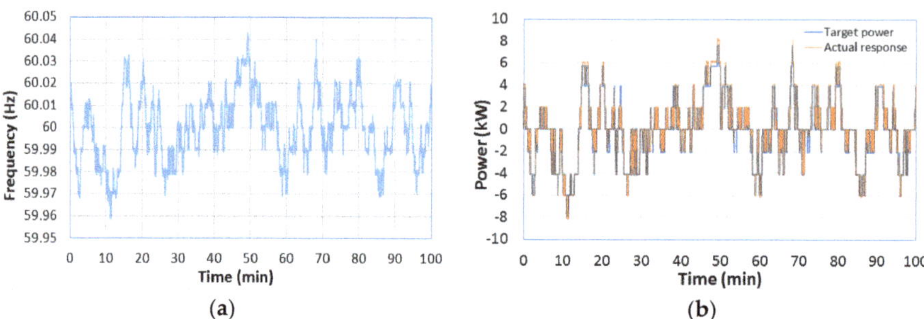

Figure 12. (a) Frequencies produced by grid simulator in AC mains and monitored by power analyzer and (b) droop control-based AC target power for DCFC/EV and AC output actual power responding from DCFC/EV (positive: charging, negative: discharging) for PFC at the CCHT-V2G facility.

Empirical measurements and Simulink-based simulation on AC output from DCFC/EV, frequency deviation, and SOC variation for PFC using grid input data, droop characteristics, and response from DCFC/EV were conducted in single-vehicle-based V2H/V2G, as shown in Figure 13. Various profiles resulting from the empirical testing and the single-vehicle-based Simulink model are compared to verify the effectiveness of the Simulink model built for single-vehicle-based V2H/V2G. The empirical and simulated profiles on AC output from DCFC/EV, frequency deviation, and SOC variation in Figure 13 are well-matched, indicating that the proposed simulation model can be used to verify electrical interactions among the integrated power system components and further for the multi-vehicle-based V2G simulation.

Figure 13c compares frequency deviation simulated in AC mains with or without bidirectional power flow provided by DCFC/EV in PFC operation. To simulate the frequency deviation variation with V2G power flow in the AC mains, it was assumed that the frequency deviation would be increased or decreased by discharging or charging EV batteries via DCFC based on the following Equations (2)–(5):

$$P_{CH} \text{ or } P_{DISCH} = \Delta f_{PFC} \times K_P \tag{2}$$

$$\text{if } \Delta f_{PFC} > 0.01 \text{ Hz}, \Delta f_{V2G} = \Delta f_{PFC} - (P_{CH}/K_P) \tag{3}$$

$$\text{if } 0.01 \text{ Hz} \geq \Delta f_{PFC} \geq -0.01 \text{ Hz}, \Delta f_{V2G} = \Delta f_{PFC} \tag{4}$$

$$\text{if } \Delta f_{PFC} < -0.01 \text{ Hz}, \Delta f_{V2G} = \Delta f_{PFC} + (P_{DISCH}/K_P) \tag{5}$$

where Δf_{PFC} = frequency deviation (Hz) simulated in AC mains for PFC operation without V2G power flow, Δf_{V2G} = frequency deviation (Hz) simulated with V2G power flow, P_{CH} = AC power input (kW) for charging DCFC/EV, P_{DISCH} = AC power output (kW) from DCFC/EV during discharging, and K_P = PFC droop gain (kW/Hz), that is 200 kW/Hz.

As shown in Figure 13c, by introducing V2G power flow from DCFC/EV in AC mains, the resulting frequency deviation would be effectively reduced; however, this simulation was conducted based on the above-mentioned assumption to utilize the droop characteristic designed for the CCHT-V2G facility. The real frequency deviation reduction depends primarily on the electrical capacities of integrated power system components and the droop gain chosen for the designed droop characteristic.

As shown in Figure 13d, the variation of SOC values was automatically monitored and collected by using both SCADA and DCFC for 100 min. The SOC change of a real EV battery pack in the empirical charging or discharging step is slower than that in the simulation because it takes time to reach an electrochemical equilibrium state in a high-capacity battery pack. In addition, SOC percentages in the utilized DCFC are monitored in rough and irregular steps, such as 76.8%, 78.0%, 78.8%, and 80.0%. However, the simulated SOC profile obtained from the Simulink model reflects instantaneous energy variations based on the theoretical energy capacity change with rapid charging and discharging power fluctuation and much smaller time steps in the simulation. The initial SOCs for the empirical PFC testing and its simulation were equally set to 78.8% for the comparative purpose. It can be seen that the initial and end SOCs, as well as SOC variation profiles in empirical testing and simulation, are well-matched.

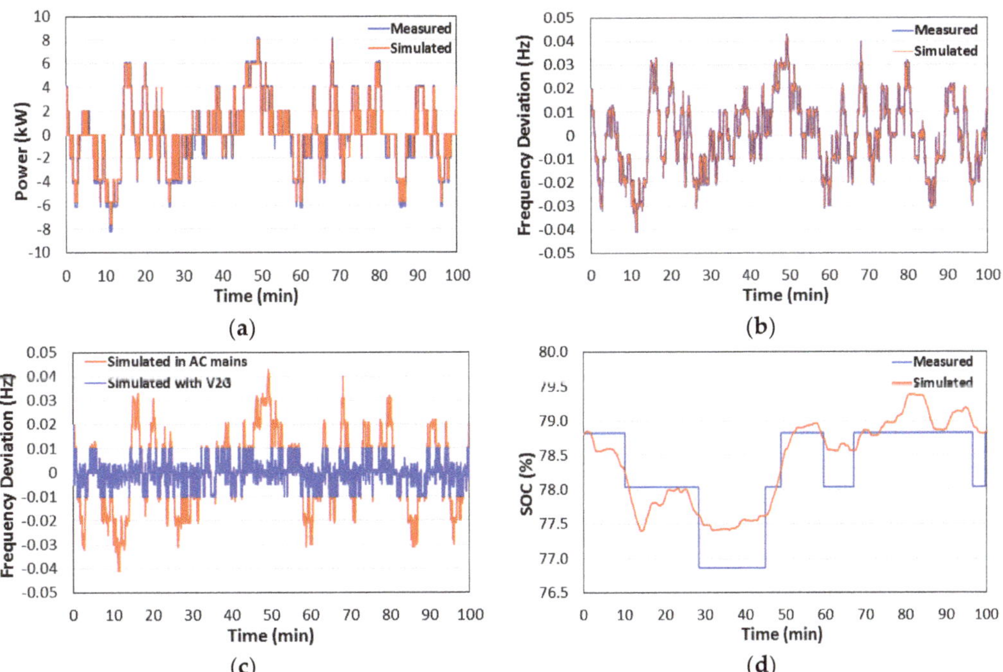

Figure 13. Comparison of (**a**) AC output from DCFC/EV (positive: charging, negative: discharging), (**b**) frequency deviation without V2G or (**c**) with V2G, and (**d**) state of charge (SOC) variation obtained from empirical testing and simulation on the single-vehicle-based PFC for 100 min (starting SOC: 78.8%).

Figure 14 shows Simulink-based simulation on AC output from DCFC/EV and SOC variation obtained from empirical testing and simulation using a normalized 2 h average standard deviation signal on single-vehicle-based SFC operation for 2 h. The initial SOCs for the empirical SFC testing and its simulation were equally set to 80% for the comparative purpose. In the case of SFC, the bidirectional power produced from DCFC/EV based on the ACE signals required for regulation up and regulation down, should be fed to the grid. The empirical and simulated profiles of AC output from DCFC/EV and SOC variation in Figure 14 are relatively well-matched, indicating that the proposed simulation model can be used to verify electrical interactions among the integrated power system components for SFC.

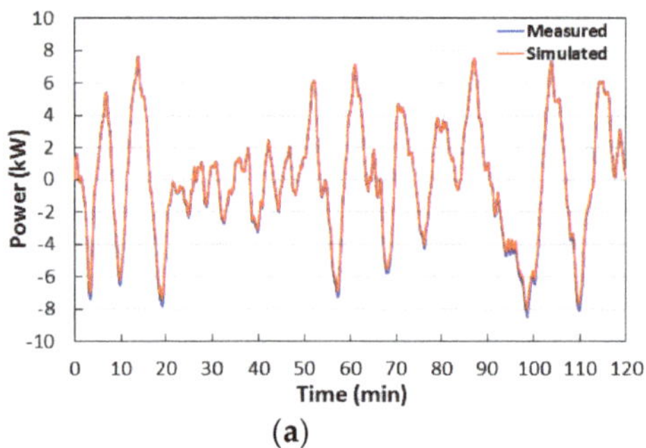
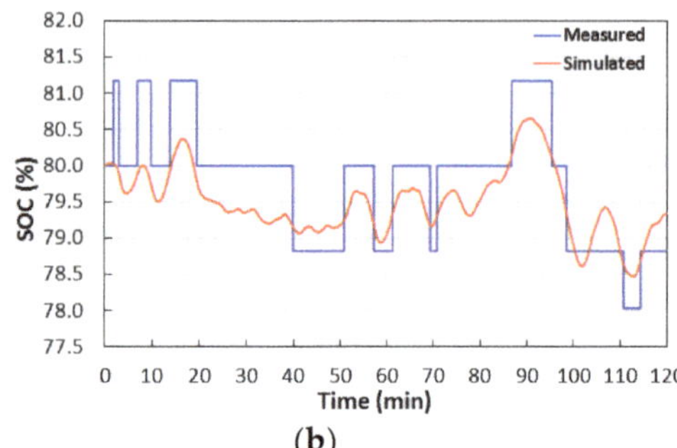

Figure 14. Simulink-based Simulation of (**a**) AC output from DC fast charger (DCFC)/EV (positive: charging, negative: discharging) and (**b**) SOC variation obtained from empirical testing and simulation using a normalized 2 h average standard deviation signal on single-vehicle-based SFC operation for 2 h (starting SOC: 80.0%).

Even if the profiles of SOC variation measured and simulated are quite similar, there is a slight discrepancy of end SOCs between measurements and simulation, as shown in Figure 14b. The energies used for the increase (regulation up) and reduction (regulation down) of active power generation are symmetric in SFC; therefore, the initial and end SOCs should be identical. However, the efficiencies of AC-to-DC (92%) and DC-to-AC (95%) conversion in the utilized DCFC, as well as charging (99%) and discharging (100%) of Li-ion EV batteries are different, resulting in a SOC that is slightly lower than the initial SOC during SFC operation. As mentioned above in Figure 13d, SOC percentages in the utilized DCFC are monitored in rough and irregular steps, such as 76.8%, 78.0%, 78.8%, and 80.0%. It seems that the SOCe of 78.8% obtained from DCFC in the empirical SFC testing could be between 78.8% and 80.0%. Therefore, it is required to utilize a DCFC with more precise SOC steps and a higher conversion efficiency over 95% in bidirectional power conversion to avoid any significant annual electrical energy loss to be potentially compensated with the financial benefits to be obtained from ancillary services.

After validating the proposed single-vehicle-based V2H/V2G model in PFC and SFC operations, as shown in Figures 13 and 14, further detailed simulation was performed to determine the impact of EV charging power and schedule on power quality and bidirectional power flow management in PFC operation at a residential house or a small facility with an integrated PV/Load and single EV system.

Simulink-based simulation on AC output from DCFC/EV, frequency deviation, SOC variation, and various power flow in PV, load, PV-to-grid, and grid-to-nanogrid was conducted for PFC operation using grid input data, droop characteristics, the response from DCFC/EV, and simulated PV power and house load demands in the single-vehicle-based V2H/V2G, as shown in Figure 15. Various parameters applied for this simulation can be found in Table 2. For each simulation scenario, the predicted PV power and simulated load data were utilized for nanogrid-level simulations. If the system frequency drops below its nominal value, the EV acts as a source by producing power to the grid to prevent further frequency drop. On the other hand, EV acts as a load by absorbing power from the grid to prevent further frequency increase.

Table 2. Operational conditions and schedule of single-vehicle-based V2H/V2G.

No	Parameters	Values	Remarks
1	Initial SOC (%)	80	- When EV is connected to DCFC
2	End SOC (%)	90	- For EV departure
3	Starting time of PFC operation	9:00	
4	Ending time of PFC operation	Same with starting time for charging EV	- Starting time for charging EV is determined by calculating minimum time to charge EV with allowable power for preparing EV departure.
5	EV departure time	16:00	
6	EV battery capacity (kWh)	40	
7	DCFC power rating (kW)	10	
8	EV charging power (kW)	5 or 10	- For SOC adjustment to prepare EV departure
9	Daily schedule on simulation	6:00 ~ 19:00	- 13 h

The scheduled charging power of the EV is estimated by the following Equation (6):

$$P_i{}^c = (SOCe - SOCi) \cdot E^r / (t^{out} - t^{in}) \tag{6}$$

where $P_i{}^c$ (in kW) is the constant charging power at the EV battery side for obtaining the charging demand to reach the expected SOCe at the plug-out time t^{out}, SOCi is the initial SOC at the plug-in time t^{in}, and E^r (in kWh) is the rated capacity of the EV battery. Both the plug-out time and the expected SOC should be provided by the EV driver in advance. It should be noted that the EV does not participate in PFC during the scheduled charging.

To prepare the end SOC of an EV for its departure at 16:00, the starting time for charging EV is determined by calculating the minimum time to charge EV with an allowable power specified in DCFC and EV batteries for preparing EV departure in time. It was assumed to use 2 kW PV power and 6 kWh house load demands for 13 h from 06:00 until 19:00, which is 10 kWh in a whole day. The initial and end SOCs are 80% and 90%, respectively, and two EV charging power levels with 5 kW and 10 kW are applied for preparing EV departure conditions.

As shown in Figure 15b, by introducing V2G power flow from DCFC/EV under PV power generation and house load demands in AC mains, the resulting frequency deviations are effectively reduced. The lower- and upper-frequency values in the stabilized period from 09:00 until 15:00 are around 59.9 Hz and 60.018 Hz that are very close to the nominal frequency, 60 Hz. It is evident that the V2G mode can contribute significantly to stabilize grid frequencies.

When the excess solar PV power is fed to the grid, over-frequency may occur in AC mains; however, the resulting frequency deviations are also well-maintained without any significant negative effect on the nanogrid. Simulation results prove that the proposed bidirectional charging strategy may contribute effectively towards frequency regulation. The frequencies are well regulated when the V2G mode is enabled, except for the period to conduct scheduled EV charging to prepare its departure at the end of PFC operation. The higher the EV charging power at the end of PFC operation, the larger the frequency deviation in AC mains, resulting in lowered power quality.

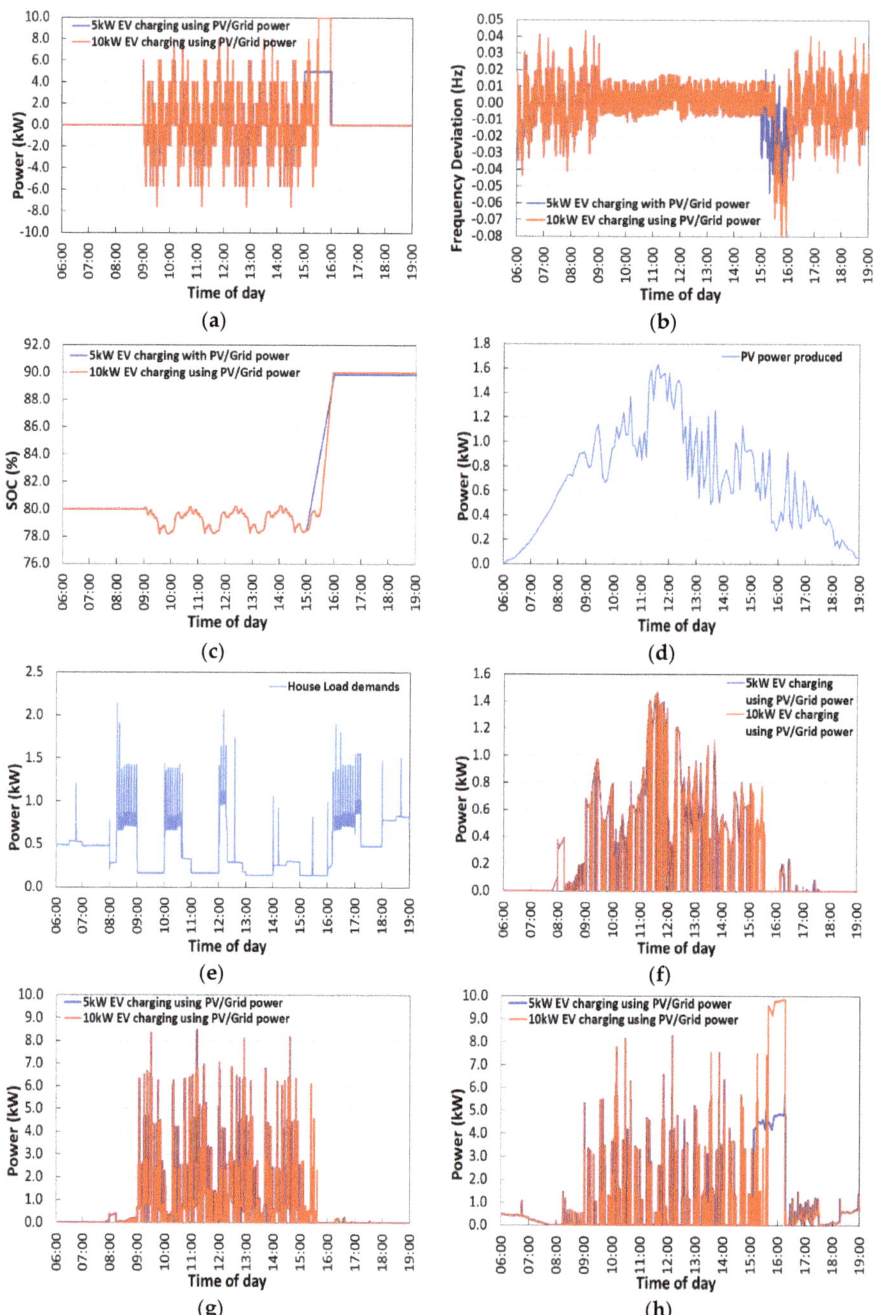

Figure 15. Simulink-based simulation of (**a**) AC output from DCFC/EV (positive: charging, negative: discharging), (**b**) frequency deviation, (**c**) SOC variation, (**d**) PV power, (**e**) house load, (**f**) power flow from PV to the external grid after the supporting load and V2G, (**g**) power flow from combined PV and V2G to external grid, and (**h**) power flow from external grid to nanogrid for PFC operation in single-vehicle-based V2H/V2G (starting SOC: 80%, duration: 13 h from 06:00 to 19:00).

As shown in Figure 15c, real-time SOC is built into the EV battery model to acquire the dynamic change of the battery energy capacity during the V2G operation and scheduled charging, where the actual charging duration is estimated based on the actual plug-in duration.

In this simulation, AC-to-AC coupled PV and EV were used; therefore, PV power generated was used first to directly charge EV during the charging steps of EV batteries in PFC operation. In addition, the PV power generated was also used for the increase of SOC to prepare EV departure and for the supporting load demands. As shown in Figure 15f, any excess PV power remaining is fed to the external grid because there is no stationary battery energy storage system (BESS) employed in this simulation. Figure 15h shows the power flow from the external grid to the nanogrid for PFC operation in single-vehicle-based V2H/V2G. The total net electrical energies supplied from the external grid are 3.13 kWh and 3.60 kWh for 5 kW and 10 kW EV charging scenarios, respectively, during the simulation period. The scenario to charge EV with 5 kW enabled to utilize more PV power because of earlier EV charging schedule than that of 10 kW EV charging scenario. To reduce the electricity cost, the increase of PV power and using a stationary BESS should be considered.

4.2. Simulation of Multi-Vehicle-based Frequency Control

A multi-vehicle V2B/V2G model based on the intelligent optimal control algorithm was developed to include a coordinated control of the SOC and charging schedule for five aggregated EVs having different departure time and SOC management profiles, as shown in Figure 10. In this model, it was assumed to utilize a commercial building with daily electricity consumption of 50 kWh and a maximum load of 10 kW. For multi-vehicle-based V2B/V2G simulation, 5 EVs with an EV battery energy capacity in the range of 24~40 kWh were individually connected to bidirectional off-board DCFCs with a power rating of 10~60 kW. It was also assumed to use the maximum renewable power of 20 kW from PV panels on the roof of the building or a PV tracker in the parking lot.

The control of power flow from integrated system components in V2B/V2G requires to reflect the preferences of EV drivers, such as the departure time, maximum allowable C-rate for charging and discharging of EV batteries, maximum usable EV battery capacity, and initial and end SOCs during V2B/V2G operations. Therefore, it is required to develop an intelligent optimal control algorithm for coordinated control of SOC and charging schedule of aggregated EVs with different departure times and SOC management profiles.

Detailed V2B/V2G operational conditions and schedule of 5 EVs based on the assumed preferences of EV drivers, shown in Table 1, were fully reflected into the proposed multi-vehicle V2B/V2G model. This simulation has 7 input parameters, including the initial SOC, end SOC, ΔSOC, departure time, EV battery energy capacity, maximum C-rate allowable, and maximum usable battery energy capacity as EV driver's preferences and operational conditions. The upper and lower limits of SOC with 90% and 30%, respectively, were applied for this simulation regardless of input parameters to prevent unnecessary battery degradation during V2B/V2G operations.

The coordinated control algorithm proposed for multi-vehicle-based V2B/V2G operations, as shown in Figure 11, was applied to verify the effectiveness of the developed model and its impact on bidirectional power flow control and power quality. Figure 16 shows Simulink-based simulation of AC output from each DCFC/EV, frequency deviation, SOC variation, PV power, building load demands, and various power flow of PV-to-grid, PV&V2G-to-grid, and external grid-to-microgrid for PFC operation using the coordinated control algorithm with 5 EVs.

The estimation on the electrical energy production from PV renewables, the energy consumption for building loads, and bidirectional power flow for frequency control as a function of time was conducted at the initial step based on predicted and simulated profiles of PV, load, and primary/secondary frequency control. The maximized EV operational time for frequency control was applied for providing economic benefits to the building energy system operator. The latest time to start charging EVs using PV renewables and/or grid power for preparing EV departure was determined by energy capacities estimated for PV/loads/frequency control and operational conditions preferred by EV drivers. The maximum C-rate allowable for charging each EV using PV and/or grid power depends on the specifications of DCFC connected and EV driver's preferences.

To prepare the SOCe of each EV with a different departure time, the starting time for charging EV is determined by calculating the minimum time to charge EV with an allowable power specified in DCFC, EV batteries, and EV driver's preferences for preparing EV departure in time. It was assumed to use 20 kW PV power and 30 kWh house load demands for 13 h from 06:00 until 19:00, which is 50 kWh in a whole day. The initial and end SOCs of each EV are specified by its own EV driver, as shown in Table 1.

The proposed decentralized coordinated control algorithm using 7 input parameters was applied for the detailed multi-vehicle-based V2B/V2G simulation with 5 EVs in Figure 16. In [12–15], various vehicle-to-grid (V2G) control strategies were proposed for both PFC and SFC. However, none of the control strategies simultaneously reflected detailed preferences of EV drivers and potential prevention of unnecessary EV battery degradation in the practical V2G operation. This included the departure time, maximum allowable C-rate for charging and discharging of EV batteries, maximum usable EV battery energy capacity, initial and end SOCs, and upper and lower SOC limits.

The scheduled charging of 5 EVs is shown in Figure 16a,c. Each EV was operated in a synchronized control from 09:00 until 14:00, and then its control was desynchronized from 14:00 until 16:00 for the intelligent coordinated control to continue PFC operation as long as possible. The fleet of 5 EVs in V2B/V2G can effectively reduce frequency deviation in a microgrid, as shown in Figure 16b. The frequencies are stabilized at around 60 Hz after starting the PFC. Small frequency vibrations resulting from the excess solar PV being fed to the grid are observed between 11:00 and 12:00. In addition, frequencies slightly decrease between 14:00 and 15:00 due to charging demands from the grid. The lower- and upper-frequency values during the PFC period and scheduled charging are around 59.99 Hz and 60.02 Hz that are very close to the nominal frequency, 60 Hz.

The frequency stabilization can be enhanced by maintaining the power balance between supply and demand in the microgrid. Further frequency deviation reduction can be achieved by using a perfectly linear PFC droop characteristic with a narrower dead-band like ±5 mHz and any DCFC with higher AC-to-DC and DC-to-AC power conversion efficiencies. To improve frequency stabilization further without significant peaks, a stationary battery energy storage system (BESS) may be considered to store the excess solar PV energy instead of injecting it into the power grid. Then, the utilization of the stored energy in the BESS for scheduled EV charging would effectively improve the under-frequency frequency values of 59.99 Hz. Interestingly, the significant under-frequency frequencies observed during scheduled charging in single-vehicle-based V2G simulation in Figure 15b disappear in the period of combined PFC operation and scheduled charging for multi-vehicle-based V2G simulation, as shown in Figure 16b. The simulation results clearly depict that the increasing number of vehicles in the EV fleet may not bring a negative impact on the power quality in the microgrid.

According to the end SOC specified by each EV driver, the controlled operation of PFC and EV charging for preparing its departure can be conducted, as shown in Figure 16c. Any PV power remaining after supporting building load demands and charging EV during both charging steps to decrease frequencies in PFC and the preparation of EV departure should be fed to the external grid because there is no stationary battery energy storage system (BESS) integrated into this simulation, as shown in Figure 16f.

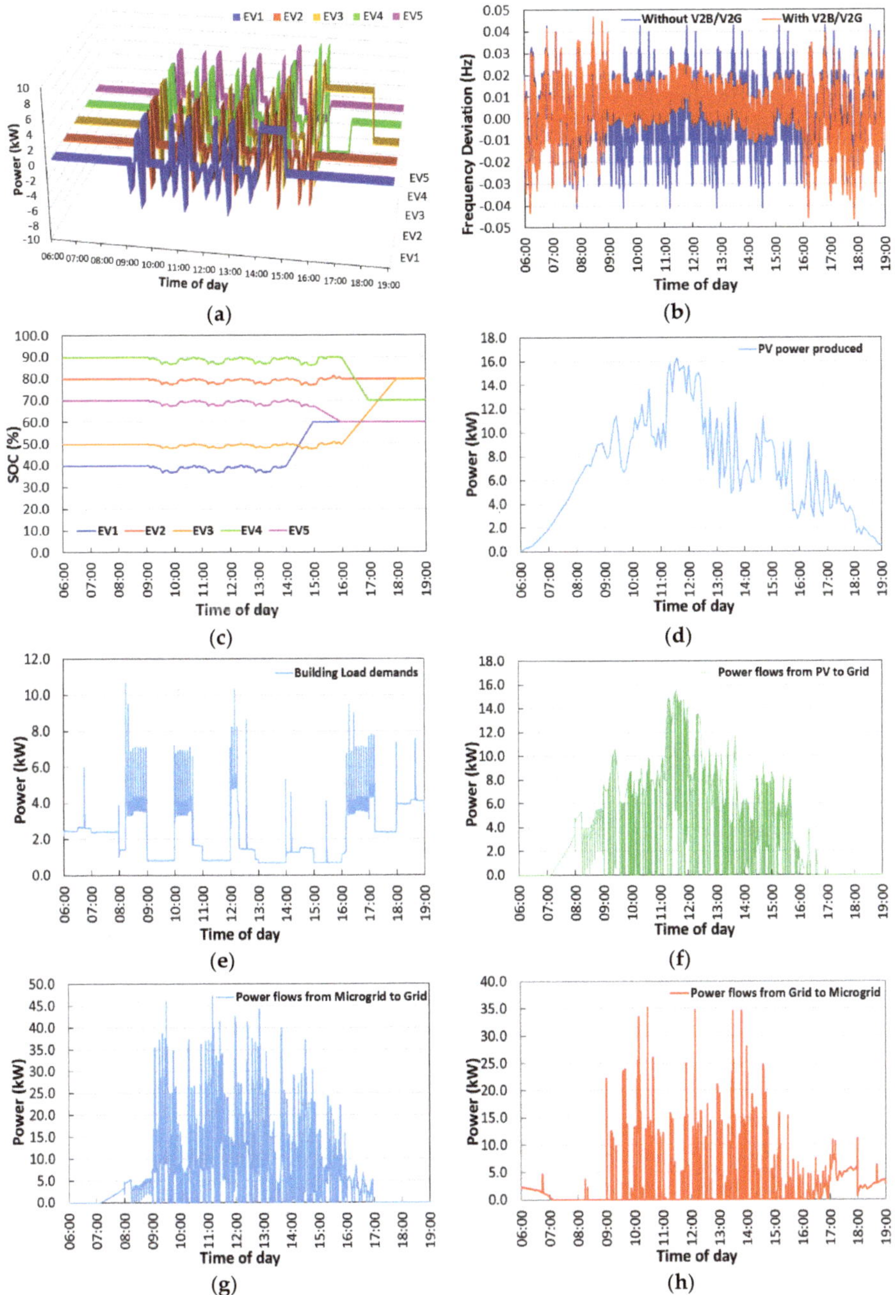

Figure 16. Simulink-based simulation of (**a**) AC output from DCFC/EV (time interval: 5 min, positive: charging, negative: discharging), (**b**) frequency deviation, (**c**) SOC variation, (**d**) PV power, (**e**) building load demands, (**f**) power flow from PV to the external grid after the supporting load and V2G, (**g**) power flow from combined PV and V2G to the external grid, and (**h**) power flow from the external grid to microgrid for PFC operation using the coordinated control algorithm with 5 EVs (duration: 13 h from 06:00 to 19:00).

Figure 16g shows power flow from the external grid to microgrid for PFC operation, EV charging, and building load demands in multi-vehicle-based V2B/V2G simulation. During the simulation period, the electrical energy produced from PV and the building load demands are 90 kWh and 30 kWh, respectively. The total electrical energies fed from the microgrid to the external grid and supplied from the external grid to the microgrid are 78.8 kWh and 28.4 kWh, respectively. The use of electricity supplied from the grid can be further reduced by more effectively utilizing PV renewables, bidirectional power flow from EVs, and any additional stationary BESS for storing excess PV renewables in the microgrid.

Figure 17 shows a Simulink-based simulation of the total V2G power flow for PFC operation with or without the proposed coordinated control algorithm with 5 EVs. The proposed coordinated control algorithm utilizes all 7 input parameters for controlling the SOC and charging schedule of 5 EVs with different departure times and SOC management profiles. There may be various comparative cases not being able to simultaneously utilize all the specified input parameters, resulting in the early termination of PFC operation, delay of EV departure, and unmatched SOCe preferred by EV drivers.

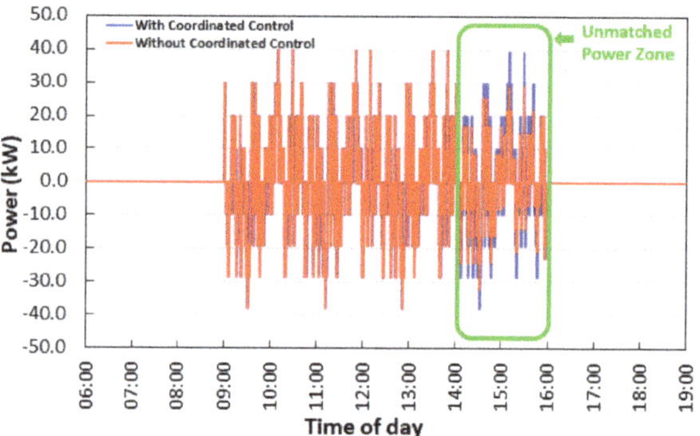

Figure 17. Simulink-based simulation of the total V2G power flow for PFC operation with or without the proposed coordinated control algorithm with 5 EVs.

One of the key objectives of the coordinated control in this simulation study is to perform PFC or SFC operation as long as possible to maximize benefits to EV and building owners, while each EV with the specified SOCe is ready for its departure in time. At the beginning of PFC operation, the V2G power flow of all 5 EVs is synchronized to provide the total V2G power based on the designed droop characteristic for the aggregated EVs. However, each EV has a different departure schedule and different charging conditions, so that intelligent desynchronized control may be required to adjust the power level of EVs participating in PFC, while other EVs prepare its departure or after the departure of any EVs based on a schedule. Otherwise, the PFC operation should be stopped when any EV needs to stop PFC and start charging for its departure.

The comparative case without the coordinated control in Figure 17 is that each EV keeps the originally assigned synchronized power flow from 09:00 until 16:00 for PFC operation. The EV charging process starts from 14:00 for the departure of EV1 at 15:00 and 15:00 for the departure of EV2 and EV5 at 16:00 as specified in Table 1. The continuous synchronized power flow from each EV for PFC operation inevitably results in the reduced V2G total power flow after 14:00. The PFC operation cannot continue after 14:00 due to the reduced total V2G power flow unmatched with the droop characteristic. PFC operation case without the coordinated control results in 2 h shorter than the PFC operation case with the intelligent decentralized coordinated control.

5. Conclusions

To verify the effectiveness of V2G-based frequency control in a microgrid, modeling and simulations for single- and multi-vehicle-based primary and secondary frequency controls were conducted to utilize the integrated components at the CCHT-V2G testing facility by using MATLAB/Simulink. The single-vehicle-based model was validated by comparing the empirical testing and simulation results of primary and secondary frequency controls. The validated conceptual model was then applied for dynamic phasor simulations on multi-vehicle-based frequency control with the proposed coordinated control algorithm for improving frequency stability and facilitating renewables integration with V2G-capable EVs in a microgrid. This proposed model includes a decentralized coordinated control of the SOC and charging schedule for five aggregated EVs with different departure times and SOC management profiles preferred by EV drivers.

The comparison of power outputs resulting from the empirical testing and simulation was conducted to validate the effectiveness of the simulation model comprising of four sub-models, including control for power quality and bidirectional power flow, PV renewables, EV/DCFC, and house loads. The empirical and simulated profiles on AC output from DCFC/EV, frequency deviation, and SOC variation were well-matched; therefore, the proposed simulation model was used to verify electrical interactions among the integrated power system components and further multi-vehicle-based V2G simulation.

After validating the conceptual models of integrated single EV and power system components, a multi-vehicle V2B/V2G model based on the coordinated control algorithm was developed and applied for Simulink-based dynamic simulations. To reflect the assumed preferences of EV drivers, detailed V2B/V2G operational conditions and schedule of 5EVs were applied for V2B/V2G simulation, including 7 input parameters of the initial SOC, end SOC, ΔSOC, departure time, EV battery energy capacity, maximum C-rate allowable, and maximum usable EV battery energy capacity. To prepare EV departure in time, each EV should be charged up to the required SOCe in a different departure time. The starting time for charging EV was determined by calculating the minimum time to charge EV with an allowable power specified in DCFC, EV batteries, and EV driver's preferences.

This multi-vehicle-based V2B/V2G simulation showed that the fleet of 5 EVs in V2B/V2G could effectively reduce frequency deviation in the microgrid. It was determined that the total net electrical energy supply from the external grid would be reduced by using PV renewables and bidirectional power flow from EVs. To utilize PV renewables more effectively, using a stationary BESS should be considered in the microgrid.

Author Contributions: Conceptualization, Y.Y. and A.T.; methodology, Y.Y. and Y.A.-S. and A.T.; software, Y.Y. and Y.A.-S. and A.T.; validation, Y.Y. and Y.A.-S. and A.T.; formal analysis, Y.Y. and Y.A.-S.; investigation, Y.Y. and Y.A.-S. and A.T.; resources, Y.Y.; data curation, Y.Y. and Y.A.-S.; writing—original draft preparation, Y.Y.; writing—review and editing, Y.A.-S. and A.T.; visualization, Y.Y. and Y.A.-S. and A.T.; supervision, Y.Y.; project administration, Y.Y.; funding acquisition, Y.Y. All authors have read and agreed to the published version of the manuscript.

Funding: This research was co-funded by the NRC's Vehicle Propulsion Technology (VPT) Program and Transport Canada—ecoTECHNOLOGY for Vehicles Program.

Institutional Review Board Statement: Not applicable.

Informed Consent Statement: Not applicable.

Data Availability Statement: Not applicable.

Acknowledgments: The authors recognize the contribution of Nameer Khan at the University of Toronto-Electrical Engineering to develop a system control platform using SCADA. The authors also recognize the contributions of Qi Liang at NRC-EME in system integration and Heather Knudsen, Patrique Tardif, Daniel Lefebvre, and Greg Burns at NRC-Construction in the maintenance of the CCHT-V2G testing facility.

Conflicts of Interest: The authors declare no conflict of interest. The funders had no role in the design of the study; in the collection, analyses, or interpretation of data; in the writing of the manuscript, or in the decision to publish the results.

Nomenclature

f_k	Grid frequency at time k
f_0	Reference (nominal) frequency
Δf_k	Frequency deviation at time k
$\eta_{ac\text{-}to\text{-}dc}$	AC-to-DC conversion efficiency
$\eta_{dc\text{-}to\text{-}ac}$	DC-to-AC conversion efficiency
E^r_n	Rated capacity of the nth EV battery
ΔE_n	Energy variation of the nth EV battery
SOC^i_n	Initial state of charge of the nth EV battery
$SOC_{n,k}$	State of charge of the nth EV battery at time k
P^c_n	Constant scheduled charging power of the nth EV for achieving the charging demand
$P_{n,k}$	V2G power at the nth EV at time k
AGC	Automatic generation control
t^{in}	Plug-in time
t^{out}	Plug-out time

References

1. Electric Vehicle Outlook 2017. Bloomberg New Energy Finance, July 2017. Available online: https://data.bloomberglp.com/bnef/sites/14/2017/07/BNEF_EVO_2017_ExecutiveSummary.pdf (accessed on 21 January 2020).
2. Pike Research a Part of Navigant. Report on Vehicle to Building Technology. 2012. Available online: https://electriccarsreport.com/2012/12/pike-research-vehicle-to-building-technologies (accessed on 21 January 2020).
3. White, C.D.; Zhang, K.M. Using vehicle-to-grid technology for frequency regulation and peak load reduction. *J. Power Sources* **2011**, *196*, 3972–3980. [CrossRef]
4. Thingvad, A.; Martinenas, S.; Anderson, P.B.; Marinelli, M.; Olesen, O.J. Economic comparison of electric vehicles performing unidirectional and bidirectional frequency control in Denmark with practical validation. In Proceedings of the 51st International Universities Power Engineering Conference IEEE, Coimbra, Portugal, 6–9 September 2016. [CrossRef]
5. Gustafsson, C.; Thurin, A. *Investigation of Business Models for Utilization of Electric Vehicles for Frequency Control*; Uppsala University: Uppsala, Sweden, 2015.
6. Høj, J.; Juhl, L.; Lindegaard, S. V2G—An Economic Gamechanger in E-Mobility? *World Electr. Veh. J.* **2018**, *9*, 35. [CrossRef]
7. Tchagang, A.; Yoo, Y. V2B/V2G on Energy Cost and Battery Degradation under Different Driving Scenarios, Peak Shaving, and Frequency Regulations. *World Electr. Veh. J.* **2020**, *11*, 14. [CrossRef]
8. EBF 483. Introduction to Electricity Markets, The Pennsylvania State University. Available online: https://www.e-education.psu.edu/ebf483/node/705 (accessed on 7 August 2020).
9. Baker, K.; Jin, X.; Vaidhynathan, D.; Jones, W.; Christensen, D.; Sparn, B.; Woods, J.; Sorensen, H.; Lunacek, M. Frequency regulation services from connected residential devices. In Proceedings of the BuildSys '16: 3rd ACM International Conference on Systems for Energy-Efficient Built Environments, Palo Alto, CA, USA, 16–17 November 2016; NREL/CP-5D00-66586; NREL: Golden, CO, USA, 2017.
10. Zarogiannis, A.; Marinelli, M.; Træholt, C.; Knezovic, K.; Andersen, P.B. A dynamic behavior analysis on the frequency control capability of electric vehicles. In Proceedings of the 2014 IEEE International Universities' Power Engineering Conference (UPEC), Cluj-Napoca, Romania, 2–5 September 2014. [CrossRef]
11. Marra, F.; Sacchetti, D.; Pedersen, A.B.; Andersen, P.B.; Træholt, C.; Larsen, E. Implementation of an Electric Vehicle Test Bed Controlled by a Virtual Power Plant for Contributing to Regulating Power Reserves. In Proceedings of the 2012 IEEE Power and Energy Society General Meeting, San Diego, CA, USA, 22–26 July 2012.
12. Pecas Lopes, J.A.; Rocha Almeida, P.M.; Soares, F.J. Using Vehicle-to-Grid to Maximize the Integration of Intermittent Renewable Energy Resources in Islanded Electric Grids. In Proceedings of the 2009 International Conference on Clean Electrical Power, Capri, Italy, 9–11 June 2009; pp. 290–295. [CrossRef]
13. Liu, H.; Hu, Z.; Song, Y.; Lin, J. Decentralized Vehicle-to-Grid Control for Primary Frequency Regulation Considering Charging Demands. *IEEE Trans. Power Syst.* **2013**, *28*, 3480–3489. [CrossRef]
14. Izadkhast, S.; Garcia-Gonzalez, P.; Frias, P. An Aggregate Model of Plug-In Electric Vehicles for Primary Frequency Control. *IEEE Trans. Power Syst.* **2015**, *30*, 1475–1482. [CrossRef]
15. Liu, H.; Yang, Y.; Qi, J.; Li, J.; Wei, H.; Li, P. Frequency droop control with scheduled charging of electric vehicles. *IET Gener. Transm. Distrib.* **2017**, *11*, 649. [CrossRef]
16. InduSoft Web Studio. Available online: https://www.indusoft.com/Products/InduSoft-Web-Studio (accessed on 22 July 2020).

17. Liu, H.; Qi, J.; Wang, J.; Li, P.; Li, C.; Wei, H. EV Dispatch Control for Supplementary Frequency Regulation Considering the Expectation of EV Owners. *IEEE Trans. Smart Grid* **2018**, *9*, 3763–3772. [CrossRef]
18. Quint, R.; Ramasubramanian, D. Impacts of Droop and Deadband on Generator Performance and Frequency Control. In Proceedings of the 2017 IEEE Power & Energy Society General Meeting, Chicago, IL, USA, 16–20 July 2017.
19. Li, P.; Hu, W.; Xu, X.; Huang, Q.; Liu, Z.; Chen, Z. A Frequency Control Strategy of Electric Vehicles in Microgrid using Virtual Synchronous Generator Control. *Energy* **2019**, *189*, 116389. [CrossRef]
20. Ferreira, S.R.; Schoenwald, D.A. Duty-Cycle Signal for Frequency Regulation Applications of ESSs. SAND2013-7315P, Sandia National Laboratories, Albuquerque, New Mexico. Available online: http://www.sandia.gov/ess/publications/SAND2013-7315P.xlsx (accessed on 25 April 2019).
21. Conover, D.R.; Crawford, A.J.; Fuller, J.; Gourisetti, S.N.; Viswanathan, V.; Ferreira, S.R.; Schoenwald, D.A.; Rosewater, D.M. *Protocol for Uniformly Measuring and Expressing the Performance of Energy Storage Systems*; PNNL-22010 Rev 2/SAND2016-3078 R; Pacific Northwest National Laboratory: Richland, WA, USA; Sandia National Laboratories: Albuquerque, NM, USA, 2016.

Article

Demand-Side Management for Improvement of the Power Quality in Smart Homes Using Non-Intrusive Identification of Appliance Usage Patterns with the True Power Factor

Hari Prasad Devarapalli [1,2,*], Venkata Samba Sesha Siva Sarma Dhanikonda [2] and Sitarama Brahmam Gunturi [3]

[1] Corporate Industry Forums and Standards Cell, A Unit of Corporate RnI, Tata Consultancy Services, Hyderabad T.S 500019, India

[2] Department of Electrical Engineering, National Institute of Technology Warangal, Warangal T.S 506004, India; sivasarma@gmail.com

[3] Component Engineering Group, Tata Consultancy Services, Hyderabad T.S 500019, India; sitaramabrahmam.gunturi@tcs.com

* Correspondence: hdevarapalli@ieee.org; Tel.: +91-94904-37035

Abstract: The proliferation of low-power consumer electronic appliances (LPCEAs) is on the rise in smart homes in order to save energy. On the flip side, the current harmonics induced due to these LPCEAs pollute low-voltage distribution systems' (LVDSs') supplies, leading to a poor power factor (PF). Further, the energy meters in an LVDS do not measure both the total harmonic distortion (THD) of the current and the PF, resulting in inaccurate billing for energy consumption. In addition, this impacts the useful lifetime of LPCEAs. A PF that takes the harmonic distortion into account is called the true power factor (TPF). It is imperative to measure it accurately. This article measures the TPF using a four-term minimal sidelobe cosine-windowed enhanced dual-spectrum line interpolated Fast Fourier Transform (FFT). The proposed method was used to measure the TPF with a National Instruments cRIO-9082 real-time (RT) system, and four different LPCEAs in a smart home were considered. The RT results exhibited that the TPF uniquely identified each usage pattern of the LPCEAs and could use them to improve the TPF by suggesting an alternative usage pattern to the consumer. A positive response behavior on the part of the consumer that is in their interest can improve the power quality in a demand-side management application.

Keywords: demand-side management; low-power consumer electronic appliances; low-voltage distribution system; non-intrusive identification of appliance usage patterns; power quality; smart home; true power factor; total harmonic distortion

Citation: Devarapalli, H.P.; Dhanikonda, V.S.S.S.; Gunturi, S.B. Demand-Side Management for Improvement of the Power Quality in Smart Homes Using Non-Intrusive Identification of Appliance Usage Patterns with the True Power Factor. *Energies* **2021**, *14*, 4837. https:// doi.org/10.3390/en14164837

Academic Editor: Surender Reddy Salkuti

Received: 13 July 2021
Accepted: 6 August 2021
Published: 8 August 2021

Publisher's Note: MDPI stays neutral with regard to jurisdictional claims in published maps and institutional affiliations.

1. Introduction

Clean and affordable energy—Goal no. 7 of United Nations' Sustainable Development Goals (SDGs)—is targeted for achievement by 2030. Demand-side management (DSM) plays a vital role in accomplishing this goal [1]. Clean energy is normally interpreted as green energy that is generated with renewable non-fossil fuels to reduce climate pollution. Interestingly, in its policies, the Government of India has articulated very well that both green energy and electrical power quality (PQ) are essential and need to be in balance without any ambiguity [2,3]. In this article, the authors consider the latter part of the interpretation of clean electrical energy.

1.1. Perspectives on Demand-Side Management

Consumers must be cognizant of unnecessary consumption and conscious of not polluting the power supply. To achieve the latter part of SDG 7—affordable energy—a wide range of technologies, such as compact fluorescent lamps (CFLs) and light-emitting diode (LED) lamps, have been deployed on a large scale by all nations across the globe, including

India. This has led to indiscriminatory usage of consumer electronic appliances (smart TVs, smart phones, smart fridges, etc.) and computers with SMPS. This pervasive usage of low-power consumer electronic appliances (LPCEAs) is defeating the very purpose of SDG 7—clean energy with quality electric power. The traditional approach to DSM is primarily focused on energy conservation, and improvements in the true power factor (TPF) could effectively be achieved by way of a human-in-the-loop DSM [4]. The purpose of this paper is to ensure affordable power and to ensure that LPCEAs do not cause issues of poor PQ in low-voltage distribution system (LVDSs).

The PQ in an LVDS is multi-dimensional and includes both the current harmonics and the power factor (PF). The ill effects of current harmonics have been discussed in detail [5–7]. While current harmonics themselves are not healthy, they also result in a poor PF due to the bidirectional exchange of reactive power between the source and the load. The current harmonics and the poor PF negatively affect the accuracy of electricity meters and the PQ of the distribution system's supply [8–10]. The existing smart energy meters in our distribution systems are not designed to measure the TPF and reactive power; hence, they are not billed to the consumers. So, the need to bill consumers for poor PQ caused by their behavior is not recognized, and hence, the requirement of compensation of reactive power has gone unnoticed [11–13]. Therefore, the authors feel strongly that the DSM for the PQ is the need of the hour, and serious attention is required in order to measure and address issues of poor PQ in LVDSs and to meet SDG 7—affordable, clean, and quality energy for all.

1.2. Non-Intrusive Monitoring of the Usage of Appliances

While homes use several electric and electronic appliances, the energy consumption is measured at a single point of common coupling (PCC) of the supply mains from the distribution system. Several techniques have been explored by various researchers since G.W. Hart explored the idea of the disaggregation of electric loads and identified the usage of individual appliances in order to prompt responsible energy consumption and healthy consumer behavior [14]. A cooperative response from consumers to the insights from this critical objective analysis will help in achieving the energy conservation targets set by the United Nations. Hart's work was performed at a time when energy availability was not in abundance. With the advent of energy-efficient smart appliances, the average energy consumption has come down significantly. Yang H. et al. articulated in detail that the DSM in smart homes is focused on the PQ issues introduced by smart appliances, including power savers, in addition to energy conservation [15]. Due to harmonics, the displacement PF is not equal to the TPF. Because household appliances are considered to have low-power consumer electronic loads, their effect on the PQ will be significant when too many of them are in use [10]. The effects of harmonics on the displacement PF and %THD are not additive, so it is not right to ignore the effects of harmonic distortion when metering energy for both consumers and utilities [13].

A typical home uses multiple appliances at different times and for different durations, thus forming different load patterns; their usage is uniform neither over a day nor over a season. As a household might have numerous load points, there are unknown loads that could be connected in open sockets, and several load patterns could possibly reach an unwieldy number (factorial N; N is the number of loads, and some of the loads are unknown), but in practice, they would usually follow a very small number of appliance usage patterns. It is not common for a modern household to use single loads at any point in time [5]. Therefore, it is clear that the focus of DSM should be on appliance usage patterns (a combination of loads), and the usage of individual loads becomes irrelevant. The recommendation to the consumer should be to switch from an appliance usage pattern with a low PQ to an appliance usage pattern with a better PQ, thereby ensuring clean power in the LVDS.

1.3. Feature Selection for NILM

The harmonic interactions among several nonlinear appliances change the PQ indexes significantly. For instance, the PF, distortion factor (DF), and current THD are directly impacted; these indexes influence the billing of the electrical energy depending on the appliance usage patterns. Therefore, from the perspective of PQ, a combination of appliances that improve the PQ can determine the amount to be billed. Two different experiences were cited: One suggested the use of fundamental signals (displacement factor) to calculate the surplus of reactive power in electrical grids. Another suggested the use of the TPF. This fact shows that this work is important for the industry in the discussion of the problems caused by nonlinear appliances on electrical energy billing systems in order to rectify PQ issues [12].

R. Gopinath et al. researched the development of robust NILM techniques for the effective management of the energy of appliances in order to support reliable and sustainable DSM [16]. A significant improvement in the system efficiency can be realized if the improvement of the PF and the elimination of harmonics can be applied in the whole network [17].

Therefore, the authors proposed the use of aggregated measures of the PQ, such as the percentage of the total harmonic distortion (THD) and the TPF, in order to understand the appliance usage patterns and for consumers to adapt to PQ-sensitive behaviors. Devarapalli et al. discussed how the percentage THD can be applied in order to disaggregate load patterns and to suggest appropriate load patterns, thus reducing the harmonic pollution produced in the system [5].

This article presents the use of the TPF as a unique feature for identifying appliance usage patterns and for helping consumers by providing them with insights in order to switch from low-PF appliance usage patterns to high-PF appliance usage patterns and to become responsible consumers. The schematic of the non-intrusive identification of appliance usage patterns in smart homes with the TPF is depicted in Figure 1.

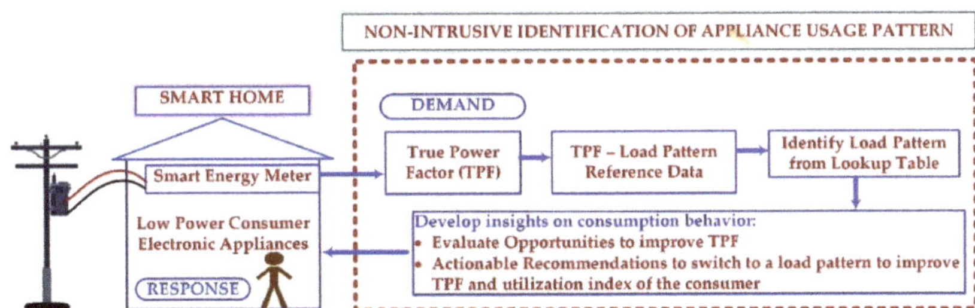

Figure 1. Schematic of demand-side management with the non-intrusive identification of appliance usage patterns in smart homes with the TPF.

1.4. The Major Contributions and Organization of the Article

- TPF measurement of real-world LPCEAs as per the IEEE 1459-2010 standard [18] by using the four-term minimal sidelobe cosine window (4MSCW)-based enhanced dual-spectrum line interpolated FFT (EDSLIFFT).
- Development of a virtual instrumentation-based measurement system for TPF measurements.
- Recommendation of appliance patterns for improvement of the TPF and DSM by using a lookup table in order to improve the utilization indexes of consumers.

This paper is organized into five sections; Section 2 describes the TPF measurement method. Section 3 elaborates on the real-time measurement of the TPF with the 4MCSW-based EDSLIFFT in the NI-cRIO (compact reconfigurable input–output)-based virtual instrumentation environment for various combinations of LPCEAs. Section 4 deliberates on the results and demonstrates that the TPF is a reliable single feature for identifying appliances' consumption patterns. Section 5 discusses the insights of the results, and Section 6 concludes with a summary of the proposed research work.

2. Measurement of the TPF Using the Four-Term MSCW-Based EDSLIFFT

In this section, the measurement of the TPF by using the 4-term MSCW-based ED-SLIFFT. The 4-term MSCW-based EDSLIFFT was proposed in [19] for the accurate estimation of the harmonics of LPCEAs. In this article, it is further extended for the computation of the TPF as per the IEEE 1459-2010 standard [18].

2.1. Overview of the Four-Term MSCW

Given the better main lobe, sidelobe, and sidelobe roll-off rate of the 4MSCW and the accuracy of the EDSLIFFT algorithm with the RT current harmonic signal analysis [19], they are further extended to measure the TPF in real time. Initially, the 4-term MSCW-based EDSLIFFT was used to measure the spectral amplitude and phase of an LPCEA's harmonic current signal. A brief overview of the 4MSCW is given in the following.

The discrete-time 4-term MSCW is expressed as:

$$w(n) = \sum_{m=0}^{M-1} (-1)^h a_h \cos\left(\frac{2\pi mn}{N}\right) \ for \ n = 0, \ 1, \ \ldots, \ N-1 \tag{1}$$

where n denotes the sample index, N denotes the total number of samples, m represents the window item index, M is the maximum window item number, and a_h denotes the window coefficients.

The spectral window corresponding to the 4MSCW is written as:

$$W(n) = \sum_{m=0}^{M-1} \frac{a_h}{2} \left[e^{\frac{-j\pi(n-m)(N-1)}{N}} \frac{\sin(n-m)\pi}{\sin\left(\frac{(n-m)\pi}{N}\right)} + e^{\frac{-j\pi(n+m)(N-1)}{N}} \frac{\sin(n+m)\pi}{\sin\left(\frac{(n+m)\pi}{N}\right)} \right]$$
$$for \ n = 0, \ 1, \ \ldots, \ N-1 \tag{2}$$

2.2. Processing with the 4MSCW-Based EDSLIFFT Algorithm

Generally, in digital signal processing, the harmonic signal is represented as [20]:

$$x(nT_s) = x(t) = \sum_{h=1}^{hmax} A_h \sin(2\pi f_h nT_s + \varphi_h) \text{ where } n = 0, \ 1, \ \ldots, \ N-1 \tag{3}$$

where the signal amplitude is denoted as A_h, and the signal frequency and phase are represented by f_h and φ_h. The harmonic signal sampling time is denoted as T_s. The harmonic order is denoted as h, which starts from 1 and reaches the maximum harmonic order of $hmax$. The traditional FFT has the issue of spectral leakage due to non-synchronous sampling because of the unstable fundamental frequency. To mitigate the spectral leakage effect, the signal is weighted by window functions.

The mathematical representation of the windowed sample signal under non-synchronous sampling is given as:

$$X(k) = \sum_{h=1}^{hmax} \frac{A_h}{2j} \left[e^{j\varphi_h} W(k - k_h) - e^{j\varphi_h} W(k + k_h) \right] \tag{4}$$

where $k = 0, 1, \ldots, (N-1)$, W indicates the 4-term MSCW function, and k_h denotes the division factor of the signal frequency and the frequency resolution, which is expressed as:

$$k_h = \frac{f_h N}{f_s} = l_h + \xi_h \tag{5}$$

where f_s is the sampling frequency of the harmonic signal, l_h is an integer value, and ξ_h $(0 \leq \xi_h \leq 1)$ is the fractional part caused by the non-synchronous sampling. The spectral line corresponding to the h_{th} harmonic lies between the two highest spectral lines—

explicitly, either the l_{eng}^{th} and the $(l_h + 1)^{th}$ or the l_h^{th} and the $(l_h - 1)^{th}$. The value of l_h is determined by the peak location index search algorithm, and the fractional part ξ_h is determined by using the EDSLIFFT algorithm, as shown in [19]. The computation process of the EDSLIFFT algorithm using 4MSCW is illustrated in Figure 2.

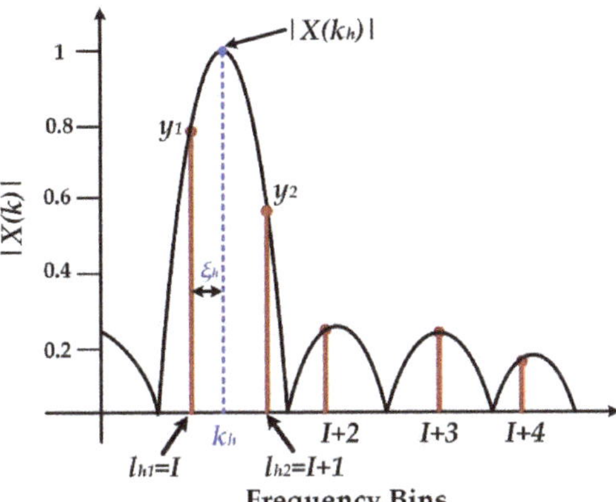

Figure 2. Computation process of the EDSLIFFT algorithm using 4MSCW.

The frequency spectrum expression used in the EDSLIFFT algorithm is given as follows:

$$X(\xi_h) = \frac{A_h}{2j}\left[e^{j\varphi_h}W(\xi_h - k_h)\right] \text{ for } \xi_h = 0, 1, \ldots, N-1 \tag{6}$$

The two spectral lines represent the hth harmonic amplitude, and they are represented by l_{h1} and l_{h2} (where $l_{h1} = I$, $l_{h2} = I + 1$, $l_{h1} < k_h < l_{h2}$). The peak locations of the harmonic amplitudes are obtained by using the peak location index search method described in [19]. Consider $y_1 = |X(I)|$ and $y_2 = |X(I + 1)|$; then, y_1 and y_2 are given as follows:

$$y_1 = |X(I)| = |A_h|\cdot|W(2\pi(I - k_h)/N| \tag{7}$$

$$y_2 = |X(I + 1)| = |A_h|\cdot|W(2\pi((I+1) - k_h)/N| \tag{8}$$

The spectral amplitudes are determined with the least-square curve-fitting technique. The symmetrical coefficient α is considered in terms of l_h and k_h as follows:

$$\alpha = k_h - l_{h1} - 0.5 \text{ for } -0.5 \le \alpha \le 0.5 \tag{9}$$

The even spectral lines in EDSLIFFT, y_1 and y_2, are represented as:

$$y_1 = |X(I)| = |A_h|\cdot|W(2\pi(-\alpha + 0.5)/N| \tag{10}$$

$$y_2 = |X(I + 1)| = |A_h|\cdot|W(2\pi(-\alpha - 0.5)/N| \tag{11}$$

A symmetrical coefficient β in terms of α is considered in order to calculate the harmonic parameters; the expression of β is as follows:

$$\beta = g(\alpha) = \frac{(y_2 - y_1)}{(y_2 + y_1)} \tag{12}$$

From Equations (10) and (11), β can be represented as:

$$\beta = g(\alpha) = \frac{|W(2\pi(-\alpha - 0.5)/N)| - |W(2\pi(-\alpha + 0.5)/N)|}{|W(2\pi(-\alpha + 0.5)/N)| + |W(2\pi(-\alpha - 0.5)/N)|} \tag{13}$$

The value of α is determined with two maximum spectral lines and the fitting polynomial $g^{-1}(\alpha)$. Thereby, the amplitude, frequency, and phase values of the given signal are obtained based upon the following interpolated formulas:

$$k_h = \alpha + I + 0.5 \tag{14}$$

$$A_h = \frac{2y_1}{|W(2\pi(I - k_h))/N|} \tag{15}$$

$$f_h = \frac{k_h f_s}{N} \tag{16}$$

$$\varphi_h = \arg(X(I)) - \arg\left[W\left(\frac{2\pi(I - k_h)}{N}\right)\right] + \frac{\pi}{2} \tag{17}$$

An LPCEA's input voltage, appliance current fundamental, and harmonic amplitudes are computed by using Equations (15)–(17). Under nonlinear loading conditions, it is necessary to measure the fundamental and harmonic content accurately; the conventional FFT has issues of spectral leakage and experiences the picket fence effect, and hence, the 4-term MSCW-based EDSLIFFT is adopted in this article. The measurement of the TPF by using the fundamental and harmonic amplitudes is discussed in Section 2.3.

2.3. TPF Measurement as per the IEEE 1459-2010 Standard

After measuring the fundamental and harmonic amplitudes of the voltage and current by using Equation (15) from the 4-term MSCW-based EDSLIFFT, the root mean square (RMS) values of the voltage and current are determined as follows:

$$V_{RMS} = \sqrt{V_{1RMS}^2 + V_{2RMS}^2 + V_{3RMS}^2 \cdots + V_{hmaxRMS}^2} \tag{18}$$

$$I_{RMS} = \sqrt{I_{1RMS}^2 + I_{2RMS}^2 + I_{3RMS}^2 \cdots + I_{hmaxRMS}^2} \tag{19}$$

The fundamental to higher-order harmonic voltages are represented by V_{1RMS}, V_{2RMS}, V_{3RMS}, $\ldots$, $V_{hmaxRMS}$ in Equation (18), and the fundamental and harmonic currents are denoted as I_{1RMS}, I_{2RMS}, I_{3RMS}, $\ldots$, $I_{hmaxRMS}$ in Equation (19). As per the IEEE 1459-2010 standard [18], the RMS of the voltage and current can be decomposed into fundamental and harmonic components as follows:

$$V_{RMS}^2 = V_{1RMS}^2 + \sum_{h>1}^{hmax} V_{hRMS}^2 = V_{1RMS}^2 + V_{HRMS}^2 \tag{20}$$

$$I_{RMS}^2 = I_{1RMS}^2 + \sum_{h>1}^{hmax} I_{hRMS}^2 = I_{1RMS}^2 + I_{HRMS}^2 \tag{21}$$

One of the most significant parameters for efficient power consumption with a good PQ is the PF. If the PF at the supply mains is at unity or close to unity, then it is designated as having a high PQ. On the contrary, a low PF means that the system is operating with a low efficiency and poor PQ. Under nonlinear appliance conditions, the PF is represented by the TPF, which is computed from the distortion factor and displacement PF. The distortion factor is determined from the current THD (THD_I).

The voltage and current THDs are determined with the following equations:

$$THD_V = \frac{\sqrt{\sum_{h>1}^{hmax} V_{hRMS}^2}}{V_{1RMS}} \tag{22}$$

$$THD_I = \frac{\sqrt{\sum_{h>1}^{hmax} I_{hRMS}^2}}{I_{1RMS}} \tag{23}$$

The voltage THD_V is negligible compared to the current THD_I; hence, the current THD is considered in the distortion factor computation, as given below:

$$DF = \frac{1}{\sqrt{1+THD_I^2}} = \frac{I_{1RMS}}{I_{RMS}} \tag{24}$$

The displacement factor is determined as follows:

$$DPF = \frac{P}{S} \tag{25}$$

where P represents the active power consumed by the LPCEAs and S denotes the total apparent power. The computation of the active, reactive, and apparent powers is discussed below.

The active power consumption of the LPCEAs given as:

$$P = V_{1RMS}I_{1RMS}\cos(\theta_1 - \delta_1) + \sum_{h>1}^{hmax} V_{hRMS}I_{hRMS}\cos(\theta_h - \delta_h) \tag{26}$$

The reactive power consumption of the LPCEAs is represented as:

$$Q = V_{1RMS}I_{1RMS}\sin(\theta_1 - \delta_1) + \sum_{h>1}^{hmax} V_{hRMS}I_{hRMS}\sin(\theta_h - \delta_h) \tag{27}$$

Based on the definitions of the voltage and current RMS values as functions of their fundamental and harmonic components, the total apparent power is expressed as:

$$S^2 = (V_{RMS}I_{RMS})^2 = \left(V_{1RMS}^2 + V_{HRMS}^2\right)\left(I_{1RMS}^2 + I_{HRMS}^2\right) \tag{28}$$

Using Equation (8), the fundamental and non-fundamental components of the apparent power are written as follows:

$$S^2 = \left(V_{1RMS}^2 I_{1RMS}^2\right) + \left(V_{1RMS}^2 I_{HRMS}^2\right) + \left(V_{HRMS}^2 I_{1RMS}^2\right) + \left(V_{HRMS}^2 I_{HRMS}^2\right) \tag{29}$$

$$S^2 = (V_{1RMS}I_{1RMS})^2 + (V_{1RMS}I_{HRMS})^2 + (V_{HRMS}I_{1RMS})^2 + (V_{HRMS}I_{HRMS})^2 \tag{30}$$

$$S^2 = S_1^2 + S_N^2 \tag{31}$$

$$S = \sqrt{S_1^2 + S_N^2} \tag{32}$$

In Equation (31), S_N is the non-fundamental apparent power because of the occurrence of harmonic frequencies in the voltage or current waveforms. The fundamental apparent power S_1 is represented by:

$$S_1 = V_{1RMS}I_{1RMS} \tag{33}$$

The non-fundamental apparent power has three distinct terms, as shown in Equation (34).

$$S_N^2 = D_I^2 + D_V^2 + S_H^2 \tag{34}$$

The term D_I denotes the current distortion power and comprises the fundamental voltage (V_{1RMS}) and the harmonic current (I_{HRMS}). It is determined as a function of the fundamental apparent power (S_1) and the current total harmonic distortion (THD_I):

$$D_I = V_{1RMS}I_{HRMS} = S_1(THD_I) \tag{35}$$

Similarly, the voltage distortion power, D_V, encompasses the harmonic voltage (V_{HRMS}) and the fundamental current (I_{1RMS}):

$$D_V = V_{HRMS} I_{1RMS} = S_1 (THD_V) \tag{36}$$

Eventually, the harmonic apparent power, S_H, is determined from both the voltage and the current terms, and it contains the total harmonic distortion of the voltage and current, as given below:

$$S_H = V_{HRMS} I_{HRMS} = S_1 (THD_V)(THD_I) \tag{37}$$

By substituting the active power given in Equation (26) and the total apparent power represented in (33), the displacement power factor is computed by using Equation (26). Then, the *TPF* is computed using the following equation:

$$TPF = DF \times DPF \tag{38}$$

Thus, the *TPF* is measured by using the 4-term MSCW-based EDLIFFT. The real-time (RT) validation of the proposed method of *TPF* measurement for identifying appliance usage patterns for DSM is presented in the next section.

3. Real-Time (RT) Measurement of the TPF

The real-time measurement of the TPF by using the 4-term MSCW-based EDSLIFFT for non-intrusive appliance usage pattern identification for DSM is described in this section. A compact reconfigurable input–output system (cRIO 9082)-based virtual instrumentation testbed from National Instruments (NI) was developed in order to deploy the proposed non-intrusive identification of appliance usage patterns. As per the authors' observation, it is prospective hardware for RT measurement of the TPF as per the requirements of international standards, such as IEEE 1459 [18], 1159 [21], and IEC 61000-4-60 [22]. Table 1. Summarizes the standards compliance of the proposed RT-based measurement method.

Table 1. Summary of the standards compliance of the proposed RT-based measurement method.

Standard	Electrical Parameters	Compliance
IEEE 1459-2010: Standard Definitions for the Measurement of Electric Power Quantities Under Sinusoidal, Non-Sinusoidal, Balanced, or Unbalanced Conditions	Instantaneous Power, Active Power, Reactive Power, Apparent power, Non-Active Power, Voltage THD, Current THD, PF	Yes
IEEE 1159-2019: IEEE Recommended Practice for Monitoring Electric Power Quality	RMS voltage, RMS Current, Frequency	Yes
IEC 61000-4-60: Testing and measurement techniques—power quality measurement methods	Power Frequency, Flicker, Voltage Magnitude, Unbalance, Harmonics, Interharmonics	Yes

The NI-cRIO 9082 was equipped with a field-programmable gate array (FPGA) architecture, which was detailed in [23]. The voltage and current signals were acquired by the NI-cRIO 9082 with the NI-9225 (voltage) [24] and NI-9227 (current) modules [25]. Then, they were processed with the 4-term MSCW-based EDSLIFFT, which was deployed in a LabVIEW-configured desktop computer system. The desktop computer was interfaced with the NI-cRIO 9082 through a TCP/IP link, as depicted in Figure 3.

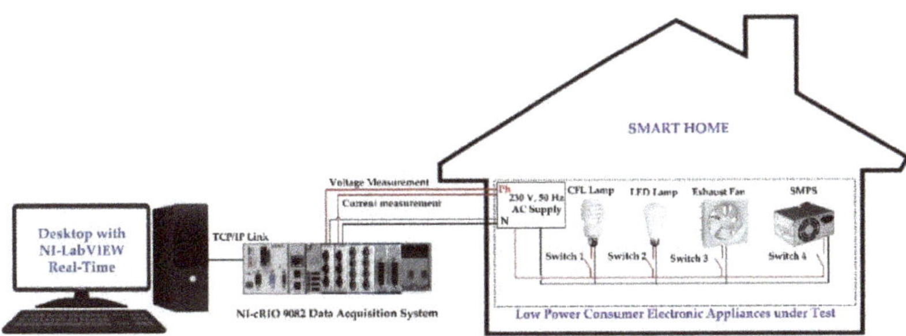

Figure 3. Hardware/lab setup for the measurement of the TPF.

CFLs, LEDs, exhaust fans, and SMPSs of personal computers are the LPCEAs that are most commonly used by consumers in developing countries; they were used here to showcase the severity of the problem for the readers, as every home that uses these appliances knowingly or unknowingly causes harmonic pollution.

The measurement hardware was coupled at the single-phase 230 V, 50 Hz utility supply mains for the TPF measurements. These TPF values were used for non-intrusive identification of appliance usage patterns. This non-intrusive appliance usage pattern identification approach took advantage of signal processing to reduce the hardware effort associated with systems intended for the identification of intrusive appliances with multiple dedicated sensors.

The TPFs of the different appliance usage patterns were measured by turning the connected appliances on or off, as shown in Figure 2. The detailed flowchart of the non-intrusive identification of appliance usage patterns with the TPF is shown in Figure 4.

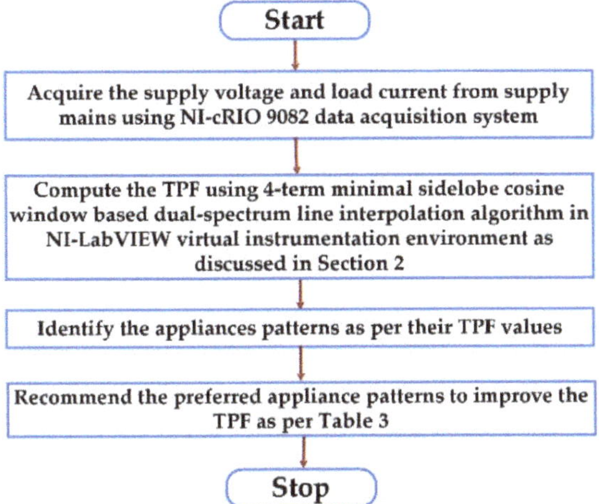

Figure 4. Flowchart for the non-intrusive identification of appliance usage patterns with the TPF.

4. RT Results

The supply voltage and appliance current waveforms acquired with the NI-cRIO 9082 for various real-world appliance combinations are illustrated in Figures 5–8. The RT values of the average TPF, active, reactive, and apparent power, and percentage THD over 24 h that were obtained from NI-cRIO at 50 kS/s are tabulated in Tables 2 and 3. The figures depict the phase differences between the input voltage and the harmonic current. The RT data values for P, Q, S, S1, SN, TPF, and %THD were observed and are tabulated in Table 4.

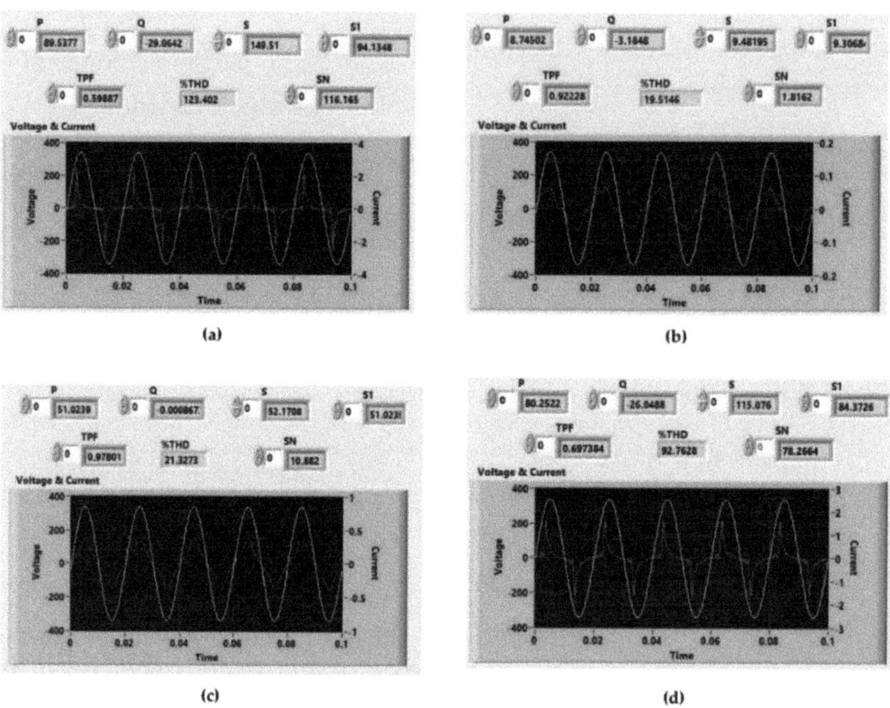

Figure 5. The voltage and current waveforms of a single real-world appliance: (**a**) CFL, (**b**) LED, (**c**) exhaust fan, and (**d**) SMPS of a PC.

Table 2. TPFs for all 15 appliance patterns in real time.

S.No	Combinations of Different Appliances	CODE	TPF
1	CFL	1 0 0 0	0.59887
2	LED	0 1 0 0	0.92228
3	Exhaust Fan	0 0 1 0	0.97801
4	SMPS of PC	0 0 0 1	0.69738
5	CFL + LED	1 1 0 0	0.63145
6	CFL + Exhaust Fan	1 0 1 0	0.75781
7	CFL + SMPS of a PC	1 0 0 1	0.64612
8	LED + Exhaust Fan	0 1 1 0	0.97925
9	LED + SMPS of a PC	0 1 0 1	0.72841
10	Exhaust FAN + SMPS of a PC	0 0 1 1	0.84232
11	CFL + LED + Exhaust Fan	1 1 1 0	0.77422
12	CFL + LED + SMPS of a PC	1 1 0 1	0.66289
13	LED + Exhaust Fan + SMPS of a PC	0 1 1 1	0.85390
14	CFL + Exhaust Fan + SMPS of a PC	1 0 1 1	0.73836
15	CFL + LED + Exhaust Fan + SMPS of a PC	1 1 1 1	0.74947

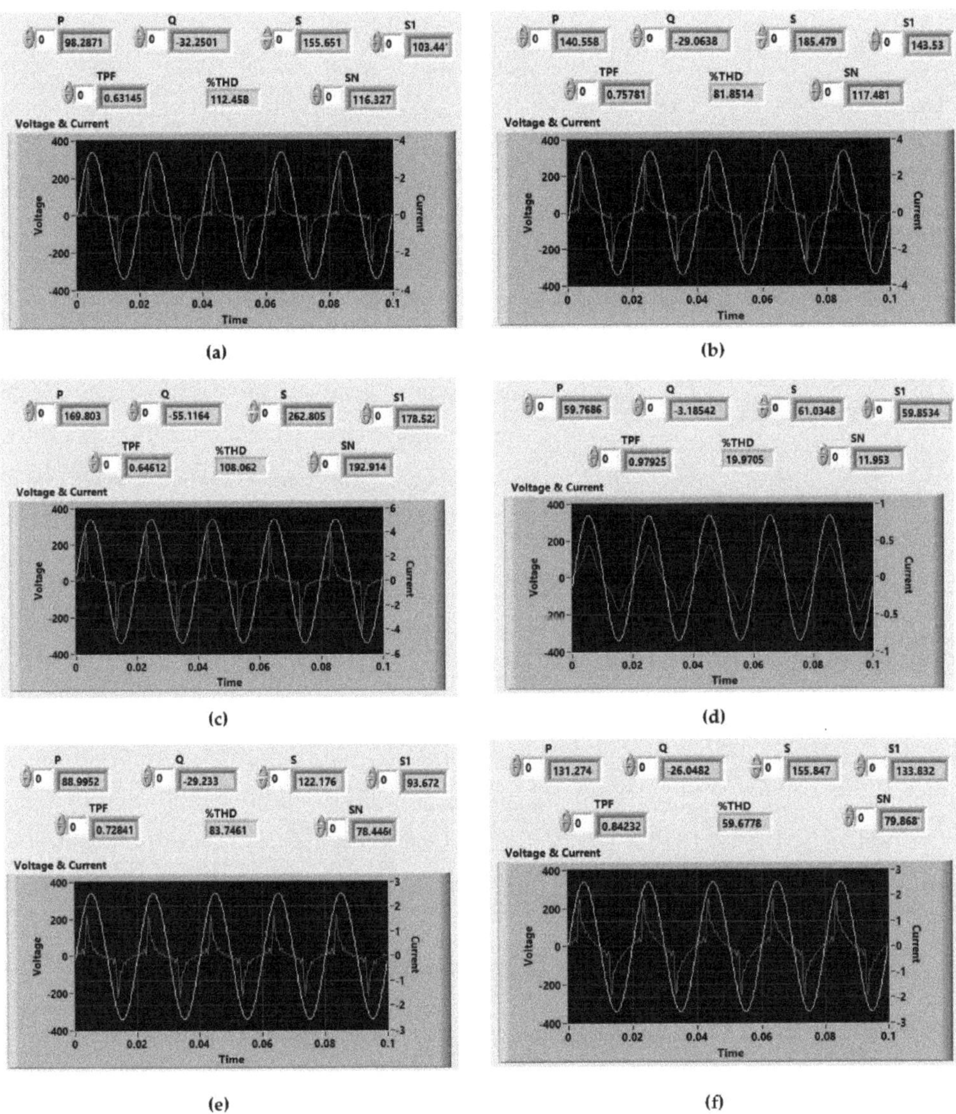

Figure 6. The voltage and current waveforms of combinations of two real-world appliances: (**a**) CFL and LED; (**b**) CFL and exhaust fan; (**c**) CFL and SMPS of a PC; (**d**) LED and exhaust fan; (**e**) LED and SMPS of a PC; (**f**) exhaust fan and SMPS of a PC.

The combinations of two real-world appliances were observed by turning on the combinations of the different two appliances. The voltage and current waveforms of the combinations of the two real-world appliances acquired with the NI-cRIO 9082 are depicted in Figure 6.

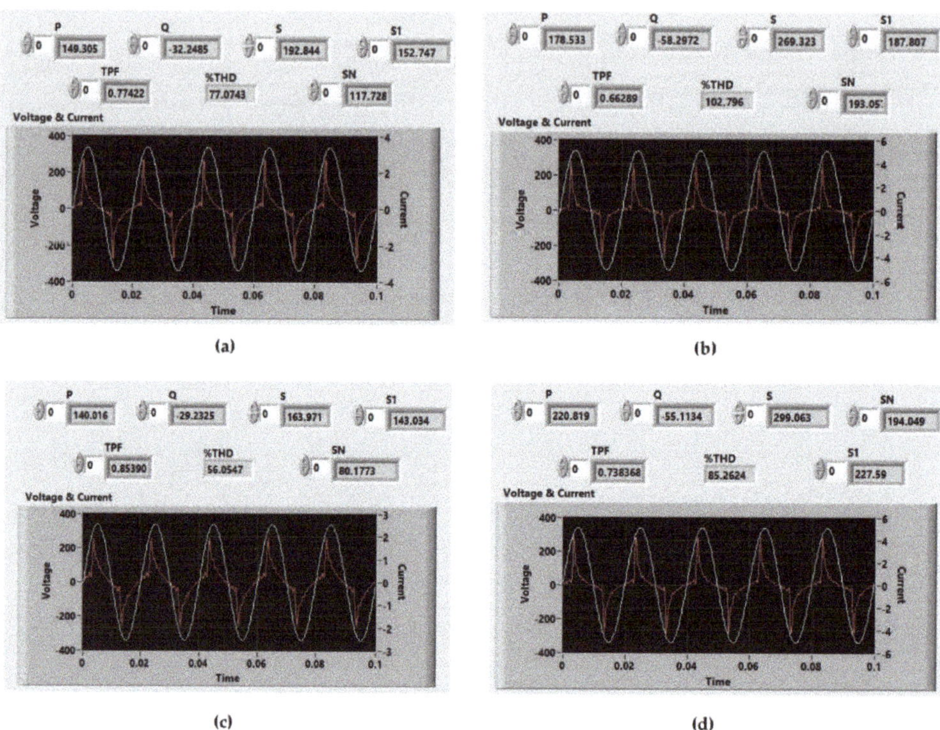

Figure 7. The voltage and current waveforms of combinations of three real-world appliances: (**a**) CFL + LED + exhaust fan;
(**b**) CFL + LED + SMPS of a PC; (**c**) LED + exhaust fan + SMPS of a PC; (**d**) CFL + exhaust fan + SMPS of a PC.

The combinations of three real-world appliances were examined by turning on the combinations of three appliances; the corresponding voltage and current waveforms of the CFL, LED, exhaust fan, and SMPS of a PC are illustrated in Figure 7.

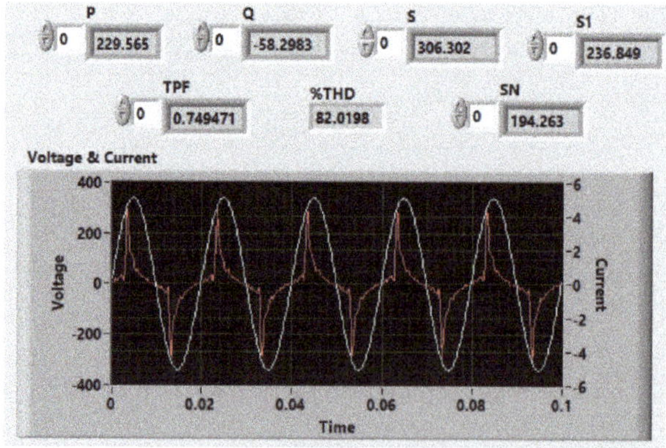

Figure 8. The voltage and current waveforms of the combination of the four real-world appliances: CFL + LED + exhaust fan + SMPS of a PC.

The voltage and current waveforms when all of the appliances were turned on are depicted in Figure 8.

Pertaining to the TPF values obtained from the operation of a single appliance to the operation of all four appliances, all were unique. Therefore, the TPF can be effectively used to identify appliance usage patterns. The TPFs measured with the four-term MSCW-based EDSLIFFT in the RT NI-cRIO 9082 system environment are depicted in Figure 4 and are tabulated in Table 2.

The performance of the experimentation demonstrated that the TPF could be uniquely identified for all combinations of the appliances. The standard deviation of the TPFs indicated that they were all unique and different. Hence, is the TPF can be used as a key feature of a lookup table in order to discern the appliances being operated.

Table 3. Actionable insights for DR management using the TPF.

Actionable Insights and Benefits							
Demand				**Response**			**Benefits**
S.No	**CODE**	**TPF**	**Actionable Insights**	**S.No**	**CODE**	**TPF**	**Change in TPF%**
1	1 0 0 0	0.59887	Turn off CFL	2	0 1 0 0	0.92228	−54.0033
2	0 1 0 0	0.92228	NR [1]	NR	0 1 0 0	0.92228	0
3	0 0 1 0	0.97801	NR	NR	0 0 1 0	0.97801	0
4	0 0 0 1	0.69738	Turn off LED for daytime	9	0 0 0 1	0.72841	−4.4495
5	1 1 0 0	0.63145	Turn off CFL	2	0 1 0 0	0.92228	−46.0574
6	1 0 1 0	0.75781	Turn off CFL	8	0 1 1 0	0.97925	−29.2210
7	1 0 0 1	0.64612	Turn off CFL for daytime	9	0 0 0 1	0.72841	−12.7360
8	0 1 1 0	0.97925	Turn off LED for daytime	3	0 0 1 0	0.97801	0.1266
9	0 1 0 1	0.72841	NR	NR	0 1 0 1	0.72841	0
10	0 0 1 1	0.84232	NR	NR	0 0 1 1	0.84232	0
11	1 1 1 0	0.77422	Turn off CFL	8	0 1 1 0	0.97925	−26.4821
12	1 1 0 1	0.66289	Turn off CFL	9	0 1 0 1	0.72841	−9.8839
13	0 1 1 1	0.85390	Turn off LED for daytime	10	0 0 1 0	0.84232	1.3561
14	1 1 0 1	0.73836	Turn off CFL	13	0 1 1 1	0.85390	−15.6481
15	1 1 1 1	0.74947	Turn off CFL	13	0 1 1 1	0.85390	−13.9338
Standard Deviation		0.1214		Standard Deviation		0.099	

[1] NR = No recommendation.

The TPF variations of the appliance usage patterns according to demand are depicted in Figure 9a. The load patterns 1, 4, 5, 6, 7, 9, 12, 14, and 15 are in the region of poor TPF values. The TPFs of the appliance usage patterns after executing the recommendations given by the actionable insights in Table 3 are illustrated in Figure 9b, which demonstrates the improvement of the TPF. Further, it is necessary to adopt mitigation devices in order to improve the TPF as per the requirements of the IEEE 519-2014 standard [26].

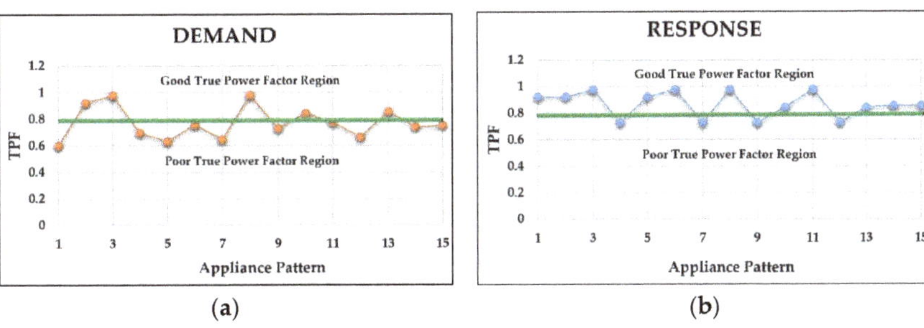

Figure 9. The TPFs of appliance usage patterns according to: (**a**) demand; (**b**) responses to the insights.

Table 4. Actionable insights for the improvement of the utilization index.

		Demand							Response					Benefits
							Actionable Insights and Benefits							
S.No	CODE	P (Watts)	Q (VARs)	S (VAs)	S_1 (VAs)	S_N (VAs)	Utilization Index = S_1/S	S.No	CODE	P (Watts)	Q (VARs)	S (VAs)	S_1/S	% Change in Utilization Index
1	1 0 0 0	89.5377	−29.0642	149.51	94.1348	116.165	0.6296	2	0 1 0 0	8.74502	−3.1848	9.48195	0.9815	35.8531
2	0 1 0 0	8.74502	−3.1848	9.48195	9.30684	1.8162	0.9815	NR	0 1 0 0	8.74502	−3.1848	9.48195	0.9815	0
3	0 0 1 0	51.0239	−0.0008	52.1708	51.0239	10.882	0.9780	NR	0 0 1 0	51.0239	−0.0008	52.1708	0.9780	0
4	0 0 0 1	80.2522	−26.0488	115.076	84.3726	78.2664	0.7331	9	0 0 0 1	88.9952	−29.233	122.176	0.7666	4.3703
5	1 1 0 0	98.2871	−32.2501	155.651	103.447	116.3270	0.6646	2	0 1 0 0	8.74502	−3.1848	9.48195	0.9815	32.2887
6	1 0 1 0	140.558	−29.0638	185.479	143.53	117.481	0.7738	8	0 1 1 0	59.7686	−3.18542	61.0348	0.9806	21.0892
7	1 0 0 1	169.803	−55.1164	262.805	178.522	192.914	0.6792	9	0 0 0 1	88.9952	−29.233	122.176	0.7666	11.3999
8	0 1 1 0	59.7686	−3.18542	61.0348	59.8534	11.953	0.9806	3	0 0 1 0	51.0239	−0.0008	52.1708	0.9780	−0.2686
9	0 1 0 1	88.9952	−29.233	122.176	93.672	78.446	0.7666	NR	0 1 0 1	88.9952	−29.233	122.176	0.7666	0
10	0 0 1 1	131.274	−26.0482	155.847	133.832	79.868	0.8587	NR	0 0 1 1	131.274	−26.0482	155.847	0.8587	0
11	1 1 1 0	149.305	−32.2485	192.844	152.747	117.728	0.7920	8	0 1 1 0	59.7686	−3.18542	61.0348	0.9806	19.2290
12	1 1 0 1	178.533	−58.2972	269.323	187.807	193.052	0.6973	9	0 1 0 1	88.9952	−29.233	122.176	0.7666	9.0475
13	0 1 1 1	140.016	−29.2325	163.971	143.034	80.1773	0.8723	10	0 0 1 0	131.274	−26.0482	155.847	0.8587	−1.5806
14	1 1 0 1	220.819	−55.1134	299.063	227.59	194.049	0.7610	13	0 1 1 1	140.016	−29.2325	163.971	0.8723	12.7595
15	1 1 1 1	229.565	−58.2983	306.302	236.849	194.263	0.7732	13	0 1 1 1	140.016	−29.2325	163.971	0.8723	11.3560

NR = No recommendation.

The actionable insights for DR management are tabulated in Table 3, which establishes the opportunity to improve the TPFs in smart homes. Utilities can benefit from the reduced malfunctioning of the equipment in their distribution systems, and consumers can benefit from the enhanced lifetimes of their LPCEAs.

5. Discussion

This section discusses the analysis of the results and the scope of future research.

5.1. Analysis of the Results

- The harmonics are non-additive, and the cumulative TPF takes both the inductive and capacitive factors of various appliance combinations into consideration.
- Individual PFs are of no consequence, and the TPF is the value to consider in order to compare the impacts of appliance usage patterns with the impacts of individual appliances, as in our article. Therefore, the TPFs of appliance combinations are nonlinear, making the appliance patterns in use uniquely identifiable.
- The TPF can be improved by selecting alternative appliance patterns, as suggested in Table 4.
- Highly nonlinear appliances, such as CFLs, SMPSs, and their combinations, consume more reactive power, which results in a poor TPF.
- The standard deviation of the TPF from the experimental results was 0.1271 for the demand, and it was reduced to 0.099 according to the recommended consumer responses. Thus, the consumer utilization index improved, as illustrated in Table 4, and the quality of the power was enhanced.
- According to the results, it is necessary to adopt current harmonic mitigation devices, such as active power filters and PF correction devices in LVDSs, in order to minimize the malfunctioning of the utility equipment, such as failures of the transformers, false tripping of the circuit breakers, etc., as well as to safeguard LPCEAs from overheating.
- It is necessary to develop smart energy meters in order to measure the THD and TPF and to recommend PQ-sensitive measures, with a provision in the power tariffs for better consumer behavior.

The quantitative metrics tabulated in Table 5 demonstrate the efficacy of the proposed deterministic method over non-deterministic methods.

Table 5. Quantitative metrics for the non-intrusive identification of appliance usage patterns in comparison with those of statistical methods.

Quantitative Metric Category	Quantitative Metrics	TPF	Statistical Methods
Feature	Sampling rate	Medium	High
Method	Process of Execution	Experimental	Empirical
Accuracy	Disaggregation percentage (D)	100	<100
	Disaggregation Error (DE)	0	>0
	Precision(P)—TP [1]/(TP + FP [2])	1	<1
	Recall (R)—TP/TP + FN	1	<1
	Accuracy = (TP + TN [3])/(TP + TN + FP + FN [4])	1	<1
	F-measure (f1) $2 \times P \times R/(P + R)$	1	<1
Training	User interface	Simple	Complex
Real-Time Implementation	Deployment capability	High	Low
Scalability	Pace of deployment	High	Low
Identification Factor	The standard deviation of the TPF	0.1271	NA
Generalization	Generalization over unseen homes	High	Medium

[1] TP = true positive; [2] FP = false positive; [3] TN = true; [4] FN = false negative.

5.2. Future Research

This paper offers a simple way of identifying appliance usage patterns, but there are a few challenges before taking this concept to market. The authors propose the following as the scope for future research.

- A smart home may have very sophisticated appliances that could be programmed and operated remotely or controlled through a mobile application. How demand-response systems can take advantage of appropriate communication is of practical interest to explore.
- The physical/thermal characteristics of electronic appliances may change over time and may, in turn, change the THD and TPF values. The accommodation of these adaptabilities in the demand-side management systems must be considered.
- The compensation of the reactive power must be studied so that a power factor at unity can be achieved at all times and all nodes in a distribution system.

6. Conclusions

The goal of this research was to save energy with good PQ in order to meet SDG 7. Due to the huge penetration of LPCEAs and their resultant current THD, the TPFs in LVDSs are poor, and consequently, the energy efficiency and PQ are heavily deteriorated. In LVDSs, poor PFs and LPCEAs are not penalized, which results in losses to the DISCOMs. This article illustrated a deterministic approach that uses the TPF in order to identify the consumption patterns of LPCEAs by using the 4-term MSCW-based EDSLIFFT in a virtual instrumentation environment for various combinations of LPCEAs in real time. It was depicted that the TPF values could be used to effectively identify various LPCEA combinations. This method of the non-intrusive identification of appliance usage patterns is essential for responsible electricity consumption with TPF that are close to unity. It is necessary to streamline the tariff structure based on the PQ. It is also necessary to mandate the correction of the PF to maintain PF that is at unity in an LVDS as per CEA regulations (2010) and to holistically comply with SDG 7. The standard deviation of the TPF is impressive, but is not sufficient to comply with the IEEE 519-2014 standard, which requires the installation of compensation devices in LVDSs when appropriate. The actionable insights recommended in this article highlight the reduction in the active power consumption and kVA requirement; they also demonstrate the improvements in the consumer utilization index. The Republic of India does not specify requirements for domestic requirements at the 230 V level, and hence, no PF penalties are imposed. Hence, the TPF is not even a concern of consumers at this moment. This work intends to highlight the deficiencies in the distribution system that must be fixed.

Author Contributions: Conceptualization, H.P.D.; Formal analysis, H.P.D. and S.B.G.; Investigation, H.P.D., and S.B.G.; Methodology, H.P.D.; Project administration, H.P.D.; Resources, V.S.S.S.S.D.; Supervision, V.S.S.S.S.D. and S.B.G.; Validation, H.P.D. and V.S.S.S.S.D.; Writing—original draft, H.P.D.; Writing—review and editing, V.S.S.S.S.D. and S.B.G. All authors have read and agreed to the published version of the manuscript.

Funding: This research received no external funding.

Acknowledgments: The authors gratefully acknowledge the Department of Electrical Engineering of the National Institute of Technology Warangal for their support.

Conflicts of Interest: The authors declare no conflict of interest.

Nomenclature

n	Sample index
m	Window item index
M	Maximum window item number
h	Harmonic order
$hmax$	Maximum harmonic order
a_h	Minimal sidelobe cosine window (MSCW) coefficients
W	Four-term MSCW function
N	Total number of samples
T_s	Sampling time
t	Time period
A_h	Amplitude of the h_{th} harmonic component
f_h	Frequency of the h_{th} harmonic component
φ_h	Phase of the h_{th} harmonic component
k	Four-term MSCW sample points ($k = 0, 1, \ldots, N-1$)
k_h	Division factor of signal frequency
f_s	Sampling frequency
l_h	Integer value of greatest spectral line
ξ_h	Fraction part of greatest spectral line
l_{h1}	Spectral line 1 representing the hth harmonic amplitude
l_{h1}	Spectral line 2 representing the hth harmonic amplitude
I	Spectral line peak index location value
y_1	Amplitude of the spectral line 1
y_2	Amplitude of the spectral line 2
α	Symmetrical coefficient 1
β	Symmetrical coefficient 2
V_{RMS}	Total input voltage RMS value
I_{RMS}	Total load current RMS value
V_{1RMS}	Input fundamental Voltage RMS value
V_{hRMS}	h^{th} order harmonic voltage RMS value
I_{1RMS}	Load fundamental current RMS value
I_{hRMS}	h^{th} order harmonic load current RMS value
P	Active power consumed by the LPCEAs
Q	Reactive power consumed by the LPCEAs
θ_1	Fundamental voltage phase angle
δ_1	Fundamental load current phase angle
θ_h	h^{th} order harmonic voltage phase angle
δ_h	h^{th} order harmonic load current phase angle
V_{HRMS}	Summation of all of the harmonic voltages
I_{HRMS}	Summation of all of the harmonic currents
S	Apparent power
S_1	Fundamental apparent power component
S_N	Non-fundamental apparent power component
D_1	Current distortion power
D_V	Voltage distortion power
S_H	Harmonic apparent power
THD_V	Voltage total harmonic distortion
THD_I	Current total harmonic distortion

Abbreviations

CEA	Central Electricity Authority
CFL	Compact fluorescent lamp
c-RIO	Compact reconfigurable input–output
DF	Distortion factor
DPF	Displacement power factor
DR	Demand response
DSM	Demand-side management
DSICOMs	Distribution companies
EDSLIFFT	Enhanced dual-spectrum line interpolated FFT
FFT	Fast Fourier transform
FPGA	Field-programmable gate array
IEEE	Institute of Electrical and Electronics Engineers
LED	Light-emitting diode
LPCEA	Low-power consumer electronic appliance
LVDS	Low-voltage distribution system
MSCW	Minimal sidelobe cosine window
NI	National Instruments
NILM	Non-intrusive load monitoring
PC	Personal computer
PCC	Point of common coupling
PF	Power factor
PQ	Power quality
RMS	Root mean square
RT	Real time
SDG 7	Sustainable Development Goal 7
SMPS	Switch-mode power supply
THD	Total harmonic distortion
TPF	True power factor

References

1. Rahman, M.A.; Islam, R.; Sharif, K.F.; Aziz, T. Developing demand side management program for commercial customers: A case study. In Proceedings of the 2016 3rd International Conference on Electrical Engineering and Information Communication Technology (ICEEICT), Dhaka, Bangladesh, 22–24 September 2016; IEEE: Piscataway, NJ, USA, 2016; pp. 1–6.
2. Srikanth, R. India's sustainable development goals–glide path for India's power sector. *Energy Policy* **2018**, *123*, 325–336. [CrossRef]
3. Hazra, S.; Bhukta, A. *Sustainable Development Goals. An Indian Perspective*, 1st ed.; Springer Nature: Cham, Switzerland, 2020; pp. 107–127.
4. Verma, P.; Patel, N.; Nair, N.-K.C. Demand side management perspective on the interaction between a non-ideal grid and residential LED lamps. *Sustain. Energy Technol. Assess.* **2017**, *23*, 93–103. [CrossRef]
5. Devarapalli, H.P.; Dhanikonda, V.S.S.S.; Gunturi, S.B. Non-Intrusive Identification of Load Patterns in Smart Homes Using Percentage Total Harmonic Distortion. *Energies* **2020**, *13*, 4628. [CrossRef]
6. Oruganti, V.S.R.V.; Bubshait, A.S.; Dhanikonda, V.S.S.S.; Simões, M.G. Real-time control of hybrid active power filter using conservative power theory in industrial power system. *IET Power Electron.* **2017**, *10*, 196–207. [CrossRef]
7. Jimenez, Y.; Cortes, J.; Duarte, C.; Petit, J.; Carrillo, G. Non-intrusive discriminant analysis of loads based on power quality data. In Proceedings of the 2019 IEEE Workshop on Power Electronics and Power Quality Applications (PEPQA), Manizales, Colombia, 30–31 May 2019; IEEE: Piscataway, NJ, USA, 2019; pp. 1–5.
8. Francisco, C. *Harmonics, Power Systems, and Smart Grids*, 2nd ed.; CRC Press: Boca Raton, FL, USA, 2015; pp. 1–278.
9. Oruganti, V.S.R.V.; Dhanikonda, V.S.S.S.; Simões, M.G. Scalable single-phase multi-functional inverter for integration of rooftop solar-PV to low-voltage ideal and weak utility grid. *Electronics* **2019**, *8*, 302. [CrossRef]
10. Mageed, H.; Nada, A.S.; Abu-Zaid, S.; Eldeen, R.S.S. Effects of waveforms distortion for household appliances on power quality. *MAPAN* **2019**, *34*, 559–572. [CrossRef]
11. da Silva, R.P.B.; Quadros, R.; Shaker, H.R.; da Silva, L.C.P. Effects of mixed electronic loads on the electrical energy systems considering different loading conditions with focus on power quality and billing issues. *Appl. Energy* **2020**, *277*, 115558. [CrossRef]
12. da Silva, R.P.B.; Quadros, R.; Shaker, H.R.; da Silva, L.C.P. A Mixed of Nonlinear Loads and their Effects on the Electrical Energy Billing. In Proceedings of the 2020 IEEE 8th International Conference on Smart Energy Grid Engineering (SEGE), Oshawa, ON, Canada, 12–14 August 2020; IEEE: Piscataway, NJ, USA, 2020; pp. 116–120.
13. da Silva, R.P.B.; Quadros, R.; Shaker, H.R.; da Silva, L.C.P. Harmonic Interaction among Electronic Loads and Its Effects on the Electrical Quantities and Billing: Case Study with Lighting Devices. In Proceedings of the 2020 7th International Conference on Electrical and Electronics Engineering (ICEEE), Antalya, Turkey, 14–16 April 2020; IEEE: Piscataway, NJ, USA, 2020; pp. 53–60.

14. Ruano, A.; Hernandez, A.; Ureña, J.; Ruano, M.; Garcia, J. NILM techniques for intelligent home energy management and ambient assisted living: A review. *Energies* **2019**, *12*, 2203. [CrossRef]
15. Yang, H.; Xue, Y.; Liu, S.; Gao, B.; Shu, Y.; Xu, Y.; Wang, J. A Judging Method of Electric Larceny in the Guise of Saving Electrical Energy. Application of Intelligent Systems in Multi-modal Information Analytics. MMIA 2019. In *Advances in Intelligent Systems and Computing*; Sugumaran, V., Xu, Z., Shankar, P., Zhou, H., Eds.; Springer: Cham, Switzerland, 2020; Volume 929, pp. 1029–1037.
16. Gopinath, R.; Kumar, M.; Joshua, C.P.C.; Srinivas, K. Energy management using non-intrusive load monitoring techniques-State-of-the-art and future research directions. *Sustain. Cities Soc.* **2020**, *62*, 102411. [CrossRef]
17. Dlamini, F.M.; Nicolae, D.V. An approach to quantify the technical impact of power quality in medium voltage distribution systems. In Proceedings of the 2016 IEEE International Power Electronics and Motion Control Conference (PEMC), Varna, Bulgaria, 25–28 September 2016; IEEE: Piscataway, NJ, USA, 2016; pp. 315–321.
18. IEEE. Standard Definitions for the Measurement of Electric Power Quantities under Sinusoidal, Nonsinusoidal, Balanced, or Unbalanced Conditions. In *IEEE Std. 1459–2010 (Revision of IEEE Std. 1459–2000)*; IEEE: New York, NY, USA, 2010.
19. Oruganti, V.S.R.V.; Dhanikonda, V.S.S.S.S.; Paredes, H.K.M.; Simões, M.G. Enhanced Dual-Spectrum Line Interpolated FFT with Four-Term Minimal Sidelobe Cosine Window for Real-Time Harmonic Estimation in Synchrophasor Smart-Grid Technology. *Electronics* **2019**, *8*, 191. [CrossRef]
20. Varaprasad, O.V.S.R.; Sarma, D.S.; Panda, R.K. Advanced windowed interpolated FFT algorithms for harmonic analysis of electrical power system. In Proceedings of the 2014 Eighteenth National Power Systems Conference (NPSC), Guwahati, India, 18–20 December 2014; IEEE: Piscataway, NJ, USA, 2014; pp. 1–6.
21. IEEE. Recommended Practice for Monitoring Electric Power Quality. In *IEEE Std. 1159–2019 (Revision of IEEE Std. 1159–2009)*; IEEE: New York, NY, USA, 2014.
22. *IEC 61000 4–30 Electromagnetic Compatibility (EMC)—Part 4–30: Testing and Measurement Techniques—Power Quality Measurement Methods*; International Electrotechnical Commission: Geneva, Switzerland, 2015.
23. User Manual, NI cRIO-9082. Available online: http://www.ni.com/pdf/manuals/376904a_03.pdf (accessed on 16 June 2021).
24. Operating Instructions and Specifications, NI 9225. Available online: https://www.ni.com/pdf/manuals/374707e.pdf (accessed on 16 June 2021).
25. Operating Instructions and Specifications, NI 9227. Available online: https://www.ni.com/pdf/manuals/375101e.pdf (accessed on 16 June 2021).
26. IEEE. Recommended practices and requirements for harmonic control in electric power systems. In *IEEE Std. 519–2014 (Revision of IEEE Std. 519–1992)*; IEEE: New York, NY, USA, 2014.

Article

Enabling Technologies for Energy Communities: Some Experimental Use Cases

Daniele Menniti [1], Anna Pinnarelli [1], Nicola Sorrentino [1], Pasquale Vizza [1,*], Giuseppe Barone [1], Giovanni Brusco [2], Stefano Mendicino [2], Luca Mendicino [1] and Gaetano Polizzi [1]

[1] Department of Mechanical, Energy and Management Engineering, University of Calabria, 87036 Rende, Italy
[2] Creta Energie Speciali S.r.l., 87036 Rende, Italy
* Correspondence: pasquale.vizza@unical.it

Abstract: It is known that the energy transition can be achieved not only with the use of renewable energy sources but also with a new conception and management of the electricity system. Renewable energy communities are then introduced as organizations for maximizing the self-consumption of energy produced from renewable energy sources. To ensure that these energy communities can operate, there is a need for enabling technologies that allow for monitoring, data and algorithms processing as well as the enabling of the same algorithms. There exists a huge confusion in the actual technologies useful to implement the energy communities. This paper first describes and groups the main enabling technologies, analyzing the services that can be offered. The scope is to emphasize the importance of having accurate, efficient and effective technologies that allow the implementation of such communities, underlining how such technologies interact with each other. Using such technologies is important to observing the possible technical and energetic results; indeed, use cases concerning the use of these enabling technologies are proposed and analyzed, showing their operating and their good environmental and energy impact.

Keywords: enabling technologies; energy community; smart meter; nanogrid; platform; power cloud

---

Citation: Menniti, D.; Pinnarelli, A.; Sorrentino, N.; Vizza, P.; Barone, G.; Brusco, G.; Mendicino, S.; Mendicino, L.; Polizzi, G. Enabling Technologies for Energy Communities: Some Experimental Use Cases. *Energies* **2022**, *15*, 6374. https://doi.org/10.3390/en15176374

Academic Editor: Surender Reddy Salkuti

Received: 21 July 2022
Accepted: 27 August 2022
Published: 31 August 2022

Publisher's Note: MDPI stays neutral with regard to jurisdictional claims in published maps and institutional affiliations.

1. Introduction

The energy transition implies a new model of social organization based on the production and consumption of energy from Renewable Energy Sources (RESs). The RESs are playing and will play an increasingly important role. It is essential to take into account their characteristics: non-programmability, uncertainty in the predictability of generation capacity, the lack of temporal coincidence between production and the final uses of energy demand and limited possibility of supplying regulation services.

Storage systems will play a central role for conventional uses (energy time shift; continuity service) and the network integration of RESs (synthetic inertia, Fast reserve, secondary and tertiary reserve, network congestion resolution, voltage regulation), system, network (transmission, distribution, local) and operator (producer/consumer) needs. The functional characteristics of storage systems can be divided into energy and power applications: the first one with a large capacity to exchange power for long periods (hours), the second one to exchange high power for short periods (seconds, minutes).

1.1. Motivation and Incitement

In this framework, different business models to better manage such sources (generation and storage) have been introduced. Among such models, the energy community or, in general, the end-user's aggregation represents a solution to optimally manage the energy production of renewable energy sources, maximizing their use.

The European Commission, in 2016, to place the consumer at the heart of the energy transition, introduced the energy communities as part of the Clean Energy for all European

---

Package (CEP). With the recent renewable energy directive (RED II) and the recent electricity market directive (EMD), a legal framework for "citizen energy communities" (CEC) and "renewable energy communities" (REC) was introduced to be interpreted and adopted into the member states' national legislation. Although energy communities constitute now a legally defined and recognized entity by the institutions of the European Union, the directives that have been promoted and voted on at an energy community level do not appear to have been transposed yet into national law in most of the member states. Even in cases where there is a sufficiently defined national legal framework, there are low rates of energy communities' development.

1.2. Literature Review and Related Works

Energy communities can be seen as an opportunity for citizens to actively participate not only in the community but also in the energy market [1].

In [2], the Local Renewable Energy Organizations are defined as organizations initiated and managed by actors from civil society, with the aim to educate or facilitate people on efficient energy use, enable the collective procurement of renewable energy or technologies or provide energy from renewable resources.

This definition, which anticipates the energy community one, highlights how the role of the citizen is central and how it is essential.

Indeed, the REDII gives a first definition of what RECs could be. Starting from the definition of renewables self-consumers and jointly acting renewables self-consumers [3], the energy Community is defined. A renewables self-consumer is defined as a final customer who generates renewable electricity for their own consumption and who may store or sell self-generated renewable electricity activity. According to the RED II, an REC can be considered as a legal entity based on open and voluntary participation, and autonomous, controlled by its shareholders or members. Such shareholders or members are in the proximity of renewable energy projects owned and developed by that legal entity. Its primary purpose is to provide environmental, economic, and social community benefits rather than financial profits.

Moreover, the CECs are defined. The differences from RECs are that they may engage in generation, including from renewable sources, distribution, supply, consumption, aggregation, energy storage, energy efficiency services or charging services for electric vehicles or provide other energy services. Another main difference seems to be the proximity aspect that characterizes the RECs [3] instead of the participation of CECs that can be spread over the territory.

The RECs aim for the participation of individuals to improve the local acceptance of renewable energy, local investment and improved participation of citizens in the energy transition [3]. The RECs will produce, consume, store and sell renewable energy, share produced renewable energy within the community and access energy markets in an integral way [3].

One of the open questions related to the RECs is the optimal number, the power plant size and the optimal mix of the several kinds of end-users to obtain the best environmental and economic results.

In [4], the main advantages of energy communities are underlined:

- cost reduction in the energy vector procurement;
- improvement of reliability and quality supply;
- active participation of citizens and use of local resources.

If an energy community, both REC and CEC, must be implemented, to obtain the aforementioned benefits, enabling technologies have to be considered. Such technologies, in some features, can be compared to the Demand Response (DR) enabling technologies.

In [5], an analysis of the enabling technologies for the DR has been carried out and has been divided into technologies for the Utility Domain and those for the End-User Domain. They are divided into four categories: metering and monitoring devices, control devices, communication systems and software programs. The same consideration about the enabling technologies can also be taken into account for the technologies used for the implementation and management of energy communities, even considering that DR (implicit or explicit) is one of the management methods of the energy communities. The technologies for both utility and end-user domains should provide a platform for the control and management of users' data with the aim to keep them informed on their behavior. There is the necessity of measuring and monitoring infrastructure for the several present devices of a control infrastructure, a communication system and of several software programs for managing the various data according to algorithms and the control of the field devices.

In [6], the different technologies that currently play a significant role in the transition to a Smart community were analyzed, not only considering energy aspects. Some of them are also important and are used as enabling technologies for energy communities. Among such technologies, there are the IoT devices, the cloud computing, the platforms and the sensors. From such devices, there is a huge quantity of data that must be analyzed as well as stored, so issues related to big data interest also affect the Energy Community implementation.

In [7], it is shown how the aspects related to enabling technologies, even in different fields, concern various sectors, ranging from sensors to security protocols, to problems relating to big data, to the information communication and technology systems.

Starting from this, both for smart cities but also in other areas such as energy communities, they lead to various challenges and solutions concerning the technical, environmental and socio-economic aspects.

In [8], a review of enabling technologies to develop a smart building is reported. Additionally, in this case, the different involved fields are analyzed, and the issues are underlined. Considering the particularity of IoT technologies, the necessity of network, cloud, system analytics, actuators and user interface is underlined.

In [9,10], a platform for energy communities is implemented, and the structure is presented. It is useful to collect and analyze the data inside the Energy Communities with the purpose to encourage the users to have aware behavior. This platform has to integrate a huge quantity of data deriving from the enabling technologies such as smart meters.

An aim of the platform is to manage the users of energy communities, providing them with useful services, as implemented in [11]. Among the services that can be provided using an opportune platform, there are those associated with blockchain that allow users to trade energy in a community [12]; when local energy trading and in particular renewable energy exchange are considered, the smart contract assumes a fundamental role, as described in [13–15]. This allows both to operate the day ahead and in a quasi-real-time way, providing services to the grid.

1.3. Contribution and Paper Organization

The enabling technologies presented in this paper are:

- the smart meter, to measure and monitor the production and the load consumption for the end-user's engagement, awareness and empowerment [16];
- the DC Nanogrid (DCNG), for real-time management of power flows among several types of generation and storage units and to maximize the shared energy in an Energy community framework [17];
- a platform for Energy Community management and so to support their creation, constitution and development.

The contribution of the paper is:

1. A review of the principal enabling technologies;
2. A specific settlement of the "advanced end-user", illustrating its technological configuration and identifying the necessary enabling technologies (smart meter, smart energy storage system, DCNG and the Energy Community Management platform (ECM));
3. The DCNG as the main enabling technology for the cooperation of more advanced end-users in the context of energy communities to maximize the shared energy;

4. The definition of some use cases and corresponding performance indexes to evaluate how a single advanced end user can operate and when it operates in an REC with the DCNG as the grid interface;

5. Demonstration test cases.

6. One of the issues of the related works is that the enabling technologies are described for a particular scope of an energy community or in general for an aggregation. There is no literature on such technologies relating to Renewable Energy Communities. Moreover, such technologies are not utilized in the context for which they are described. In the present paper, such technologies are described from a system point of view, then an example of such technologies is implemented and utilized, showing the performance both of such technologies and of their operation in an energy community.

2. Principal Enabling Technologies for the Rec Implementation

To meet the objectives of decarbonization and to allow the energy transition to be effective, cultural changes based on energy saving and consumption efficiency must be started.

A change in producing, storing, consuming and exchanging energy is needed to be self-sufficient; at the same time, the energy is shared, and citizens are involved in the sustainable development of their city.

Therefore, a series of enabling technologies that allow all of this to be possible become necessary, allowing the integration of different resources that would otherwise be difficult to integrate. At the same time, they allow and facilitate the implementation and management of energy communities.

In the paper, the attention is focused on the smart meter (SM), energy storage system (ESS), DCNG and Energy Community management platform (ECM).

2.1. Smart Meter Description and Peculiarities

Among the various enabling technologies for the Energy community, there are technologies that allow us to measure and monitor the power exchanged by the end-user with the grid; this is allowed by the smart meter (SM).

It consists of an energy monitoring system, which allows us to analyze electricity consumption and production both for residential and industrial end-users. This can allow the end-user to optimize the use of RES generation and change their behavior to reduce energy consumption.

The SM, in addition to having the purpose of making the end-user aware of their behavior, is used to monitor and acquire the various data that will be used for the energy management and optimization platform for the REC.

The end-users are monitored, including consumers, producers and prosumers, and, according to these data, it is possible to proceed to manage energy to maximize the self-consumption to provide flexibility services to the grid, as well as to balance the energy within the aggregation.

In [18], it is highlighted how, with the introduction of SMs, it is possible to offer new services: providing more detailed information on electricity, using ad hoc applications or simply informing how to save energy.

At the level of DSO, and in particular of the energy provider, the use of the SM is an essential tool to limit energy fraud, alongside that of the development of smart grids, including a variety of different components.

In [19], an analysis of the SMs and their different uses is carried out.

In [20], the role of SMs in monitoring the network and providing services for energy communities is analyzed, highlighting how time resolution is a parameter that must be taken necessarily into consideration according to the services and the purpose of such technologies.

In [21], it is highlighted how the use of an efficient monitoring system, especially of reactive power, is also useful in compensating for the power factor in a system where reactive power represents a problem. It is also highlighted how the use of the saved data allows for the improvement in the compensation operations of this power factor. The utilization of such data allows in fact to carry out an accurate analysis of the users.

In [22], it is highlighted that a high temporal resolution is necessary to be able to develop models for the management of domestic users. This is underlined also in this paper, where a particular SM with a high temporal resolution for management and monitoring purposes is utilized.

The SM has brought advanced functionalities oriented to smart grid and end-user needs such as the Demand Side Management (DSM) for unregulated market trading and real-time information. The proposed SM, as depicted in Figure 1, is realized to change a normal citizen into an active smart citizen with its energy behavior.

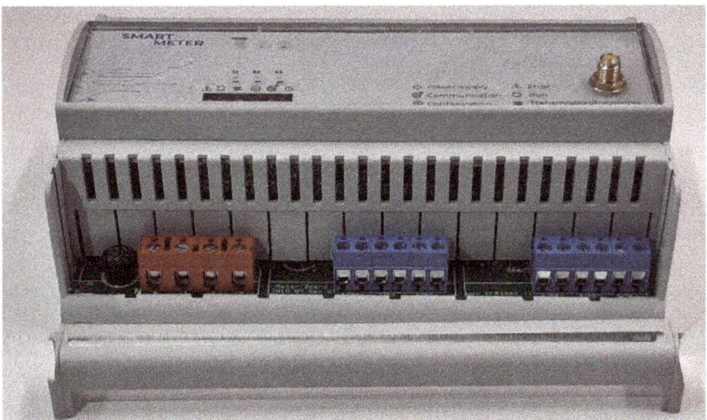

Figure 1. Smart meter.

Smart citizens changing their lifestyle routines and energy consumption can provide energy services and consumption flexibility, also reducing their energy costs. A smart citizen becomes part of a 'smart energy community' and helps to ensure the quality of supply and environment preservation.

The SM's functionalities are understandable, user-friendly and standardized and provide a broad range of technical functionalities to inform end-users about their consumption, helping to increase awareness of energy consumption and, furthermore, engage them actively to participate in the electricity (and gas, water, heat) supply market.

Smart Meter Communication

To allow the acquisition of measurement data recorded by SMs, a software prototype has been developed on the cloud and stores the data on a MySQL relational Data Base (DB). The software communicates with the SMs from which it acquires electrical measurements.

The data stored on a DB are appropriately aggregated and analyzed to be used as input for any dashboard for the users or for the aggregator.

Such SMs obviously need a reliable internet connection to send the measured data every required time step (5 s in this case) and thus avoid their dispersion in order not to affect the operation of the same optimization and energy management models.

This Smart Meter is a combination of both hardware and software. The hardware part was designed and built on an electronic board that must condition electrical quantities to create an intelligent monitoring tool. The software management of the smart meter was implemented using a TM4C1294 Texas Instruments microcontroller.

All the electrical quantities to be monitored are suitably transduced and conditioned to be acquired by the ADCs of the microcontroller.

For simplicity, an HTTP-based communication has been adopted. APIs that the smart meter invokes have been developed.

The data stored on the DB are appropriately aggregated and analyzed to be used as input for any dashboard for users or for the aggregator.

The data are measured with a granularity of 5 s, which allows both to see details that otherwise would not have been visible and to perform a more precise power control, as explained in [16].

In Figure 2 the difference between a profile with a time step of 5 s and 15 min is reported, where it is possible to see how the 5 s profile allows to provide more detailed control and management.

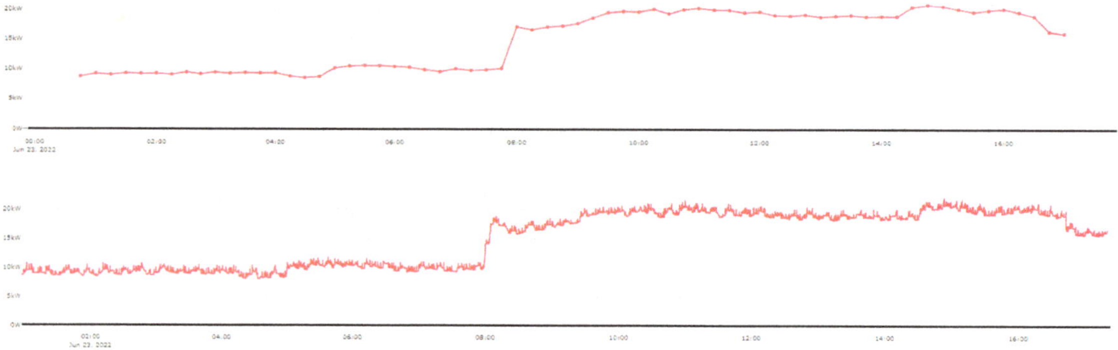

Figure 2. Comparison between a 15 min time step profile and a 5 s profile.

2.2. Energy Storage Systems

Electric storage systems include a broad category of devices. A classification of ESSs frequently adopted in the literature refers to the specific form of energy and distinguishes the storage systems in:

- electrochemical storage (lead acid batteries, lithium-ion batteries, zebra, nickel-metal hydride etc.).
- mechanical-type storage (CAES compressed air storage, high- and low-speed mechanical flywheels, pumping hydroelectric basins).
- electrostatic storage (supercapacitors).
- electromagnetic storage (superconductive magnetic energy storage—SMES).
- chemical storage (hydrogen).

The storage systems are generally chosen based on the service they are required to perform, the power service and/or the energy service.

Energy storage systems can serve at various locations in which electricity is produced, transported, consumed and held in reserve (backup). Depending on the location, storage can be large-scale (GW), medium-sized (MW) or it can use micro and local systems (kW).

The most used technologies are the Li-ion batteries, super-capacitor and PEM-based power-to-hydrogen.

These ESSs can achieve very high performance in terms of response speed and the modulation of power. The impact will be as incisive as possible to implement an aggregate and coordinated management as Virtual Storage. In this perspective, the virtual ESSs represent an important resource for the national electricity system, with the aim of providing synthetic inertia services (very intense response lasting a few seconds) and fast reserve (delivery time up to fractions of an hour), i.e., regulation services to the frequency variation, as well as peak shaving and load leveling functions, i.e., functions for the reduction in power peaks and flattening of the load curve.

The need for some of these services, in particular, that of the fast reserve, has led European TSOs, including TERNA, to promote pilot projects. To date, the storage technologies present on the market, in particular, the electrochemical ESS, are interfaced with the network through converters and control and management systems that are not easy to integrate and coordinate with each other.

The new storage capacity, appropriately controlled and managed, would allow the supply of synthetic inertia services to the network to restore the system inertia of fast reserve to speed up the response of the primary regulation to restore the nominal frequency value in the network, peak shaving and/or the load-leveling services necessary to support the penetration of the RES into the electrical system.

Therefore, the installation of storage systems is a promising solution to support the integration of RESs, particularly when devices are intelligently coordinated, such as virtual storage plants (VSPs), to provide a wide range of energy services.

Energy storage can supply more flexibility and balance to the grid, providing a backup to intermittent renewable energy. Locally, it can improve the management of distribution networks, reducing costs and improving efficiency. More in general, energy storage can provide many valuable services across the whole energy system. Indeed, energy storage is essential to balance supply and demand. Peaks and troughs in demand can often be anticipated and satisfied by increasing or decreasing generation at short notice. In a low-carbon system, intermittent RESs mean it is more difficult to vary output, and rises in demand do not necessarily correspond to rises in RES generation. Higher levels of energy storage are required for grid flexibility and grid stability and to cope with the increasing use of intermittent wind and solar electricity [23].

2.3. DC Nanogrid

To support this energy transition, it will be necessary to use system management, exploiting new flexible resources such as the previous mentioned smart storage systems to ensure adequate levels of system stability, safety and resilience, and so a new paradigm in terms of small-scale hybrid systems to integrate and coordinate decentralized distributed flexible resources is needed.

The DCNG is a hybrid system to maximize the self-consumption, accumulation and sharing of energy from renewable sources.

It can integrate different power sources and storage systems supplying uninterruptible and interruptible loads. It can operate when interconnected to the electricity grid but at the same time also in islanded mode (see Figure 3).

It represents a change in the plant engineering paradigm, interposing a DCNG as an interface between the end-user loads and the grid, behind the meter.

It is a complex and modular hybrid system of conversion and energy control systems capable of simultaneously managing multiple types of generation sources and/or storage systems of different technologies, as well as exchanging power with the electricity grid.

Under regular operating conditions, the DCNG operates to satisfy the critical load's demand, maximizing the use of RES generation when it is not required to follow a given power profile to be exchanged with the electricity grid (required by a central energy management system).

On the contrary, when it is required to exchange a given power with the electricity grid, the DCNG operates to exploit the integrated RESs and storage units, but at the same time guaranteeing continuity in supplying the critical loads.

The DCNG energy management is based on the DC Bus Signaling (DBS) strategy. The voltage of the DC bus on which the various RES and storage units are integrated is used as the only control signal. Indeed, there is no communication between the different sources, generation, storage system and loads. They operate independently with the voltage of the DC bus as a reference, through which the power flows are managed [17].

This hardware and software configuration, therefore, allows the management of the DCNG in an automated manner, reducing the interaction with other components. This enables the automated participation of the end-users.

The DCNG has a modular architecture capable of being extended both "locally", by connecting various devices to the base module (see Figure 4), consisting of a PV plant and a conventional battery ESS, and in a widespread manner, by physically connecting multiple DCNGs via their DC bus or even by virtually interconnecting different DCNGs through an internet connection (Figure 3), managed by the EMP based on the Power Cloud approach [23].

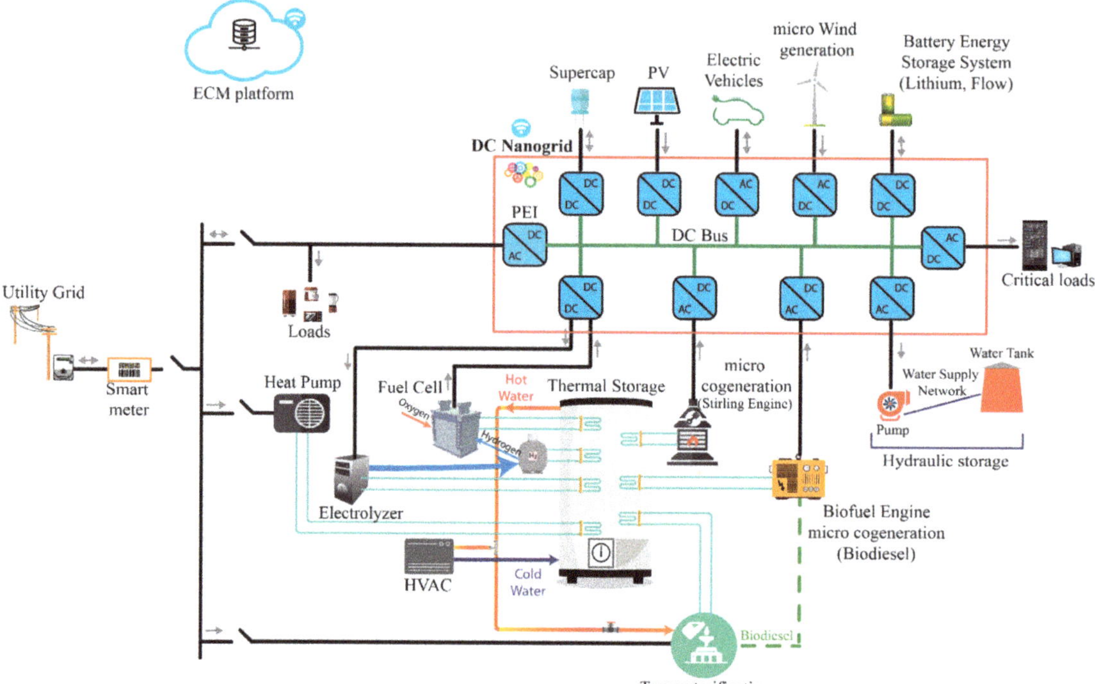

Figure 3. DCNG general configuration.

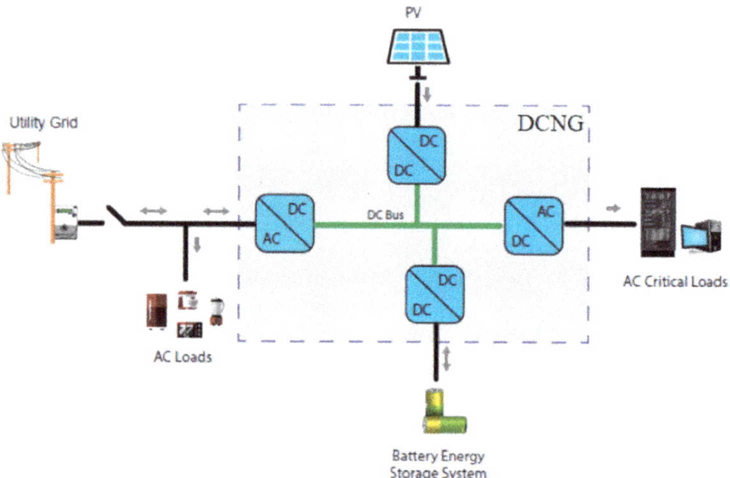

Figure 4. DCNG base module.

In this way, an advanced end-user (AEU) is defined: an aware and active end-user, which is a consumer with an installed smart meter, a PV plant (prosumer) and a smart storage system (prosumager) coordinated by the DCNG, configured as illustrated in Figure 5.

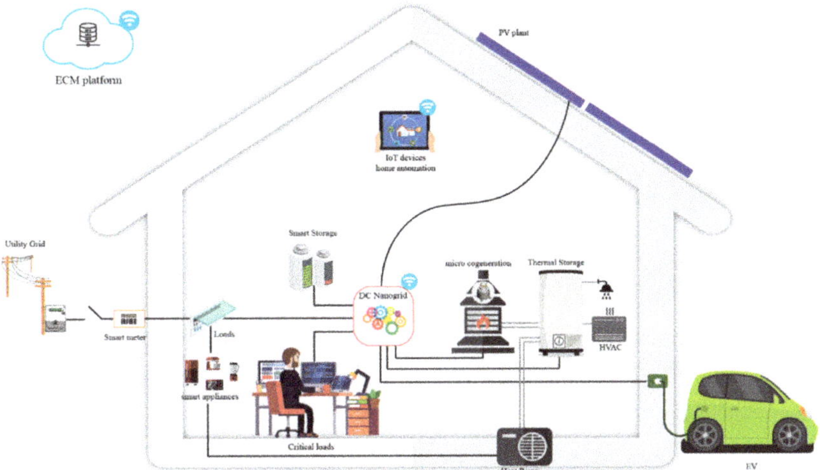

Figure 5. Advanced end-user configuration.

Several AEUs can cooperate among them so as to create an energy community (Figure 6) in compliance with actual and future regulations and technical and market rules. Coordinating the various AEUs, it is possible to maximize the economic return of the members of the whole energy community, maximizing the self-consumption rate until it tends to assume the behavior of a "Nonsumer" and at the same time supports the network when required to provide flexibility services.

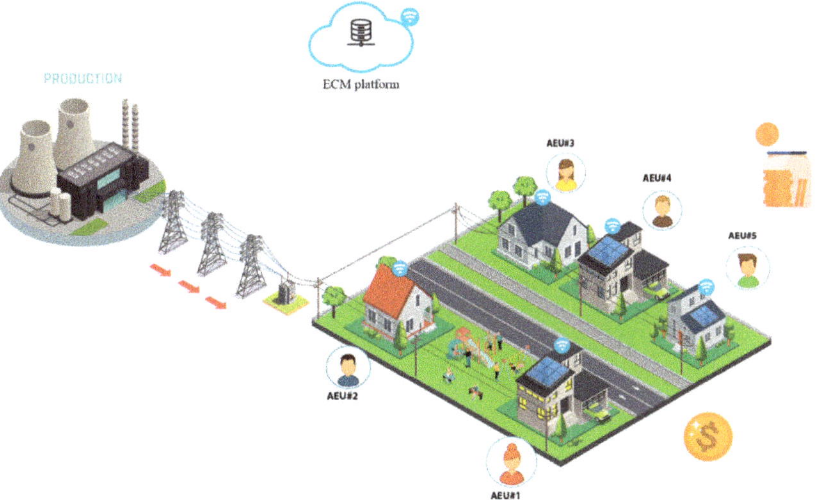

Figure 6. Advanced end-user energy community configuration.

2.4. Energy Community Management Platform

To manage the various AEUs in a coordinated manner, it is necessary to implement an Energy Community Management (ECM) platform that allows the exchange of data and, therefore, allows monitoring, communication and control.

2.4.1. The Communication Architecture

The data flow scheme for communication between the ECM platform, the DCNG and the smart meters is illustrated in Figure 7. In addition to the integration and management of PV generation and storage systems, the DCNG can communicate with the ECM platform. An aggregator, with the help of the energy management platform, manages the REC. The aggregator, in real time, monitors the power exchanged with the grid by end-users with SMs (User Data) and the status of the DGNG (NG Data). In addition, the aggregator can send to each DCNG a set of values, as set points, which translates into the power profile exchanged with the grid, which the DCNG must try to follow.

Figure 7. Communication architecture.

The communication by the aggregator takes place on a broadcast channel, so the AEUs, "members" of the REC, are enabled to receive the set points.

2.4.2. Platform Communication Protocol

The data communication protocol between DCNGs and SMs with the ECM platform is a Machine-To-Machine communication, using Message Queue Telemetry Transport (MQTT), assuming that the channel is encrypted using SSL, for which an MQTTS protocol was used.

The SM is only able to issue messages, while the DCNG can issue messages and receive/issue commands. Two types of messages have been foreseen in the communication: active and reactive. A message is said to be active if it is transmitted by the DCNG without a request from the aggregator (for example, the transmission of a malfunction alarm), while a message is reactive (synchronous type) when it follows a request–response-type scheme, the aggregator requests specific information from the DCNG or when sending specific commands.

The communication architecture, in terms of the MQTT protocol, is represented in Figure 8, where the Broker is represented by the energy management platform while the Server is the DCNG.

It should be emphasized that the exchange of information takes place on a broadcast application channel, which is 1:n. Each DCNG and SM enabled to communicate on the broadcast channel represents the members of the community.

The communication architecture used is the Publishers/Subscribers type and includes:

-   a broker server that queues messages from various community members;
-   a client able to subscribe and publish messages on a particular channel made available by the Broker;
-   one or more "listeners" such as DCNGs which receive messages published on the Broker's channels and perform certain functions.

The communication provides for a flow control for authentication based on a username and password registered in the DCNG and in the SM. Both DCNGs and SMs send messages with state-type operations containing information on the status of the DCNG (#NG/data) or the measurements taken by the SM installed by the user (#User/data). Finally, in the communication flow, there are also messages with action-type operations, where the aggregator sends the setpoint to be followed to the individual DCNGs.

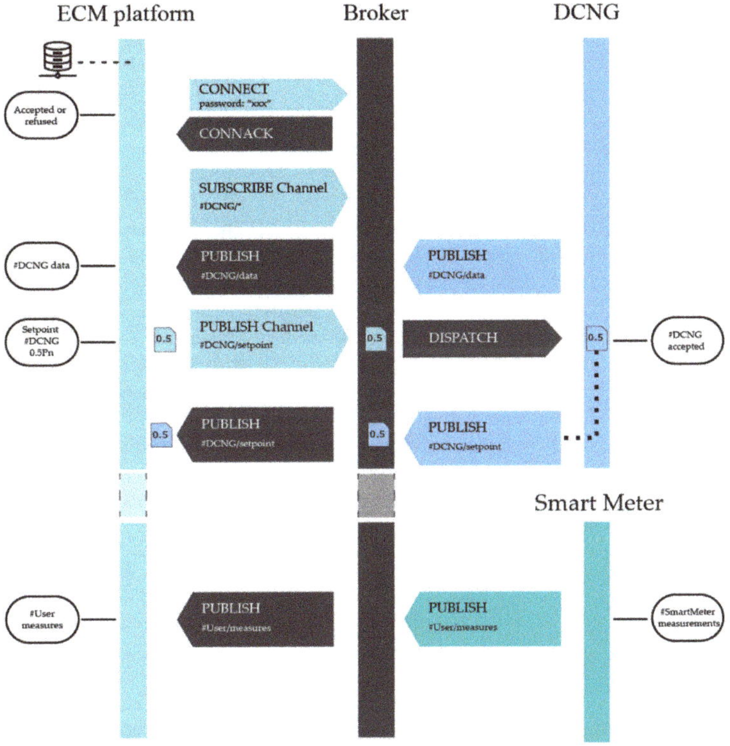

Figure 8. Data flow control summary of the communication protocol among the DCNG, SM and aggregator.

The data are encapsulated in a string message encoded according to the JavaScript Object Notation (JSON) format, and its content is divided into a header and a body. In the header, in addition to useful information for flow control (Sequence number, Request ID), there is the "operation" key, which indicates the type of operation to which the body of the message refers. There are two types of operation, type "action" and type "state". An action-type operation is used to send commands to the DCNGs (for example, the aggregator sends the set points to the DCNGs), while a state-type operation is used by the DCNGs themselves and by the SMs to send data (for example, measurements taken by the SMs and/or other information such as the state of charge of the ESS).

2.4.3. ECM Platform Description and Operation

The ECM platform allows the management of the data coming from the various AEUs (DCNGs and SM), the provision of services and the elaboration of any management algorithms implementing the ECM platform processes, as follows.

ECM Platform processes

1. Day-ahead self-consumption optimization: The process aims to optimize the power flows exchanged with the energy storage system to maximize self-consumption using deterministic equations that also consider production and load forecasts such as those relating to the storage system. This algorithm, using the SOC forecasts for the ESS,

divides the energy surplus among the different ESSs and at the same time uses the stored energy to supply the loads. Power swaps are scheduled the day before.

2. Day-ahead dispatching services: Through this process, the scheduled availability of the day before is provided for individual AEUs (equipped with an energy storage system) to change their profile, therefore, to provide services to the network if required. Starting from the previous process, the SOC of the various ESS is evaluated and, if it is greater than a reference SOC, the availability to provide services to the network is scheduled/communicated. Using the availability of individual AEUs determines the availability for the entire REC.

3. The real-time management of requests to modify the exchange power profile: starting from what was programmed the day before, this process calculates in real time the power that the AEUs must exchange with the storage systems to provide the services requested if necessary [8]. If the grid operator sends a new request in terms of the power profile to exchange with the grid, according to the communication of the day before, this must be provided by the REC. This request is distributed among the AEUs, as REC members, according to the previously determined availability.

4. Real-time balancing profile process: since the previous processes are based on forecasts, there could be errors in the forecasts. For this purpose, through this process, it is possible to balance the power between what is programmed and the real power profiles, opportunely charging and/or discharging the energy storage system [8]. In particular, the comparison was drawn between the forecasts and the power measurements and, according to the real SOC ESS, the power is modified.

ECM Platform Services

1. Day-Ahead User Load Prediction: This service predicts the day-before hourly load for the different end-users in the aggregation.

2. Forecast of the day-ahead AEU production: This service allows the forecast of the hourly production of the day ahead for end-users to be equipped with a photovoltaic system.

3. Forecast of the day-ahead ESS state of charge: This service is used for the forecast of the SOC of the different ESS which is useful for providing the energy planning of the day ahead.

4. Day-ahead ESS charge/discharge profile: through this service, the energy that the ESS must exchange is scheduled the day ahead.

5. Forecast of the day-ahead AEU trading power profiles: it determines the power (injection and absorption) that the AEU will exchange with the network.

6. Day-ahead AEU dispatching service availability: this service is useful for calculating the availability of the individual AEUs to modify your profile and provide flexibility services to the grid.

7. Availability of day-ahead dispatching services: starting from the AEUs' availability to provide services to the grid, the willingness to provide services of the REC is determined.

8. Real-time forecast of the ESS state of charge: it determines the power that the aggregation must exchange to provide services to the grid in real time.

9. Real-time state of charge of the user's energy storage system: this service divides the power to provide services to the grid among the various end-users, previously determined for the REC.

10. Real-time AEU energy storage system charge/discharge: using the day-ahead schedule and real-time services, the power that the ESSs need to exchange is determined.

11. Real-time AEU exchange power profile: it determines the power that the end-user must exchange with the network, considering both the real-time and the previous day's planning.

12. End-users real-time power balance services: the last service determines the power that the AEUs must exchange to limit the imbalance between the programmed day-ahead power profiles (injection and absorption) and the real-time ones.

2.4.4. Utilized Management Method

In [17,24], the authors present the model to optimally manage the power flow, providing services to the grid. The proposed control strategy is utilized not only to locally control the energy needs of a single AEU but also to manage more than one AEU, which constitutes a Renewable Energy Community. They also provide services to the grid, in particular, balancing services. To this scope, the power surplus has been injected into the grid or into the storage system, as requested by the operator.

The method can be used for more days and different time intervals. The objective function (*OF*) allows several optimization aims for the exchanged power with the grid. The minimization of the overall exchanged energy, maximum power and power peak for a defined time interval can be requested:

$$
OF : \min \left(\sum_{\substack{t=1 \\ d=1}}^{\substack{\frac{D}{T}}} f\left(P_{gt}^{d}\right) \right)
$$

where P_{gt}^{d} is the exchanged power with the grid for a single user, for the time interval t and for the day d. This method is subject to several constraints that concern the exchanged power with the grid, the stored energy, the power exchanged with the storage systems and other conditions, as defined in [17,24].

3. REC Performance Indexes

The REC performances can be evaluated using some specific indexes. They have been defined considering both the H2020 research projects on this theme, the Italian national directives [25–27], what has been proposed in the literature [28] and based on the data stored in the ECM platform.

3.1. Indexes for the Single AEU and for the REC Performance: Self-Consumption and Self-Sufficiency Indexes

The aim is to identify indexes, in addition to the AEU's local energy self-sufficiency, to have information regarding the quality of the self-consumption carried out inside the REC.

The indexes used to evaluate the degree of individual or shared/collective self-consumption can be measured with reference to different time intervals (hour, day, month, year), although the most useful is the hourly, as considered in this paper.

The collective self-consumption index must be "normalized", comparing the amount of energy consumed in a time interval with the amount of renewable energy produced and available to evaluate the benefits related to the shared self-consumption in the energy communities.

The aim is to individuate and reward the AEUs most virtuous in terms of self-consumption and shared self-consumption, that is, concentrated in the hours of maximum RES production. It is necessary to build an index that returns information that is not only quantitative but also qualitative.

For this reason, in the calculation of shared self-consumption, it was decided to compare local consumption with the total consumption of the aggregation, normalizing the index.

The following variables will be used in the formulation of the proposed indexes:

- h: 1, 2, ... , 24 index representing hours;
- i: 1, 2, ... , n index representing the REC AEU number;
- g: 1, 2, ... , Ng index representing the month day;
- $E_{p,i}^{h}$: energy produced by the AEU i in the hour h;
- E_{p}^{h}: energy produced from all RES generation plants inside the REC in the hour h;
- E_{i}^{h}: energy absorbed by the AEU i in the hour h;
- $E_{L}^{h} = \sum_{i=1}^{n} E_{i}^{h}$: energy absorbed by all REC AEUs;

- $E^h_{AC} = min\left(E^h_p; E^h_L\right)$: energy shared by the REC, according to the ARERA definition, provided in the regulation 318/2020/R/eel. It is defined as the minimum between the hourly value of energy produced by renewable sources and the sum of the energy adsorbed by all the loads of the community.
- $E^h_{AL,\,i} = min\left(E^h_i; E^h_{p,i}\right)$: locally self-consumed energy or energy self-sufficiency, it is represented by the minimum value between the AEU's production and consumption on the same site.

3.1.1. Hourly Shared Community Self-Consumption Index (IAC)

The index refers to the shared self-consumption of the REC in relation to the total amount of energy produced in the same hour (E^h_p) from all RES generation plants. It expresses the relationship between the energy shared in the hour and all the energy produced in the same hour. The index varies between 0 and 1 and is formulated as follows:

$$IAC^h = \frac{E^h_{AC}}{E^h_p},\tag{1}$$

3.1.2. Individual AEU Hourly Local Self-Consumption Index (IAS)

It is the ratio between the hourly, locally self-consumed energy and the energy produced in the same hour by the same AEU (when production is greater than zero).

$$IAS^h = \frac{E^h_{AL,i}}{E^h_{p,i}},\tag{2}$$

3.1.3. Single AEU Hourly Shared Self-Consumption Index (IAC)

The index refers to the self-consumption of the single AEU and corresponds to the percentage of the collective shared energy that is consumed by the AEU when it is produced (the energy stored in local storage for AEU is also counted as instantly self-consumed energy).

Considering the renewable self-produced energy given by the self-consumption index IAC^h and the locally consumed energy compared to the total energy (E^h_i/E^h_L), the index is normalized with the hourly production $E^{\,h}_{p,i}$ compared to the maximum generation contribution of the user ($max_g\ E^{\,h}_{p,i}$). The index varies between 0 and 1 and can be calculated as:

$$IAC^h_i = \frac{E^h_i}{E^h_L} \times \left(IAC^h \times \frac{E^h_{p,\,i}}{max_g\left(E^h_{p,\,i}\right)} \right),\tag{3}$$

The first term is the load power of the AEU in the hour compared to the total REC load in the same hour. The second term represents the normalized self-consumption index, that is, the product between the collective self-consumption index and the normalized AEU's production. It is possible to determine the shared self-consumption indexes, both individual and collective, on other time scales (daily, monthly or yearly) using the same formulas reported above appropriately scaled over the correct time horizon.

3.1.4. Services Request Reliability Index (ISRR)

The index refers to the ability of the AEU to satisfy a flexibility service request. It can be defined for each day, for the community or for the single AEU. It is the ratio between the error valuated as the difference between the averaged actual exchanged power (P^h_{act}) and the averaged request power (P^h_{sched}), and the averaged actual exchanged power (P^h_{act}).

$$ISRR = \frac{P^h_{act} - P^h_{sched}}{P^h_{act}} \times 100\tag{4}$$

4. Experimental Use Cases

The aim of the experimental use cases is to demonstrate the need for accurate, efficient and effective technologies for the implementation of RECs. In particular, it is shown how these technologies can interact with each other to achieve different results. For example, maximizing the self-consumption of the single AEU with or without the provision of flexibility services. These goals are achieved by exploiting local generation from RESs and ESSs.

The implemented and analyzed use cases consider a REC of four AEUs, as illustrated in Figure 9. Each AEU is equipped with a 3 kW PV plant, an ESS, a converter to supply the critical loads and a converter for the interface with the grid; other external loads are monitored through an SM. The difference consists of the capacities of the ESS, which, respectively, are: 18 kWh, 6 kWh, 14 kWh and 26 kWh.

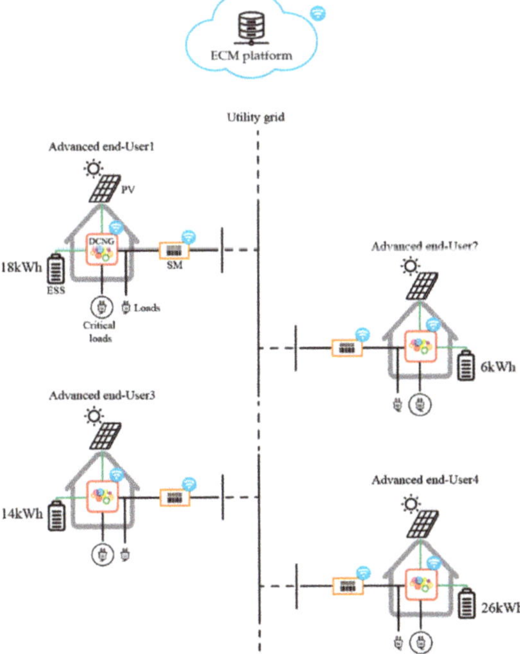

Figure 9. Community of four AEUs.

The AEUs are interfaced through the ECM platform using the communication architecture before described.

Firstly, the analysis is carried out considering only one AEU operation and after the REC one. Three use cases are identified and analyzed, as illustrated in Table 1 and below described.

Table 1. Experimental Use Cases.

Use Case	Description	Goal of the Use Case
UC_1	AEU self-consumption without any flexibility service request	To demonstrate that the AEU can maximize the use of local generation and the local self-consumption, therefore minimizing the CO_2 production and the energy costs at the single AEU level.
UC_2	AEU self-consumption with a flexibility service request	To demonstrate that the AEU can maximize the use of local generation and the local self-consumption and promptly satisfy a flexibility service request.
UC_3	REC self-consumption without and with a flexibility service request	To show the AEU behavior without and with a flexibility service request when different AEUs work in an REC.

(a) Use case 1(UC_1): AEU self-consumption without any flexibility service request.

This use case interests the single AEU. Its aim is to demonstrate that the AEU, through the DCNG with its control and management system of the RES generation and storage units, can maximize the use of local generation and the local self-consumption, therefore minimizing the CO2 production and the energy costs at the single AEU level.

For the performance evaluation of this use case, the IAS, hourly local self-consumption index, is used.

(b) Use case 2 (UC_2): AEU self-consumption with a flexibility service request.

This use case interests the single AEU. Its aim is to demonstrate that the AEU, through the DCNG with its control and management system of the different RES generation and storage units, can maximize the use of local generation and the local self-consumption and promptly satisfy a flexibility service request.

In this use case, the grid operator (TSO/DSO) or an intermediary operator (such as Balancing Service Provider—BSP) sends the flexibility request to the AEU, which operates through the DCNG to follow such a request. The DCNG operates by controlling its internal flexibility resources to follow the required power profile, minimizing the difference between the requested profile and the real-time power profile.

In this use case, the IAS and ISRR indexes are evaluated.

(c) Use case 3 (UC_3): REC self-consumption without and with a flexibility service request.

In this use case, we observe the real-time operation of the AEUs operating as an REC.

In this case, we first observe how each AEU behaves when a flexibility service is not requested and, therefore, when they operate in the community simply to increase the self-consumption of the same.

Subsequently, a request for a flexibility service is assumed, and then the behavior of each AEU is analyzed.

In this use case, the hourly IAC and ISRR indexes are evaluated.

5. Numerical Results

The numerical results have been obtained using the AEU laboratory prototypes implemented in a LASEER laboratory (see Figure 10). Each AEU is configured as illustrated in Section 4 and uses the DCNG with its own control and management system for the RES generation and storage units.

5.1. Test–UC_1

In its operation, without any external flexibility request, thanks to the DBS strategy, the DCNG maximizes the use of local generation and at the same time minimizes the CO_2 emissions, reducing the energy exchanged with the grid [17].

The PV, the load power profiles and the power exchanged between the grid and the AEU are reported in Figures 11–13. The power exchanged between the grid and the grid should be equal to zero (Figure 13). In some time instances, such a net power profile is different from zero, due in particular to the exchanged ESS power. The ESS power profile and SOC are reported in Figures 14 and 15.

In Figure 16 the net power profile that the AEU would have exchanged with the grid without the use of the DCNG is reported. It has been numerically evaluated. It is lower than zero when there is a power surplus that has to be injected into the grid and vice versa when there is a power deficit. Therefore, power is required to supply the loads.

From Figure 13, it can be observed that for almost the whole day, the AEU exchanges a power equal to zero with the grid. Only at the beginning of the day and at the end of the day are there withdrawals from the grid due to insufficient energy in the ESS.

In addition, there is a peak of absorption from the grid around 8.30 a.m. due to the high load power in that time interval and the reduced PV production, as the power that the ESS can supply is also limited.

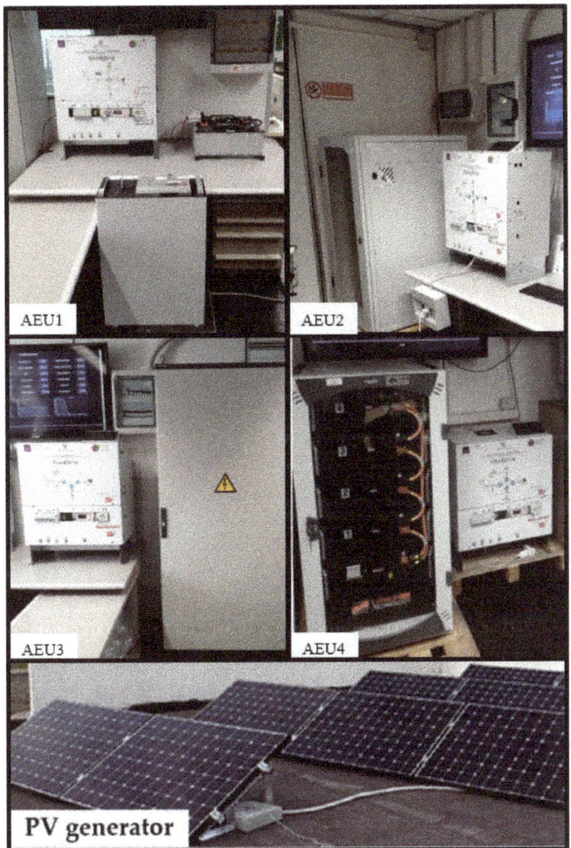

Figure 10. AEUs Laboratory prototypes.

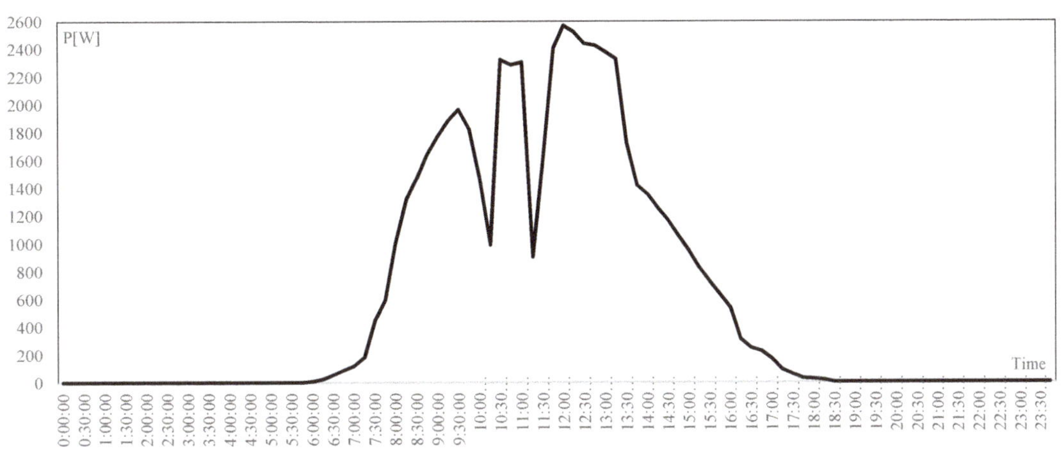

Figure 11. PV power production.

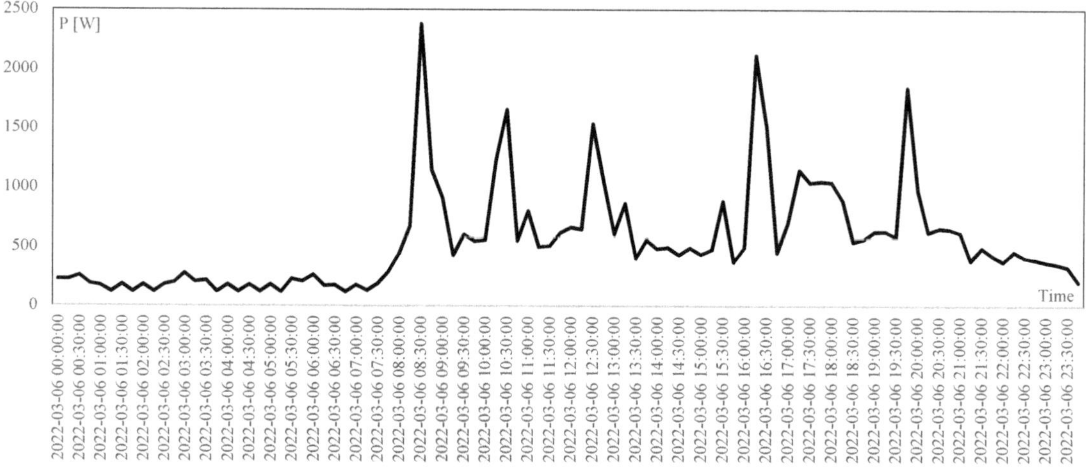

Figure 12. Load power profile.

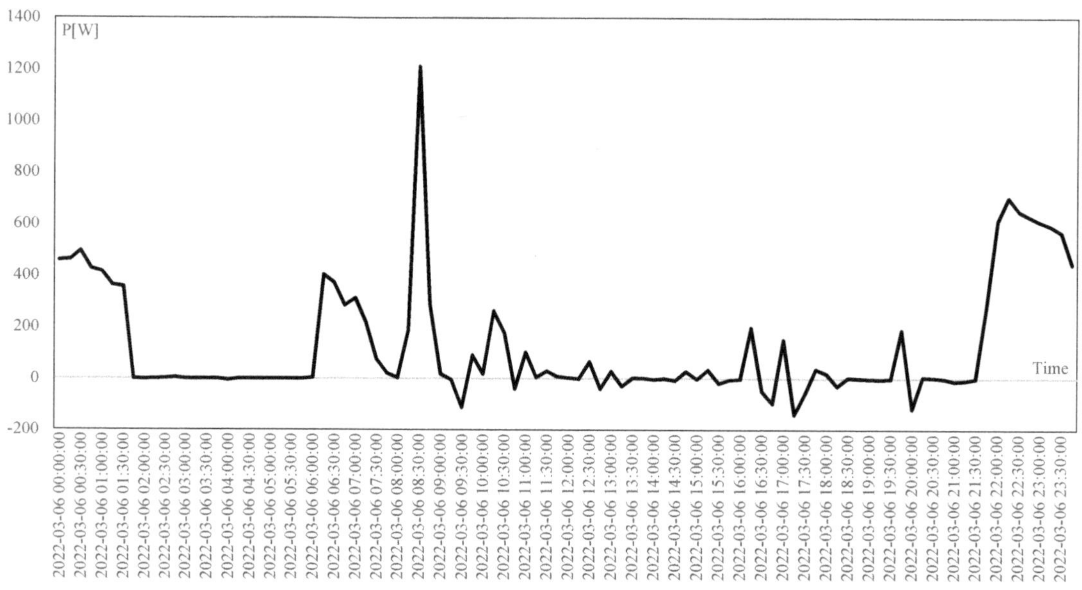

Figure 13. Exchanged grid net power.

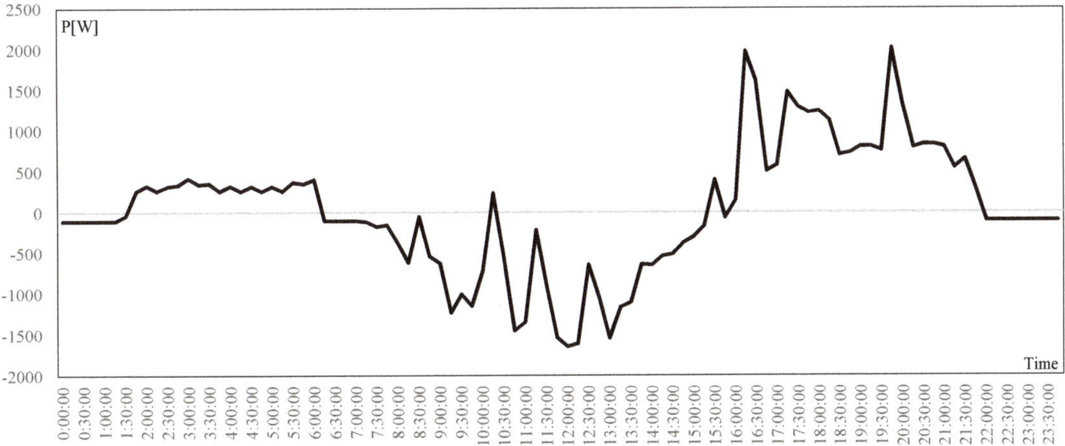

Figure 14. ESS exchanged power profile.

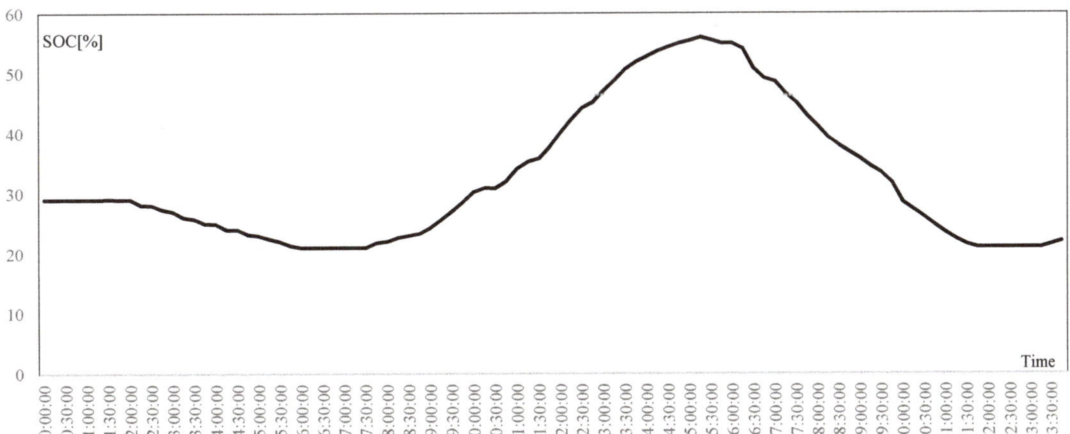

Figure 15. ESS state of charge.

During the day, the PV-produced energy is equal to 14.04 kWh. The stored energy is equal to 6.66 kWh, while the supplied energy from the battery is equal to 7.37 kWh. The energy absorbed by the loads is equal to 13.17 kWh. The energy absorbed from the grid that is not supplied by the DCNG is equal to 3.23 kWh. The surplus of PV-produced energy that is not stored in the ESS or used to supply the load is equal to 0.19 kWh.

With the proposed AEU configuration and the control and management performed by the DCNG, considering the actual grid energy generation mix, the emitted equivalent CO2 is equal only to 1.32 kg with respect to 5.4 kg emitted without the use of DCNG.

If the IAS is evaluated for the hours when there is PV generation, its average value is equal to 67%, with hourly peak values also equal to 100%.

Therefore, we can affirm that, when comparing Figures 13 and 16, the AEU equipped with DCNG is able to minimize energy exchanges with the grid and take advantage of the available resources, and CO_2 emissions can also be reduced.

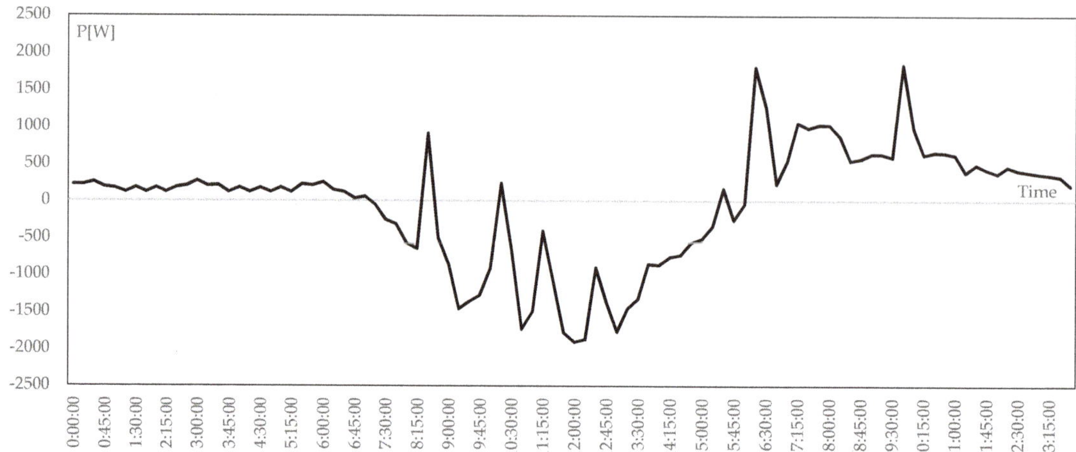

Figure 16. Exchanged power with the grid without the use of the DCNG.

5.2. Test–UC_2

The UC_2 starts considering the AEU operating as described in the UC_1: the first aim is to maximize the self-production, but at the same time, the AEU communicates a flexibility availability equal to 200 W for every time (a quarter of an hour) of the day.

Based on that AEU availability, the grid operator sends the flexibility day requests as reported in Table 2, where the reduction power request (upward flexibility) is reported with a negative sign, and vice versa for the increased power requests (downward flexibility).

Table 2. Flexibility services requested to the AEU.

Time (Considered Quarter)	Flexibility Request [W]
0:45	−150
1:45	200
2:00	200
3:30	−100
3:45	200
5:30	−200
8:45	200
9:00	200
11:15	−200
11:30	−150
14:00	−200
17:00	100
20:45	−100
21:00	−50
23:15	100

In this UC, the AEU must operate accordingly, modifying its grid exchange power profile to satisfy as much as possible the flexibility service requests.

In Figure 17, the power profile exchanged is reported. It can be observed that in the first part of the day, the net exchanged power is different from zero. This is due to the low SOC of the ESS, which does not allow to supply all loads. The flexibility requests are satisfied using the ESS; in particular, the required power is injected or absorbed by the ESS.

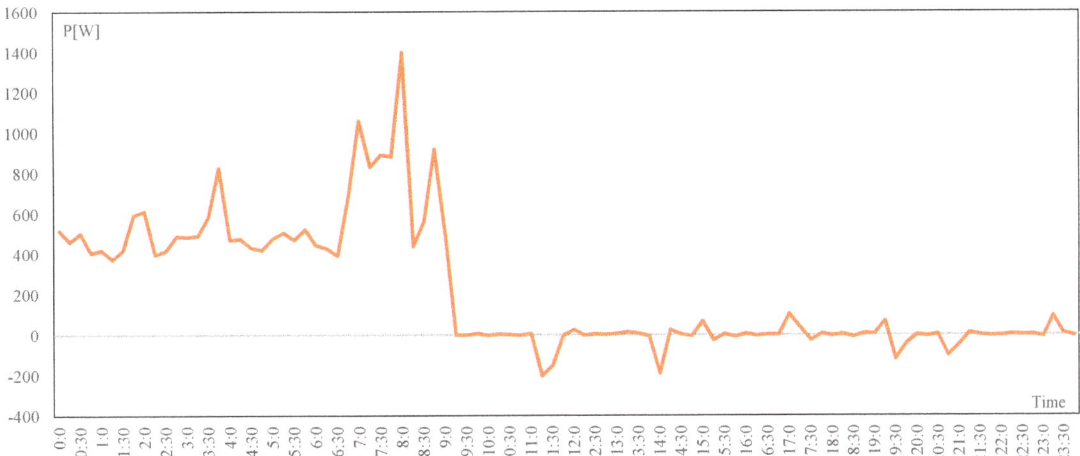

Figure 17. Exchanged grid net power with flexibility service.

In Figures 18 and 19, the ESS power and the SOC profile are reported. The ESS is unable to follow the load profile at the beginning of the day because of the low SOC; in the same way, it is unable to perform any requests for flexibility.

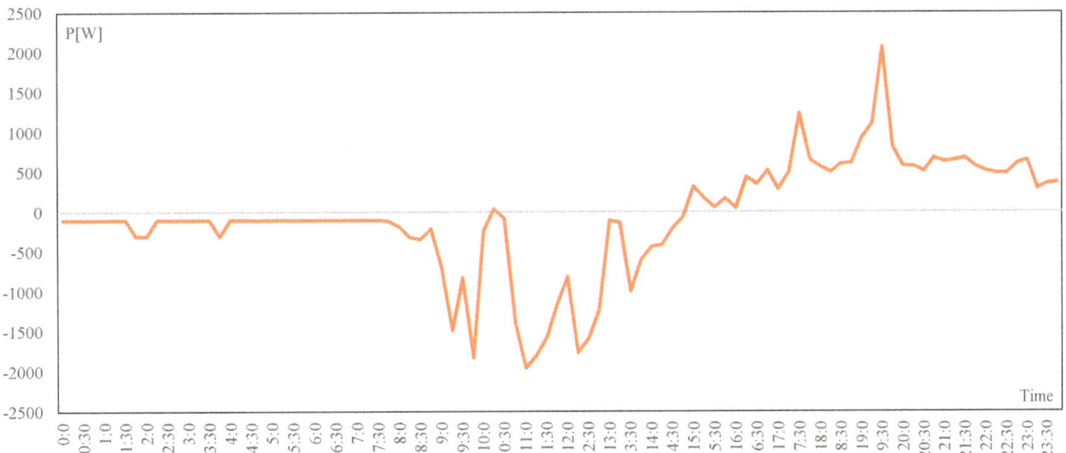

Figure 18. ESS power profile with flexibility service.

During the day, the total power for downward and upward flexibility requests are, respectively, 0.3 kWh and 0.287 kWh. All the upward flexibility requests have been satisfied, while the downward flexibility requests are satisfied only at about 61%. This is an extreme case, but still possible, in which the AEU storage system is not ready to be able to satisfy the flexibility requests.

A way to avoid such problems is to reserve a capacity of ESSs only for these services or to use several AEUs operating in a REC as illustrated in UC3. In this way, the different requests can be provided by the other AEUs.

In this use case, evaluating the IAS for the hours when there is PV production, there is not an evident difference between the case where services are required and where services are not required; its average value is about 75%, with hourly peak values also equal to 100%.

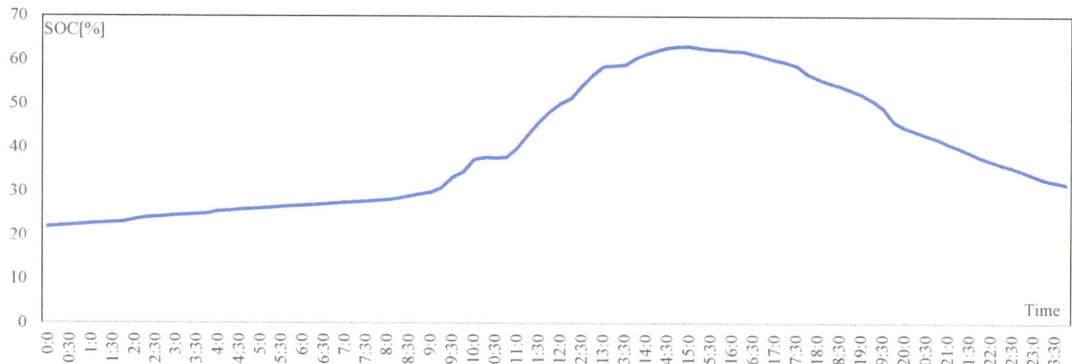

Figure 19. ESS SOC with flexibility service.

The ISRR has been also calculated and it is equal to about 1.2%, which means that most of the flexibility requests have been satisfied.

Therefore, with this use case, the authors want to demonstrate that an AEU equipped with DCNG that manages an RES plant and a storage system is able to provide flexibility services. In fact, to demonstrate it, we can compare Figure 17 starting from 11:00 am onwards with Table 2. It is observed that the power exchange with the grid is zero (or almost) when there is no flexibility service request, while it is equal to the flexibility service requested in the other instances of time. However, to be able to respond to these services, it is necessary to have a storage system that does not have to be completely charged or discharged (see first hours of Figure 17 and compare it with Table 2; the flexibility service is not satisfied).

5.3. Test–UC_3

In this use case, four AEUs are assumed to be part of an REC. The main goal is to maximize the self-consumption. It is assumed that flexibility services are also requested from the grid operator, as illustrated in Table 3.

Table 3. Flexibility services requested to the REC.

Time (Considered Quarter)	Flexibility Request [W]
0:45	400
3:30	−550
9:30	−700
13:0	200
15:45	600
17:30	−350
20:45	−200
22:45	−250

In Figures 20–23, the load, the PV generation, the ESS power profiles of the four AEUs and the REC net power profile exchanged with the grid are reported. The last one is equal to zero when no flexibility request is present. Otherwise, it is equal to the request.

Each specific flexibility request is distributed to each AEU by dividing the power request in proportion to the capacity of the ESS associated with each single AEU.

As can be seen from the numerical results, in terms of the power exchanged with the grid, the operation of AEUs as an REC allows to maximize the self-consumption as well as to satisfy flexibility requests using the ESSs appropriately.

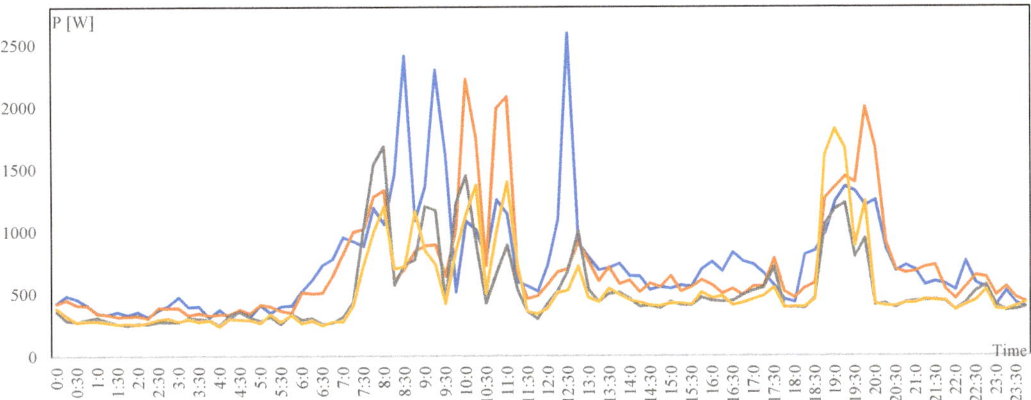

Figure 20. Load power profile for the different AEUs: AEU1 (blue), AEU2 (orange), AEU3 (grey), AEU4 (yellow).

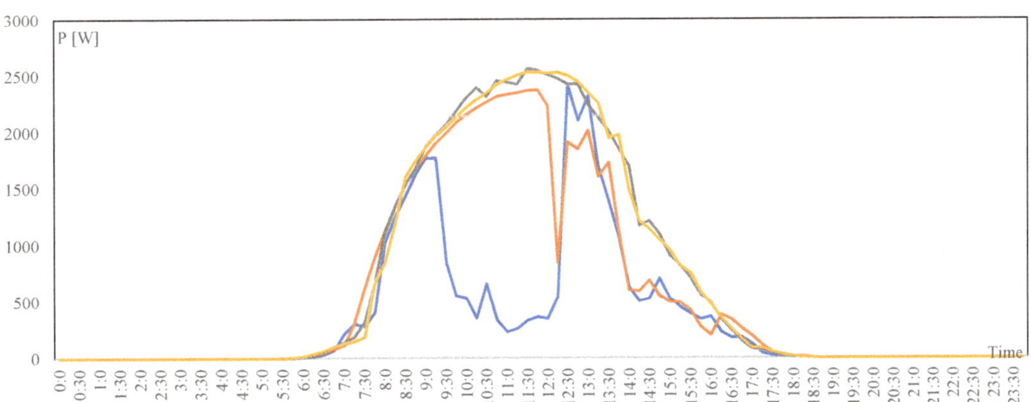

Figure 21. PV power production profiles for the different AEUs: AEU1 (blue), AEU2 (orange), AEU3 (grey), AEU4 (yellow).

This is evident from Figure 23. In fact, this profile is null (or almost) when there is no request for a flexibility service, as the REC maximizes self-consumption, while the profile is non-zero when there is a service request, and the profile is equal to the request (see Table 3). This is possible by exploiting all the resources, RES Figure 21 and ESS Figure 22, of the REC.

The hourly IAC has been evaluated for the hours where the PV production is greater than zero: it ranges between 93.7% and 100%; its average value is about 99.3%. These values are due to the optimal operation of the storage systems.

The ISRR is quite null, which means overall flexibility requests at the REC level have been satisfied.

Figure 22. Exchanged power with the storage system for the different AEUs: AEU1 (blue), AEU2 (orange), AEU3 (grey), AEU4 (yellow).

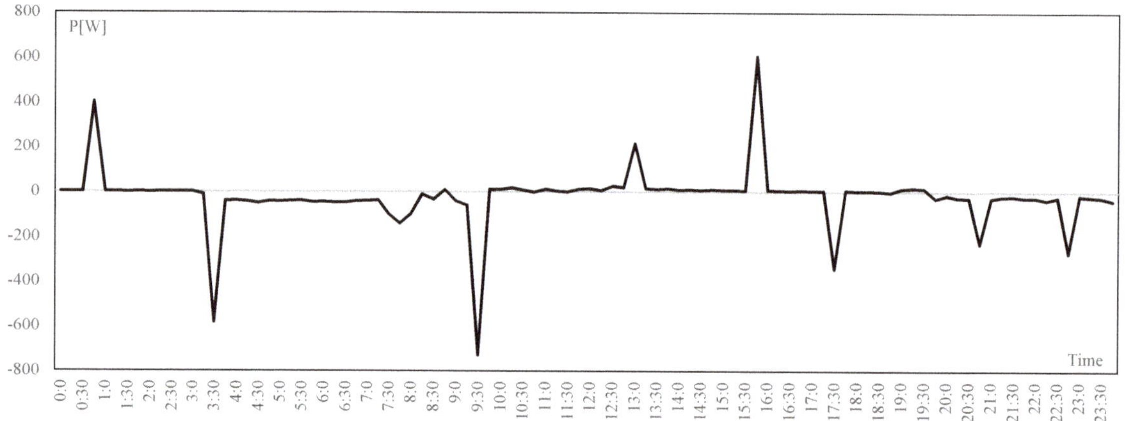

Figure 23. Exchanged net power for the community with flexibility services requests.

6. Conclusions

Globally, many states, including Italy, are committed to undertaking an effective and sustainable path of deep decarbonization in all sectors through a profound restructuring of energy systems. This has led to the need for new management models to maximize the use of energy produced from renewable sources, ensuring safety conditions, continuity of service and resilience of the electricity system itself.

Possible reference models are that of the Renewable Energy Communities (REC) and the Citizen Energy Communities (CEC). In particular, the model based on CERs was taken as a reference by the European Commission—with the Directive (EU) 2018/2001 (called RED II) recently implemented by the Italian Government on 15 December 2021 with Legislative Decree 199—to allow participation active consumer in the energy transition through the sharing of electricity produced from renewable sources, the promotion of energy efficiency and, therefore, the spread of electric mobility. CERs have been introduced to maximize the use of energy from renewable sources and minimize energy costs for users.

This paper, in this contest, illustrates the main enabling technologies, smart meter, ESS, DCNG and ECM platform, that must be used to support the growth of RECs. The innovative settlement of the Advanced End-User is introduced, defined as an active and pro-active end-user equipped with the above-mentioned enabling technologies.

Finally, some use cases and the associated performance indexes have been identified and defined, respectively. The numerical results performed both, considering the operation of a single AEU and four AEUs operating as an REC, are illustrated, and discussing when flexibility service requests are sent or not, highlighting how it is possible to both maximize the self-consumption and satisfy the flexibility service request in the case of an REC consisting of AEUs as members.

Through the analysis and comparison of the different considered UCs, it was possible to demonstrate the importance of the individual technologies and their combined use as enabling elements for RECs. UC1 shows how the use of the proposed technologies increases self-consumption. This increase is measured through IAS which goes from a virtual average value of 15% to that of 67%. On the other hand, the comparison between UC1 and UC2 shows that, in the case of a single AEU, there is not an obvious difference between the cases where services are required and where services are not required. Finally, UC3 demonstrates that the community use of the above technologies enables further performance measured by means of IAC, of which the average value is about 99.3%.

Author Contributions: Conceptualization, A.P., G.B. (Giovanni Brusco) and P.V.; methodology, D.M. and A.P.; software, S.M. and G.B. (Giuseppe Barone); validation, N.S., G.B. (Giovanni Brusco) and A.P.; formal analysis, P.V.; investigation, L.M.; resources, D.M., L.M. and S.M.; data curation, G.B. (Giuseppe Barone), G.P. and S.M.; writing—original draft preparation, P.V. and G.B. (Giovanni Brusco); writing—review and editing, A.P. and G.P.; visualization, G.B. (Giovanni Brusco); supervision, D.M., A.P. and P.V.; project administration, P.V. and A.P.; funding acquisition, D.M., A.P. and N.S. All authors have read and agreed to the published version of the manuscript.

Funding: This research was funded by the European Union's Horizon 2020 research and innovation program under grant agreement n° 864283, by the National Operational Programme for Research and Innovation 2014–2020 "Fondo Sociale Europeo, Azione 1.2, Attraction and International Mobility n° AIM1857122-2 and by the National Project "ComESto-Community Energy Storage" PON ricerca e innovazione 2014–2020 MIUR-ARS01_01259.

Institutional Review Board Statement: Not applicable.

Informed Consent Statement: Not applicable.

Data Availability Statement: Not applicable.

Conflicts of Interest: The authors declare no conflict of interest.

References

1. Soeiro, S.; Dias, M.F. Renewable energy community and the European energy market: Main motivations. *Heliyon* **2020**, *6*, e04511. [CrossRef]
2. Boon, F.P.; Dieperink, C. Local civil society based renewable energy organisations in the Netherlands: Exploring the factors that stimulate their emergence and development. *Energy Policy* **2014**, *69*, 297–307. [CrossRef]
3. Cejka, S.; Zeilinger, F.; Veseli, A.; Holzleitner, M.T.; Stefan, M. A Blockchain-based Privacy-friendly Renewable Energy Community. In Proceedings of the 9th International Conference on Smart Cities and Green ICT Systems—Smartgreens, Siemens, Germany, 2–4 May 2020; pp. 95–103.
4. Ceglia, F.; Esposito, P.; Marrasso, E.; Sasso, M. From smart energy community to smart energy municipalities: Literature review, agendas and pathways. *J. Clean. Prod.* **2020**, *254*, 120118. [CrossRef]
5. Honarmand, M.E.; Hosseinnezhad, V.; Hayes, B.; Shafie-khah, M.; Siano, P. An overview of demand response: From its origins to the smart energy community. *IEEE Access* **2021**, *9*, 96851–96876. [CrossRef]
6. Iqbal, A.; Olariu, S. A survey of enabling technologies for smart communities. *Smart Cities* **2020**, *4*, 54–77. [CrossRef]
7. Ahad, M.A.; Paiva, S.; Tripathi, G.; Feroz, N. Enabling technologies and sustainable smart cities. *Sustain. Cities Soc.* **2020**, *61*, 102301. [CrossRef]
8. Jia, M.; Komeily, A.; Wang, Y.; Srinivasan, R.S. Adopting Internet of Things for the development of smart buildings: A review of enabling technologies and applications. *Autom. Constr.* **2019**, *101*, 111–126. [CrossRef]
9. Gagliardelli, L.; Zecchini, L.; Beneventano, D.; Simonini, G.; Bergamaschi, S.; Orsini, M.; Fabio, M. ECDP: A Big Data Platform for the Smart Monitoring of Local Energy Communities. In Proceedings of the 1st International Workshop on Data Platform Design, Management, and Optimization, Online, 29 March 2022.

10. Hill, M.; Duffy, A. A Digital Support Platform for Community Energy: One-Stop-Shop Architecture, Development and Evaluation. *Energies* **2022**, *15*, 4763. [CrossRef]

11. Pereira, H.; Gomes, L.; Faria, P.; Vale, Z.; Coelho, C. Web-based platform for the management of citizen energy communities and their members. *Energy Inform.* **2021**, *4*, 43. [CrossRef]

12. Van Leeuwen, G.; AlSkaif, T.; Gibescu, M.; Van Sark, W. An integrated blockchain-based energy management platform with bilateral trading for microgrid communities. *Appl. Energy* **2020**, *263*, 114613. [CrossRef]

13. Górski, T. Reconfigurable Smart Contracts for Renewable Energy Exchange with Re-Use of Verification Rules. *Appl. Sci.* **2022**, *12*, 5339. [CrossRef]

14. Yahaya, A.S.; Javaid, N.; Alzahrani, F.A.; Rehman, A.; Ullah, I.; Shahid, A.; Shafiq, M. Blockchain based sustainable local energy trading considering home energy management and demurrage mechanism. *Sustainability* **2020**, *12*, 3385. [CrossRef]

15. Zou, W.; Lo, D.; Kochhar, P.S.; Le, X.B.D.; Xia, X.; Feng, Y.; Xu, B. Smart contract development: Challenges and opportunities. *IEEE Trans. Softw. Eng.* **2019**, *47*, 2084–2106. [CrossRef]

16. Barone, G.; Brusco, G.; Menniti, D.; Pinnarelli, A.; Polizzi, G.; Sorrentino, N.; Vizza, P.; Burgio, A. How smart metering and smart charging may help a local energy community in collective self-consumption in presence of electric vehicles. *Energies* **2020**, *13*, 4163. [CrossRef]

17. Barone, G.; Brusco, G.; Menniti, D.; Pinnarelli, A.; Sorrentino, N.; Vizza, P.; Burgio, A.; Bayod-Rújula, Á.A. A Renewable Energy Community of DC Nanogrids for Providing Balancing Services. *Energies* **2021**, *14*, 7261. [CrossRef]

18. Van Aubel, P.; Poll, E. Smart metering in the Netherlands: What, how, and why. *Int. J. Electr. Power Energy Syst.* **2019**, *109*, 719–725. [CrossRef]

19. Gerasopoulos, S.I.; Manousakis, N.M.; Psomopoulos, C.S. Smart metering in EU and the energy theft problem. *Energy Effic.* **2022**, *15*, 12. [CrossRef]

20. Sanduleac, M.; Ciornei, V.I.; Toma, L.; Plamanescu, R.; Dumitrescu, A.M.; Albu, M. High reporting rate smart metering data for enhanced grid monitoring and services for energy communities. *IEEE Trans. Ind. Inform.* **2021**, *18*, 4039–4048. [CrossRef]

21. Cano Ortega, A.; Sánchez Sutil, F.J.; De la Casa Hernández, J. Power factor compensation using teaching learning based optimization and monitoring system by cloud data logger. *Sensors* **2019**, *19*, 2172. [CrossRef]

22. Sanchez-Sutil, F.; Cano-Ortega, A.; Hernandez, J.C.; Rus-Casas, C. Development and calibration of an open source, low-cost power smart meter prototype for PV household-prosumers. *Electronics* **2019**, *8*, 878. [CrossRef]

23. Pinnarelli, A.; Menniti, D.; Sorrentino, N.; Bayod, A. Chapter 8: Optimal management of energy storage systems integrated in nanogrids for virtual "nonsumer" community. In *Elsevier Book Distributed Energy Resources in Local Integrated Energy Systems: Optimal Operation and Planning*, 1st ed.; Optimal Operation and Planning; Elsevier: Amsterdam, The Netherlands, 2021; eBook; ISBN1 9780128242148. Paperback; ISBN2 9780128238998.

24. Mendicino, S.; Menniti, D.; Palumbo, F.; Pinnarelli, A.; Sorrentino, N.; Viggiano, L.; Vizza, P. An IT platform for the management of a Power Cloud community leveraging IoT, data ingestion, data analytics and blockchain notarization. In Proceedings of the 2021 IEEE PES Innovative Smart Grid Technologies Europe (ISGT Europe), Espoo, Finland, 18–21 October 2021; pp. 1–6.

25. DMEA—Wholesale Energy Markets and Environmental Sustainability Department. Self-Consumption and Energy Communities. Resolution ARERA 318/2020/R/eel. Available online: https://www.arera.it/it/docs/20/318-20.htm (accessed on 4 August 2020).

26. Ministerial Decree. Identification of the Incentive Tariff for the Remuneration of Renewable Energy Plants Included in the Experimental Configurations of Collective Self-Consumption and Renewable Energy Communities. Available online: https://www.mise.gov.it/index.php/it/normativa/decreti-ministeriali/decreto-ministeriale-16-settembre-2020-individuazione-della-tariffa-incentivante-per-la-remunerazione-degli-impianti-a-fonti-rinnovabili-inseriti-nelle-configurazioni-sperimentali-di-autoconsumo-collettivo-e-comunita-energetiche-rinnovabili (accessed on 16 September 2020).

27. GSE. Technical Rules for Access to the Shared Electricity Enhancement and Incentive Service. Available online: https://www.arera.it/allegati/docs/22/003-22dmea.pdf (accessed on 4 April 2022).

28. Bianco, G.M.; Bonvini, B.; Bracco, S.; Delfino, F.; Laiolo, P.; Piazza, G. Key Performance Indicators for an Energy Community Based on Sustainable Technologies. *Sustainability* **2021**, *13*, 8789. [CrossRef]

Article

Spectral Kurtosis Based Methodology for the Identification of Stationary Load Signatures in Electrical Signals from a Sustainable Building

Luis A. Romero-Ramirez [1], David A. Elvira-Ortiz [1], Rene de J. Romero-Troncoso [1], Roque A. Osornio-Rios [1], Angel L. Zorita-Lamadrid [2], Sergio L. Gonzalez-Gonzalez [3] and Daniel Morinigo-Sotelo [2,*]

[1]   HSPdigital–CA Mecatronica, Facultad de Ingenieria, Universidad Autonoma de Queretaro,
      Campus San Juan del Rio, Rio Moctezuma 249, Col. San Cayetano, C. P., San Juan del Rio 76807, Mexico;
      lromero@hspdigital.org (L.A.R.-R.); delvira@hspdigital.org (D.A.E.-O.); troncoso@hspdigital.org (R.d.J.R.-T.);
      raosornio@hspdigital.org (R.A.O.-R.)

[2]   Research Group HSPdigital-ADIRE, Institute of Advanced Production Technologies (ITAP),
      University of Valladolid, 47011 Valladolid, Spain; zorita@eii.uva.es

[3]   Research Group Termotecnia, University of Valladolid, 47011 Valladolid, Spain;
      sergiolorenzo.gonzalez@uva.es

*   Correspondence: daniel.morinigo@eii.uva.es

Citation: Romero-Ramirez, L.A.;
Elvira-Ortiz, D.A.;
Romero-Troncoso, R.d.J.;
Osornio-Rios, R.A.;
Zorita-Lamadrid, A.L.;
Gonzalez-Gonzalez, S.L.;
Morinigo-Sotelo, D. Spectral Kurtosis
Based Methodology for the
Identification of Stationary Load
Signatures in Electrical Signals from a
Sustainable Building. *Energies* **2022**,
15, 2373. https://doi.org/10.3390/
en15072373

Academic Editor: Surender Reddy
Salkuti

Received: 15 February 2022
Accepted: 22 March 2022
Published: 24 March 2022

Publisher's Note: MDPI stays neutral
with regard to jurisdictional claims in
published maps and institutional affiliations.

Abstract: The increasing use of nonlinear loads in the power grid introduces some unwanted effects, such as harmonic and interharmonic contamination. Since the existence of spectral contamination causes waveform distortion that may be harmful to the loads that are connected to the grid, it is important to identify the frequency components that are related to specific loads in order to determine how relevant their contribution is to the waveform distortion levels. Due to the diversity of frequency components that are merged in an electrical signal, it is a challenging task to discriminate the relevant frequencies from those that are not. Therefore, it is necessary to develop techniques that allow performing this selection in an efficient way. This paper proposes the use of spectral kurtosis for the identification of stationary frequency components in electrical signals along the day in a sustainable building. Then, the behavior of the identified frequencies is analyzed to determine which of the loads connected to the grid are introducing them. Experimentation is performed in a sustainable building where, besides the loads associated with the normal operation of the building, there are several power electronics equipment that is used for the electric generation process from renewable sources. Results prove that using the proposed methodology it is possible to detect the behavior of specific loads, such as office equipment and air conditioning.

Keywords: digital signal processing; green buildings; total harmonic distortion; spectral analysis; spectral kurtosis

1. Introduction

In a world that needs to evolve towards energy sustainability, smart buildings that are energy efficient are a necessity. These buildings are characterized by energy flexibility, renewable energy production and user interaction [1], and must be designed for near-zero energy, which is accomplished by managing renewable energy sources, advanced monitoring and control systems, energy storage and demand flexibility. Climate and user responses along with monitoring and supervision are some of the basic functions of smart buildings [1], which adapt to climatic conditions to minimize energy demand and generate energy to supply their energy consumption [2], and monitoring and supervision are necessary to control loads and set comfort settings [3]. However, to achieve the aim of adequate energy efficiency it is necessary to use modern lighting systems (LED and compact fluorescent lamps) [4,5], as well as numerous types of non-linear loads or DC appliances that need individual rectifiers to facilitate the connection of the equipment to the AC grid [6], and of course, to integrate the use of renewable energies, especially photovoltaic technology.

However, despite the advantages that the introduction of these elements brings when it comes to achieving a smart building, power quality (PQ) problems can arise in the electrical networks of these buildings that generate distortions, overloads, unbalances and voltage fluctuations [7–13] Within these PQ problems, the generation of harmonics and interharmonics are a particularly important concern in the electrical grid, as they arise precisely due to the operation of interconnected electricity producers (cogeneration, photovoltaics, etc.) [14] and the use of converters based on power electronics as an interface between distributed generation and the electricity grid [15], as well as the numerous non-linear loads existing in this type of buildings, thus causing energy losses, reduction of the life cycle and malfunctioning of equipment and installations. For this reason, research into power quality has increased a lot in recent times, especially in its detection, and since a disturbance can appear in the power grid at any time, it is important to develop techniques that allow continuous monitoring of electrical signals to determine the existence of anomalous behavior. In this sense, techniques that focus on the detection of PQ disturbances can be classified into two sets: transform-based (non-parametric) methodologies and model-based (parametric) methodologies [16], with the first group being the most observed in recent literature. Among all the non-parametric methodologies, the fast Fourier transform (FFT) is yet the most used technique when it is necessary to carry out an analysis in the frequency domain [17]. Notwithstanding, the FFT presents some drawbacks: it can only be used when the signal to analyze is stationary, and the spectral leakage that appears when a frequency component is not an integer multiple of the resolution of the transform. Thus, several techniques have been proposed for dealing with the issues related to the use of the FFT. Most of these alternative techniques extend their analysis to the time-frequency domain allowing studying both stationary and transient disturbances. Examples of these techniques are the short-time Fourier transform (STFT), wavelet transform (WT), S-transform (ST), Hilbert–Huang transform (HHT), empirical mode decomposition (EMD), multiple signal classification (MUSIC), among others [12,18,19]. Although these techniques deal with some of the concerns regarding FFT, they present other non-desired effects as mode mixing and high computational effort.

On the other hand, the parametric approaches aim to develop an accurate mathematical model that describes the behavior of any PQ disturbance [20,21]. These methods require an a priori knowledge of the parameters that describe the disturbance (severity, duration, spectral content, among others); however, they do not provide information regarding how these parameters are obtained. Thus, the use of methodologies that involve time and frequency domain high order statistic (HOS) have gained popularity, not only for parameter identification but also for working along with artificial intelligence tools, such as artificial neural networks (ANN) and support vector machines (SVM) to detect and classify a large set of PQ disturbances [22]. Within parametric approaches and given that spectral contamination with harmonics and interharmonics is probably the most common issue related to the existence of non-linear loads and distributed generation, the use of frequency-domain features, such as spectral kurtosis (SK), has gained popularity. Recently, the SK has been widely explored to perform a feature extraction that provides information about the existence of specific spectral components in electric signals [23,24]. The effectiveness of the spectral kurtosis relies on the fact that it is insensitive to Gaussian noise, and the computational burden and execution time associated with this feature are low. The works reported so far, using SK [25,26] or other techniques [27], have been demonstrated to be effective for detecting the presence of harmonics and interharmonics with a constant amplitude trend, even when their energy is low. However, they do not provide a time tracking of the detected spectral components to see their behavior along the day. The presence of non-linear loads is inherent in every smart building, not only for the existence of office or residential equipment but also for the use of self-generation systems mainly based on renewable energies. Therefore, smart buildings are likely to experience high levels of spectral distortion due to the large number of non-linear loads they have to handle. In this sense, it is important to develop techniques that allow performing a complete detection, quantification, and tracking of the most significant spectral components in the grid to determine how they evolve and be able to take actions when required to guarantee a proper quality for the final users.

The contribution of this work is developed in this line, introducing a methodology for the detection and quantification of stationary frequency components (harmonics and interharmonics) in the electrical signals of a smart building, and fusing the SK with the FFT. The use of the proposed methodology presents some advantages against the conventional approaches, for instance, the ease of its implementation since the mathematics behind this methodology are simple. This situation results in a technique that is efficient and demands a low computational burden. Moreover, it is possible to perform a time tracking of a specific stationary frequency component (SFC) without problems, such as the mode mixing introduced by time-frequency transforms, such as WT and EMD, this way it is possible to perform a quantification of how every SFC contributes to detriment the quality of the power grid. Finally, it is important to mention that, although SK is a widely explored technique, its use is more extended in the identification of transient events. Additionally, the combination of SK and FFT represents a novel solution for the time tracking and quantification of SFC. The proposed methodology aims to be a helpful tool to perform an estimation of the types of loads that appear in the grid and how they impact the quality of the supply. This way it is possible to propose actions in order to mitigate the undesired effects associated with a specific type of load. To validate this methodology, measurements have been carried out in a smart building located in the northeast of Spain, using two weeks of data from two different years. Since the proposed methodology performs a time tracking of the harmonics and interharmonics in the electric signals, it can be used as a tool for monitoring the levels of distortion associated with any frequency component along the day. This is helpful to determine when the distortion levels are beyond the admissible levels and take corrective actions when required. Moreover, with the proposed methodology, it is possible to perform an analysis of the power consumption habits that are presented in the building and how they influence the PQ of the building.

2. LUCIA Nearly Zero Energy Building

The building under analysis is named LUCIA (Lanzadera Universitaria de Centros de Investigación Aplicada, in Spanish for University Shuttle of Applied Research Centers) [28]. Founded in 2015 by the University of Valladolid, Spain, with the purpose of working as an applied research building that incorporates scientific centers from different disciplines and applied research laboratories through the creation of spin-off companies from enterprises with a technological base. The building presents a flexible distribution that allows several simultaneous uses to meet a changing demand from the users.

LUCIA is a Nearly Zero Energy Building (NZEB) and a ZERO CO_2 building that has obtained some certifications that endorse it as the most sustainable building in Europe and in the entire northern hemisphere, and it is also ranked second in the same category all over the world. It is the second building in the world with the best score in the LEED certification (the first in the northern hemisphere), and it has the LEED Platinum certificate with 98 points. LEED is a worldwide voluntary certification system for sustainable buildings; it was developed by the U.S. Green Building Council and it is the certification in matter of sustainable buildings with the highest global recognition [29]. Additionally, the LUCIA building holds the highest rating (5 leaves) in the VERDE certification, a voluntary national certification that evaluates the reduction of the environmental impacts regarding the building under test, compared with those associated with another hypothetical building that is limited to meet the requirements specified by the local regulations [30]. Moreover, the LUCIA building has been awarded and recognized in different competitions and forums for the sustainable solutions that have been implemented in it. This building has been designed as a base for investigating the social aspects of sustainable buildings; it is a prototype for verifying the hypothesis that sustains the methods for environmental evaluation and one of its main tasks is the investigation on topics that have not been parameterized so far by using only renewable energies.

The building has an area of 7500 m^2, distributed over 4 floors: three of them are located above the ground level and the fourth consists of a basement with a fully open parking space with natural lighting and ventilation (see Figure 1).

Figure 1. General view of the LUCIA building.

With the aim of minimizing the water and energy consumption, as well as the environmental impact, a series of systems and strategies have been implemented in the building to incorporate renewable energies in order to satisfy the requirements of air conditioning, heating and all the electrical services and lighting. These strategies can be categorized into two groups: active and passive strategies. The passive strategies are related to the building design, such as quality thermal isolation, sunlight control through skylights, a geothermal energy system for ventilation among others. In the case of the active strategies, they are described in more detail below since their importance is bigger due to the purpose of the present work:

(a) Solar photovoltaic energy: A part of the electric energy consumption is covered by two photovoltaic generation systems that are installed in the building. One of them is compounded by several photovoltaic modules that are located on the outer face of the double curtain wall in the central area of the southeast side of the building (Figure 2a). This first system (herein called the south system) delivers a rated power of 10 kW. The second system consists of photovoltaic glasses incorporated in two skylights (Figure 2b) above each one of the stairwells in the two communication blocks of the building; therefore, this system fulfills a double function: it provides natural lighting as a part of the roof structure, and it also produces energy delivering a rated power of about 5 kW.

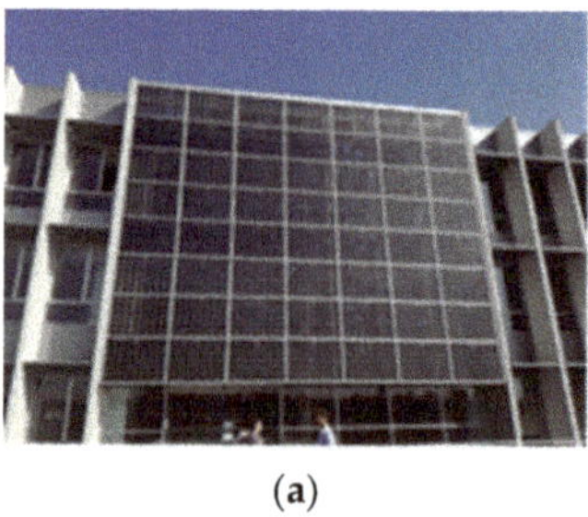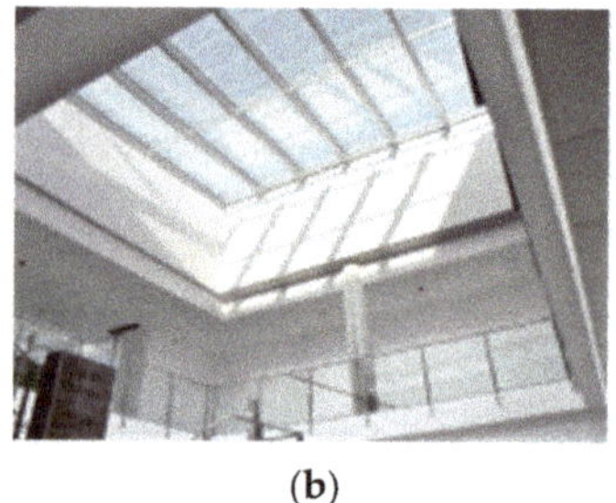

(a) (b)

Figure 2. View of the photovoltaic installations (**a**) south system; and (**b**) skylight.

(b) Trigeneration system: There is a biomass cogeneration installation with a nominal power of 100 kWe and about 180 kWt. It is a gasifier that transforms biomass and wood chips, into syngas that feeds some internal combustion engines. The thermal use of the system, when there is a demand for cooling, is completed with the installation of an absorption chiller that allows the air conditioning installation to provide cold.

(c) Installation of air conditioning and ventilation: it is a mixed air-water system with four-pipe fan coils as terminal units that allow to provide heating and cooling simultaneously in different parts of the building. The air is treated in the primary air conditioning system, equipped with a high-efficiency adiabatic heat recovery unit before it is delivered to the different locations inside the building through the fan coils.

(d) Both, the motors of the pumps from the hydraulic circuits and the motors from the fans of the air conditioner, are connected to variable frequency drives (VFD) that allow achieving the maximum efficiency by adjusting the operating conditions to the instantaneous needs. This way, only the required water flow is handled.

(e) Intelligent building management through a supervision, control and monitoring system of the facilities that allow configuring the different elements of the air conditioning and lighting systems for an efficient operation.

- For lighting control, there are light intensity sensors that regulate and adjust the luminosity of the lamps in the workspaces when they need to be turned on. In the common areas, there are also presence detectors to limit the lamp activation to the necessary moments and when the levels of natural lighting are not sufficient.
- For air conditioning, there are thermostats in each space that allow to independently regulate and control the contribution of heath or cold in each room.
- The building is fully monitored in terms of thermal and electrical parameters; it counts with 97 grid analyzers that allow knowing the consumption of each workspace in the facility as well as the energy that is being produced by generation systems located in the building. It also integrates seven thermal energy meters at different locations within the air conditioning installation; temperature sensors in offices, laboratories and common spaces and even a weather station on the roof of the building to know different environmental and meteorological parameters.

Additionally, it must be mentioned that the LUCIA building incorporates its own electrical substation to manage the building energy consumption. This substation uses an 800 kVA dry-type transformer from Merlin Gerin. The main specifications of the substation are presented in Table 1.

Table 1. Specifications for the substation in the LUCIA building.

Specification	Value
Voltage ratio	13.2–20 kV/0.42 kV
Secondary voltage	420 V
Short Circuit Voltage	6%

Although all the aforementioned technology allows the building to generate its own energy and ensures more efficient performance of the overall tasks that must be carried out, the use of cogeneration with renewable energies, such as solar photovoltaic and biomass, along with the intelligent systems for the building management, introduces some challenges for the power grid. Since all these systems require nonlinear elements for proper functioning, a considerable number of spectral components (harmonics and interharmonics) are introduced to the power grid. Therefore, it is important to perform a continuous tracking of these components in order to guarantee that the power quality of the grid remains at acceptable levels that allow the proper operation of the numerous sensitive equipment operating in the technology laboratories that make up this building. Moreover, since the building possesses its own substation, a degradation in the power quality of the grid must be associated only with loads and processes that are performed inside the building, because the transformer acts as a filter for all the disturbances that may come from the outside.

3. Methodology

This paper presents a methodology for the detection and quantification of stationary frequency components (harmonics and interharmonics) in the electrical signals of a smart building, fusing SK with FFT, and the assessment of their impact on the electrical network of a building.

Figure 3 shows a general block diagram of the proposed methodology that is intended to develop two main tasks:

- A spectral analysis, which is in charge of the identification and quantification of station-ary frequency components (SFC) in electrical signals and will, in turn, be performed in three stages. First, the SK is used for the detection of SFC in current signals in order to identify specific frequency components that are related to consumption habits in the building. Then, since there are frequency components that are irrelevant to the study, discrimination of non-significant SFC is performed based on characteristics, such as amplitude variability throughout the day, and persistence during working hours. Finally, FFT is performed on the current signal to quantify the SFC detected by the SK.

- A PQ analysis, which aims to assess the impact of SFCs on the smart building network. A PQ analysis is performed with the current and voltage signals to calculate the power consumption and THD associated with the loads inside the building.

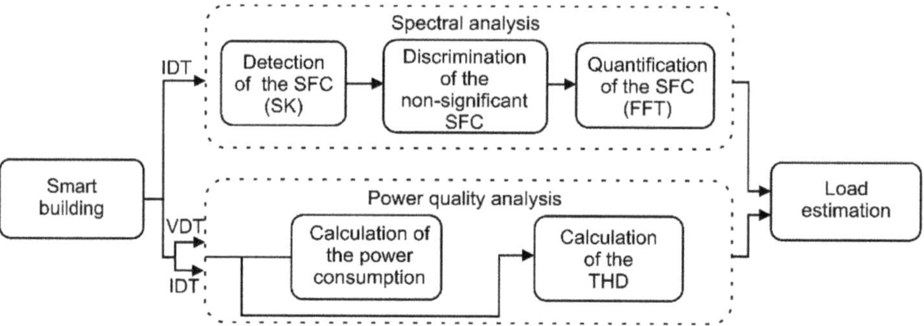

Figure 3. Description of the proposed methodology.

These two analyses are designed to provide a reliable estimate of the nature of the loads introducing pollution into the network to assist network maintainers. The voltage and current signals from the smart building are measured at the low voltage panel and are used in both the spectral analysis and the PQ analysis.

In a first step, the SK assesses all the spectral components that appear in the electric signals and separate the stationary components from those that sporadically appear. Then the FFT is computed to measure the energy of the frequency components that are selected by the SK. Additionally, a general PQ analysis is performed to complement the evaluation of the smart building. Finally, results are analyzed to determine the impact of harmonics, inter-harmonics, and PQ in the electric grid of the smart building. This process is itera-tively carried out to obtain a timestamp of the daily behavior of the stationary frequency components (SFC) that are present in the electric signals.

3.1. Spectral Analysis

As indicated above, the main task of the spectral analysis consists of the detection and quantification of the SFCs appearing in the current signals of the building, considering the fusion of two different techniques: SK and FFT. This analysis has been set up in three phases.

3.1.1. Detection of the SFC

The first step in the spectral analysis consists of the detection of the SFC and the SK is the technique that allows achieving this goal. SK is a technique that uses statistical tools for identifying spectral components that may be transient or stationary. These last are those that concern this work. To perform the computation of the SK the Welch estimation is used according to (1):

$$SK(f) = \frac{M4(f)}{M2(f)^2} - 2 \tag{1}$$

where $SK(f)$ is the frequency domain series that contain the SK values; $M4(f)$ and $M2(f)$ are the spectra of the fourth and the second order moment of the signal, respectively, and they are obtained as described by (2) and (3):

$$M4(f) = E\{|X(f)|^4\} \tag{2}$$

$$M2(f) = E\{|X(f)|^2\} \tag{3}$$

where $E\{\ \}$ is the expected value of time-series from the signal spectrum $X(f)$.

Although the estimation of the SK presented in (1) is mainly used for the identification and characterization of non-stationary series [31], it can also be suitable for the identification of stationary components since it considers the presence of both: Gaussian and non-Gaussian components in a signal. The parameters for computing the algorithm are selected as follows: 8192 samples for computing FFT with an overlap of 3072 points. Additionally, a Hann window of 4082 points is used to reduce the spectral leakage. The obtained spectra are averaged every 10 min and the fourth and the second moment of the signal are computed over these averaged values. The resulting values are substituted in (1) and a total of 144 values are obtained to represent the behavior of everyday analysis. It is important to mention, that the average spectrum is stored to be used later to obtain the energy of the identified stationary components.

As aforementioned, SK is a powerful tool for detecting non-stationary and stationary frequency components as well. This detection is characterized by positive values for non-stationary frequencies and negative values for stationary frequencies, the rest of spectral components, such as Gaussian noise, present a zero (or close to zero) value. Thus, to start with the identification task, all the frequency components with an SK value lower than -0.89 are isolated and considered as preliminary results. This process is carried out on the 144 SK values obtained for each day of analysis.

3.1.2. Discrimination of the Non-Significant SFC

Since the purpose of this work is to identify those frequencies that show a consistent amplitude behavior along the day, it is important to track those frequencies that repeatedly appear as stationary in the SK computation. Moreover, it is desirable to identify the frequency components that can be related to the building occupation during working hours. In order to separate the SFC that significantly contributes to the harmonic and interharmonic contamination of the grid, a discrimination of the non-significant SFC is conducted. Therefore, the frequencies that are present in at least 8 h of the day (i.e., frequencies that appear in 48 of the SK computation), are selected. The rest of the frequency components are discarded because they appear just for short time periods during the day. Furthermore, in this analysis, it is expected that frequencies, such as the fundamental frequency and some of its harmonics appear as stationary frequencies.

Even though a Hann window is applied to reduce the spectral leakage related to the calculation of the FFT, this effect can still be present in the results and it may lead to a misinterpretation of the stationary frequencies. Thus, the tracking of the stationary frequencies is performed searching for frequency values within a threshold instead of looking for a specific frequency. All the frequencies within the threshold are considered to be the same and they are addressed with the central value of the threshold.

3.1.3. Quantification of the SFC

For the quantification of the SFC, the spectrum calculated for the SK values is used. However, only the energy of those frequencies selected in the previous stage is considered. The energy of each frequency is stored to visualize its behavior and trend amplitude during each day. Furthermore, the quantification presents a form to visualize the results of the identification of the SFC in graphs showing amplitude variability along each day.

3.2. Power Quality Analysis

To complement the results delivered by the SK-FFT methodology and to make an overview of the occupation patterns in the sustainable building, a PQ analysis is performed. Since the main purpose of this work is related to the existence of frequency components different than the fundamental one, harmonic analysis is performed using the THD as the parameter to measure the levels of distortion during the days of analysis. The THD is calculated according to the standard IEEE 1159-2019 [32] using (4):

$$THD = 100\frac{\sqrt{\sum_{i=2}^{50} X_i^2}}{X_1},\tag{4}$$

where THD represents the total harmonic distortion as a percentage of the energy of the fundamental frequency component; X represents the electric magnitude under analysis (voltage or current signal); X_1 is the energy of the fundamental frequency component, and; X_i is the energy of the i-th harmonic. The THD index is obtained for the current and for the voltage signals. The THD is obtained for time windows of 200 ms and then the average per minute is calculated and reported.

Additionally, the power consumption related to the building is calculated to show the difference in the occupation level for the two years of analysis. This way it is also possible to determine how the increase of the loads attached to the grid influence the quality of the power supply. In this work, the instantaneous power is computed as the product of the instantaneous values of the voltage and current signals as presented in (5):

$$S(t) = v(t)i(t),\tag{5}$$

where $S(t)$ is the instantaneous apparent power as a time function; $v(t)$ and $i(t)$ are the instantaneous value of the voltage and current signals, respectively. Since the data acquisition systems work at a sampling rate of 8000 Hz, a total of 8000 values per second are computed for the apparent power. However, the obtained power values are separated into one-minute windows and the average per minute is delivered as result. Additionally, to obtain a more reliable analysis of the power that is effectively used in the building, the active power is obtained considering (6):

$$P(t) = \frac{1}{T}\int_0^T S(t)dt,\tag{6}$$

where: P is the active power and T is the time interval over the active power computation is carried out. In order to keep the consistency of the data, T is also selected as 1 min. Since the calculation of the active power is performed over discrete signals, the discretization of (6) presented in (7) is used:

$$P(t) = \frac{1}{T}\sum_{j=1}^n v_n i_n,\tag{7}$$

where n is the number of samples and v_n and i_n are the n-th samples of the voltage and current signals, respectively. Finally, the reactive power $Q(t)$ is calculated using (8) to provide an overview of the energy that is not effectively used in the building.

$$Q(t) = \sqrt{S(t)^2 - P(t)^2} \tag{8}$$

The 1-min window used in the power consumption computation is selected so there exists consistency between the values for the *THD* and the power consumption. *THD* and power provide information about the change in the consumption habits in the building and determine the existence of an increment in the PQ contamination, and distortion levels between the days and years analyzed. This information gives a general picture of the status of the building each year, but it does not provide specific information, such as which harmonics or interharmonics are stationary or how they behave along the day. Then, the co-operation between the SK-FFT methodology for the identification of stationary frequencies and the PQ analysis delivers a more complete study of the building consumption along the time.

4. Experimental Setup

4.1. Description of the Data Acquisition System

Experimentation is performed in the facilities of the LUCIA building. Here, the consumptions of the complete building are measured at the low voltage distribution panel using a proprietary data acquisition system (DAS). The DAS uses a Field Programmable Gate Array (FPGA) device as the main processor and it is able to acquire signals from seven simultaneous channels (three for voltage signals and four for current signals) with a 16-bit resolution and at a sampling frequency of 8000 Hz. The data acquisition is carried out following a series of steps as depicted in Figure 4. In the first stage, the current and voltage from the probes are received by the DAS. Then, the signals are carried through a conditioning and isolation process. The conditioning mainly consists of an impedance coupling; therefore, a resistive element of 10 kΩ is selected to work along with the inductive elements associated with current probes, whereas a 250 kΩ resistive element is incorporated for the voltage probes. It must be mentioned that the values of the voltage and current signals must remain within ± 2 VDC. In order to avoid security risks, the physical ground must be decoupled; thus, an isolation amplifier is used to accomplish this purpose. Additionally, the same amplifier allows obtaining a differential amplification with a gain of 8. Next and with the aim of avoiding signal aliasing, a filtering stage is performed. Here, a passive low-pass filter with a cutoff frequency of 3.2 kHz is implemented. At this point, the signals have been properly conditioned so they can be used in the data conversion process. To perform this conversion, an 8-channel analog to digital converter (ADC) is used. The selected ADC includes a programmable gain amplifier (PGA) and an electromagnetic interference (EMI) filter specially designed for PQ applications. The responsible for the configuration and the proper operation of the ADC is an FPGA-based processor. This processor is also in charge of establishing the sampling frequency, that in this particular case is selected to be 8000 Hz. Finally, the data storage is performed using an external micro SD module that allows easily changing the memory card once it is full. The DAS incorporates Bluetooth communication; thus, a personal tablet can be used as a user interface (see Figure 5a). Voltage signals are directly measured with cables from the secondary of the power transformer in the low voltage panel of the building whereas the current measurements are performed using the SCT-013-010 sensors by YHDC (see Figure 5b) installed in the secondary winding of measurement current transformers. The data are collected for two years with different consumptions. During the first year, the building works at about 50% of its capacity, whereas in the second year the installation operates at 100% of its capacity. One week of each year is selected for applying the proposed methodology to determine the existence of spectral components that constantly appear in the power grid and that may be related to the building operation. Additionally, a power quality analysis is performed on the voltage and current signals for the two weeks to determine how the SFC identified by the methodology impacts the grid.

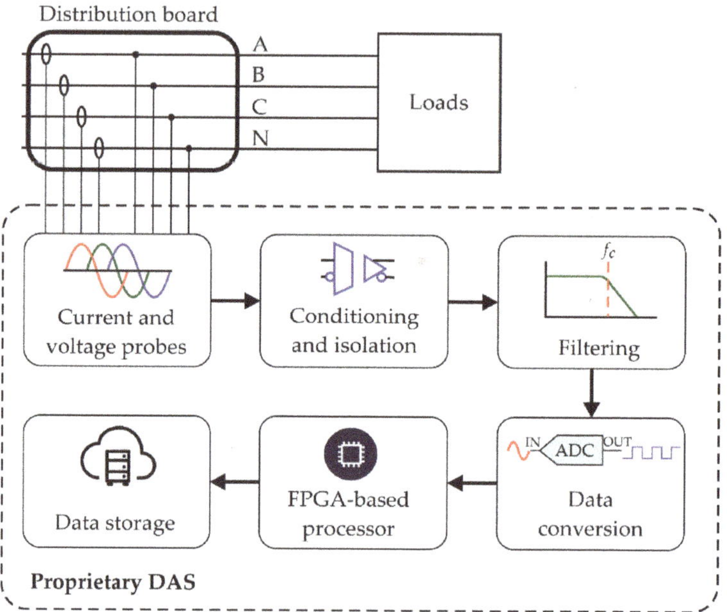

Figure 4. Functional diagram for the DAS.

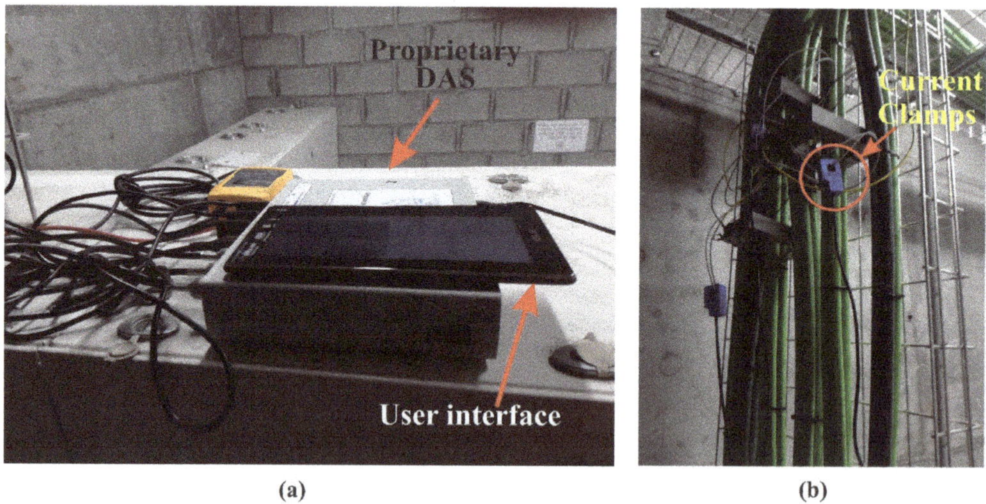

Figure 5. Description of the (**a**) Proprietary DAS with user interface; and (**b**) placement of the current clamps in the secondary winding of measurement current transformers.

4.2. Description of the Experimentation Dates

Two different data sets are selected in this work. The first data set is formed by the voltage and current signals from a complete week comprised between the months of February and March during the year 2018. The selected week goes from 24 February to 2 March. These days are selected because they are winter days but they are also working days; therefore, it is possible to observe the effect of important loads associated with climatization. The second data set is selected to be a week from the year 2019. In this case, the selected dates start on 15 May and end on 21 May. This week was chosen to take into account the effects of air-conditioning systems. The one-year separation between the two data sets is intended to present a view of the change in the consumption patterns inside the building from one year to another.

5. Results and Discussion

As mentioned in the experimental setup, a week of data is acquired in different years (2018 and 2019) from the described smart building. The proposed methodology is used for detecting stationary frequency components (SFC) and performing the power quality (PQ) analysis on the acquired data. Then, an analysis is performed to estimate the relationship between the SFC detected and loads of the smart building. Due to the nature of the loads in the smart building and the implemented strategies for control and energy generation, it is expected that the signals present harmonic and interharmonic contamination; therefore, the proposed methodology is suitable for finding the spectral component that is contaminating the power supply. Several SFC are identified by the methodology in the two years of analysis. The methodology identifies SFC until the range of 4 kHz (half the sampling frequency); however, this work analyzes frequencies until 1 kHz. A total of 39 SFC are identified in 2018 whereas 44 SFC are found in 2019. It is worth noticing that most of the SFC detected by the methodology are interharmonics; notwithstanding, the fundamental frequency and its harmonics are also detected as SFC. Figure 6 shows the amplitude behavior of the 3rd harmonic for each monitored day in the years 2018 and 2019. The 3rd harmonic in both years shows a constant amplitude all the week, but a different amplitude between the two years. This difference is related to the weather, 2018 data are acquired in winter and 2019 data are acquired in spring. Therefore, most of the loads used in the smart building are weather dependent, for instance, climatization.

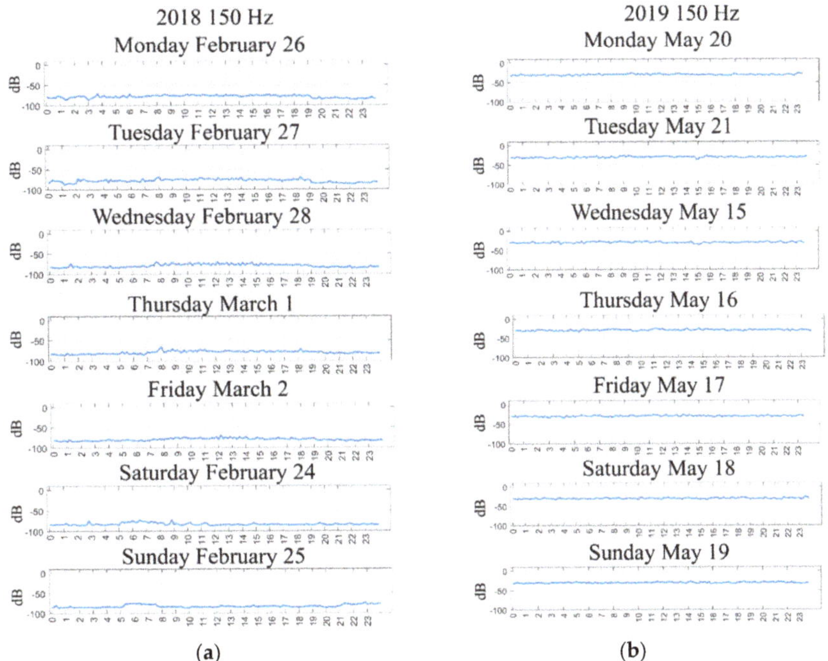

Figure 6. Amplitude behavior for the 3rd harmonic of the current signal (**a**) in 2018; and (**b**) in 2019.

The result of the spectral analysis and the PQ analysis are shown below.

5.1. Stationary Frequency Components and Loads Estimation

The amplitude behavior of the SFC detected in the current signals has a direct relationship with the loads connected to the electric grid in the smart building. Figure 6 shows some of the detected SFC in the electric current; these frequencies present changes in their energy values at specific hours during the weekdays. For the week taken from the year 2018, it is observed that the appearance of the frequencies of 32 Hz, 133 Hz and 167 Hz is consistent for different workdays (see Figure 7a,b). Moreover, the behavior of these three components is very similar so it can be inferred that all of them are produced for loads with similar characteristics. It is observed that the amplitude of the fluctuations in these frequency components presents a magnitude between 30 and the 70 dB and they occur at the beginning of the day, around 7:00, and at the end of the day, nearly 19:00 h. This schedule represents the period of the maximum occupancy of the building and it also corresponds with the start-up and the shutdown of the climatization systems. It is expected that these types of loads introduce a noticeable SFC since around 80% of the building consumption, because of the climatization (fan coils) and a Data Processing Centre (DPC). On the other hand, during the weekend the energy values of these spectral components remain constant (see Figure 7c,f). Additionally, it is worth noticing the fact that during the weekend, the amplitude values for the 32 Hz, 132 Hz and 167 Hz components are not only constant but low, indicating that the loads that cause the existence of such components are not operating during these days. Moreover, the climatization system is not started during the weekend; therefore, it cannot introduce SFC to the power grid. Some of the SFC that appear in the year 2018, still remain in the year 2019. This is the case for the 32 Hz and the 132 Hz components (see Figure 7d,e). Although in 2018 the identified SFC corresponds with a 133 Hz frequency and in the year 2019 the reported value is 132 Hz, these frequencies are so close to each other that it can be assumed that they are the same. Another important situation is the fact that harmonics always appear as SFC because they present a constant behavior for all the days. Notwithstanding, Figure 7d–f presents the temporal behavior of the 7th harmonic and it is observed that it behaves the same during weekdays as on weekends. This is an important situation because it shows that most loads of the building impact the generation of interharmonics instead of harmonics. This is an expected situation since many of the loads associated with sustainable buildings are nonlinear. In contrast, in Figure 7e it is observed that an increment in the amplitude of the 132 Hz starts at 0:00 h and finishes at 5:00 h. This time lapse coincides with the operation of the air purification system. At this point, it is worth noticing that the existence of frequency components different from the fundamental component causes a waveform distortion of the voltage and current signals. Such waveform distortion is harmful to the loads attached to the power grid because it reduces the lifetime of the internal components of every electric and electronic device. In industrial facilities, this kind of situation may lead to unexpected stops in the production process representing financial losses. On the other hand, at residential facilities, the presence of waveform distortions can be traduced in the malfunctioning of home appliances. Therefore, in order to propose solutions for mitigating the detrimental effects associated with harmonic and interharmonic components, it is necessary to develop robust and reliable methodologies that allow quantifying and identifying the specific components that cause waveform distortion. This turns out to be the main purpose of the proposed methodology.

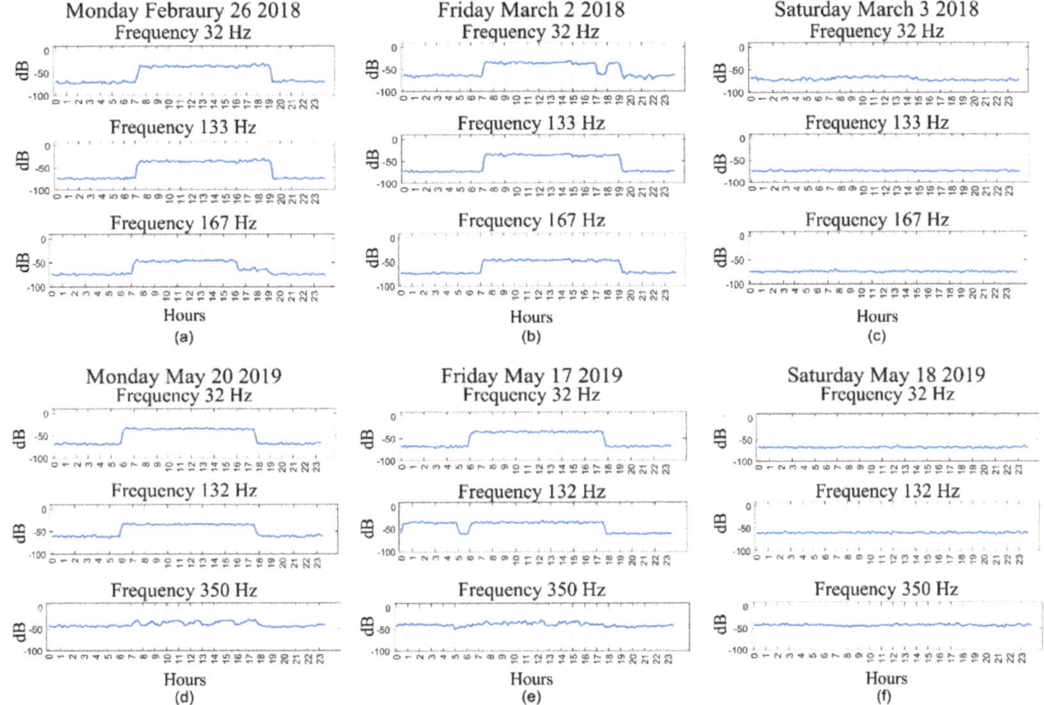

Figure 7. SFC for the current signals: (**a**) frequencies 32 Hz, 133 Hz and 167 Hz for Monday 26 February 2018, (**b**) frequencies 32 Hz, 133 Hz and 167 Hz for Friday 2 March 2018, (**c**) frequencies 32 Hz, 133 Hz and 167 Hz for Saturday 3 March 2018, (**d**) frequencies 32 Hz, 132 Hz, and 350 Hz for Monday 20 May 2019, (**e**) frequencies 32 Hz, 132 Hz, and 350 Hz for Friday 17 May 2019, and (**f**) frequencies 32 Hz, 132 Hz, and 350 Hz for Saturday 18 May 2019.

It is reported that freezers and cool rooms represent between 20% to 25% of the building power consumption, whereas the consumption related to loads, such as illumination and office equipment, is almost imperceptible. The freezers and cool rooms can turn on at any time of the day. Figure 8a,d,g shows the 850 Hz component detected as SFC on different days of the year 2018, and the 340 Hz component detected as SFC on some other days of the year 2019 (see Figure 8b,e,h). These frequencies show a random variability in their amplitude that can be related to the operation of freezers and cool rooms. At this point is important to mention that these two frequencies present similar behavior during all the workdays. However, some frequency components behave differently depending on the weekday. For instance, during the year 2019, it is found that 668 Hz is an SFC. On Monday 20 May 2019 this frequency component presents a rising at 6:00 and then a falling at around 18:00 (see Figure 8c). Notwithstanding, on Friday 17 May the amplitude behavior is different since it presents several minor variations at night (see Figure 8f). Thus, it can be inferred that an atypical load was started during the night of 17 May 2019. During the weekend these SFC remain in very low values, but the 850 Hz component found on 3 March 2018 shows several variations (see Figure 8g). This situation can be easily explained due to the fact that the vending machines and some of the freezers in the building remain working even at weekend.

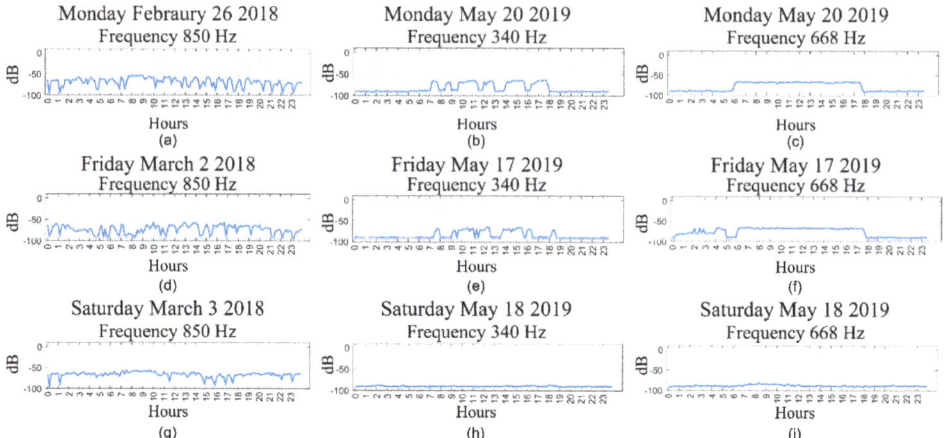

Figure 8. SFC for the current signals: (**a**) frequency 850 Hz Monday 26 February 2018, (**b**) frequency 340 Hz Monday 20 May 2019, (**c**) frequency 668 Hz Monday 20 May 2019, (**d**) frequency 850 Hz Friday 2 March 2018, (**e**) frequency 340 Hz Friday 17 May 2019, (**f**) frequency 668 Hz Friday 17 May 2019, (**g**) frequency 850 Hz Saturday 3 March 2018, (**h**) frequency 340 Hz Saturday 18 May 2019, (**i**) frequency 668 Hz Saturday 18 May 2019.

In addition, the detected SFC can cause long-term problems to the grid and the loads attached to it, because these frequencies cause waveform distortion that compromises the correct functioning of the loads attached to the grid. The proposed methodology brings information that makes it possible to take more specific corrective actions to mitigate the undesired effects. For instance, by knowing the frequencies that are mainly contributing to the detriment of the power quality in the grid, it is possible to correctly tune a filter that helps to attenuate the amplitude of the harmonic and interharmonic components. This way, the waveform distortion due to harmonic and interharmonic contamination can be reduced and the power quality can be improved.

It is important to mention that there are some other techniques that can also perform identification of SFC in an electrical signal, for instance, the FFT can perform this identification by itself. However, by using only the FFT it is not possible to perform the time tracking that is achieved by the proposed methodology. Some other techniques, such as WT and EMD could provide a solution that brings information regarding the temporal behavior of certain frequencies. Notwithstanding, these techniques deliver information for frequency bands rather than for specific frequencies. The proposed approach identifies and quantifies single frequency components delivering more accurate data since it avoids the mode mixing that can cause the contribution of a frequency component to be considered more than one time for the same analysis. Finally, there is some commercial equipment that allows performing an analysis of the harmonic distortion in an electrical signal. However, this equipment provides information regarding the total contribution of all the existent harmonics and it is not possible to obtain data about any specific harmonic. Moreover, only harmonics are considered and interharmonics are left aside. By using the methodology proposed in this work, it is possible to perform an analysis of both: harmonics and interharmonics. Additionally, information regarding specific frequencies is available for being used as a tool to cope with the methodologies for harmonic and interharmonic mitigation.

Since there are some other works and methodologies that try to address the same or similar issues to the methodology proposed in this work, Table 2 presents a comparison of the previously reported works against this work (The x indicates the capabilities of the different methodologies).

Table 2. Comparison of the proposed methodology with previous works.

Methodology	Identification of Stationary Harmonics	Identification of Stationary Interharmonics	Tracking Behavior over Time	Relationship with Possible Loads
SK [25]	x	-	1 h	-
AVMD + HT [27]	x	-	-	-
RLS-IEKF [32]	x	x	-	-
WT [33]	x	-	-	x
SK-FFT (this work)	x	x	Weekly	x

From Table 2, it is observed that the SK by itself has been previously used for the identification of stationary frequency components; however, its use has been reported only for the identification of harmonics, and interharmonics are left aside. Moreover, the SK also performs a time tracking of the frequency components but only on a 1-h basis and it does not provide information regarding the nature of the load that causes the existence of certain harmonics. On the other hand, the technique reported in [27] fuses adaptive variational mode decomposition (AVMD) with the Hilbert transform (HT). Although this methodology delivers good results on the identification of stationary harmonics, interharmonics are not analyzed and no time tracking of the behavior of the harmonic is presented. An explanation of the loads that may cause the detected harmonics is not provided either. Besides the proposed methodology, in [32] it is proposed the use of recursive least squares (RLS) and an iterated extended Kalman filter (IEKF) for the identification of both: harmonic and interharmonic components. Notwithstanding, a time tracking of the frequency components is not presented and the type of load that may be related to a specific frequency component is not analyzed. In [33] the use of wavelet transform (WT) is introduced for identifying harmonic components and the authors of this work also try to introduce an explanation of the loads that generate such harmonics. However, interharmonics are not studied and the time evolution of the found frequencies is not presented. Finally, the SK-FFT methodology proposed in this particular work is able to identify and quantify any SFC considering harmonics and interharmonics as well. Moreover, the time evolution of every SFC is tracked and presented and this behavior over time is used for analyzing the type of load that causes each SFC. Thus, the methodology here proposed is the only one that is capable of fulfilling these four tasks: identification of stationary harmonics, identification of stationary interharmonics, tracking behavior over time and presenting a relationship with possible loads.

5.2. PQ Analysis

The PQ analysis is performed on data from both years. This analysis is used as a complement to the results obtained by the SK-FFT methodology. Figure 9 shows the voltage THD calculated for both years, the behavior of this index is similar from 2018 to 2019. The register of the loads from the smart building shows that in both years there is almost the same number of loads, and this is reflected in the THD. Furthermore, the THD remains within acceptable levels during the two years of analysis. At this point, it is important to mention that the THD index is computed considering only the influence of harmonics. As it can be observed in the previous sections, most of the SFC that appear are interharmonics; therefore, their contribution is not considered. This is an important situation because most of the international standards that regulate PQ issues consider the THD as the main index for quantifying waveform distortion. The results found in this work show that interharmonics can also have a significant contribution to waveform distortion; then, their contribution should be considered. In this sense, the proposed methodology can inform about the interharmonics that significantly contribute to the waveform distortion so they can be considered in a new index for quantifying this waveform distortion.

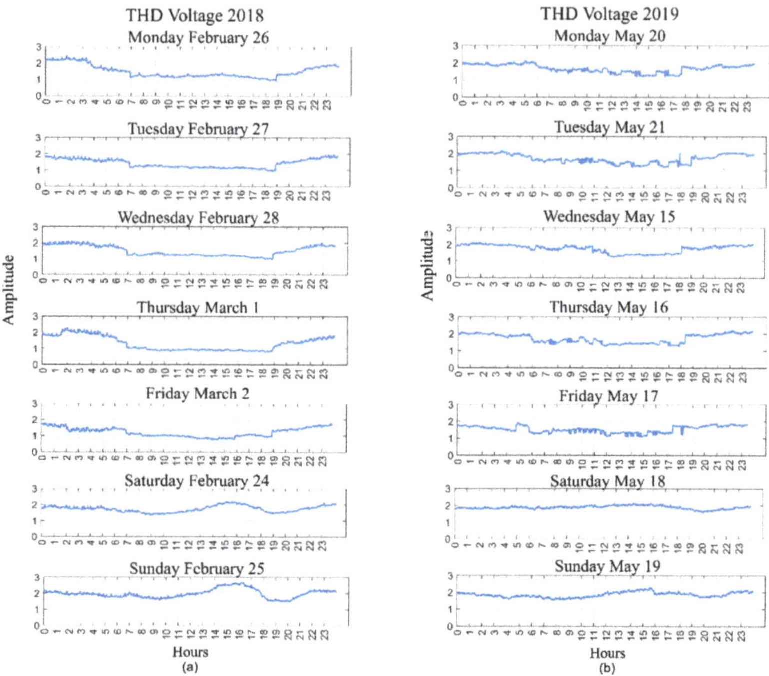

Figure 9. Voltage THD for the: (**a**) week from 2018, (**b**) week from 2019.

As part of the PQ analysis, the active power is calculated. Figure 10 shows the active power from both years. The levels of active power in 2018 have a maximus in 100 kW, and the levels of active power in 2019 have a maximus in 200 kW. This difference is related to the weather because the data from 2018 is acquired in winter and the data from 2019 is acquired in spring and it is observed that more energy is required for cooling the building than for warming it.

Finally, it is important to mention that this methodology is suitable to be applied in facilities and installations different than the smart building presented here. For instance, this methodology may result helpful at industrial facilities where an important number of loads, such as electric motors, are operating simultaneously. Identifying unexpected frequency components can be useful for avoiding damage to sensitive equipment and reducing losses for the enterprise related to unexpected stops and poor PQ. Thus, the application of the proposed methodology at industrial facilities is left as a perspective for future work.

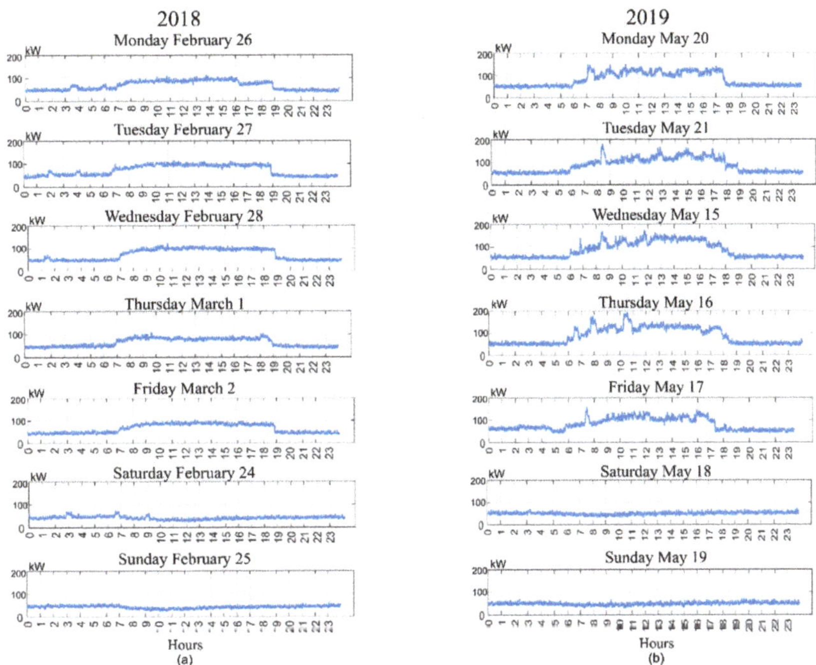

Figure 10. Active Power (**a**) week from 2018 (**b**) week from 2019.

6. Conclusions

In this work, the fusion of two well-known techniques is proposed for the identification and isolation of stationary frequency components in electric signals. These techniques are the SK and the FFT. By combining these two techniques it is possible to use SK for tracking specific frequency components with stationary behavior, and with the FFT it is possible to perform an accurate estimation of each stationary component. This allows the identification and quantification of the SFC that cause waveform distortion in a robust and reliable way.

In addition, the described methodology has some advantages over some of the conventional approaches, as explained below. First, this proposed approach identifies and quantifies individual frequency components rather than frequency bands. This situation makes it possible to provide more accurate data, as it avoids mode mixing. Secondly, this methodology also allows observing the behavior of each frequency component over time; therefore, it is possible to determine the causes of specific frequencies to determine if any corrective action needs to be implemented.

Author Contributions: Conceptualization, L.A.R.-R., D.A.E.-O. and R.d.J.R.-T.; methodology, L.A.R.-R., D.A.E.-O. and D.M.-S.; validation, L.A.R.-R., D.A.E.-O. and R.A.O.-R.; formal analysis, A.L.Z.-L., S.L.G.-G. and D.M.-S.; investigation, L.A.R.-R., D.A.E.-O., and S.L.G.-G.; resources, A.L.Z.-L., D.M.-S. and R.A.O.-R.; data curation, L.A.R.-R., D.A.E.-O. and D.M.-S.; writing—original draft preparation, L.A.R.-R., D.A.E.-O. and D.M.-S.; writing—review and editing, A.L.Z.-L., R.A.O.-R. and R.d.J.R.-T.; visualization, D.A.E.-O. and D.M.-S.; supervision, D.M.-S., R.d.J.R.-T and A.L.Z.-L.; project administration, R.d.J.R.-T. and D.M.-S.; funding acquisition, A.L.Z.-L., D.M.-S. and R.A.O.-R. All authors have read and agreed to the published version of the manuscript.

Funding: This work was partially funded by the Universidad de Valladolid; by the Mexican council of science and technology (CONACYT) under grant 743842; and by FONDEC-UAQ 2020 FIN202011 project.

Institutional Review Board Statement: Not applicable.

Informed Consent Statement: Not applicable.

Conflicts of Interest: The authors declare no conflict of interest.

References

1. Al Dakheel, J.; Del Pero, C.; Aste, N.; Leonforte, F. Smart buildings features and key performance indicators: A review. *Sustain. Cities Soc.* **2020**, *61*, 102328. [CrossRef]
2. Golpîra, H.; Khan, S.A.R. A multi-objective risk-based robust optimization approach to energy management in smart residential buildings under combined demand and supply uncertainty. *Energy* **2019**, *170*, 1113–1129. [CrossRef]
3. Marinakis, V.; Karakosta, C.; Doukas, H.; Androulaki, S.; Psarras, J. A building automation and control tool for remote and real time monitoring of energy consumption. *Sustain. Cities Soc.* **2013**, *6*, 11–15. [CrossRef]
4. Ciugudeanu, C.; Buzdugan, M.; Beu, D.; Campianu, A.; Galatanu, C.D. Sustainable Lighting-Retrofit Versus Dedicated Luminaires-Light Versus Power Quality. *Sustainability* **2019**, *11*, 7125. [CrossRef]
5. Lucia, O.; Cvetkovic, I.; Sarnago, H.; Boroyevich, D.; Mattavelli, P.; Lee, F.C. Design of Home Appliances for a DC-Based Nanogrid System: An Induction Range Study Case. *IEEE J. Emerg. Sel. Top. Power Electron.* **2013**, *1*, 315–326. [CrossRef]
6. Pena-Alzola, R.; Bianchi, M.A.; Ordonez, M. Control Design of a PFC with Harmonic Mitigation Function for Small Hybrid AC/DC Buildings. *IEEE Trans. Power Electron.* **2015**, *31*, 6607–6620. [CrossRef]
7. Nallusamy, S.; Velayutham, D.; Govindarajan, U.; Parvathyshankar, D. Power quality improvement in a low-voltage DC ceiling grid powered system. *IET Power Electron.* **2015**, *8*, 1902–1911. [CrossRef]
8. Wunder, B.; Ott, L.; Szpek, M.; Boeke, U.; Weiß, R. Energy efficient DC-grids for commercial buildings. In Proceedings of the 2014 IEEE 36th International Telecommunications Energy Conference (IN TEL EC), Sao Paulo, Brazil, 17–20 August 2014; pp. 1–8.
9. Parchure, A.; Tyler, S.J.; Peskin, M.A.; Rahimi, K.; Broadwater, R.P.; Dilek, M. Investigating PV Generation Induced Voltage Volatility for Customers Sharing a Distribution Service Transformer. *IEEE Trans. Ind. Appl.* **2017**, *53*, 71–79. [CrossRef]
10. Alam, M.J.E.; Muttaqi, K.M.; Sutanto, D. Battery Energy Storage to Mitigate Rapid Voltage/Power Fluctuations in Power Grids Due to Fast Variations of Solar/Wind Outputs. *IEEE Access* **2021**, *9*, 12191–12202. [CrossRef]
11. Chamana, M.; Chowdhury, B.H.; Jahanbakhsh, F. Distributed Control of Voltage Regulating Devices in the Presence of High PV Penetration to Mitigate Ramp-Rate Issues. *IEEE Trans. Smart Grid* **2018**, *9*, 1086–1095. [CrossRef]
12. Elvira-Ortiz, D.A.; Morinigo-Sotelo, D.; Duque-Perez, O.; Jaen-Cuellar, A.Y.; Osornio-Rios, R.A.; Romero-Troncoso, R.D.J. Methodology for Flicker Estimation and Its Correlation to Environmental Factors in Photovoltaic Generation. *IEEE Access* **2018**, *6*, 24035–24047. [CrossRef]
13. Zahedmanesh, A.; Muttaqi, K.M.; Sutanto, D. Alleviation of Voltage Variations Introduced by Unbalanced Allocation of Single-phase Loads in a Distribution Network Integrated with PVs and PEVs. In Proceedings of the 2020 IEEE International Conference on Power Electronics, Smart Grid and Renewable Energy (PESGRE2020), Kerala, India, 2–4 January 2020; pp. 1–6. [CrossRef]
14. Raj, N.T.; Iniyan, S.; Goic, R. A review of renewable energy based cogeneration technologies. *Renew. Sustain. Energy Rev.* **2011**, *15*, 3640–3648. [CrossRef]
15. Mancasi, M.; Vatu, R.; Ceaki, O.; Porumb, R.; Seritan, G. Evolution of smart buildings. A Romanian case. In Proceedings of the 2015 50th International Universities Power Engineering Conference (UPEC), Stoke-on-Trent, UK, 1–4 September 2015; pp. 1–4.
16. Mahela, O.P.; Shaik, A.G.; Gupta, N. A critical review of detection and classification of power quality events. *Renew. Sustain. Energy Rev.* **2015**, *41*, 495–505. [CrossRef]
17. Deng, H.; Gao, Y.; Chen, X.; Zhang, Y.; Wu, Q.; Zhao, H. Harmonic Analysis of Power Grid Based on FFT Algorithm. In Proceedings of the 2020 IEEE International Conference on Smart Cloud, Washington, DC, USA, 6–8 November 2020; pp. 161–164.
18. Gupta, N.; Seethalekshmi, K.; Datta, S.S. Wavelet based real-time monitoring of electrical signals in Distributed Generation (DG) integrated system. *Eng. Sci. Technol. Int. J.* **2020**, *24*, 218–228. [CrossRef]
19. Thirumala, K.; Pal, S.; Jain, T.; Umarikar, A.C. A classification method for multiple power quality disturbances using EWT based adaptive filtering and multiclass SVM. *Neurocomputing* **2019**, *334*, 265–274. [CrossRef]
20. Rodriguez-Guerrero, M.A.; Carranza-Lopez-Padilla, R.; Osornio-Rios, R.A.; Romero-Troncoso, R.D.J. A novel methodology for modeling waveforms for power quality disturbance analysis. *Electr. Power Syst. Res.* **2017**, *143*, 14–24. [CrossRef]
21. Elkholy, A. Harmonics assessment and mathematical modeling of power quality parameters for low voltage grid connected photovoltaic systems. *Sol. Energy* **2019**, *183*, 315–326. [CrossRef]
22. Mishra, M. Power quality disturbance detection and classification using signal processing and soft computing techniques: A comprehensive review. *Int. Trans. Electr. Energy Syst.* **2019**, *29*. [CrossRef]
23. Liu, Z.; Zhang, Q.; Han, Z.; Chen, G. A new classification method for transient power quality combining spectral kurtosis with neural network. *Neurocomputing* **2014**, *125*, 95–101. [CrossRef]
24. Ray, P.K.; Eddy, Y.F.; Krishnan, A.; Dubey, H.C.; Gooi, H.B.; Amaratunga, G.A.J. Wavelet Trans-form-Spectral Kurtosis Based Hybrid Technique for Disturbance Detection in a Microgrid. In Proceedings of the 2018 IEEE Power & Energy Society General Meeting (PESGM), Portland, OR, USA, 5–10 August 2018; pp. 1–5.
25. Sierra-Fernández, J.-M.; Rönnberg, S.; De La Rosa, J.-J.G.; Bollen, M.H.J.; Palomares-Salas, J.-C. Application of Spectral Kurtosis to Characterize Amplitude Variability in Power Systems' Harmonics. *Energies* **2019**, *12*, 194. [CrossRef]

26. Cifredo-Chacón, M.Á.; Perez-Peña, F.; Quirós-Olozábal, Á.; González-de-la-Rosa, J.J. Implementation of pro-cessing functions for autonomous power quality measurement equipment: A performance evaluation of CPU and FPGA-based embedded system. *Energies* **2019**, *12*, 914.
27. Cai, G.; Wang, L.; Yang, D.; Sun, Z.; Wang, B. Harmonic Detection for Power Grids Using Adaptive Variational Mode Decomposition. *Energies* **2019**, *12*, 232. [CrossRef]
28. Edificio LUCIA-Universidad de Valladolid. Available online: http://edificio-lucia.blogspot.com/ (accessed on 16 October 2021).
29. LEED Rating System. Available online: https://www.usgbc.org/leed (accessed on 27 December 2021).
30. Green Building Council España-Certificación Verde. Available online: https://gbce.es/certificacion-verde/ (accessed on 16 October 2021).
31. Antoni, J. The spectral kurtosis: A useful tool for characterizing nonstationary signals. *Mech. Syst. Signal Process.* **2006**, *20*, 282–307.
32. Enayati, J.; Moravej, Z. Real-time harmonics estimation in power systems using a novel hybrid algorithm. *IET Gener. Transm. Distrib.* **2017**, *11*, 3532–3538. [CrossRef]
33. Nath, S.; Sinha, P.; Goswami, S.K. A wavelet based novel method for the detection of harmonic sources in power systems. *Int. J. Electr. Power Energy Syst.* **2012**, *40*, 54–61. [CrossRef]

Article

DC Fault Current Analyzing, Limiting, and Clearing in DC Microgrid Clusters

Navid Bayati [1], Hamid Reza Baghaee [2], Mehdi Savaghebi [1,*], Amin Hajizadeh [3], Mohsen N. Soltani [3] and Zhengyu Lin [4]

1   Electrical Engineering Section, Department of Mechanical and Electrical Engineering, University of Southern Denmark, 5230 Odense, Denmark; navib@sdu.dk
2   Department of Electrical Engineering, Amirkabir University of Technology, Tehran 15875-4413, Iran; hrbaghaee@aut.ac.ir
3   Department of Energy Technology, Aalborg University, 6700 Esbjerg, Denmark; aha@energy.aau.dk (A.H.); sms@energy.aau.dk (M.N.S.)
4   Wolfson School of Mechanical, Electrical and Manufacturing Engineering, Loughborough University, Loughborough LE11 3TU, UK; Z.Lin@ieee.org
*   Correspondence: mesa@sdu.dk

Abstract: A new DC fault current limiter (FCL)-based circuit breaker (CB) for DC microgrid (MG) clusters is proposed in this paper. The analytical expressions of the DC fault current of a bidirectional interlink DC/DC converter in the interconnection line of two nearby DC MGs are analyzed in detail. Meanwhile, a DC fault clearing solution (based on using a DC FCL in series with a DC circuit breaker) is proposed. This structure offers low complexity, cost, and power losses. To assess the performance of the proposed method, time-domain simulation studies are carried out on a test DC MG cluster in a MATLAB/Simulink environment. The results of the proposed analytical expressions are compared with simulation results. The obtained results verify the analytical expression of the fault current and prove the effectiveness of the proposed DC fault current limiting and clearing strategy.

Keywords: DC microgrid; fault; cluster; DC/DC converter; fault current limiter (FCL)

Citation: Bayati, N.; Baghaee, H.R.; Savaghebi, M.; Hajizadeh, A.; N. Soltani, M.; Lin, Z. DC Fault Current Analyzing, Limiting, and Clearing in DC Microgrid Clusters. *Energies* **2021**, *14*, 6337. https://doi.org/10.3390/en14196337

Academic Editor: Surender Reddy Salkuti

Received: 1 September 2021
Accepted: 1 October 2021
Published: 4 October 2021

Publisher's Note: MDPI stays neutral with regard to jurisdictional claims in published maps and institutional affiliations.

1. Introduction

DC microgrids (MGs) have attracted wide attentions over the last years in both industry and academia. At the point of practical applications, DC MGs make sense to be used because important types of distributed energy resources (DERs), such as fuel cells (FCs), photovoltaic (PV) arrays, wind turbines (WTs), and battery energy storage systems (BESSs), and also many types of electronics loads are DC [1]. The DC MGs have demonstrated superiority over the alternating current (AC) MGs from the viewpoints of reliability, efficiency, control complexity, penetration of renewable energy resources, and connection of DC loads [2–4]. Despite these significant advantages, designing and implementing a suitable protection scheme for DC MGs remains an important challenge [5]. The protection challenges in the rapid rise of DC fault current and the absence of a naturally occurring zero crossing point potentially lead to sustained arcs. To address the DC MG protection challenges, proper grounding architectures, fast fault current limiting methods, efficient fault detection/location strategies, and well-designed DC circuit breakers are required [6].

As a solution for better power support, the DC MG clusters are designed with the help of bidirectional interlink DC/DC converters [7]. One or several DC MGs can be connected to the AC grid by bidirectional interlink AC/DC converters. Due to the bidirectional current in the interconnected link between DC MGs, a bidirectional DC/DC converter should be implemented in these lines. In [8], a bidirectional DC/DC converter has been presented for connecting two adjacent DC MGs. This converter has two functions: operating as a boost converter during the power flow from the low voltage (LV) to the high voltage (HV) side, and as a buck converter during the power flow from HV to the LV side. DC MGs can be connected to the main utility grid via a modular multilevel converter (MMC) [9–11]. Due to the modular structure of MMCs, it is not necessary to connect the power electronic devices in series, and so the difficulties of manufacturing are reduced [12]. The MMC system provides a suitable output for modulated current and voltage. Therefore, these systems are widely used in AC/DC converter applications [13].

Up to now, most of the reported studies on the DC MG clusters have been focused on the control [14–17], switching network topology [18], power scheduling mechanism [19], management [20], and resiliency [21] of these systems. When the DC MG clusters consist of several DC systems and the connection to the utility grid, the DC fault characteristics and corresponding fault clearing solutions should be analyzed in depth. In Table 1, the focus of the reported studies on DC MG clusters is summarized to highlight the lack of enough investigations on the protection of these systems. In [22], an LV DC circuit breaker (CB) has been suggested for the protection of autonomous DC MG clusters. However, the interlink converter between individual DC MGs was not discussed in the paper. Moreover, the system under study has only been operated in islanded mode. In [23], the performance of a multiport DC/DC converter in a DC network has been analyzed to find solutions for protecting a DC network equipped with multiport converters and avoid the long clearing time of traditional DC CBs. A hybrid MMC (HMMC) with integrated BESS has been proposed in [24], and its performance during AC and DC grid faults has been analyzed. In [25], the fault analysis method has been presented for a simple AC/DC voltage source converter for a DC MG application.

Table 1. Comparison of the reported studies on DC MG clusters.

Reported Study	Focusing Area	Operation Mode
[14]	Communication-based control	Grid-connected
[15]	Coordinated control	Islanded and grid-connected
[16]	Power flow control	Islanded
[17]	Modeling and control	Islanded
[18]	Switching network topology	Islanded
[19]	Power scheduling mechanism	Islanded
[20]	Power exchange management	Islanded
[21]	Resiliency	Grid-connected

DC fault isolation and limitation are the most essential technical obstacles during the fault in interconnection lines of DC MG clusters. Generally, there have been four solutions for solving these problems:

- Implementing AC CB at the AC side of the AC/DC converter for faults at the interconnection line between the AC grid and DC MG cluster: Several AC CB strategies have been investigated for LV AC MGs protection systems [26,27]. However, the slow response time of the AC CBs makes them unsuitable for implementation in the clustered DC MGs.
- Integrating FCL functionality into the converter of interconnection lines: FCL is deactivated during the normal operation of the system; however, during the fault, it adds a high resistance to the fault path to limit the value of the fault current [28]. In the literature, both AC FCL [29] and DC FCL [30] have been proposed for converter-based LV systems. However, the implementation of conventional FCLs in DC MG clusters suffers from high response time, cost, size, and installation complexity [31].
- Implementing DC CB in series to the DC/DC converter for faults at the interconnection line between the two adjacent DC MGs: In [32], a hybrid DC CB has been presented to break the DC fault current up to 9 kA within 5 ms. However, existing DC CBs have some disadvantages, such as low technology maturity and high manufacturing cost.
- Implementing fault limiting capability in the converter control: Some of the reported strategies have suggested active FCL strategies for the AC/DC converters to limit the fault current and enhance the fault-ride-through (FRT) capability of MGs [33,34]. Con-

verter limits the fault current to two times the nominal value to prevent overheating. Therefore, for adopting this option in the converters, more powerful electronic devices are required, and the power losses and the cost of the system increase [35].

By reviewing the previous literature, it has been found that there is a lack of study on the fault analysis of the bidirectional DC/DC converter as the interlink of DC MG clusters. This paper proposes an analytical expression of the current of a DC MG cluster under fault conditions. First, the analytical model of the fault current of the AC/DC converter between the grid and DC MG is presented. Second, the analytical model of a bidirectional interlink DC/DC converter of the interconnection line between two adjacent DC microgrids in a clustered DC MG is derived in detail. Then, a DC fault clearing strategy is proposed based on using a DC FCL in series with a DC circuit breaker. Finally, the performance of the proposed method is evaluated based on time-domain simulation studies on a test DC MG cluster in MATLAB/Simulink environment. The simulation results are also compared with the results of the proposed analytical model. The obtained results verify the analytical expression of the fault current and prove the effectiveness of the proposed DC fault current limiting and clearing strategy. The main salient features of this presented study are as follows:

i.   A detailed analysis of converters during short circuit faults is investigated, which allows the better design of converters during faults. It can also model the converter behavior during different stages of fault by transient equations.
ii.   An accurate transient analysis of DC MG clusters during fault by considering the characteristic of converters is presented. Therefore, the importance of the current limiting of interconnected lines between converters is highlighted and investigated.
iii. A DC FCL is proposed for interconnected DC MG clusters, which has a higher speed, lower coordination problems, and power losses make it different from existing FCL strategies. Moreover, the proposed method is designed and specified for DC MG clusters, which have a few studies on the protection of these systems.

The rest of the paper is organized as follows: The structure of the DC MG cluster is discussed in Section 2. The analytical expression of the AC/DC converter is presented in Section 3. Section 4 proposes the detailed analytical expression of the bidirectional DC/DC converters in the DC MG cluster. In Section 5, a strategy is presented for limiting and clearing the DC fault in the DC MG cluster. Section 6 is dedicated to the simulation results and discussion to verify the performance and demonstrate the accuracy of the presented analytical equations. Finally, conclusions are stated in Section 7.

2. DC MG Cluster

The structure of a sample DC MG cluster consisting of two DC MGs is depicted in Figure 1, and the parameters are presented in Table 2. Neighboring DC MGs can be connected by a bidirectional DC/DC converter. Because a DC MG cluster can operate in grid-connected mode, one or several DC MGs are connected to the upstream AC utility grid by AC/DC converters. Both DC MGs in this cluster are connected to the AC grid by AC/DC MMCs. Increasing the number of connections increases the probability and the current level of a fault. Therefore, detailed fault analysis is an important task in these systems. In this paper, only the faults in the interconnected lines between one of the DC MGs to the AC/DC MMC are investigated, as mentioned by *F1* and *F2* in Figure 1. For example, *F1* is a candidate scenario for a fault at the interconnected link between grid and DC microgrids. The highest possible fault current usually belongs to the interconnected lines because typically, these lines have the highest power transfer capability and connect two main buses. Figure 2a,b illustrate the structure of an AC/DC MMC. Each sub-module (SM), as shown in Figure 2c, consists of two IGBTs, two freewheeling diodes, and a cell capacitor [36]. The modulation strategy of this scheme is different from the two-level voltage source converters (VSCs), which makes high-quality waveforms and low switching losses. Section 3 discusses the fault characteristics of this type of converter in a DC MG cluster.

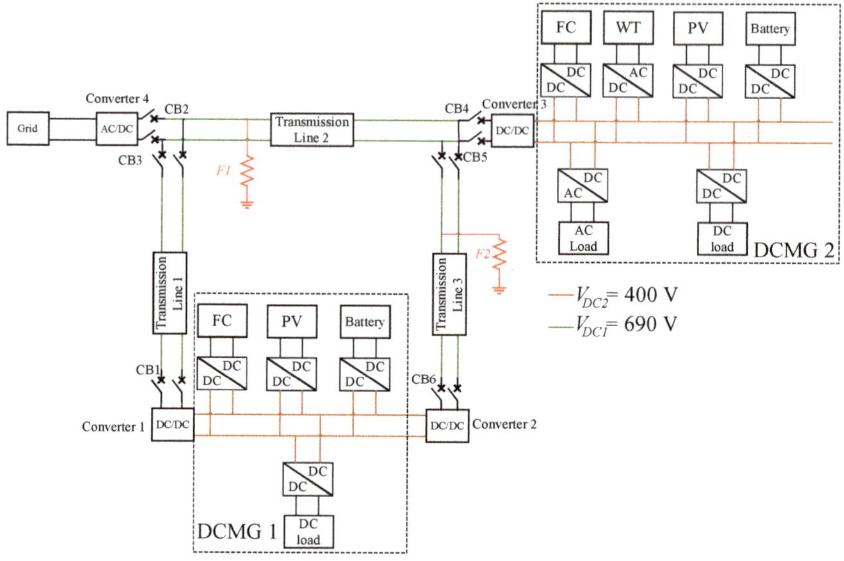

Figure 1. Structure of a DC MG cluster.

Table 2. Main parameters of the case study.

Component	Rated Value
Inductor of DC/DC converter	5 mH
Capacitor of DC/DC converter, C1	500 μF
Capacitor of DC/DC converter, C2	5000 μF
Resistance of DC/DC converter	1.358 mΩ
Capacitor of AC/DC converter	600 μF
Inductance of AC/DC converter	76 mH
Resistance of DC/DC converter	1.5 mΩ
Nominal voltage of VDC1	690 V
Nominal voltage of VDC2	400 V
Line resistance	1.6 mΩ/m
Line inductance	0.1 mH/m

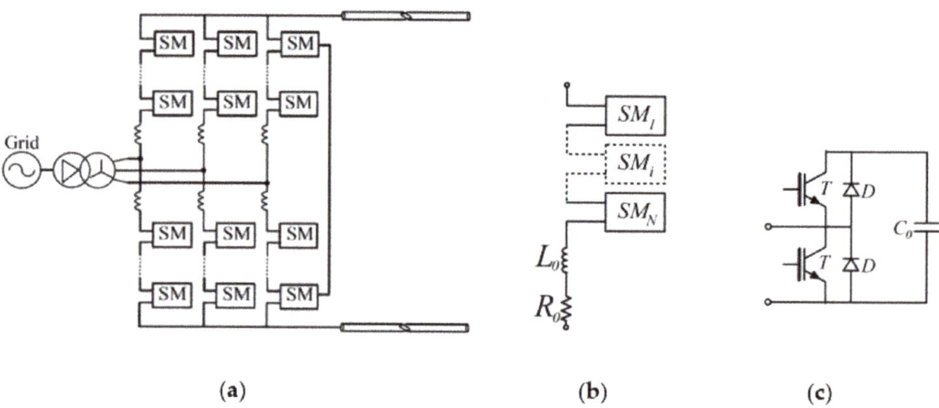

Figure 2. (**a**) Structure of an AC/DC MMC, (**b**) structure of each arm, (**c**) equivalent model of SM.

In Figure 3, a bidirectional DC/DC converter for use in the interconnection line between two adjacent DC MGs is represented, as shown in Figure 1 for converters one to three. This type of converter for managing the bidirectional power flow between two DC MGs has been introduced in [9]. This converter is a boost converter when power flow is from the LV (V_{DC2}) side to the HV (V_{DC1}) side, and a buck converter for power flow from the HV to the LV. Therefore, during the normal operation mode, this converter meets an appropriate bidirectional power sharing and voltage regulation among the DC MGs. The detailed fault analysis of this converter in the DC MG cluster is discussed in Section 4.

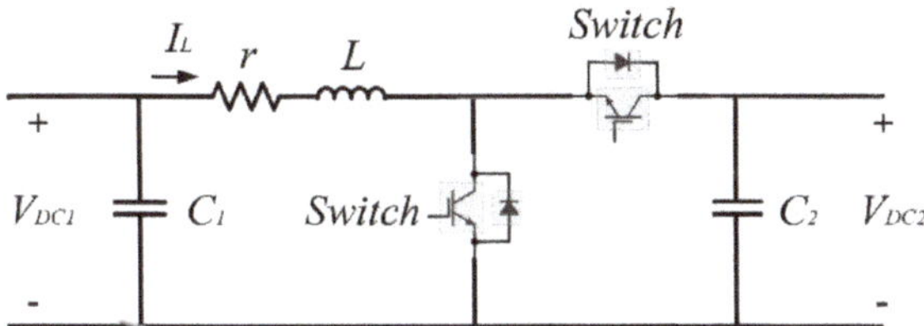

Figure 3. Structure of a bidirectional DC/DC converter.

3. DC Fault Analysis of AC/DC Converter

The previous literature has presented the two stages of a fault in the converters, namely capacitor-discharge and freewheeling diode operation. With the onset of a fault, the capacitors are discharged into the cable, and a high-rise current flows through it. Then, after a few milliseconds, the voltage of the terminal of the converter gets reversed, and the diodes start conducting. It should be noted that due to the surge energy in converters during the capacitor discharge stage, caused by high di/dt, the main damages to power electronic converters happen during the capacitor discharge stage [37].

The fault current starts with the capacitor discharge state, which has the highest magnitude of the current. Each SM of the interface AC/DC converter between the grid and the DC MGs includes a capacitor, inductance, and resistance, as shown in Figure 2b. The values of R_0, C_0, and L_0 are assumed to be the equivalent resistance, capacitor, and inductance of all SMs in each arm, respectively. According to [38], the value of the fault current of the DC side, F_1 fault in Figure 1, during the capacitor discharge state is given by

$$I_{DC}(s) = \frac{1}{s^2(\frac{2}{3}L_0 + L_{DC}) + s(\frac{2}{3}R_0 + R_{DC}) + \frac{N}{6C_0}} \tag{1}$$

by using inverse Laplace transform of Equation (1), the time domain equation of fault current is obtained as:

$$i_{DC}(t) = -\frac{1}{sin\theta}Ae^{-\frac{t}{\sigma}}sin(\omega_{DC}t - \theta) + Be^{-\frac{t}{\sigma}}sin(\omega_{DC}t) \tag{2}$$

where, L_{DC} and R_{DC} are the inductance and resistance of the DC line, respectively. N shows the number of SMs in each arm. The values of A and B are calculated by the initial values of the current and voltage of the AC/DC converter and

$$\theta = arctan(\sigma\omega_{DC}) \tag{3}$$

$$\sigma = \frac{4L_0 + 6L_{DC}}{2R_0 + 3R_{DC}} \tag{4}$$

$$\omega_{DC} = \sqrt{\frac{2N(2L_0 + 3L_{DC}) - C_0(2R_0 + 3R_{DC})^2}{4C_0(2L_0 + 3L_{DC})^2}} \tag{5}$$

After the DC line's voltage reaches zero, the voltage of the terminal gets reversed, and diodes start to conduct. Therefore, each SM could be modeled by a resistance, an inductance, and a capacitor. According to [39], the lower and upper arm currents of MMC could be calculated by

$$\begin{cases} i_{pa}(t) = A_0 + \sum_{n=1}^{\infty} A_n sin(n\omega t + \varphi_n) \\ i_{na}(t) = A_0 + \sum_{n=1}^{\infty} (-1)^n A_n sin(n\omega t + \varphi_n) \\ i_a = i_{pa}(t) - i_{na}(t) = \sum_{n=2k-1}^{\infty} 2A_n sin(n\omega t + \varphi_n) \end{cases} \tag{6}$$

where, i_{pa} and i_{na} are upper and lower arm currents, respectively. i_a is the current from the AC side, A_0 and A_n are the DC component and nth harmonic of the current of the arm, and k is a positive integer. The harmonic orders in i_{pa} higher than A_0 and A_1 are assumed to be zero, and by applying Kirchhoff's voltage law to Figure 2, the following equation can be obtained

$$\begin{cases} u_{AC} = L_{ac}\frac{di_a}{dt} + L_0\frac{di_{pa}}{dt} \\ u_{AC} = Usin(\omega t + \beta) \end{cases} \tag{7}$$

where, u_{ac} is the voltage of the AC side, and U is the magnitude of this voltage, L_{ac} is the inductance of the AC line, and β is the angular displacement of the harmonics in the upper and lower arm currents. Therefore, by comparing Equations (6) and (7), the $A_n = 0$ for $n = 2, 3, 4$. Consequently, the value of the DC fault current during the freewheeling stage is obtained as

$$i_{pa}(t) = \frac{U}{2\omega L_{ac} + \omega L_0}(1 - cos(\omega t + \beta)) \tag{8}$$

It should be noted that Equation (8) represents the steady-state performance of the converter after the blocking stage. It takes some time for DC current to vary from the blocking moment to the steady state. This process can generally be modelled by a first-order inertia behavior. Therefore, the complete expression of DC current can be shown by

$$i_{dc}(t) = I_{dc\infty} + (I_{dcBlock} - I_{dc\infty})e^{-\frac{t}{\tau_{dcBlock}}} \tag{9}$$

where, $I_{dc\infty}$ is the peak of (8), $I_{dcBlock}$ is the initial DC current after MMC blocking, and $\tau_{dcBlock}$ is the first-order inertia time constant. For practical systems, $\tau_{dcBlock}$ is between 10 ms to 200 ms [39].

4. DC Fault Analysis of DC/DC Converter

A DC fault in the interlink DC/DC converter between two adjacent DC MGs (F_2 fault in Figure 1) is a severe condition for near to fault converters. During a fault, the fault analysis should be done in two states. The first state is the response of the RLC equivalent circuit due to the DC link capacitor discharge of the bidirectional DC/DC converter. The second stage starts after the current magnitude reaches the maximum value and consequently, the voltage of the capacitor reaches zero.

4.1. Analysis of Capacitor Discharge

After a fault, at the first step, capacitors start discharging through the cable impedance and the equivalent circuit of the bidirectional DC/DC converter in a DC MG cluster, as shown in Figure 4. R_L and L_L are the resistance and inductance of the cable, respectively, and R_f is the fault resistance.

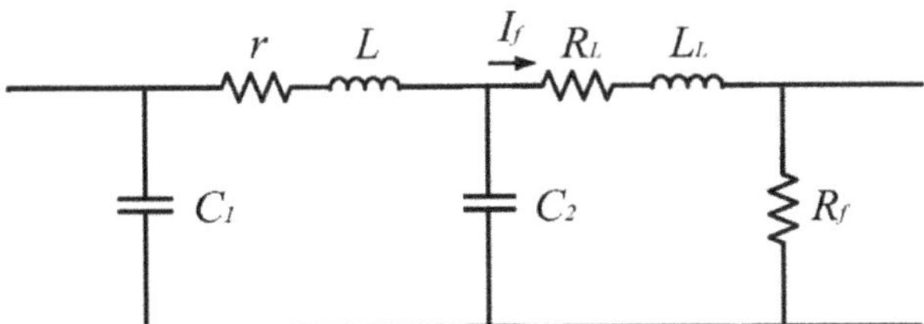

Figure 4. Equivalent circuit of DC/DC converter during capacitor discharge.

The RLC circuit response in the frequency domain can be obtained by Equation (11).

$$I_f(s) = \frac{LL_L C_2 i s^3 + s^2(C_2 u_2 L + L_L r C_2 i) + s(Li + L_L i + C_2 u_2 r + \frac{C_2}{C_1} L_L i) + u_1 + \frac{C_2}{C_1} u_2}{LL_L C_2 s^4 + s^3(RC_2 L + rL_L C_2) + s^2(RrC_2 + L + L_L(1 + \frac{C_2}{C_1})) + s(r + \frac{RC_2}{C_1} + R) + \frac{1}{C_1}} \tag{10}$$

where, u_1 and u_2 are the initial voltages of the C_1 and C_2, respectively, and i is the initial value of the current of the converter. r and L are the internal resistance and inductance of the DC/DC converter, respectively, and $R = R_f + R_L$. The fault current in the time domain can be calculated by solving the differential Equation of (11).

$$LL_L C_2 \frac{d^4 I_f}{dt^4} + (RC_2 L + rL_L C_2)\frac{d^3 I_f}{dt^3} + (RrC_2 + L_L + L + L_L\frac{C_2}{C_1})\frac{d^2 I_f}{dt^2} + (r + R + R\frac{C_2}{C_1})\frac{dI_f}{dt} + \frac{1}{C_1}I_f = 0 \tag{11}$$

Due to the low length of the cable between two adjacent DC MGs, the value of line inductance is very small. Therefore, assuming $L_L = 0$, Equation (11) is simplified as Equation (12)

$$(RC_2 L)\frac{d^3 I_f}{dt^3} + (RrC_2 + L)\frac{d^2 I_f}{dt^2} + (r + R + R\frac{C_2}{C_1})\frac{dI_f}{dt} + \frac{1}{C_1}I_f = 0 \tag{12}$$

And the fault current is calculated as

$$I_f(t) = Ae^{-\alpha t} + Be^{-\lambda t}\cos(\mu t) + De^{-\lambda t}\sin(\mu t) \tag{13}$$

where, the value of A, B, and D are calculated by the initial values of capacitors voltages. α, μ, and λ are obtained by the Equations (14)–(16)

$$\Delta_0 = R^2 r^2 C_2^2 + L^2 - LRrC_2 - 3R^2 C_2 L \tag{14}$$

$$\Delta_1 = 2R^3 r^3 C_2^3 - 3R^2 r^2 C_2^2 L - 3RC_2 rL^2 + 2L^3 - 9R^3 rC_2^2 L - 9R^2 C_2 L^2$$
$$+ \frac{27R^2 C_2^2 L^2}{C_1} \tag{15}$$

$$\zeta = \sqrt[3]{\frac{\Delta_1 \pm \sqrt{\Delta_1^2 - 4\Delta_0^3}}{2}} \tag{16}$$

Then, Equation (16) has three different roots, one of them is a real number, and two of them are complex. The real value is α, the real part of the complex values is λ, and the imaginary parts are $\pm\mu$.

$$\tau = \frac{1}{3RC_2 L}(RrC_2 + L + \delta^k \zeta + \frac{\Delta_0}{\delta^k \zeta}) \tag{17}$$

where δ is

$$\delta = -\frac{1}{2} \pm \frac{\sqrt{3}}{2}j \tag{18}$$

where, j is the imaginary unit, δ defines the three roots of Equation (16), and k is 0, 1, and 2.

4.2. Freewheeling Diode Operation

The second step of the fault current analysis is the freewheeling diode stage. This stage starts after DC-link capacitors' (C_1 and C_2 in the equivalent circuit of Figure 5) voltages reach zero. Then, the converter terminal voltage is reversed and makes a diode-capacitor equivalent circuit, as depicted in Figure 5.

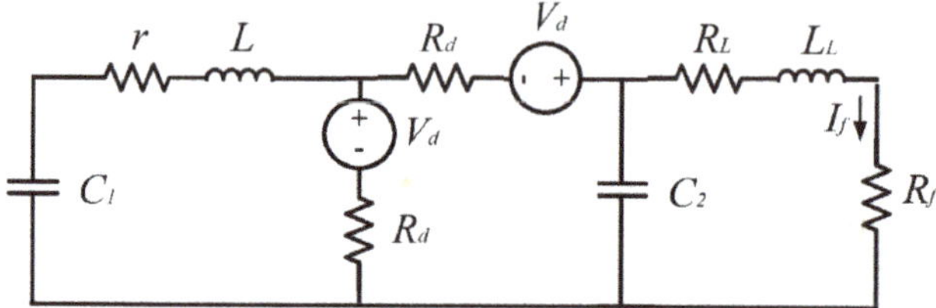

Figure 5. Equivalent circuit of DC/DC converter during the freewheeling diode operation.

The differential equation of the fault current during the freewheeling diode stage for calculating the fault current in the time domain is shown in Equation (19).

$$(2R_d RC_2 L)\frac{d^3 I_f}{dt^3} + \left((R + 2R_d)L + 2R_d RC_2 r + R_d^2 RC_2\right)\frac{d^2 I_f}{dt^2}$$
$$+ (Rr + (R + 2r)R_d + \frac{2R_d RC_2}{C_1} + R_d^2)\frac{dI_f}{dt} + (\frac{R + 2R_d}{C_1})I_f = 0 \tag{19}$$

R_d is the diode resistance, and consequently, the fault current in the time domain is obtained by

$$I_f(t) = \frac{2V_d}{2R_d + R} + Ee^{-vt} + Fe^{-\eta t}cos(\Psi t) + Ge^{-\eta t}sin(\Psi t) \tag{20}$$

where, V_d is the voltage of the diode, and E, F, and G are defined by using initial values of the freewheeling diode stage. η and Ψ are the real and imaginary parts of Equation (23), respectively, and v is the real value of Equation (23).

$$\begin{cases} a_1 = 2R_d RC_2 L \\ a_2 = RL + 2R_d RC_2 r + R_d^2 RC_2 + 2R_d L \\ a_3 = Rr + RR_d + \frac{2R_d RC_2}{C_1} + 2rR_d + R_d^2 \\ a_4 = \frac{R + 2R_d}{C_1} \end{cases} \tag{21}$$

$$\begin{cases} b_1 = a_2{}^2 - 3a_1 a_3 \\ b_2 = 2a_2^3 - 9a_1 a_2 a_3 + 27a_1^2 a_4 \\ b_3 = \sqrt[3]{\frac{b_2 \pm \sqrt{b_2^2 - 4b_1^3}}{2}} \end{cases} \tag{22}$$

$$\Gamma = \frac{1}{6R_d RC_2 L}(a_2 + \delta^k b_3 + \frac{b_1}{\delta^k b}) \tag{23}$$

5. DC Fault Clearing and Limiting Solution

In this section, a novel DC FCL-based CB configuration is presented as a solution for installing a lower-rate CB and limiting the fault current. Figure 6 depicts the schematic connection of the proposed DC FCL-based CB including a rectifier bridge, a resistor, an inductance, a capacitor, and also a DC CB which is connected in series with the DC/DC converter of Figure 1 in the interconnected line. Herein, the FCL is installed before the converter, which is used to achieve the energy transmission and conversion between DC MGs. Note that the proposed FCL can operate in both the normal and fault operation mode of the system. The basic concept of the proposed DC FCL-based CB operation uses a novel FCL in series with DC CB to reduce the fault current before clearing the fault by CB. It offers a soft current clearing and a lower fault current.

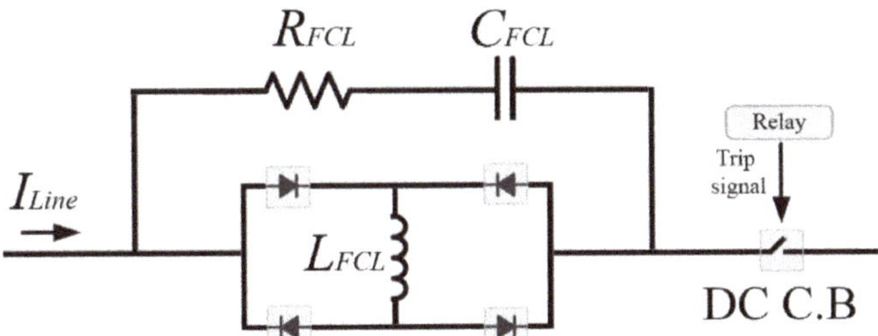

Figure 6. Proposed DC FCL configuration.

It should be noted that in Figure 6, it is necessary to use a diode bridge in the DC link; otherwise, the ripple of current generates a voltage drop, which will have an impact on the performance of the normal operation of connected converters. The DC FCL is composed of three main parts:

i. A diode rectifier bridge;
ii. A series inductor;
iii. A shunt RC branch for fault current reduction.

Furthermore, in the proposed DC FCL, the diode bridge composed of four diodes is utilized to enable the bidirectional current-flowing function, replacing the anti-parallel structure of the half and full-controlled switches. It also improves the economic considerations, due to lower price of diodes than the half and full-controlled switches. However, the forward voltage of diodes altogether causes a small voltage drop, but, in comparison to the total voltage drop in line, this voltage drop is negligible and has an insignificant impact on the normal operation [40].

The DC FCL-based CB operation is divided into two modes, namely normal and fault operation mode. During the normal operation mode, as shown in Figure 7a, the line current only flows through the inductor, and since the normal current of the system is DC, the overall power losses of inductance are relatively low, and they depend on the reactor copper loss. As shown in Figure 7b, during the fault, depending on the frequency components of the fault current, the capacitor branch absorbs some part of the fault current, and the inductance makes an impedance path for this current. The equivalent impedance of the DC FCL-based CB during the fault is obtained as

$$Z_{FCL} = \frac{s(LRCs + L)}{LCs^2 + RCs + 1} \tag{24}$$

where, L, R, and C are the inductance, resistance, and capacitance of the DC FCL-based CB, respectively. In the frequency domain, the size of the impedance of FCL for the fault current limiting stage is calculated as

$$|Z_{FCL}| = \frac{(L\Psi)(\sqrt{R^2C^2\Psi^2 + 1})(\sqrt{(1 - LC\Psi^2) + R^2C^2\Psi^2})}{(1 - LC\Psi^2)^2 + R^2C^2\Psi^2} \tag{25}$$

where, Ψ is the frequency of the current during the freewheeling diode operation mode, and then during the capacitor discharge stage, it will be replaced by μ, as explained in Equation (16). Based on Equation (25), the fault current magnitude is reduced by the coefficient of $1/|Z_{FCL}|$, as the total impedance of FCL during fault conditions. Therefore, the value of Z_{FCL} should be selected as higher than 1 Ω, to operate as a fault current reducing component. To determine the parameters of FCL, the thermal resistance of the equipment should be considered [3] to find the reduction value. Therefore, the objective value of Z_{FCL} can be determined by:

$$\frac{I_{obj}}{I_{peak}} = \frac{1}{|Z_{FCL}|}$$
$$\text{Subject to :}$$
$$L_{min} \leq L \leq L_{max}$$
$$R_{min} \leq R \leq R_{max}$$
$$C_{min} \leq C \leq C_{max}$$

(26)

where, I_{obj} is the final value of fault current after reduction, and I_{peak} is the peak of fault current before reduction. The values of R, C, and L should be in the range of minimum and maximum values of existed components. Therefore, the proposed DC FCL parameters can have different values based on cost and existing components, which causes a better availability of the design of this FCL.

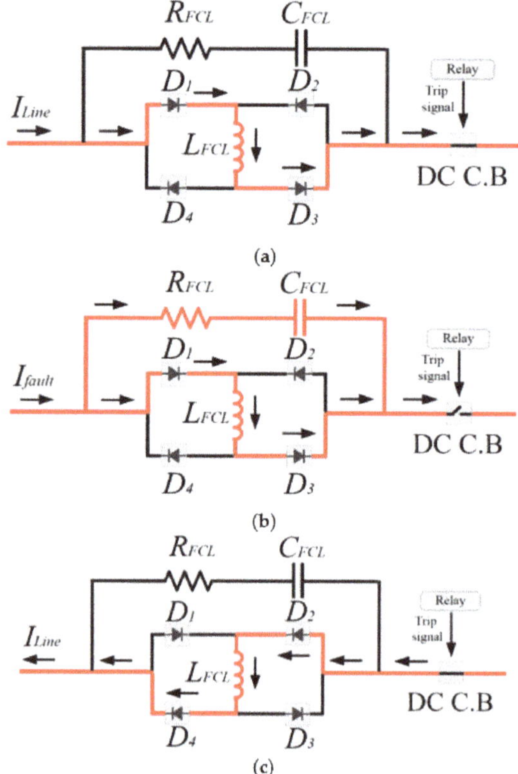

Figure 7. DC FCL-based CB and current flow (**a**) normal, (**b**) fault, (**c**) bidirectional condition.

During normal operation, as long as the DC current direction is positive, as shown in Figure 7a, the diodes D_1-D_3 are in conducting state. On the other hand, during the negative DC current direction, as shown in Figure 7c, the diodes D_2-D_4 are in conducting state. In this situation, the DC reactor L_{FCL} is bypassed from the DC line, thus, having no impact on the DC grid during normal operations. After a fault, the DC fault current rises rapidly to a high magnitude. As shown in Figure 7b, due to the high frequency of the DC fault current, the RC circuit will be paralleled with inductance to limit the fault current to the desired value. Moreover, this bridge-type FCL can deal with the bidirectional fault current limitation in the DC systems.

The modeling of the DC FCL-based CB has been based on the consideration of the transient impedance during the fault, since they are effective parameters in the study of the behaviors of the fault current. From Figure 1, an AC-DC converter is utilized to convert the AC grid power to the DC MG cluster power. Due to the existence of the power electronic components and controllers, the transient impedance of the proposed method is obtained by (25), which is different from the normal operation mode. The access of DC FCL-based CB helps to increase the impedance of the system and the damping factor, for decreasing the fault current from the two aspects. Meanwhile, this current reduction will help relays to keep their settings during variations of fault resistances.

In order to avoid the negative impacts of the directly installed DC reactors, the bridge-type FCLs are implemented in the DC grids. The implemented bridge-type FCL in a DC system consists of an H-bridge composed of four series diodes, and a branch composed of the series-connected DC reactor. Therefore, it can avoid the negative impact of the DC reactor on a system in normal operation. However, during normal operation, the rated current will flow through diodes in the bridge-type FCL, thus leading to a small power loss [30].

During normal operation, the current of the DC inductor is an almost ripple-free DC current with a magnitude equal to the peak of the DC-link current. Therefore, the ripple current in the DC inductor of FCL is approximately zero, and then, the average current in FCL in normal operation is equal to the peak value of DC-link current in steady-state. Consequently, the total power losses of DC FCL (P_{Loss}) is the sum of power losses of DC inductor ($P_{Inductor}$) and diode-bridge (P_{Bridge}), and can be determined as follows [28]

$$\begin{cases} P_{Bridge} = 2V_d I_{Line} \\ P_{Inductor} = r_d I_{max}^2 \\ P_{Loss} = I_{Line}(r_d I_{Line} + 2V_d) \end{cases} \tag{27}$$

where, I_{line} is normal current, V_d is the voltage drop of diodes, I_{max} is the peak value of DC-link current in steady-state, r_d is the resistance of the inductor. On the other hand, in a DC system without FCL, the normal flowing power can be defined by P_{DC}. Thus, the ratio of power losses to active flowing power is defined by K, and can be calculated by:

$$K = \frac{P_{Loss}}{P_{DC}} = \frac{I_{Line}(r_d I_{Line} + 2V_d)}{P_{DC}} \tag{28}$$

For example, consider the under-study system by V_{DC} = 690 V, r_d = 0.005 Ω, V_d = 2 V, I_{line} = 20 A. For this case, the value of K = 0.5%. This shows that in the presence of the proposed DC FCL, the total power loss has a very small percentage of the overall rated power of the system, and it will be acceptable in most practical applications. Furthermore, by using a superconductor inductor in DC FCL, the power loss of the inductor will be cancelled out, however, it increases the cost and weight of the system. Obviously, it is also possible to use a switch parallel with DC FCL to bypass it during steady-state conditions and avoid any power losses, however, it will increase the overall operation time of the proposed DC FCL.

6. Simulation and Real-Time Validation Results

In this section, the simulation results and discussion are presented for the case study system of Figure 1 with the parameters provided in Table 2. To validate the proposed fault analysis method, two cases are considered to obtain the fault current behavior for both AC/DC and DC/DC converters in the interconnected lines. In addition, the effectiveness of the proposed DC FCL-based CB is evaluated by using time-domain simulation studies in MATLAB/Simulink software environment and verified by comparing the proposed DC FCL with other reported FCL techniques.

6.1. Behavior of AC/DC Converter during the Fault

In the first part, the presented fault analysis of the AC/DC converter between the grid and the DC MG cluster, as represented in Section 3, is compared with the simulation results. Figure 8 shows the behavior of the fault current for a fault with a resistance of 0.1 Ω for both analytical results and simulation results. The results of Figure 8 show a small error between the analytical model and simulation results.

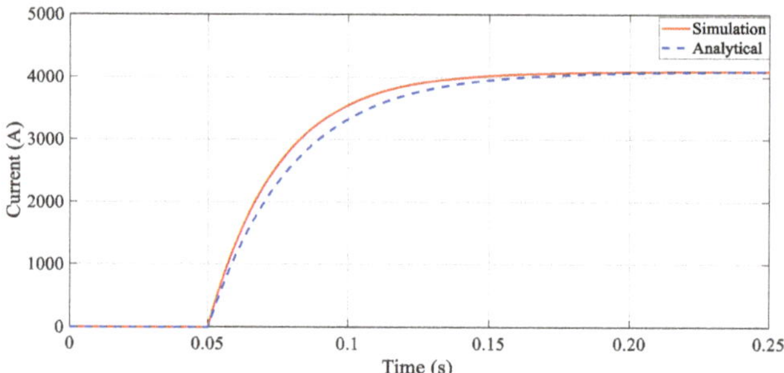

Figure 8. DC current of MMC during the fault.

6.2. Behavior of DC/DC Converter during the Fault

In Figure 9, the behavior of the DC/DC converter during a fault with a fault resistance of 0.1 Ω is presented for both analytical and simulation cases. During the capacitor discharge, the fault current reaches 6485 A in less than 0.03775 ms. One of the main parameters used for detecting the fault, as a threshold of the switching and relay settings, is the slope of the fault current. Therefore, this threshold is 171,788 kA/s and 158,616 kA/s by calculation using Equation (13) and simulation, respectively, which shows a small error of 7%. Thus, selecting the threshold by the proposed analytical equations is suitable for adjusting the setting of the relay. The parameters of Equations (13) and (20) for this case study are shown in Table 3.

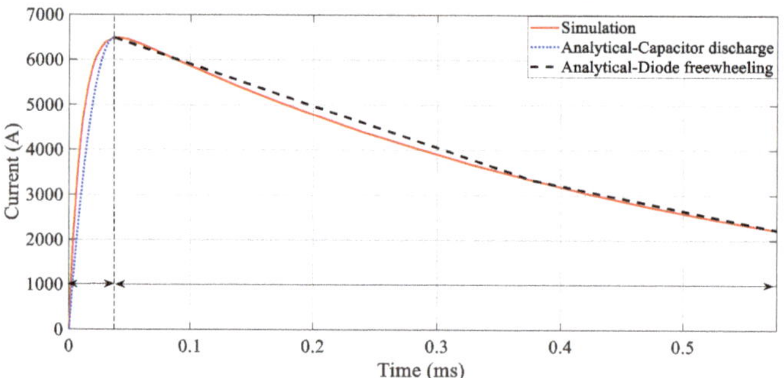

Figure 9. Comparison between analytical and simulation results.

Table 3. Parameters of the analytical model.

	Parameter	Value (s^{-1})
	α	2153
Capacitor discharge	λ	76
	μ	604
	υ	102×10^3
Freewheeling diode	η	2.157
	Ψ	88.5

6.3. Characteristics of DC FCL-Based CB during the Fault

By installing a DC FCL-based CB in series with a bidirectional DC/DC converter in the interconnected line between two DC MGs in a DC MG cluster, the desired maximum fault current magnitude is controlled by selecting a suitable value for resistance, inductance, and capacitance parameters of the DC FCL-based CB. According to the thermal resistance of the converter diodes [3], the maximum magnitude of the fault current is selected as 1100 A, and the equivalent impedance of FCL is calculated by Equation (25). The values of R_{FCL}, L_{FCL}, C_{FCL} are selected as 0.5 Ω, 0.1 H, and 2 mF, respectively. The fault current characteristic of the system equipped with DC FCL-based CB is shown in Figure 10. In Figure 11, the impact of the fault resistance on the performance of the proposed FCL is evaluated. The fault current peak of DC MGs reduces by increasing the fault resistance, and therefore, by using the proposed DC FCL scheme, fault current magnitude remains approximately constant. Furthermore, for low fault resistances, a considerable reduction in fault current is caused by using the proposed DC FCL scheme.

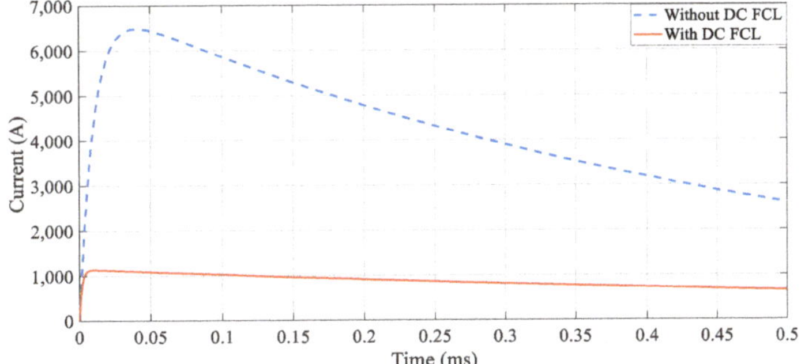

Figure 10. Impact of the DC FCL on the fault current.

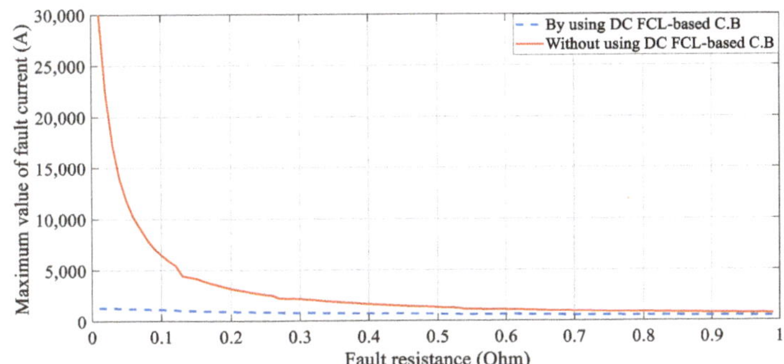

Figure 11. Performance of proposed DC FCL-based CB on the fault current.

In DC MG clusters, the fault current is injected from different sources. For example, in Figure 1, a fault at F1 with a fault resistance of 0.15 Ω causes a high-rise current from the DC MG and grid, as shown in Figure 12. Consequently, due to the converters' capacitor discharge, the fault current injected into the faulty point also has a high-rise peak, as represented in Figure 13. This high-rise current can damage the converters; therefore, installing the proposed DC FCL reduces the fault current to a lower level, as presented in Figure 13.

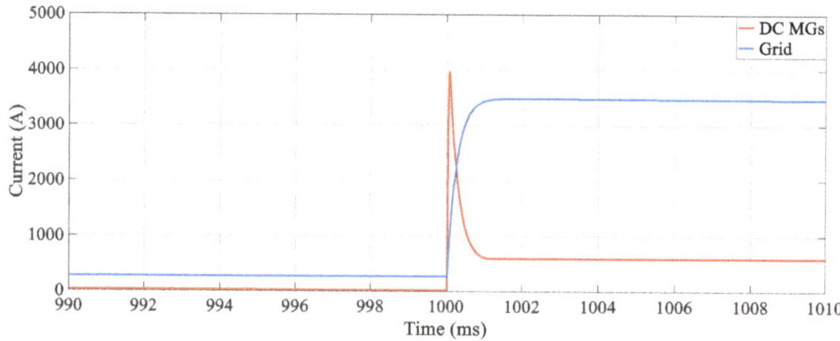

Figure 12. Fault current contributions without FCL, from grid and DC MGs.

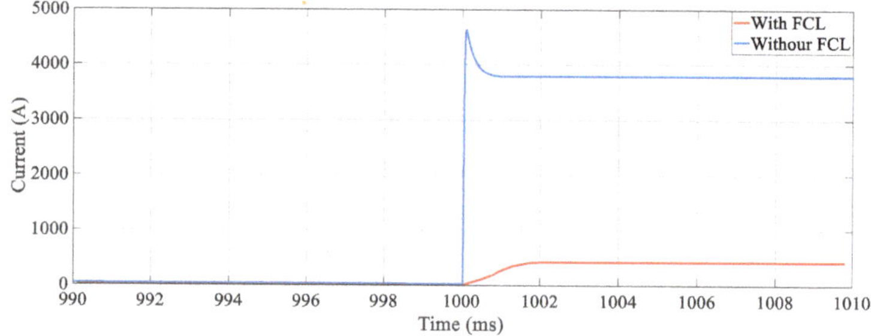

Figure 13. Fault current with and without FCL.

6.4. Real-Time Validation

The proposed DC FCL scheme has been modeled using an OPAL-RT simulator for validating the performance of the proposed method using real-time simulation. The experimental setup with oscilloscope, personal computer (PC), and OPAL-RT simulator is shown in Figure 14. Figure 15a,b presents the per-unit fault current waveforms for a fault in MG at F1, which occurred at $t = 1$ s with fault resistances of 0.45 Ω and 1.5 Ω, respectively. This verification using OPAL-RT assures the operation and effectiveness of the proposed scheme during different fault conditions.

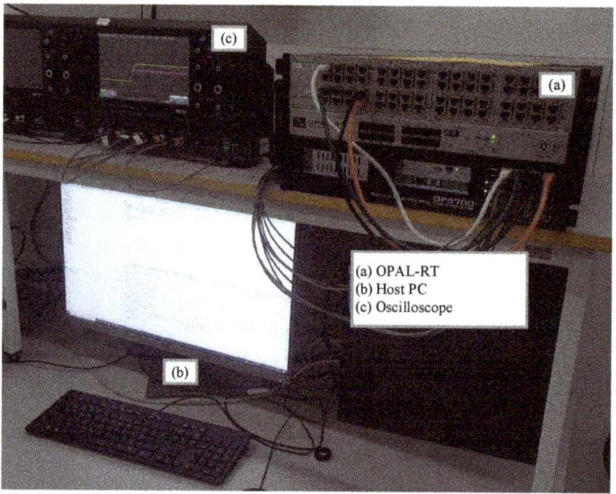

Figure 14. Real-time setup.

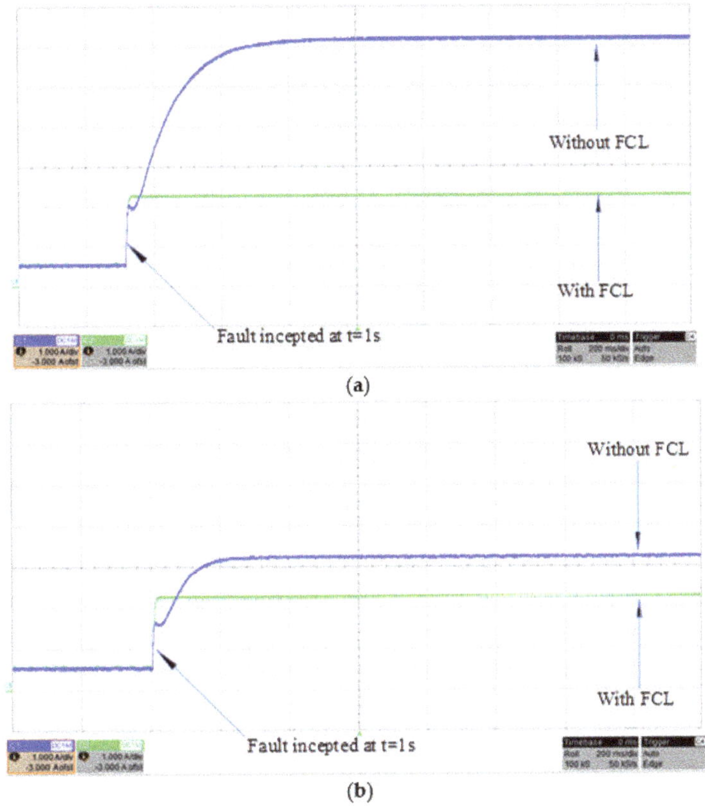

Figure 15. Fault current waveform with and without FCL for (**a**) 0.45 Ω and (**b**) 1.5 Ω faults.

6.5. Comparison between the Proposed DC FCL and Existing FCL Strategies

Finally, the proposed DC FCL-based CB is compared with other methods from several points of view, including cost, size, bidirectional operation, speed, complexity, and impact on the coordination of overcurrent relays. These comparison points are summarized in Table 4. Moreover, to better clarify the comparison of the existing methods and proposed scheme, detailed installed components for each DC FCLs, and the range of implemented voltage are summarized in Table 5.

Table 4. Comparison of the proposed DC FCL with other existing methods.

Type	Cost	Size	Speed	Complexity	Coordination Problem	Power Losses
[32]	High	Large	High	High	Medium	Low
[41]	High	Large	Low	High	Low	Low
[42]	High	Large	Medium	High	Miscoordination	Medium
[43]	High	Large	Medium	High	Miscoordination	Medium
[44]	Low	Large	Medium	High	Miscoordination	High
[45]	Low	Large	Low	Low	Medium	Low
[46]	Low	Small	High	High	Medium	High
proposed method	Moderate	Small	High	Low	Low	Low

Table 5. Comparison of the proposed DC FCL with the existing methods.

Type	Fault Current Limiting Component	Installed Power System
[32]	Arrester bank	High voltage
[41]	Ultrafast switch, diode bridge, a resistor with a high positive temperature	Medium Voltage
[42]	Switch, diode bridge, inductor capacitor branch	Low Voltage
[43]	Superconductor resistance bank	Medium Voltage
[44]	Diode bridge, two parallel RC thyristors, two series inductor diodes	Medium Voltage
[45]	Resistor, two IGBTs, switch	Low Voltage
[46]	Metal oxide varistor, diode bridge, IGBT, parallel RC	Medium Voltage
proposed method	Diode bridge, parallel RC, inductor	Low and Medium Voltage

Cost: The overall cost of an FCL strategy is estimated based on utilizing semiconductor and mechanical components and auxiliary circuits, such as communication and cooling systems. In order to investigate the overall cost, it is supposed that the mechanical switches are from the same type in the considered structures, and the cost of the semiconductor elements (including MOSFETs, silicon-controlled rectifiers, and diodes) are determined for a certain current and voltage rating. Among the existing FCL schemes, due to the implementation of arrester or superconductor resistance banks, the methods presented in [32,41–43] have the highest cost, and the strategies proposed in [44–46] have the lowest cost, due to the utilization of only a few elements.

Size: The size of an FCL is determined by several required control and power elements. The methods proposed in [45,46] have a small size, however, its size increases by increasing the rating of the FCL. On the other hand, the semiconductor and mechanical branches in [32,41,44] result in the large size of their suggested FCLs. Implementing the capacitor charging circuit and superconductor with a cooling system in [42,43], respectively, resulted in bulky FCLs. The installed components in each method are also shown in Table 5.

Speed: The speed of an FCL is determined by the breaking and operating time of mechanical switches. In [41,45], the multistage current limiting and positive temperature coefficient resulted in a low-speed operation for these structures, and these methods limited the current by approximately 2 ms. Implementing one stage FCL using PWM in [46] caused a fast FCL operation. The methods of [42–44], with an operation time of approximately 1 ms, due to the switching in zero current and the fact that they require charging time, are categorized as medium-speed FCLs.

Complexity: Control, power circuits and the capability of bidirectional conducting are three parameters that determine the complexity of an FCL scheme. Therefore, [39–45] have the lowest complexity, while [32,41–43,46] have the highest complexity. In references [32,41–43,46], the systems are relatively complicated because of employing several operating sequences, PWM function, complicated power circuitry including the considerable number of mechanical switches and semiconductor devices, active circuitry to charge their commutation capacitors, and super-conductor elements.

Impact on the Coordination with Other Protection Devices: Most protection devices are coordinated based on the magnitude of the fault current of the system. The method suggested in [41] can only be used in the upstream network, which causes better coordination for overcurrent relay only in the upstream. Furthermore, due to the transient behavior in [42–44], these schemes are not appropriate for the coordination of fast protection devices. On the other hand, due to the expected value of the fault current and suitability of suggested FCLs in [45,46], these FCL strategies help to coordinate the protection devices.

Power Losses: The overall power losses of an FCL are evaluated during the normal condition. The method in [41] uses a fast-tripping switch with zero resistance during normal conditions, however, it was unusable for MV DC systems. In [42,43], FCL included two diodes series with a capacitor. Therefore, the power losses include the losses of the internal resistance of diodes and the ESR of the capacitor. The FCL of [44–46] has three diodes in a normal operation mode that cause a power loss. In [32,45], during the normal condition, only one IGBT is conducting, and the power losses of IGBTs are normally less than diodes.

The proposed DC FCL-based CB does not require a communication system, capacitor charging circuit, and has a low number of elements, but the rating of inductance and diode, and the high value of current tolerance of diodes increase the cost. Then, this method is categorized as a moderate-cost DC FCL strategy. Moreover, there is no need for a capacitor discharge circuit or any controlling circuit for DC FCL; thus, the size of this structure is low. In the proposed scheme, due to the passive structure of FCL and its connection in series with the DC/DC converter, the speed of fault current limiting of the proposed structure is high. On the other hand, in terms of the coordination problem, as the proposed FCL provides an approximately constant fault current in the protected line by changing the fault resistance, the value of fault current will be constant. Therefore, the overcurrent relays do not require new settings, and it can be a reliable and secure coordination of current-based protection devices. The proposed method uses two diodes in the current path during normal conditions. Therefore, the total power losses are only the sum of power losses of these diodes.

The qualitative performance evaluation of each method based on six performance parameters is depicted as spider diagrams in Figure 16. The axes in the spider charts present qualitative performance parameters in the range of low to high, starting from the center point. For example, in the category of cost, the low-cost and high-cost are marked with 1 and 3, respectively.

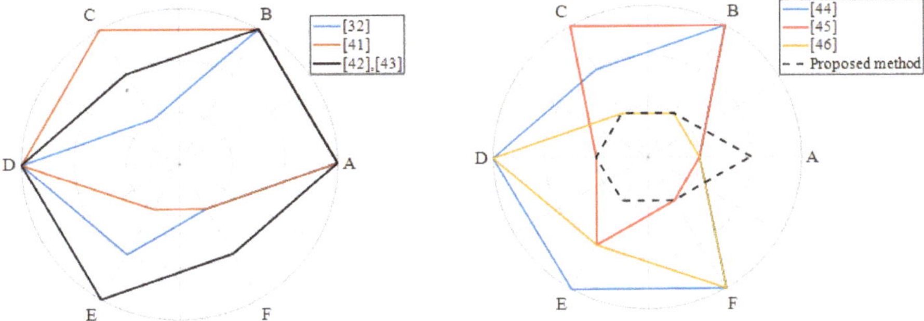

Figure 16. Comparative evaluation of existed and proposed scheme, (A) cost, (B) size, (C) operation time, (D) complexity, (E) coordination problem, (F) power losses.

7. Conclusions

In this paper, the DC fault currents of a DC MG cluster were analyzed. The analytical DC fault current expressions of a bidirectional DC/DC converter and AC/DC MMC converter were presented. After the fault occurrence, the capacitor discharge of the converters causes a high-rise fault current within several milliseconds, and the main damage is caused by the *di/dt* level in the power electronic devices. Then, the freewheeling stage starts, and the fault current reduces to the steady-state value. Consequently, a new DC FCL-based CB was proposed to limit the fault current and provide a soft clearing for CB. The proposed DC FCL scheme causes an approximately constant maximum fault current in a lower level of fault current before installation of FCL in the protected line, and it makes the setting adjustment of other protection devices more convenient and reduces the required current rating of the CBs. The simulation results and the comparison with several previously reported FCL strategies proved the accuracy of the analytical expressions and the proper operation of the proposed DC FCL-based CB during the fault condition.

Author Contributions: Methodology, validation, writing—original draft preparation, N.B.; writing—review, H.R.B. and Z.L.; supervision, A.H. and M.N.S.; editing, M.S. All authors have read and agreed to the published version of the manuscript.

Funding: This research received no external funding except for the real-time simulation setup as mentioned in the acknowledgement below.

Institutional Review Board Statement: Not applicable.

Informed Consent Statement: Not applicable.

Data Availability Statement: Data sharing not applicable.

Acknowledgments: The authors would like to express their appreciation for the support of Energi Fyn's Development Foundation for the OPAL-RT real-time simulation setup used for extracting the results.

Conflicts of Interest: The authors declare no conflict of interest.

References

1. Bayati, N.; Hajizadeh, A.; Soltani, M. Protection in DC microgrids: A comparative review. *IET Smart Grid* **2018**, *1*, 66–75. [CrossRef]
2. Martirano, L.; Rotondo, S.; Kermani, M.; Massarella, F.; Gravina, R. Power Sharing Model for Energy Communities of Buildings. *IEEE Trans. Ind. Appl.* **2020**, *57*, 170–178. [CrossRef]
3. Bayati, N.; Baghaee, H.R.; Hajizadeh, A.; Soltani, M. A Fuse Saving Scheme for DC Microgrids with High Penetration of Renew-able Energy Resources. *IEEE Access* **2020**, *8*, 137407–137417. [CrossRef]
4. Beheshtaein, S.; Cuzner, R.M.; Forouzesh, M.; Savaghebi, M.; Guerrero, J.M. DC microgrid protection: A comprehensive review. *IEEE J. Emerg. Sel. Top. Power Electron.* **2019**, *12*. [CrossRef]
5. Aghaee, F.; Dehkordi, N.M.; Bayati, N.; Hajizadeh, A. Distributed Control Methods and Impact of Communication Failure in AC Microgrids: A Comparative Review. *Electronics* **2019**, *8*, 1265. [CrossRef]
6. Bayati, N.; Baghaee, H.R.; Hajizadeh, A.; Soltani, M. Localized protection of radial DC microgrids with high penetration of con-stant power loads. *IEEE Syst. J.* **2020**, *8*, 4145–4156.
7. Bayati, N.; Baghaee, H.R.; Hajizadeh, A.; Soltani, M.; Lin, Z. Mathematical morphology-based local fault detection in DC Mi-crogrid clusters. *Electr. Power Syst. Res.* **2020**, *1*, 106981.
8. Kumar, M.; Srivastava, S.C.; Singh, S.N.; Ramamoorty, M. Development of a control strategy for interconnection of islanded di-rect current microgrids. *IET Renew. Power Gener.* **2014**, *9*, 284–296. [CrossRef]
9. Marquardt, R. Stromrichterschaltungen Mit Verteilten Energiespeichern. German Patent DE10103031A1, 1 December 2011.
10. Yang, H.; Saeedifard, M.; Yazdani, A. An Enhanced Closed-Loop Control Strategy with Capacitor Voltage Elevation for the DC–DC Modular Multilevel Converter. *IEEE Trans. Ind. Electron.* **2019**, *66*, 2366–2375. [CrossRef]
11. Bayat, H.; Yazdani, A. A Hybrid MMC-Based Photovoltaic and Battery Energy Storage System. *IEEE Power Energy Techn. Syst. J.* **2019**, *6*, 32–40. [CrossRef]
12. Saeedifard, M.; Iravani, R. Dynamic performance of a modular multilevel back-to-back HVDC system. *IEEE Trans. Power Deliv.* **2010**, *25*, 2903–2912. [CrossRef]
13. Westerweller, T.; Friedrich, K.; Armonies, U.; Orini, A.; Parquet, D.; Wehn, S. Trans bay cable-world's first HVDC system using multilevel voltage-sourced converter. In Proceedings of the CIGRE Session, Paris, France, 22–27 August 2010.
14. Tu, C.; Xiao, F.; Lan, Z.; Guo, Q.; Shuai, Z. Analysis and Control of a Novel Modular-Based Energy Router for DC Microgrid Cluster. *IEEE J. Emerg. Sel. Top. Power Electron.* **2018**, *7*, 331–342. [CrossRef]
15. Qiao, L.; Li, X.; Di, H.; Guo, L.; Xu, Y.; Tan, Z.; Wang, C. Coordinated control for medium voltage DC dis-tribution centers with flexibly interlinked multiple microgrids. *J. Mod. Power Syst. Clean Energy* **2019**, *7*, 599–611.
16. Mudaliyar, S.; Duggal, B.; Mishra, S. Distributed Tie-Line Power Flow Control of Autonomous DC Microgrid Clusters. *IEEE Trans. Power Electron.* **2020**, *35*, 11250–11266. [CrossRef]
17. Chen, Z.; Yu, X.; Xu, W.; Wen, G. Modeling and Control of Islanded DC Microgrid Clusters With Hierarchical Event-Triggered Consensus Algorithm. *IEEE Trans. Circuits Syst. I Regul. Pap.* **2020**, *68*, 376–386. [CrossRef]
18. Lu, X.; Lai, J. Distributed Cluster Cooperation for Multiple DC MGs Over Two-Layer Switching Topologies. *IEEE Trans. Smart Grid* **2020**, *11*, 4676–4687. [CrossRef]
19. Wahid, A.; Iqbal, J.; Qamar, A.; Ahmed, S.; Basit, A.; Ali, H.; Aldossary, O. A Novel Power Scheduling Mechanism for Islanded DC Microgrid Cluster. *Sustainability* **2020**, *12*, 6918. [CrossRef]
20. John, B.; Ghosh, A.; Goyal, M.; Zare, F. A DC Power Exchange Highway Based Power Flow Management for Interconnected Microgrid Clusters. *IEEE Syst. J.* **2019**, *13*, 3347–3357. [CrossRef]
21. Simonov, M. Dynamic Partitioning of DC Microgrid in Resilient Clusters Using Event-Driven Approach. *IEEE Trans. Smart Grid* **2014**, *5*, 2618–2625. [CrossRef]
22. Yaqobi, M.A.; Matayoshi, H.; Danish, M.S.S.; Lotfy, M.E.; Howlader, A.M.; Tomonobu, S. Low-Voltage Solid-State DC Breaker for Fault Protection Applications in Isolated DC Microgrid Cluster. *Appl. Sci.* **2019**, *9*, 723. [CrossRef]

23. Corti, M.; Tironi, E.; Ubezio, G. DC networks including multiport DC/DC converters: Fault analysis. *IEEE Trans. Dustry Appl.* **2016**, *52*, 3655–3662. [CrossRef]
24. Chen, Q.; Li, R.; Cai, X. Analysis and Fault Control of Hybrid Modular Multilevel Converter with Integrated Battery Energy Storage System. *IEEE J. Emerg. Sel. Top. Power Electron.* **2016**, *5*, 64–78. [CrossRef]
25. Mohanty, R.; Pradhan, A.K. Protection of Smart DC Microgrid With Ring Configuration Using Parameter Estimation Approach. *IEEE Trans. Smart Grid* **2017**, *9*, 6328–6337. [CrossRef]
26. Tang, G.; Xu, Z.; Zhou, Y. Impacts of three MMC-HVDC configurations on AC system stability under DC line faults. *IEEE Trans. Power Syst.* **2014**, *29*, 3030–3040. [CrossRef]
27. Song, S.-M.; Kim, J.-Y.; Choi, S.-S.; Kim, I.-D.; Choi, S.-K. New Simple-Structured AC Solid-State Circuit Breaker. *IEEE Trans. Ind. Electron.* **2018**, *65*, 8455–8463. [CrossRef]
28. Jalilian, A.; Hagh, M.T.; Abapour, M.; Muttaqi, K.M. DC-link fault current limiter-based fault ride-through scheme for invert-er-based distributed generation. *IET Renew. Power Gener.* **2015**, *9*, 690–699. [CrossRef]
29. Alam, S.; Abido, M.A.Y.; Hussein, A.E.-D. Non-Linear Control for Variable Resistive Bridge Type Fault Current Limiter in AC-DC Systems. *Energies* **2019**, *12*, 713. [CrossRef]
30. Li, B.; He, J.; Li, Y.; Wen, W.; Li, B. A Novel Current-Commutation-Based FCL for the Flexible DC Grid. *IEEE Trans. Power Electron.* **2019**, *35*, 591–606. [CrossRef]
31. Arani, A.A.; Bayati, N.; Mohammadi, R.; Gharehpetian, G.B.; Sadeghi, S.H. Fault Current Limiter optimal sizing considering dif-ferent Microgrid operational modes using Bat and Cuckoo Search Algorithm. *Arch. Electr. Eng.* **2018**, *67*, 321–332.
32. Callavik, M.; Blomberg, A.; Häfner, J.; Jacobson, B. *The Hybrid HVDC Breaker*; ABB Grid Systems Technical Paper; ABB: Zurich, Switzerland, 2012.
33. Lin, W.; Jovcic, D.; Nguefeu, S.; Saad, H. Full-bridge MMC converter optimal design to HVDC operational requirements. *IEEE Trans. Power Deliv.* **2015**, *31*, 1342–1350. [CrossRef]
34. Baghaee, H.R.; Mirsalim, M.; Gharehpetian, G.B.; Talebi, H.A. A new current limiting strategy and fault model to improve fault ride-through capability of inverter interfaced DERs in autonomous microgrids. *Sustain. Energy Technol. Assess.* **2017**, *24*, 71–81. [CrossRef]
35. Modeer, T.; Nee, H.P.; Norrga, S. Loss comparison of different sub-module implementations for modular multilevel con-verters in HVDC applications. In Proceedings of the 2011 14th European Conference on Power Electronics and Applications, Birmingham, UK, 30 August–1 September 2011.
36. He, Y.; Zheng, X.; Nengling, T.; Wang, J.; Nadeem, M.H.; Liu, J. A DC Line Protection Scheme for MMC-Based DC Grids Based on AC/DC Transient Information. *IEEE Trans. Power Deliv.* **2020**, *35*, 2800–2811. [CrossRef]
37. Gowaid, I.A. A Low-Loss Hybrid Bypass for DC Fault Protection of Modular Multilevel Converters. *IEEE Trans. Power Deliv.* **2016**, *32*, 599–608. [CrossRef]
38. Xu, Z.; Xiao, H.; Xiao, L.; Zhang, Z. DC Fault Analysis and Clearance Solutions of MMC-HVDC Systems. *Energies* **2018**, *11*, 941. [CrossRef]
39. Ilves, K.; Antonopoulos, A.; Norrga, S.; Nee, H.-P. Steady-State Analysis of Interaction Between Harmonic Components of Arm and Line Quantities of Modular Multilevel Converters. *IEEE Trans. Power Electron.* **2011**, *27*, 57–68. [CrossRef]
40. Rashid, G.; Ali, M.H. Nonlinear Control-Based Modified BFCL for LVRT Capacity Enhancement of DFIG-Based Wind Farm. *IEEE Trans. Energy Convers.* **2016**, *32*, 284–295. [CrossRef]
41. Steurer, M.; Frohlich, K.; Holaus, W.; Kaltenegger, K. A novel hybrid current-limiting circuit breaker for medium volt-age: Principle and test results. *IEEE Trans. Power Deliv.* **2003**, *18*, 460–467. [CrossRef]
42. Zyborski, J.; Lipski, T.; Czucha, J.; Hasan, S. Hybrid arcless low-voltage AC/DC current limiting interrupting device. *IEEE Trans. Power Deliv.* **2000**, *15*, 1182–1187. [CrossRef]
43. Lee, B.W. Complex Superconducting Fault Current Limiter. U.S. Patent 11/616 458, 2006.
44. Keshavarzi, D.; Farjah, E.; Ghanbari, T. Hybrid DC circuit breaker and fault current limiter with optional interruption capa-bility. *IEEE Trans. Power Electron.* **2017**, *33*, 2330–2338. [CrossRef]
45. Tang, Y.; Duarte, J.L.; Smeets, R.P.P.; Motoasca, T.E.; Lomonova, E.A. Multi-stage dc hybrid switch with slow switching. In Proceedings of the 37th Annual Conference of the IEEE Industrial Electronics Society, Melbourne, VIC, Australia, 7–10 November 2011; pp. 1462–1467.
46. Maitra, A.; McGranaghan, M.; Lai, J.S.; Short, T.; Goodman, F. Multifunction Hybrid Solid-State Switchgear. U.S. Patent 7 405 910, 29 July 2008.

energies

Article

Discrimination of Transformer Inrush Currents and Internal Fault Currents Using Extended Kalman Filter Algorithm (EKF)

Sunil Kumar Gunda [1,*] and Venkata Samba Sesha Siva Sarma Dhanikonda [2]

[1]   Department of Electrical Engineering, National Institute of Technology, Warangal 506004, India
[2]   Department of Electrical and Electronics Engineering, Kakatiya Institute of Technology and Science, Warangal 506015, India; dvss@nitw.ac.in
*   Correspondence: gsunil24@student.nitw.ac.in; Tel.: +91-96-1893-1612 or +91-85-5509-2408

Abstract: The discrimination of inrush currents and internal fault currents in transformers is an important feature of a transformer protection scheme. The harmonic current restrained feature is used in conventional differential relay protection of transformers. A literature survey shows that the discrimination between the inrush currents and internal fault currents is still an area that is open to research. In this paper, the classification of internal fault currents and magnetic inrush currents in the transformer is performed by using an extended Kalman filter (EKF) algorithm. When a transformer is energized under normal conditions, the EKF estimates the primary side winding current and, hence, the absolute residual signal (ARS) value is zero. The ARS value will not be equal to zero for internal fault and inrush phenomena conditions; hence, the EKF algorithm will be used for discriminating the internal faults and inrush faults by keeping the threshold level to the ARS value. The simulation results are compared with the theoretical analysis under various conditions. It is also observed that the detection time of internal faults decreases with the severity of the fault. The results of various test cases using the EKF algorithm are presented. This scheme provides fast protection of the transformer for severe faults.

Keywords: transformer; internal fault currents; magnetic inrush currents; extended Kalman filter (EKF) algorithm; harmonic estimation

Citation: Gunda, S.K.; Dhanikonda, V.S.S.S.S. Discrimination of Transformer Inrush Currents and Internal Fault Currents Using Extended Kalman Filter Algorithm (EKF). *Energies* **2021**, *14*, 6020. https://doi.org/10.3390/en14196020

Academic Editor: Surender Reddy Salkuti

Received: 20 July 2021
Accepted: 9 September 2021
Published: 22 September 2021

Publisher's Note: MDPI stays neutral with regard to jurisdictional claims in published maps and institutional affiliations.

1. Introduction

Transformers require an efficient protection system from faulty conditions in the power system network. The classification of magnetic inrush currents and internal fault currents is a challenging issue for proper relay design. The magnetizing inrush current results in maloperation of the differential relay in transformer protection. The transformer may become saturated with inrush currents and internal faults due to the presence of zero-sequence components, which may lead to maloperation of differential current protection. The waveform correlation technique is suggested for differential current protection of transformers. The analysis of zero sequence currents due to inrush currents and internal faults can be performed using the waveform correlation technique [1].

An accelerated Convolutional neural network (CNN) was suggested for estimating the transformer magnetizing current from internal faults on a 230 kV transmission network. It was observed that the CNN can have feature extraction and fault detection blocks in a single deep neural network block by enabling the system to provide important features automatically [2]. The fundamental theory and principles of magnetizing the inrush current of the transformer are essential when it is energized. The analysis of the magnetizing current is important for transformer protection [3]. A new technique was proposed with two moving windows to predict the magnitude of differential currents of the transformer for the discrimination of inrush currents from internal faults [4]. The least error square (LES) was proposed to classify the fault currents and the magnitude of magnetic inrush currents with the two moving windows technique. The LES method has a very fast response capability with good estimation. A new method was proposed for the classification of fault currents and magnetic inrush currents of the transformer by using measurement of similarity index between the updated new sequence and a prescribed template. The proposed technique was used to sort the sample points of differential currents and fault currents to classify the fault currents and magnetic inrush currents [5].

The discrimination of the inrush currents from the internal faults in power transformers was suggested based on the runs test method, which is derived from the inherent differences between the waveforms [6]. A short time correlation transform was used to discriminate the internal fault current and inrush current by extracting the dead angle from the differential current. The effect of current-transformer (CT) saturation on the inrush current was also considered for proper discrimination of the faults [7].

Discrete wavelet transform (DWT) was suggested for the separation of magnetic inrush currents and fault currents by extracting the energy functions. DWT as used with different wavelet functions to achieve higher accuracy and reliability [8]. The saturation of the transformer will affect the classification of magnetic inrush currents and fault currents of a single-phase transformer. An efficient compensation algorithm was suggested for the discrimination of magnetic inrush currents and fault currents, which are affected by CT saturation [9].

The EKF algorithm as used to predict the primary winding current for the classification of magnetic inrush currents and fault currents. The EKF algorithm was used to estimate the primary side winding current of the transformer based on the residual current [10]. The novel method was used for the classification of internal fault currents and inrush currents based on the two-terminal network method. The absolute difference of the active power (ADOAP) flowing into and consumed by the two-terminal network was considered, so that the internal faults were identified [11]. The high-temperature resistive Superconducting Fault Current Limiters (SFCL) is suggested for discrimination of an internal fault from inrush current of transformer. The SFCL will operate for high fault currents, which gives more sensitivity than the differential protection [12]. The internal and external electrical faults and inrush current of the transformer are can be analysed using Maximal Overlap Discrete Wavelet Transform (MODWT) with Daubechies4 wavelet function by extracting their features. These extracted features are considered for training the classifiers of Decision Tree (DT) and Artificial Neural Network (ANN) [13].

The wavelet transform is one of the signal processing techniques that are used for feature extraction and fault analysis. The discrimination of electromagnetic transients and internal faults were also estimated by using wavelet transform analysis [14]. The wavelet pocket transform (WPT) has more features than the fast wavelet transform (FWT) to discriminate signals in both time and frequency domains. The WPT is an accurate method for the characterization of signals for both the frequency and time domains [15].

The sine wave curve-fitting was suggested for classifying the magnetic inrush currents and internal fault currents for three-phase transformers. The sinusoidal wave was fit to the normalized differential current (I_{diff}) by utilizing least squares techniques (LSQ) for every individual phase. The calculation of residual signals is based on the difference between normalized differential currents and fitted signals. The prediction of internal fault currents and magnetic inrush currents by using the residual signals may be possible in less than 10 ms, i.e., half cycle of power frequency [16]. A review of the reliability and safety of high-speed trains with advanced branches of intelligent transportation was performed by using fault detection and diagnosis techniques. This literature review will help with the fault detection of high-speed trains, and it suggests an advanced method for future investigation [17]. The fault detection and diagnosis (FDD) of high-speed trains in electric traction is suggested based on modified principal component analysis and a broad learning system. The broad learning system can be used to extract fault information without requiring mathematical models or a control mechanism for high-speed trains [18]. The hybrid KNN-GA is used for differential protection of the transformer, i.e., discrimination of magnetic inrush current and fault current with improved performance. The proposed algorithm also increases the performance of the deferential relay [19]. The inrush currents and fault currents are distinguished by the novel method, i.e., generalized delayed signal cancelation (GDSC), based on the sequence components along with the fundamental component. The fundamental, second harmonic positive and negative sequence components are calculated for the estimation of internal faults and magnetic inrush currents [20].

The discrimination of transformer fault currents and magnetic inrush currents is achieved by multi-scale multivariate fuzzy entropy (MMFE), by applying the fuzzy membership function. The MMFE renders better results than the WT method for classification of fault currents and inrush currents, based on the comparison of numerical values of the fixed value of current [21]. The extraction of features with the measured data from the differential currents of a power transformer is suggested by using random the forest-based fault discrimination (RFBFD) technique [22]. New techniques, such as LES, DWT and EKF, are suggested for analysis and estimation of the current from the measured waveform. It is observed that the DWT is given detailed coefficients for current estimation with the varying window technique and EKF is suggested for current estimation with the prediction and correction method.

From the above analysis of the literature survey, it is observed that discrimination of internal fault currents and inrush currents is still a problem for researchers. In this paper, a novel technique is presented for the classification of internal fault currents from the inrush currents using the EKF algorithm.

In this paper, an EKF algorithm is explored to predict primary side currents of the transformer, using the nonlinear state-space transformer model. A residual signal is derived from the difference of measured and estimated currents on the primary side winding of the transformer. When the transformer is under normal conditions, the residual signal becomes zero based on the EKF estimation. If the transformer is in a faulty condition, large residual currents are generated. The residual signal amplitude is compared with the threshold value in order to classify the magnetic inrush current and internal fault current. In this paper, the diagonal elements of the covariance matrix are utilized to analyze the severity of the fault occurrence in the transformer. The nonlinear state-space model of the one-phase transformer is developed under steady-state conditions and the proposed EKF algorithm in the next sections. The simulation results are presented to evaluate the EKF for different conditions of the transformer.

2. Transformer State-Space Nonlinear Model

According to the literature [10], the nonlinear state-space model for the transformer is formulated by the proposed method, i.e., the EKF. In this regard, the nonlinear equations are derived in state-space form for the transformer. The transformer equivalent circuit is presented in Figure 1 [10].

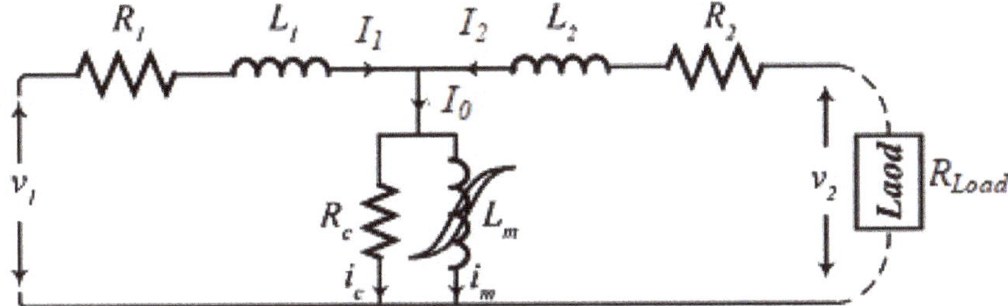

Figure 1. Single-phase transformer equivalent circuit under steady-state conditions.

By applying Kirchhoff's current law (KCL) at the single node of the circuit shown in Figure 1,

$$I_1 + I_2 = i_c + i_m \tag{1}$$

By applying Kirchhoff's voltage law (KVL), the loop equations are applied at the input side and output side of the equivalent circuit of the transformer,

$$v_1 = R_1 I_1 + \frac{dI_1}{dt} \tag{2}$$

$$v_2 = R_2 I_2 + \frac{dI_2}{dt} \tag{3}$$

where $\lambda_1 = \lambda_{l1} + \lambda_m$ and $\lambda_2 = \lambda_{l2} + \lambda_m$. We know that L_1 and L_2 are considered linear inductances; the flux linkages of L_1 and L_2 can be written as $\lambda_{l1} = L_1 I_1$ and $\lambda_{l2} = L_2 I_2$. Therefore, I_1 and I_2 are given as

$$I_1 = \frac{\lambda_1 - \lambda_m}{L_1} \tag{4}$$

$$I_2 = \frac{\lambda_2 - \lambda_m}{L_2} \tag{5}$$

Applying the KVL in the middle loop of the above circuit is given as:

$$R_c i_c = \frac{d\lambda_m}{dt} \tag{6}$$

Substituting Equation (4) in Equation (2), the first state-space equation is achieved, which is

$$\frac{d\lambda_1}{dt} = -\frac{R_1}{L_1}\lambda_1 + \frac{R_1}{L_1}\lambda_m + v_1 \tag{7}$$

(First state-space equation)

Substituting Equation (5) in Equation (3), and considering $V_2 = -R_L I_2$, the second state-space equation is modified as

$$\frac{d\lambda_2}{dt} = -\frac{R_L + R_2}{L_2}\lambda_2 + \frac{R_L + R_2}{L_2}\lambda_m \tag{8}$$

(Second state-space equation)

Considering [1,4–6], the third state-space equation, which associates the nonlinear nature of the transformer magnetic core, is given by

$$\frac{d\lambda_m}{dt} = \frac{R_c}{L_1}\lambda_1 + \frac{R_c}{L_2}\lambda_2 - R_c\left(\frac{1}{L_1} + \frac{1}{L_2}\right)\lambda_m - R_c i_m \tag{9}$$

(Third state-space equation)

To predict the current in the primary side winding with the EKF method, the nonlinear state-space model output is derived as:

$$y = I_1 = \frac{1}{L_1}\lambda_1 - \frac{1}{L_1}\lambda_m \tag{10}$$

Equations (7)–(10) give the state-space formulation of the nonlinear model of the transformer. In Equation (9), the inrush current i_m is not determined. Therefore, it is required to determine a nonlinear model of transformer, $i_m = g(\lambda_m)$, in which g describes the nonlinear nature of the hysteresis andmagnetic saturation of the core.

According to the literature [10], various approaches are suggested for modeling this nonlinear behavior such as tangent, polynomial, and hyperbolic tangent functions, etc. Based on a trial and error approach, it was concluded that the combination of two Gaussian functions will best describe the nonlinear behavior of the 1-phase transformer that is studied in this paper. The criterion for assessment of the different models is the root mean square of the error (RMSE). The Gaussian model is given as:

$$i_m = g(\lambda_m) = sign(\lambda_m) \times \left[ae^{-\left(\frac{|\lambda_m|-b}{c}\right)^2} + de^{-\left(\frac{|\lambda_m|-e}{f}\right)^2} \right] \tag{11}$$

According to the literature [10], the main coefficients of the Gaussian model are estimated by using nonlinear LS (NLS) optimization and the Levenberg-Marquardt algorithm technique. Different Gaussian functions are used to predict the magnetization curve of a transformer, as presented in Table 1 [10].

Table 1. Different functions used for the estimation of the magnetization curve of the transformer.

No	Function	RMSE
I	$i_m = \lambda_m \left[-2 + 4\tan\left(0.65\lambda_m\right)^2 \right]$	1.1082
II	$i_m = 16\lambda_m^5 - 47\lambda_m^4 + 43\lambda_m^3 - 11.5\lambda_m^2 - 0.5\lambda_m + 0.02$	1.3214
III	$i_m = \text{sgn}(\lambda_m)\left[44e^{-\left(\frac{|\lambda_m|-6}{2}\right)^2} + 170e^{-\left(\frac{|\lambda_m|-10}{3.5}\right)^2} \right]$	0.7042

Functions I, II, and III of the above Table 1 are used for various estimations of the magnetization curve, i.e., magnetizing current vs. magnetization and transformer inrush currents, as shown in Figure 2a,b, respectively [10]. The magnetic inrush currents are provided based on the various functions presented in the above Table 1 and it shows the behavior of the magnetic core [10].

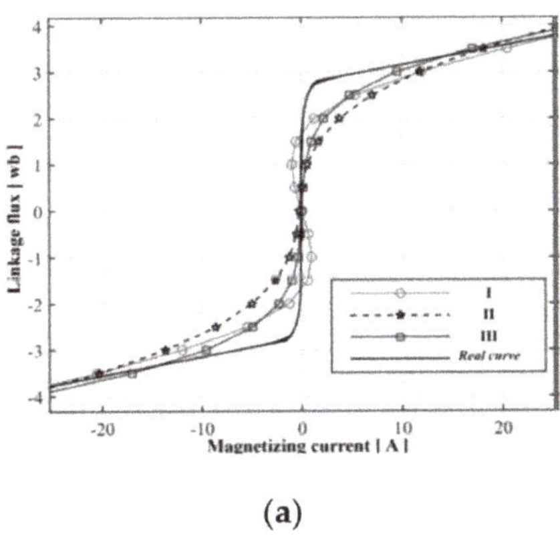

(**a**)

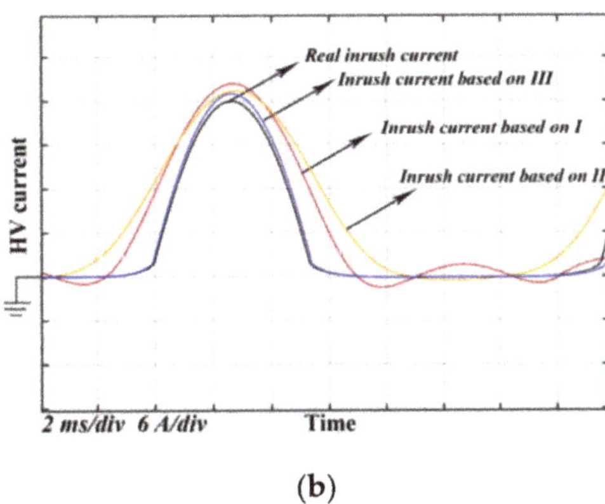

(**b**)

Figure 2. Various estimated magnetization curves (**a**) with their respective inrush currents (**b**).

Different functions from Table 1 are used for modelling the magnetic core behavior of the transformer. Different estimations of the magnetization curve are compared in Figure 2a. Moreover, with different fulfilled estimations, the expected waveforms relevant to the transformer inrush current under various conditions are shown in Figure 2b.

3. EKF Formulation and Its Working Principle

The proposed working principle of EKF algorithm is given below.

3.1. Proposed Working Principle of the EKF

According to the literature [10], the nonlinear state-space model's equation of the 1-phase transformer considered was derived in the previous section. The transformer primary side winding current using EKF is estimated for the model considered [10]. If the transformer is under a fault condition, the EKF will accurately predict the input current. If the considered transformer is in a faulty period, the model would not give proper nonlinear behavior due to the existing internal faults. Therefore, the estimation error of the EKF is increased, which depends on the accuracy of the model. In this regard, the estimation error is accountably large during the incipient cycles. Therefore, the absolute residual signal (ARS), i.e., the difference between the estimated and measured primary side currents, is increased at the incipient cycle while energizing a transformer in a faulty condition. This feature is used for the classification of transformer inrush currents from internal faults. Therefore, the ARS compares with the predefined threshold value for confirming or rejecting the protection systems [10].

3.2. Formation of the EKF

The nonlinear behavior of the state-space model can be given as follows:

$$\begin{aligned}
\lambda\cdot &= f(\lambda, v_1, \omega, t) \\
I_1 &= h(\lambda, v, t) \\
\omega &\approx (0, Q) \\
v &\approx (0, R)
\end{aligned} \tag{12}$$

where v and ω are uncorrelated white noises with zero mean and $\lambda^{3\times1}$ is the state vector. The complete algorithm of the proposed EKF is given by the following flowchart in Figure 3 [10].

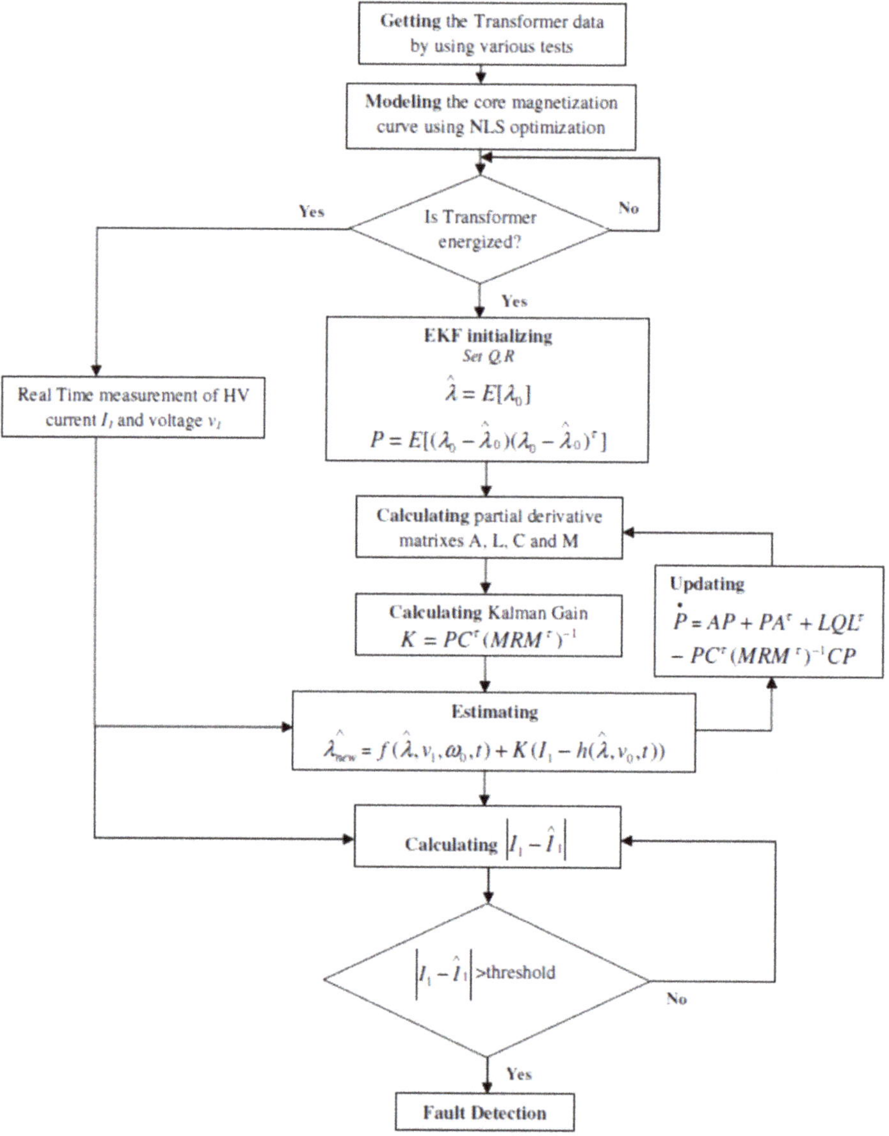

Figure 3. Flowchart of the proposed EKF method for estimation of different fault currents.

The nonlinear vectors of Equation (12) for the transformer modeled in earlier sections are formulated in Equation (13).

$$\begin{bmatrix} \dot{\lambda}_1 \\ \dot{\lambda}_2 \\ \dot{\lambda}_m \end{bmatrix} = f(\lambda, v_1, \omega, t) = \begin{bmatrix} \frac{-R_1}{L_1}\lambda_1 + \frac{R_1}{L_1}\lambda_m + v_1 + \omega_1 \\ \frac{-R_L+R_2}{L_2}\lambda_2 + \frac{R_L+R_2}{L_2}\lambda_m + \omega_2 \\ \frac{R_C}{L_1}\lambda_1 + \frac{R_C}{L_2}\lambda_2 - R_C\left(\frac{1}{L_1} + \frac{1}{L_2}\right)\lambda_m - R_C(\mathrm{sgn}(\lambda_m)[ae^{-(\frac{|\lambda_m|-b}{c})^2} + de^{-(\frac{|\lambda_m|-e}{c})^2}] + \omega_3 \end{bmatrix} \tag{13}$$
$$y = I_1 = h(\lambda, v, t) = \frac{1}{L_1}\lambda_1 - \frac{1}{L_1}\lambda_m + v$$

In Equation (13), y and v_1 are the measured magnetic inrush currents of the transformer and input voltage of each sample datum, respectively. Moreover, Q and R matrices are considered as

$$Q = \begin{bmatrix} 0.2 & 0 & 0 \\ 0 & 0.05 & 0 \\ 0 & 0 & 0.5 \end{bmatrix}$$
$$R = 1$$

The Q and R matrices were chosen based on the present model and measurement uncertainties, and the matrices A, L, C, and M were also obtained, as given in Equations (14)–(17).

$$A = \left.\frac{\partial f}{\partial \lambda}\right|_{\hat{\lambda}} =$$

$$\begin{bmatrix} -\frac{R_1}{L_1} & 0 & \frac{R_1}{L_1} \\[6pt] 0 & -\frac{R_L+R_2}{L_2} & \frac{R_L+R_2}{L_2} \\[6pt] \frac{R_C}{L_1} & \frac{R_C}{L_2} & \begin{cases} -R_C\left(\frac{1}{L_1}+\frac{1}{L_2}\right) - R_C\left(\left(ae^{-(\frac{\hat{\lambda}_m-b}{c})^2}\right)\left(-\frac{2}{c}\left(\frac{\hat{\lambda}_m-b}{c}\right)\right)\right) + \left(de^{-(\frac{\hat{\lambda}_m-e}{f})^2}\right)\left(-\frac{2}{f}\left(\frac{\hat{\lambda}_m-e}{f}\right)\right) \\ \qquad\qquad\qquad\qquad \text{when } \hat{\lambda}_m \geq 0 \\[8pt] R_C\left(\frac{1}{L_1}+\frac{1}{L_2}\right) - R_C\left(\left(ae^{-(\frac{\hat{\lambda}_m+b}{c})^2}\right)\left(-\frac{2}{c}\left(\frac{\hat{\lambda}_m-b}{c}\right)\right)\right) + \left(de^{-(\frac{\hat{\lambda}_m+e}{f})^2}\right)\left(-\frac{2}{f}\left(\frac{\hat{\lambda}_m-e}{f}\right)\right) \\ \qquad\qquad\qquad\qquad \text{when } \hat{\lambda}_m \leq 0 \end{cases} \end{bmatrix} \tag{14}$$

$$L = \left.\frac{\partial f}{\partial \omega}\right|_{\hat{\lambda}} = \begin{bmatrix} 1 & 0 & 0 \\ 0 & 1 & 0 \\ 0 & 0 & 1 \end{bmatrix} \tag{15}$$

$$C = \left.\frac{\partial h}{\partial \lambda}\right|_{\hat{\lambda}} = \begin{bmatrix} \frac{1}{L_1} & 0 & -\frac{1}{L_1} \end{bmatrix} \tag{16}$$

$$M = \left.\frac{\partial h}{\partial v}\right|_{\hat{\lambda}} = 1 \tag{17}$$

3.3. Operating Criteria

The residual signal instantaneous magnitude value is determined to be the criterion for differentiating fault currents of the transformer. The residual signal can be shown as:

$$R = y - \hat{y} \tag{18}$$

where y and $\hat{y}$ are the measured value (from the current transformer) and estimated (with EKF) currents, respectively. If the ARS is greater than the threshold value, the fault condition is discriminated. The selection of the threshold value is very important, because it may affect the operating time of estimation and accuracy percentage of the algorithm.

4. Analysis of Simulation Results of Transformer under Various Faults

The proposed method was implemented by using simulated data of the transformer to distinguish the different inrush and fault conditions. The inrush current may be present in the system when the transformer is energized under no-load or lightly loaded conditions. The equivalent circuit of the respective transformer under such conditions is shown in

Figure 4. Consider R_s, R_p, and R_t are a series, core losses and source resistances and L_s and L_m are series and magnetizing inductances, respectively.

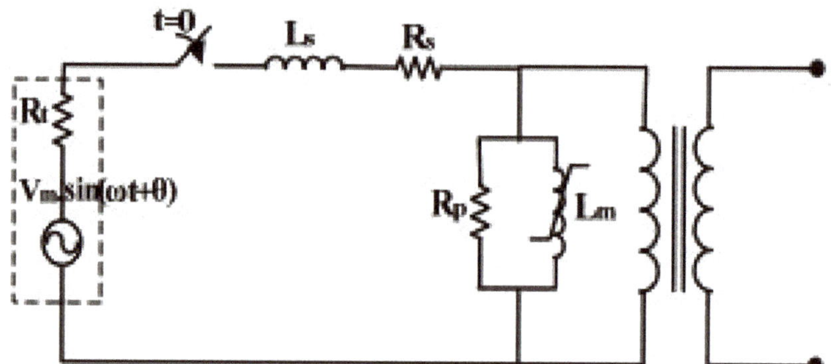

Figure 4. Transformer equivalent circuit under the no-load condition.

4.1. Transformer Parameters

The single-phase transformer equivalent model was designed and simulated using MATLAB/Simulink, 2018a version by MathWorks, Natick, USA for different internal fault conditions and inrush currents conditions.

The transformer's parameters are tabulated in Table 2.

Table 2. Specifications of the single-phase transformer.

Specification No.	Parameters	Value
1	KVA rating	15 KVA
2	Frequency	50 Hz
3	Voltage Rating	2300/230
4	Rated Primary Current	6.53 amps
5	Winding Resistance on Primary	7.334 mΩ
6	Winding Resistance on Secondary	4.634×10^{-5} Ω
7	Winding Reactance on Primary	3.428 Ω
8	Winding Reactance on Secondary	6.194 Ω

4.2. Types of Inrush Currents, Modeling, and Simulation of Inrush Currents

1. Inrush current energization
2. Recovery inrush current
3. Sympathetic inrush current

4.2.1. Factors Affecting Inrush Currents

The inrush phenomenon occurs due to over-fluxing of the transformer core for a temporary period. This phenomenon may depend on:

1. The switching time of the voltage waveform, when the transformer is energized.
2. The magnitude and polarity of the flux (residual) persisting in the transformer at the re-energization time.
3. Resistance on the primary side winding.

4.2.2. Effects of Switching Angle

The phase angle of starting voltage depends on the transformer when it is switched on. Therefore, the flux in the core also starts from zero at the time of switching. As per Faraday's law of electromagnetic induction, the voltage induced across the winding is given as $e = \frac{d\varphi}{dt}$, where φ is the flux in the core. Hence, the flux φ is the integral of the voltage wave, which can be determined and it is shown in Figure 5 [3].

$$e = E\sin\omega t = \frac{d\varphi}{dt} \Rightarrow \varphi = \int edt = E\int \sin\omega t dt$$

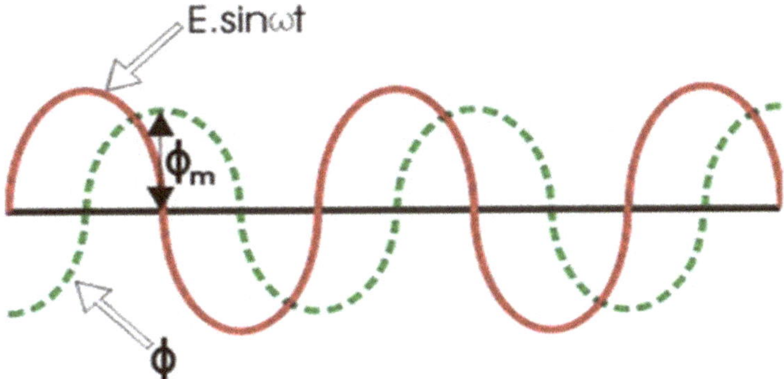

Figure 5. Calculation of the induced emf.

The saturation of the transformer starts when it is crossed above the steady-state value of the flux, i.e., when the transformer is switching, the maximum value of flux will double the steady-state maximum value. After the occurrence of the steady-state maximum value of flux, the core becomes saturated and the respective current required for producing the rest of flux when it is high is shown in Figure 6 [3]. Let φ_m be the maximum value of the steady-state flux. In this regard, the transformer primary side will draw a high peak current from the source side. This phenomenon is treated as the inrush current or magnetizing inrush current of the transformer.

$$\varphi_m^{'} = (E/\omega)\int_0^{\pi} \omega\sin\omega t dt = \varphi_m\int_0^{\pi} \sin\omega t d(\omega t) = 2\varphi_m$$

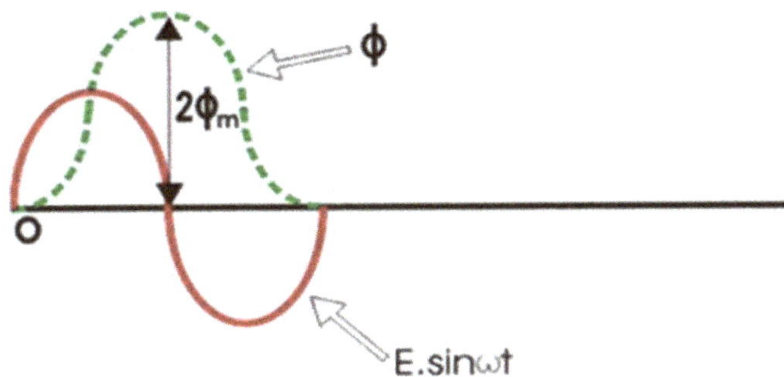

Figure 6. Transformer flux doubling effect.

During the voltage switching at 90°, the produced flux is the minimum and hence the current drawn. If the switching angle increases, it leads to a decrease in the amplitude of inrush current, which is shown in Figure 7 [3]. The representation of inrush current energization is shown in Figure 8, and inrush currents for various switching angles are shown in Figure 9.

The magnitudes of magnetic inrush currents for various switching angles are given in Table 3.

Hence, the highest inrush current magnitude is treated for a zero-degree switching angle and the least inrush current magnitude is treated for a 90-degree switching angle. It is observed that increasing the switching angle decreases the amplitude of inrush current for different materials, as shown in Figures 10 and 11.

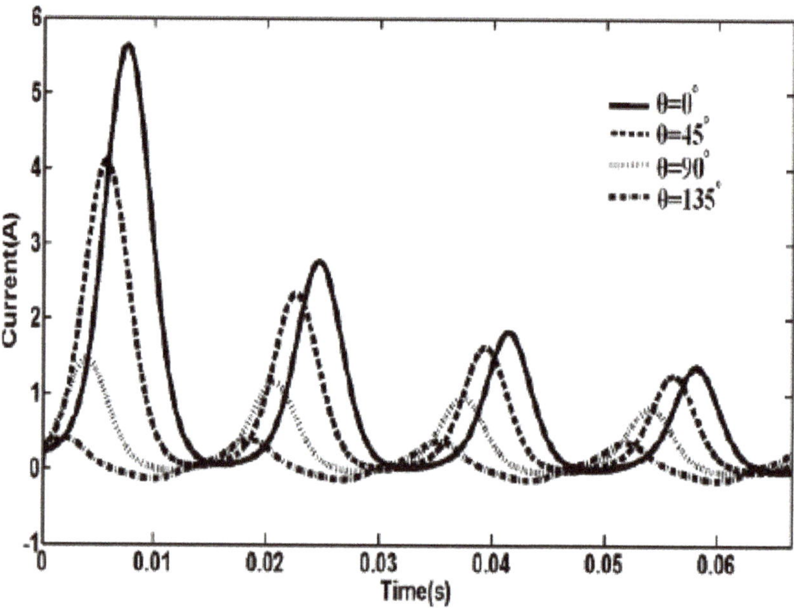

Figure 7. Variation in switching angle effects on inrush current amplitude.

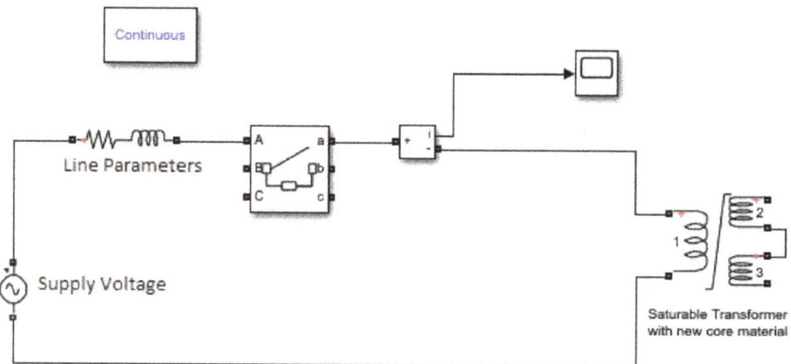

Figure 8. Representation of energization inrush current.

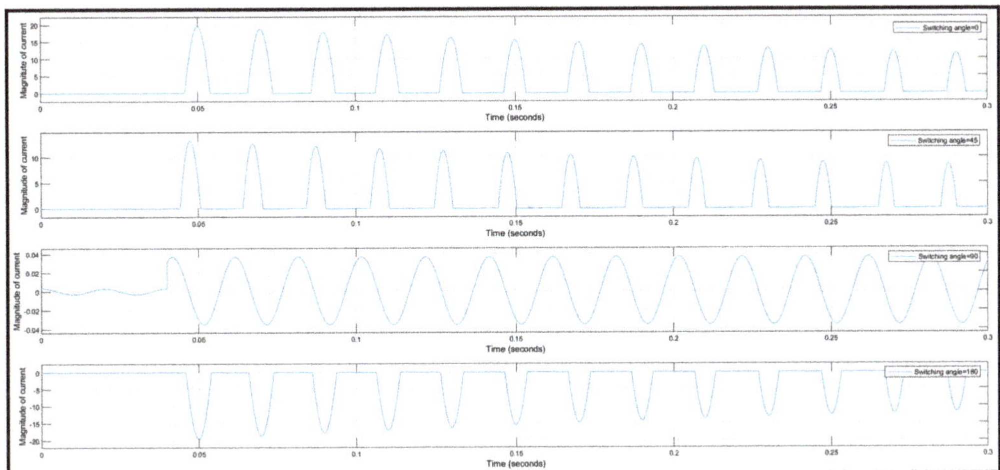

Figure 9. Inrush currents for various switching angles.

Table 3. Magnitudes of magnetic inrush current for various switching angles.

SNo.	Switching Angle	Inrush Current Magnitude
1	0 degree switching	19.688 A
2	45 degree switching	13.170 A
3	90 degree switching	0.037 A
4	180 degree switching	−19.573 A

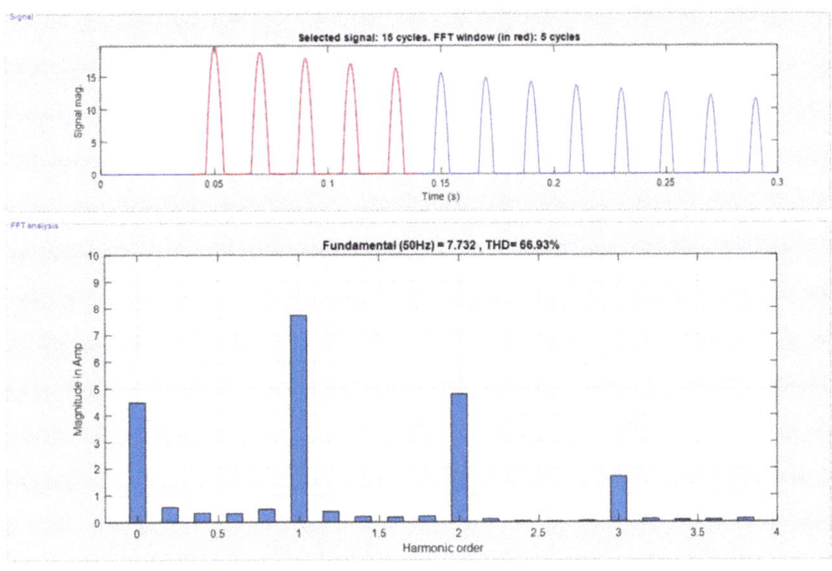

Figure 10. Analysis of inrush current using FFT for old materials.

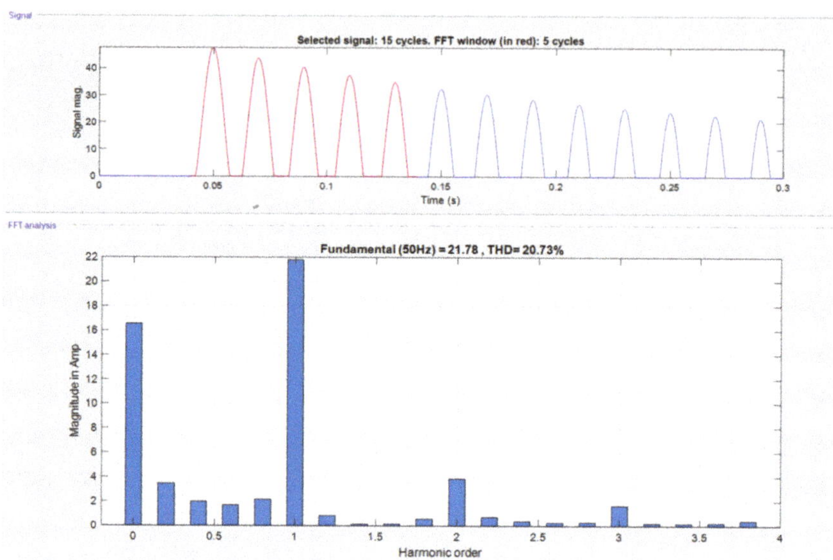

Figure 11. Analysis of inrush currents using FFT for new materials (metglass).

Table 4 shows the magnitude of harmonic currents for old metals and new metals (metglass) of transformer core.

Table 4. Magnitudes of harmonic currents for different metals of the transformer.

Sl.No.	Harmonic Oder	Old Materials	New Materials (Metglass)
1	Fundamental harmonic (1st) magnitude	7.73 A	21.78 A
2	2nd harmonic magnitude	4.78 A	3.86 A
3	3rd harmonic magnitude	1.69 A	1.65 A
4	Total harmonic distortion (%THD)	66.93%	20.73%

4.3. Modeling and Simulation of Internal Faults

The generation of internal faults, i.e., transformer internal turn faults along with current waveforms, is shown in Figures 12 and 13, respectively.

Table 5 shows the magnitude of internal fault current for different conditions of the transformer.

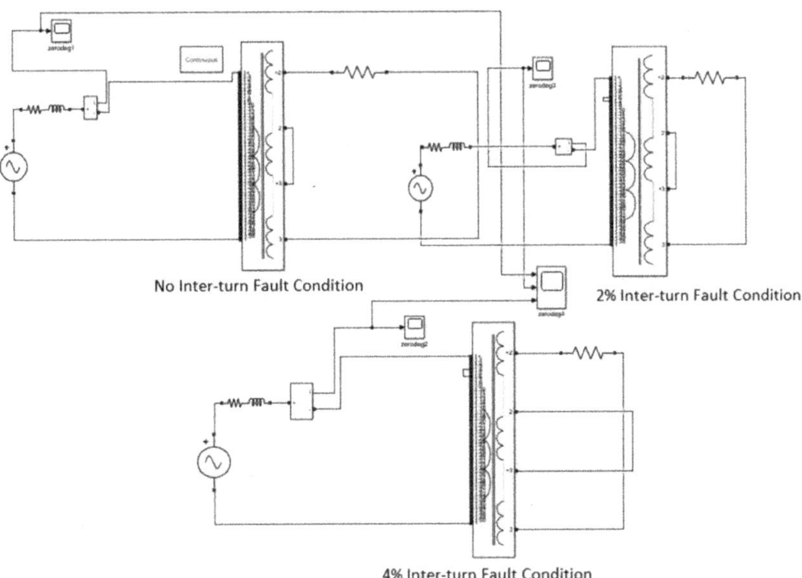

Figure 12. Simulink model for various transformer internal (turn-to-turn) faults (no fault, 2% fault and 4% fault).

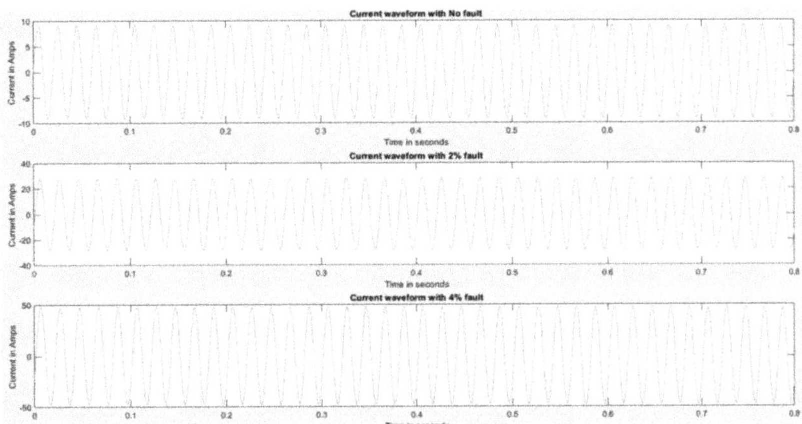

Figure 13. Internal faults with energization for different percentages of winding shorted.

Table 5. Magnitudes of internal fault currents for different cases.

Sl.No.	Particulars	No Fault	2% Fault	4% Fault
1	Current Magnitude (RMS value) (Simulation analysis)	6.52 A	20.23 A	34.48 A
2	Current Magnitude (Peak = RMS × 1.414 value) (Simulation analysis)	9.151 A	28.226 A	47.666 A
3	Current Magnitude (Peak = RMS × 1.414 value) (Mathematical analysis)	9.21 A	28.61 A	48.854 A

Hence, the simulation results are validated with the mathematical analysis under various conditions i.e., no fault, 2% fault, and 4% fault.

4.4. Implementation of the EKF for the Specified Transformer

The simulation of transformer with the EKF for primary current estimation with the specific data of the transformer is shown in Figure 14.

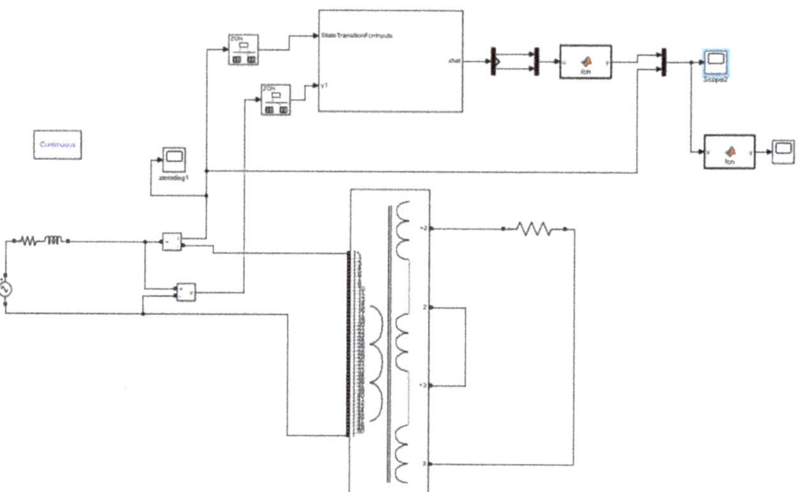

Figure 14. Simulink model of the EKF for primary current estimation.

The functions used in the EKF block are:

```
1. State transition function
function x = myStateTransitionFcn(x,v)
    dt = 0.0001;
    ind1 = 0.012256;
    ind2 = 0.000106;
    R1 = 2.59;
    R2 = 2×4.6344e-05;
    RC = 105,000;
    x = x + [(-R1×x(1)/ind1) + (R1× x(3)/ind1) + v;
    (-R2× x(2)/ind2) + (R2× x(3)/ind2);RC× x(1)/ind1 + RC× x(2)/ind2 – (RC× x(3)) ×(1/ind1
    + 1/ind2)-RC× sign(x(3)) ×(13.89×
exp(-((abs(x(3))-6.17)/2.58)^2)) – RC× sign(x(3)) ×(54.55× exp(-((abs(x(3))-6)/1.92)^2))] ×
    dt;
    end
2. Measurement function
function y = myMeasurementFcn1(x)
    ind1 = 0.012256;
    ind2 = 2× 5.55368e-5;
    R1 = 2.59;
    R2 = 2× 4.6344e-05;
    RC = 105,000;
    y = (1/ind1) × x(1) – (1/ind1) × x(3);
    end
```

4.5. Simulation Results of EKF Algorithm for Various Conditions

When the transformer is energized in various healthy conditions, the transformer is able to estimate it almost perfectly with a low residual signal value. The various estimated magnetization curves with their respective inrush currents mentioned in Section 2 are more useful for analysis of the EKF algorithm for estimation of inrush current [10].

The transformer is now energized after it has a 2% inter-turn fault. The combination of inrush and fault currents is not perfectly estimated by the EKF algorithm with a residue larger than in Figure 15; it can be observed in a close-up view of the EKF estimation during the inrush case in Figure 16. Similarly, the EKF was tested over the algorithms with various fault conditions; again, a high residual signal was generated, as shown in Figures 17 and 18.

The simulation model for discrimination of magnetic inrush current, 2% internal fault, 4% internal fault, and 6% internal fault along with current waveforms is given in Figure 19a,b and the respective ARS signals obtained from various sources are depicted in Figure 20, which indicates that the residual signal keeps increasing with the severity of the fault.

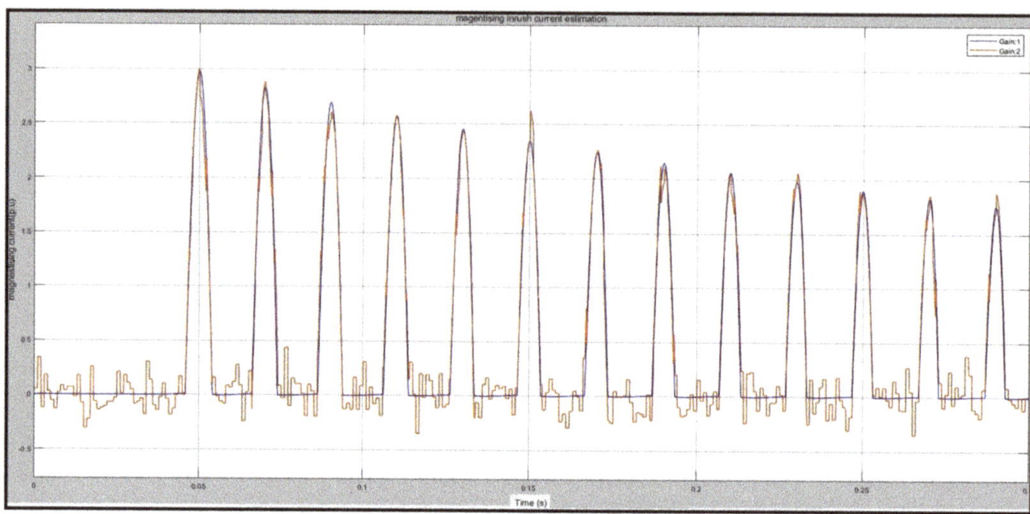

Figure 15. Estimation of inrush current by EKF algorithm (blue line is actual current and red line is the EKF estimation current).

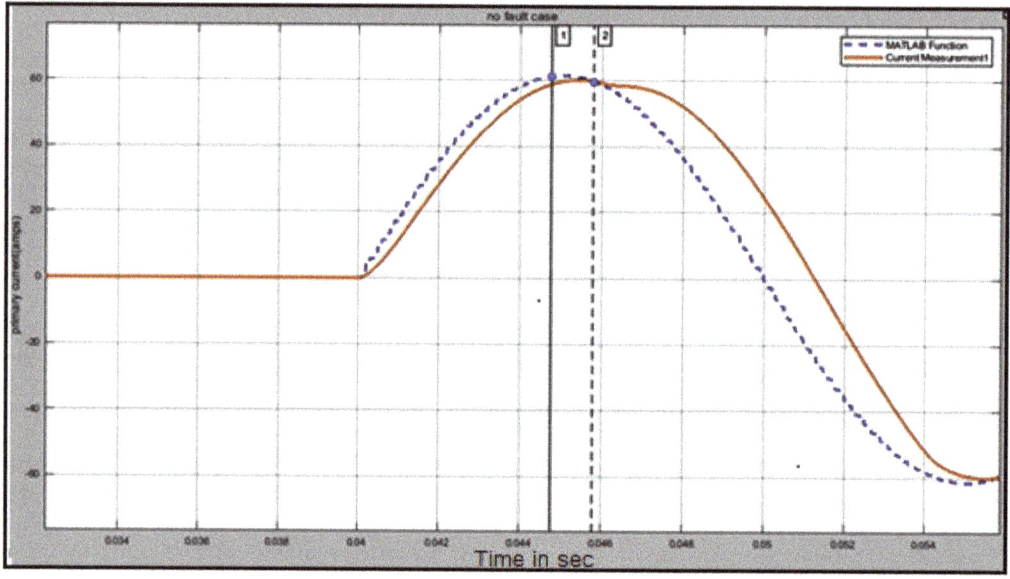

Figure 16. Close up view of EKF estimation during inrush case (red line depicts actual current where the blue line is the estimated current waveform).

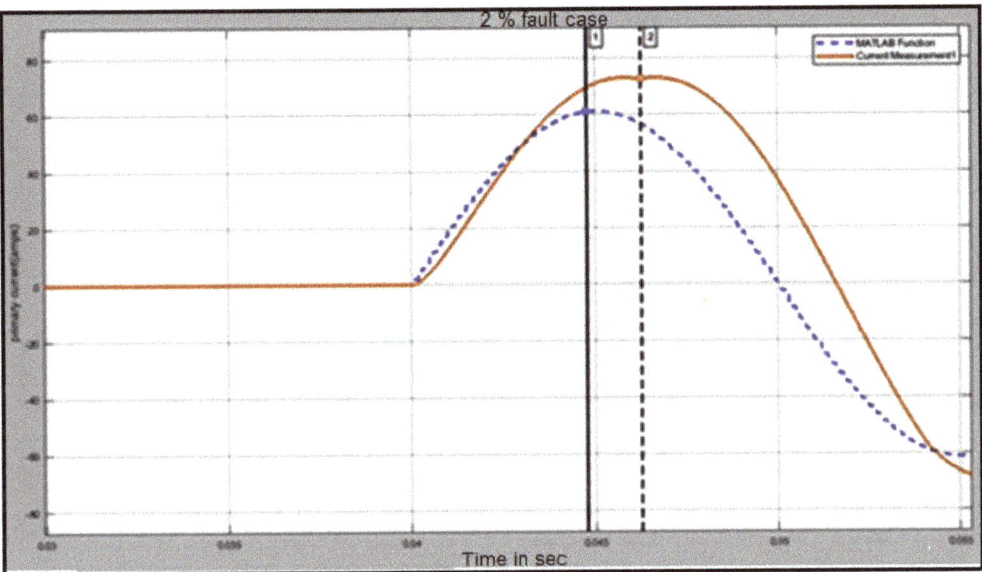

Figure 17. EKF estimation during energization with a 2% internal fault.

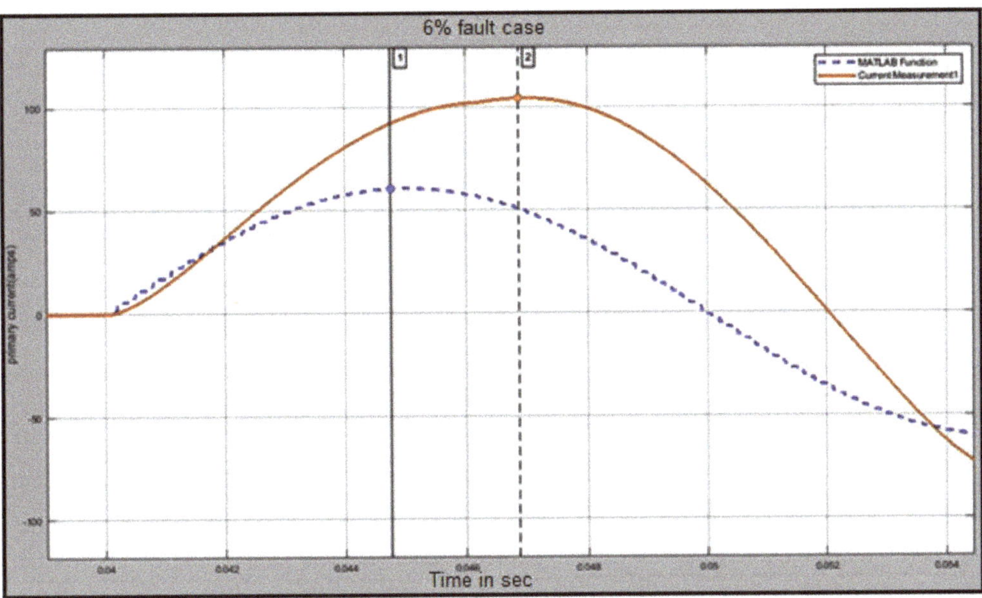

Figure 18. EKF estimation during energization with a 6% internal fault.

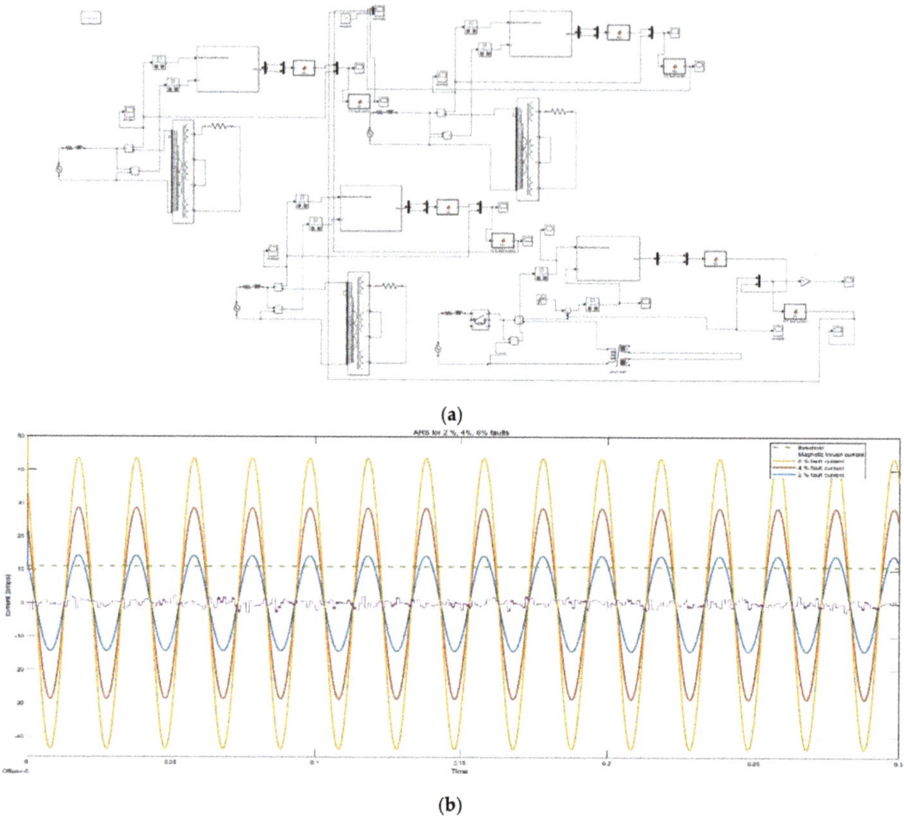

(a)

(b)

Figure 19. (**a**) Simulation model with various types of internal faults. (**b**) ARS signals of magnetic inrush current along with fault current waveforms.

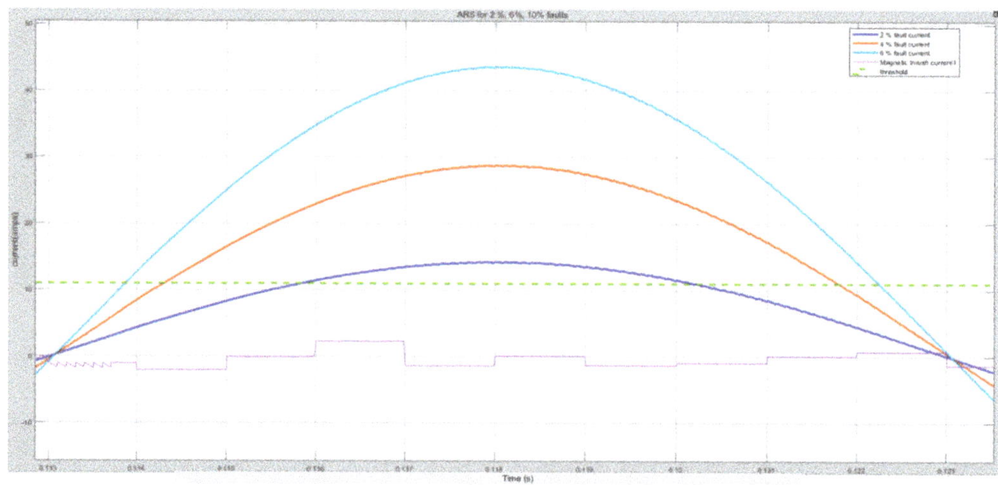

Figure 20. ARS signals for various cases.

The Table 6 indicates the fault detection time with respect to the severity of the faults along with ARS signal amplitudes for different cases.

Table 6. ARS signals and the fault detection time for various cases.

S.No.	Different Cases	ARS Signal Amplitude (amps)	Detection Time (ms)
1	For all inrush currents	less than 5 amps	-
2	Threshold value	11 amps	-
3	Energization with 2% fault	14.3 amps	116
4	Energization with 4% fault	28.82 amps	115
5	Energization with 6% fault	43.82 amps	114

4.6. Observations

- The EKF algorithm was simulated for cases of inrush current with various switching angle cases, and the residue signal was found to be low and within limits (less than the selected threshold value in all cases).
- The internal fault current could not be estimated by the EKF filter and all the ARS values were found to be greater than the selected threshold value of 11 amperes (all fault conditions had an ARS of 11A or more, so this was selected as the threshold). If ARS > 11A, it is an internal fault; otherwise, it is not.
- While we prefer a shorter operating time, some fault cases may be perceived as inrush because the ARS value might not have crossed the threshold by then. Nevertheless, the operating point was always low (max 116 milliseconds, which is much less than a cycle). As the severity of the fault increases, the detection time decreases as the required residual threshold is achieved earlier.
- EKF is independent of the low second harmonic problem as the filter takes into account the BH curve of the core. If the core material or other parameters of the transformers are changed, the equations for the EKF algorithm will have to be changed as well and, hence, the algorithm is able to predict the inrush currents in every case provided the equations are derived and implemented carefully.
- The state-space model and all the coefficients must be derived very carefully. Even a slight mistake can cause the EKF to diverge, which is bound to give the wrong output.

5. Conclusions

In this paper, the EKF was implemented for the discrimination of inrush currents and internal faults current of a single-phase transformer. The state-space equations of the power transformer were simplified using simple circuit theory and curve fitting (for saturation curve). By using the two-step predictive-corrective mechanism of the EKF algorithm, the estimation of the primary winding current of the transformer was achieved. The transformer primary winding current was estimated for various switching angles and faults using the EKF. It was observed that the ARS value is zero under normal conditions, and it will not be equal to zero for internal fault and inrush phenomena conditions. Hence, the EKF algorithm was implemented for discriminating the internal faults and inrush faults by keeping the threshold level at the ARS value. The simulation results were validated with the theoretical analysis under various fault conditions. It is also concluded that the detection time of internal faults decreases with the severity of the fault. Hence, this scheme provides fast protection of the transformer for severe faults. The results of various test cases using the EKF algorithm were presented. The implementation of the EKF can be extended to protect other power system equipments for classification of faults.

Author Contributions: Conceptualization, S.K.G. and V.S.S.S.S.D.; data curation, S.K.G.; formal analysis, S.K.G.; methodology, S.K.G. and V.S.S.S.S.D.; supervision, V.S.S.S.S.D.; writing—original draft, S.K.G. and V.S.S.S.S.D.; writing—review and editing, S.K.G. and V.S.S.S.S.D. Both authors have read and agreed to the published version of the manuscript.

Funding: This research received no external funding.

Institutional Review Board Statement: Not applicable.

Informed Consent Statement: Not applicable.

Acknowledgments: The authors gratefully acknowledge the Department of Electrical Engineering of the National Institute of Technology, Warangal for their support.

Conflicts of Interest: The authors declare no conflict of interest.

References

1. Zheng, T.; Yang, X.; Guo, X.; Wang, X.; Zhang, C. Zero-Sequence Differential Current Protection Scheme for Converter Transformer Based on Waveform Correlation Analysis. *Energies* **2020**, *13*, 1814. [CrossRef]
2. Shahabodin, A.; Mousa, A.; Benyamin, P.; Mohammad, M. Integration of Accelerated Deep Neural Network into Power Transformer Differential Protection. *IEEE Trans. Ind. Inform.* **2020**, *16*, 865–876.
3. Gopika, R.; Deepa, S. Study on Power Transformer Inrush Current. *IOSR J. Electr. Electron. Eng.* **2017**, *2*, 59–63.
4. Hassan, A.; Majid, S. A novel technique for internal fault detection of power transformers based on moving windows. *Int. Trans.Electr. Energy Syst.* **2014**, *24*, 1263–1278.
5. Luliang, Z.; Mengshi, L.; Tianyao, J.; Jie, Z. Identifying magnetizing inrush in power transformers based on symmetry of current waveforms. *IEEJ Trans. Electr. Electron. Eng.* **2017**, *12*, 959–960.
6. Bahador, F.; Golshan, M.; Abyaneh, H.; Saghaian-Nejad, M. A runs test-based method for discrimination between internal faults and inrush currents in power transformers. *Eur. Trans. Electr. Power* **2011**, *21*, 1392–1408.
7. Zhang, H.; Wen, J.; Liu, P.; Malik, O. Discrimination between fault and magnetizing inrush current in transformers using short-time correlation transform. *Electr. Power Energy Syst.* **2002**, *24*, 557–562. [CrossRef]
8. Youssef, O. A Wavelet-Based Technique for Discrimination between Faults and Magnetizing Inrush Currents in Transformers. *IEEE Trans. Power Deliv.* **2003**, *18*, 170–176. [CrossRef]
9. Hajipour, E.; Vakilian, M.; Majid, S. Current-Transformer Saturation Compensation for Transformer Differential Relays. *IEEE Trans. Power Deliv.* **2015**, *30*, 2293–2302. [CrossRef]
10. Naseri, F.; Kazemi, Z.; Arefi, M.; Farjah, E. Fast Discrimination of Transformer Magnetizing Current from Internal Faults: An Extended Kalman Filter-Based Approach. *IEEE Trans. Power Deliv.* **2018**, *33*, 110–118. [CrossRef]
11. Ma, J.; Wang, Z.; Yang, Q.; Liu, Y. A Two Terminal Network-Based Method for Discrimination between Internal Faults and Inrush Currents. *IEEE Trans. Power Deliv.* **2010**, *25*, 1599–1605. [CrossRef]
12. Sahebi, A.; Samet, H. Discrimination between internal fault and magnetising inrush currents of power transformers in the presence of a superconducting fault current limiter applied to the neutral point. *IET Sci. Meas. Technol.* **2016**, *10*, 537–544. [CrossRef]
13. Bagheri, S.; Moravej, Z.; Gharehpetian, G. Classification and Discrimination among Winding Mechanical Defects, Internal and External Electrical Faults, and Inrush Current of Transformer. *IEEE Trans. Ind. Inform.* **2018**, *14*, 484–493. [CrossRef]
14. Faiz, J.; Lotfi-Fard, S. A Novel Wavelet-Based Algorithm for Discrimination of Internal Faults from Magnetizing Inrush Currents in Power Transformers. *IEEE Trans. Power Deliv.* **2006**, *21*, 1989–1996. [CrossRef]
15. Eissa, M. A Novel Digital Directional Transformer Protection Technique Based on Wavelet Packet. *IEEE Trans. Power Deliv.* **2005**, *20*, 1830–1836. [CrossRef]
16. Ahmadi, M.; Samet, H.; Ghanbari, T. Discrimination of internal fault from magnetizing inrush current in power transformers based on sine-wave least-squares curve fitting method. *IET Sci. Meas. Technol.* **2014**, *9*, 73–84. [CrossRef]
17. Hongtian, C.; Bin, J. A Review of Fault Detection and Diagnosis for the Traction System in High-Speed Trains. *IEEE Trans. Intell. Transp. Syst.* **2020**, *21*, 450–465.
18. Chao, C.; Wang, W.; Chen, H.; Zhang, B.; Shao, J.; Teng, W. Enhanced Fault Diagnosis Using Broad Learning for Traction Systems in High-Speed Trains. *IEEE Trans. Power Electron.* **2021**, *36*, 7461–7469.
19. Thote, P.; Daigavane, M.; Daigavane, P.; Gawande, S. An Intelligent Hybrid Approach Using KNN-GA to Enhance the Performance of Digital Protection Transformer Scheme. *Can. J. Electr. Comput. Eng.* **2017**, *40*, 151–161.
20. Batista, Y.; Souza, H.; Santos Neves, F.; Dias Filho, R. A GDSC-Based Technique to Distinguish Transformer Magnetizing from Fault Currents. *IEEE Trans. Power Deliv.* **2018**, *33*, 589–599. [CrossRef]
21. Li, C.; Zhou, N.; Liao, J.; Wang, Q. Multiscale multivariate fuzzy entropy-based technique to distinguish transformer magnetising from fault currents. *IET Gener. Transm. Distrib.* **2019**, *13*, 2319–2327. [CrossRef]
22. Shah, A.; Bhalja, B. Fault discrimination scheme for power transformer using random forest technique. *IET Gener. Transm. Distrib.* **2016**, *10*, 1431–1439. [CrossRef]

Article

Third-Order Sliding Mode Applied to the Direct Field-Oriented Control of the Asynchronous Generator for Variable-Speed Contra-Rotating Wind Turbine Generation Systems

Habib Benbouhenni [1] and Nicu Bizon [2,3,4,*]

[1] Department of Electrical & Electronics Engineering, Faculty of Engineering and Architecture, Nisantasi University, Istanbul 34481742, Turkey; habib.benbouenni@nisantasi.edu.tr
[2] Doctoral School, Polytechnic University of Bucharest, 313 SplaiulIndependentei, 060042 Bucharest, Romania
[3] Faculty of Electronics, Communication and Computers, University of Pitesti, 110040 Pitesti, Romania
[4] ICSI Energy, National Research and Development Institute for Cryogenic and Isotopic Technologies, 240050 Ramnicu Valcea, Romania
* Correspondence: nicu.bizon@upit.ro

Abstract: Traditional direct field-oriented control (DFOC) techniques with integral-proportional (PI) controllers have undesirable effects on the power quality and performance of variable speed contra-rotating wind power (CRWP) plants based on asynchronous generators (ASGs). In this work, a commanding technique based on the DFOC technique for ASG is presented on variable speed conditions to minimize the output power ripples and the total harmonic distortion (THD) of the grid current. A new DFOC strategy was designed based on third-order sliding mode (TOSM) control to minimize oscillations and the THD value of the current and active power of the ASG; the designed technique decreases the current THD from ASG and does not impose any additional undulations in different parts of ASG. The designed technique is simply implemented on traditional DFOC techniques in variable speed DRWP systems to ameliorate its effectiveness. Also, the results show that by using the designed TOSM controllers, in addition to regulating the active and reactive powers of the ASG-based variable speed CRWP system, the THD current and active power undulations of the traditional inverters can be minimized simultaneously, and the stator current became more like a sinusoidal form.

Keywords: third-order sliding mode control; asynchronous generators; variable speed dual-rotor wind turbine; direct field-oriented control; integral-proportional

---

Citation: Benbouhenni, H.; Bizon, N. Third-Order Sliding Mode Applied to the Direct Field-Oriented Control of the Asynchronous Generator for Variable-Speed Contra-Rotating Wind Turbine Generation Systems. *Energies* **2021**, *14*, 5887. https://doi.org/10.3390/en14185877

Academic Editors: Surender Reddy Salkuti and Davide Astolfi

Received: 22 August 2021
Accepted: 13 September 2021
Published: 17 September 2021

Publisher's Note: MDPI stays neutral with regard to jurisdictional claims in published maps and institutional affiliations.

1. Introduction

A sliding-mode command (SMC) is a robust command technique that forces the system to slide along a prescribed switching surface and changes the dynamics of a system by using a discontinuous command signal [1]. Compared to that of other command strategies, SMC attracted significant interest due to its high robustness to external disturbances, simplicity, ease of implementation, and low sensitivity to the system parameter variations [2]. Similar to other methods, we can find this method in all fields of applied sciences, such as robotics, process command, machine command, and motion command. On the other hand, the instrumentation is among the most widely used applications of the SMC technique. There are new applications of the SMC method, for example, anomaly detection and congestion command. SMC technique was used for AC machine drive in wind turbines. In [3], the author uses the SMC technique to regulate the power of the induction generator. In [4], direct power command (DPC) was designed based on the SMC method of the induction generator. Active power undulations were reduced by using the neural SMC (NSMC) technique [5]. However, the NSMC technique was able to offer better effectiveness and performance compared to that of classical SMC strategy. Benbouhenni et al. proposed the fuzzy SMC technique to reduces stator current and torque undulations of an asynchronous generators-based wind turbine [6]. This technique combines the advantages of fuzzy logic and the classical SMC technique so the chattering phenomena are eliminated. In [7], a robust SMC technique based on neuro-fuzzy controllers was developed for ASG.

Despite the many advantages and numerical simulation results, as well as the experimental results, the SMC method is still characterized by several problems, such as the chattering issue. When this method is used to produce electricity from a wind farm, it presents ripples at the level of effective electric current, and this is undesirable because high-quality current and energy can be obtained only if the ripple ratio is very small. There are several solutions proposed to ameliorate the characteristics and effectiveness of the SMC technique, and thus, to improve the quality of the electric current. When the current is high quality, the life of the devices is extended to a longer period, thus reducing the cost of purchasing the devices. In addition, the cost of periodic equipment maintenance is reduced. Among the proposed solutions to reduce chattering and improve SMC performance, we find terminal SMC (TSMC) technique [8], synergetic control [9], integral SMC (ISMC) strategy [10], fast TSMC technique [11], adaptive global SMC control [12], total SMC [13], nonsingular TSMC [14], fast TSMC technique [15], and fast integral TSMC [16].

Traditionally, the terminal SMC (TSMC) command was proposed in 1996 by Yu et al. [17]. Compared to that of the classical SMC technique, TSMC offers some superior characteristics such as high steady-state tracking precision and fast and finite-time convergence, such as motor control, robot command, multiagent systems, stochastic nonlinear systems, second-order SMC command, TSMC observer, and distributed command. Nevertheless, the TSMC command technique has a singularity problem. The latter is among the biggest obstacles that limit the use of the TSMC method. In [18], a nonsingular fast TSMC technique was proposed for position tracking of the electric cylinder. ASG control through rotor current controllers proposed by the TSMC technique is designed in [19], and in this work, the current and torque undulations are reduced compared to the classical method. A super twisting fractional-order terminal SMC technique was designed to command the rotor converter of ASG [20]. In [21], the rotor side converter of ASG is controlled by using the second-order integral TSMC method. The TSMC approach improves the performance of the ASG-based wind turbine compared to a standard SMC method [22].

In recent decades, several new methods were published with the goal of overcoming the chattering phenomenon of classical SMC techniques. In [23], an approach of the high-order SMC technique was designed. In [24], another approach was designed to overcome the problem of the high-order SMC technique. These two techniques adopted indirect methods to overcome the problem of the chattering phenomenon. In [25], the author uses the super twisting algorithm (STA) to control the induction generator. The neural algorithm and the STA technique were combined to reduces voltage and stator current ripples of ASG-based wind power [26]. The combination of the STA algorithm and the neuro-fuzzy methods was completed to obtain a more robust method, as well as to reduce oscilations at the level of electromagnetic torque and active power [27]. STA algorithm with fuzzy logic was used to control doubly fed induction machines [28]. In [29], a direct power control (DPC) with the STA algorithm minimized the ripples of active power compared to that of DPC with a look-up table. In [30], the author uses the STA algorithm to command the power of the synchronous generator (SG). In [31], the fractional order-based super twisting algorithm (FOSTA) improves the performances of the BLDC compared to that of proportional-integral (PI) controllers. The STA technique minimizes the chattering effect, which is inherent in the traditional SMC method [32,33]. In [34], the adaptive intelligent global SMC method was proposed to minimize the current ripple of a DC-DC buck inverter. In [35], the DPC strategy was designed based on the variable gain STA technique to command ASG-based wind turbines. In [36], the STA improves the performance of the DFIG and the quality of output power compared to classical SMC control.

There is another method that was suggested in several scientific works to ameliorate the characteristics of the SMC method under the name of second-order continuous SMC method. This method is characterized by simplicity and durability, and it can be accomplished easily. This method is used to ameliorate the effectiveness and efficiency of the DPC and DTC of the asynchronous generator [37,38]. Synergetic control and the SMC method are combined to minimize the chattering phenomenon and ameliorate the characteristics of the quality current of the ASG [39]. This designed strategy is a simple nonlinear technique that can reduce the total harmonic distortion (THD) of current compared to that of the traditional strategy. To obtain lower current ripples and high-quality energy efficiency, a new nonlinear strategy is designed in [40]. This designed new strategy is simple and gives very excellent results compared with that of the PI controller. This method was called terminal synergetic control.

There are many nonlinear methods proposed to ameliorate the purity of current produced from renewable sources. All these methods have one goal: to obtain good results in terms of ripples of effective power, electric current, and THD value.

In this work, we attempt to deliver a new method for SMC to reduce the chattering phenomenon as well as improve the performance and efficiency of the asynchronous generator-based contra-rotating wind turbine power (CRWP). This method will be called below as the third-order sliding mode (TOSM) command. This new method is the main contribution made in this paper.

In this work, the TOSM command was designed to control and minimize the ripples of current, reactive power, torque, and active power of ASG-based variable- speed CRWP systems. A TOSM technique to overcome the undulations power problem is designed for the direct field-oriented control (DFOC) technique of variable-speed CRWP systems. A DFOC control strategy with TOSM controllers is employed to enable avoidance of the power undulations and improve the response time. The system states can be ensured to improve in variation parameters. The principle of the proposed TOSM controllers is detailed in the work. Validation of the designed technology is carried out by digital simulations using Matlab software.

In summary, the novelties and main findings of this paper are as follows:

-   A new TOSM controller based on the DFOC control scheme is proposed to reduce ripples of both reactive and active powers.
-   TOSM controllers minimize the tracking error for reactive and active powers towards the references of ASG-based variable speed CRWP systems.
-   The DFOC-TOSM control scheme with the PWM technique minimizes the THD value of torque, voltage, and active power of ASG-based variable speed CRWP.

Thus, the combination of the work is as follows. In Section 2, contra-rotating wind turbine system models are presented. In Section 3, the proposed nonlinear controller is presented using the STA controller. Section 4 includes the DFOC control scheme with designed controllers. Section 5 presents and discusses the numerical results of the research carried out.

2. CRWP Model

Wind energy is inexhaustible, clean, does not require much maintenance, and is inexpensive. The use of wind energy is now common in the world in the field of electric energy generation. The most widely used generation system for wind power conversion is the single-turbine, which is capable of converting 59% of the total wind power into useful electrical power [41]. This is still a very small percentage and needs to be increased. In order to ameliorate the characteristics and development of a single-rotor system, there are several scientific types of research in this regard to increase the efficiency of power recovery. Among these most effective solutions, we find the two-rotor system or contra-rotating wind power. The latter can increase the power conversion efficiency by 40% compared to that of a single turbine system (STS) [42]. This new method is explained in [43–47]. In this new technology, two turbines were used to raise and increase the ability to collect energy from the wind. Experiments proved the effectiveness of this new technology. However, the first is a small turbine and the second is a large turbine. The difference between new and old technology lies in the number of mechanical components and the amount of energy gained from the wind. As is known through the scientific experiments conducted on this new technology, all the results are in favor of the new technology. Besides, this new turbine has excellent characteristics in regions with high and low wind speeds. Among its advantages is that it operates at lower tip speed ratios compared to that of the traditional STS [47]. However, this method has several disadvantages, such as high costs, control difficulties, and risk of subsynchronous resonance. On the other hand, the new technology has more mechanical components than the classic wind turbine. The mechanical energy obtained from this new technology is shown in Equation (1). Through this equation, the total energy of this technology is the sum of the two energies of the small and large turbines:

$$P_{CRWP} = P_T = P_{ST} + P_{LT} \tag{1}$$

where P_{ST} and P_{LT} are the mechanical power of the small and large turbines.

In the CRWP system, the resulting aerodynamic torque is the sum of the torques of the small and large turbines and is represented by the Equation (2).

$$T_{CRWP} = T_T = T_{ST} + T_{LT} \tag{2}$$

where T_{ST} and T_{LT} are the aerodynamic torque of the small and large turbines.

Equations (3) and (4) represent the aerodynamic torque for of the large and small turbines [45].

$$T_{LT} = \frac{C_p}{2 \cdot \lambda_{LT}^3} \cdot \rho \cdot \pi \cdot R_{LT}^5 \cdot w_{LT}^2 \tag{3}$$

$$T_{ST} = \frac{C_p}{2 \cdot \lambda_{ST}^3} \cdot \rho \cdot \pi \cdot R_{ST}^5 \cdot w_{ST}^2 \tag{4}$$

where: R_{ST} and R_{LT} are the blade radius of the small and large turbines; ρ: the air density; λ_{ST}, λ_{LT}: the tip speed ration of the small and large turbines; w_{ST}, w_{LT} the mechanical speed of the small and large turbines.

Equations (5) and (6) represents the tip speed ratios of the small and large turbines, respectively.

$$\lambda_{ST} = \frac{w_{ST} \cdot R_{ST}}{V_{ST}} \tag{5}$$

$$\lambda_{LT} = \frac{w_{LT} \cdot R_{LT}}{V_{LT}} \tag{6}$$

where V_{ST} and V_{LT} are the speed of the unified wind on small and large turbines.

To calculate the wind speed at a point between a large and a small turbine, we use Equation (7). Fifteen m is the distance (x) between the center of the large turbine and the center of the small turbine [48].

$$V_x = V_{LT} \cdot \left(1 - \frac{1 - \sqrt{1 - C_T}}{2} \cdot \left(1 + \frac{2 \cdot x}{\sqrt{1 + 4 \cdot x^2}}\right)\right) \tag{7}$$

where C_T is the trust coefficient ($C_T = 0.9$) and Vx is the wind speed at a point between a large and small turbine.

Cp in terms of the pitch angle (β) is given by Equation (8) [49].

$$C_p(\beta, \lambda) = 0.517 \cdot \left(\frac{116}{\lambda_i} - 0.4 \cdot \beta - 5\right) \cdot e^{\frac{-21}{\lambda_i}} + 0.0068 \cdot \lambda \tag{8}$$

3. Third-Order Sliding Mode Control

There are many types of SMC techniques in the literature. All these proposed methods aim to reduce chattering phenomena. The one with the simplest algorithm and the easiest experimentally is the STA method [50,51]. This method was applied in several fields, for example, electronic and electrical control. In [52], there are two types of STA which are as follows: super twisting extended state observer and super twisting SMC. These two methods were applied to wind turbines to compare them. In [53], the characteristic of DFOC was improved by the STA method. Experimental results showed the characteristics of the STA technique in improving the response time of a six-phase motor. STA technique was proposed to control the DC-link of the NPC inverter [54]. Internal permanent-magnet synchronous motors (IPMSMs) were controlled by an adaptive STA observer [55]. In [56], the author proposes to use the STA technique to command the output current of the multilevel converter. Permanent-magnet synchronous motors (PMSMs) were controlled by two different types of nonlinear controllers (STA and SMC) [57]. Experimental results showed that the STA improved the motor properties compared to that of SMC. In [58], the author proposes to use the STA technique to control the photovoltaic system. The STA technique has a very high efficacy compared to that of the classical method, as seen in the results of [58]. Synchronous motors (SMs) were controlled by the STA method [59]. STA observer with adaptive parameters estimation was proposed to command the IPMSMs [60]. In [61], the author proposes to use the STA technique to control the DC motor. The numerical simulation results showed that the characteristics of the STAR technique improve the performance and efficiency of the DC motor.

In this section, a new technique to reduce the chattering phenomena and minimizes the ripples of active power and current for ASG-CRWP was designed. The designed technique, named third-order sliding mode (TOSM), is an effective method for uncertain systems to overcome the main drawbacks of the classical SMC technique and STA described in the literature. The proposed strategy is a nonlinear method based on the STA method. The command input of the designed TOSM controller comprises three inputs given by Equation (9):

$$u(t) = u_1(t) + u_2(t) + u_3(t) \tag{9}$$

$$u_1(t) = \lambda_1 \cdot \sqrt{|S|} \cdot Sign(S) \tag{10}$$

$$u_2(t) = \lambda_2 \cdot \int Sign(S)dt \tag{11}$$

$$u_3(t) = \lambda_3 \cdot Sign(S) \tag{12}$$

Equation (13) shows the output of the proposed TOSM method:

$$u(t) = \lambda_1 \cdot \sqrt{|S|} \cdot Sign(S) + \lambda_2 \cdot \int Sign(S)dt + \lambda_3 \cdot Sign(S) \tag{13}$$

The tuning constants λ_1, λ_2, and λ_3 are used to refine the TOSM strategy for controller smoothing.

This suggested method was used to improve the performance of the DFOC method. Figure 1 shows the structure of the proposed TOSM controller.

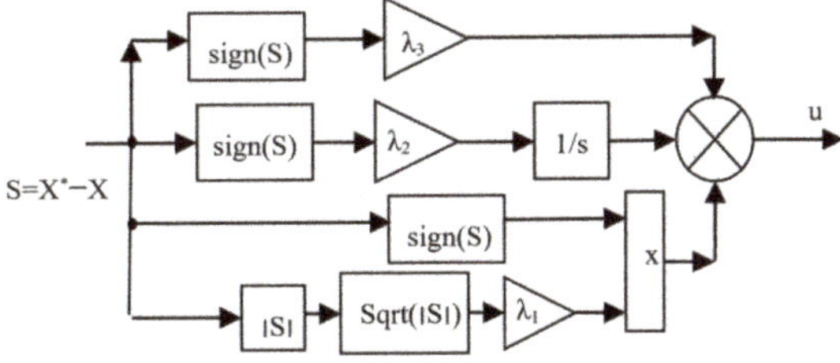

Figure 1. Structure of proposed TOSM controller.

The stability condition is given by:

$$S \cdot \dot{S} < 0 \tag{14}$$

This TOSM technique is designed in this work to minimize the ripple of stator current, active power, electromagnetic torque, and reactive power in an ASG-based CRWP system using the DFOC strategy and an inverter controlled by the classical PWM technique.

4. DFOC-TOSM Control

In the literature, the FOC technique was widely used for AC machines. The principle of this control is to orient the stator flux along the axis of the rotating frame. Two types of FOC strategies were studied in literature: direct and indirect FOC control. Traditional FOC control was proposed to control the induction generator [62]. In [63], the indirect FOC (IFOC) technique improved the performance of the induction generator compared with that of the DFOC technique. In [64], the FOC strategy was designed based on the neuro-fuzzy (NF) controllers to minimize the current and flux of the ASG. The experimental result shows the superiority of the designed FOC-NF technique. In [65], the FOC strategy was designed based on SVM technique and hysteresis current to minimize the reactive power ripples of ASG-based traditional wind power.

In this section, we propose to ameliorate the performances of the DFOC strategy of the asynchronous generator integrated into the CRWP system. This method is based on the following principle [64]:

$$\psi_{ds} = 0 \text{ and } \psi_{qs} = \psi_s \tag{15}$$

Hence, direct stator voltage and quadrature stator voltage can be written as:

$$\begin{cases} V_{qs} = V_s = w_s \cdot \psi_s \\ \quad\ V_{ds} = 0 \end{cases} \tag{16}$$

Equation (17) expresses each of direct stator current and quadrature stator current of the induction generator.

$$\begin{cases} I_{qs} = -\frac{M}{L_s} \cdot I_{qr} \\ I_{ds} = -\frac{M}{L_s} \cdot I_{dr} + \frac{\psi_s}{L_s} \end{cases} \tag{17}$$

The active and reactive power are obtained from Equation (18).

$$\begin{cases} Q_s = -1.5 \left(\frac{\omega_s \cdot \psi_s \cdot M}{L_s} \cdot I_{dr} - \frac{\omega_s \cdot \psi_s^2}{L_s} \right) \\ P_s = -1.5 \frac{\omega_s \cdot \psi_s \cdot M}{L_s} \cdot I_{qr} \end{cases} \tag{18}$$

Equation (19) represents the torque.

$$T_{em} = -1.5 \, I_{qr} \cdot \psi_{ds} \cdot p \cdot \frac{M}{L_s} \tag{19}$$

Equation (20) expresses each of direct rotor voltage and quadrature rotor voltage of the induction generator.

$$\begin{cases} V_{qr} = R_{dr} \cdot I_{qr} + \left(L_r - \frac{M^2}{L_s} \right) p \cdot I_{qr} - g \cdot w_s \left(L_r - \frac{M^2}{L_s} \right) I_{dr} + g \cdot \frac{M \cdot V_s}{L_s} \\ V_{dr} = R_{dr} \cdot I_{dr} + \left(L_r - \frac{M^2}{L_s} \right) \cdot p \cdot I_{dr} - g \cdot w_s \cdot \left(L_r - \frac{M^2}{L_s} \right) \cdot I_{qr} \end{cases} \tag{20}$$

Equation (21) represents direct and quadrature rotor current of the ASG.

$$\begin{cases} I_{qr} = \left(V_{qr} - g \cdot w_s \cdot \left(L_r - \frac{M^2}{L_s} \right) \cdot I_{qr} - g \cdot \frac{M \cdot V_s}{L_s} \right) \cdot \frac{1}{R_r + \left(L_r - \frac{M^2}{L_s} \right) \cdot p} \\ I_{dr} = \left(V_{dr} - g \cdot w_s \left(L_r - \frac{M^2}{L_s} \right) \cdot I_{qr} \right) \cdot \frac{1}{R_r + \left(L_r - \frac{M^2}{L_s} \right) \cdot p} \end{cases} \tag{21}$$

The DFOC method is a very simple and easy to implement method, and it can be applied to any electric machine. In addition, it is inexpensive compared to that of other methods. The principle of this method can be expressed as in Figure 2. From this figure, it is observed that two independent PI controllers to control the direct and quadrature rotor voltages are used.

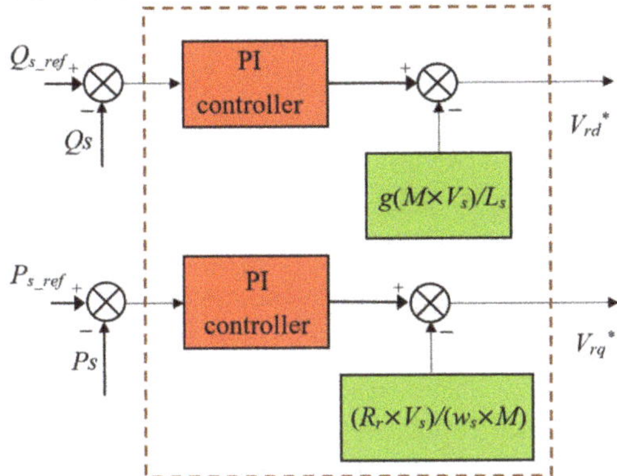

Figure 2. Classical DFOC strategy.

This method is simple but provides greater ripples for both current and active power. Also, the response time is rather large compared, for example, with direct power command (DPC) and direct torque command (DTC), and this is due to the use of PI controllers.

In this paper, we propose to ameliorate the characteristics and effectiveness of this method by the proposed nonlinear TOSM method. Thus, by improving the quality of the current produced, the life of the generator and the system as a whole is extended as well.

The principle of the TOSM-DFOC strategy is the direct regulation of the current, active power, voltage, torque, and reactive power of the ASG-based CRWP system by using two TOSM techniques. The two regulated variables are reactive and active powers, which are usually controlled by TOSM controllers. The idea is to keep the active and reactive power quantities within the designed sliding surfaces.

The DFOC with TOSM controllers (TOSM-DFOC) is a modification of the traditional DFOC strategy, where the PI regulators were replaced by TOSM controllers, as shown in Figure 3.

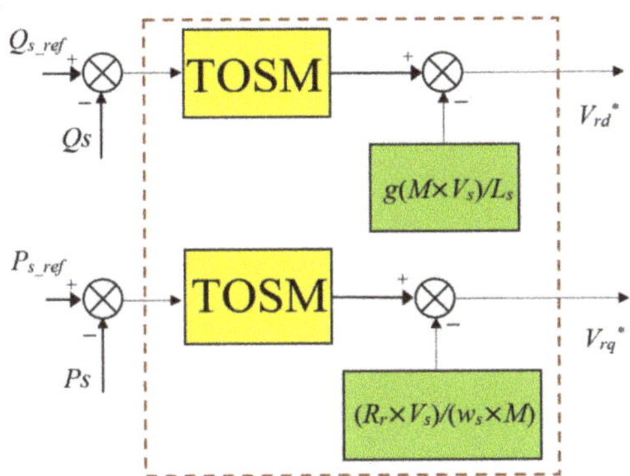

Figure 3. Structure of DFOC-TOSM strategy.

The TOSM technique was designed to force the regulated dynamics towards manifolds and keeps them there. To force the ASG active and reactive powers to track their corresponding references, the sliding surfaces S_{Qs} and S_{Ps} of the reactive and active powers are selected as the error between the desired and real dynamics, being given by Equations (22) and (23), respectively:

$$S_{Ps} = P_s^* - P_s \tag{22}$$

$$S_{Qs} = Q_s^* - Q_s \tag{23}$$

The sliding surfaces shown in Equations (22) and (23) are used as inputs for the TOSM command law. Thus, TOSM regulators for the active and reactive power are used to influence the two rotor voltage components, as in Equations (24) and (25), respectively:

$$V_{dr}^* = \lambda_1 \cdot \sqrt{|S_{Qs}|} \cdot Sign(S_{Qs}) + \lambda_2 \cdot \int Sign(S_{Qs})dt + \lambda_3 \cdot Sign(S_{Qs}) \tag{24}$$

$$V_{qr}^* = \lambda_1 \cdot \sqrt{|S_{Ps}|} \cdot Sign(S_{Ps}) + \lambda_2 \cdot \int Sign(S_{Ps})dt + \lambda_3 \cdot Sign(S_{Ps}) \tag{25}$$

This proposed technique is implemented for a DFOC strategy based on the TOSM controllers to obtain minimum active power undulations and to minimize the chattering phenomenon. The controller structure of the TOSM technique for the reactive and active powers of the DFOC strategy is presented in Figures 4 and 5, respectively.

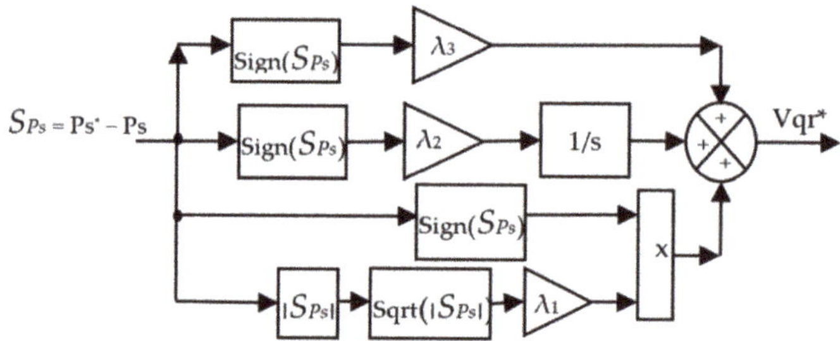

Figure 4. Proposed TOSM active power controller.

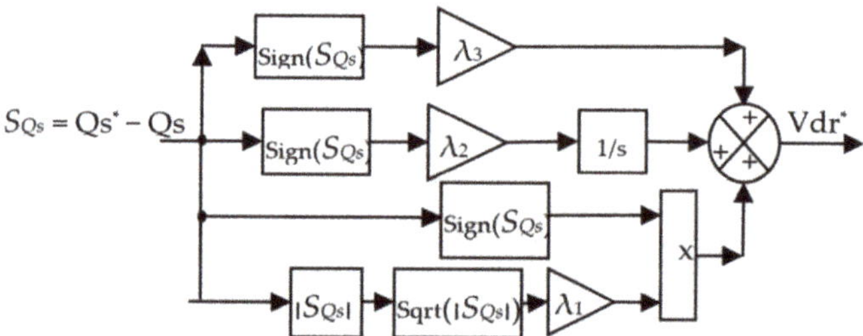

Figure 5. Proposed TOSM reactive power controller.

The basic block diagram of the DFOC-TOSM strategy is shown in Figure 6. This strategy is designed as a simple algorithm that provides a robust command. However, this proposed strategy gives a fast response dynamic compared to DFOC, DPC, and DTC. This designed technique can better minimize the active and reactive power undulations compared to that of classic DPC, DFOC, and DTC strategies.

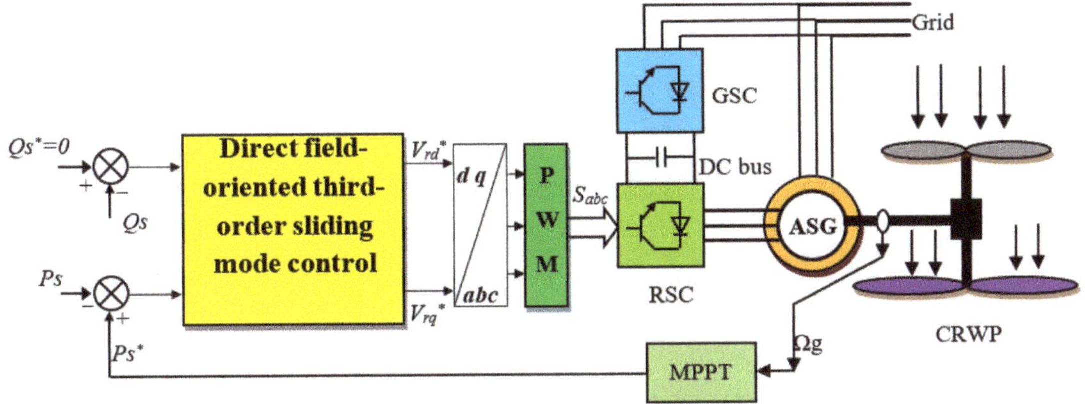

Figure 6. DFOC-TOSM control of ASG-CRWP.

The proposed method is one of the best methods available today, and this is due to the use of TOSM. The following table (Table 1) shows a comparative study between the proposed method and some of most famous and widely used methods, such as DTC, DPC, and FOC control scheme. From Table 1, we note that the proposed method is more robust than the rest of the methods in terms of improving the dynamic response and THD value. But this method is rather complicated compared to that of the classic DTC and DPC.

Table 1. A comparative study between DFOC-TOSM with some techniques.

Control Techniques	Controller	Complexity	Current Oscillations	Reference Tracking	Dynamic Responses	Sensitivity to Parameter Change	THD (%) of Current
DTC control	Hysteresis controller	Low	High	Good	Good	High	High
DPC control	Hysteresis controller	Low	High	Good	Good	High	High
FOC control	PI	High	High	Acceptable	Acceptable	High	High
Proposed technique (DFOC-TOSMC)	TOSM controller	High	Low	Excellent	Excellent	Medium	Low

5. Numerical Simulation

The proposed TOSM controller as applied for the ASG reactive and active powers was implemented using MATLAB® software. Comparison between the proposed method and the classical PI controller in terms of current, active power, torque and reactive powers undulations reduction, wind speed changing, trajectory tracking, and robustness to ASG parameter variations.

The generator used in the digital simulation has the following characteristics: f = 50 Hz, p = 2, Rs = 0.012 Ω, Lr = 0.0136 H, Pn = 1.5 MW, Rr = 0.021 Ω, Ls = 0.0137 H, Vn = 380 V, J = 1000 K×gm², M = 0.0135 H, F = 0.0024 N×m.

The grid used in the simulation has the following parameters: nominal grid voltage (Vg = 389 V), nominal grid frequency (fs = 50 Hz), DC-link voltage (E = 1200 V), filter inductance (Lg = 6 mH), and filter resistance (Rg = 0.15 Ω).

DFOC with PI controllers (DFOC-PI) and DFOC-TOSM are simulated and compared in terms of current, reactive power, torque and active power undulations, reference tracking, and THD value of current.

To test the effectiveness and robustness of the proposed method, we proposed two different tests: the first is related to tracking the trajectory and the second to assess the robustness to changes in some parameters of the machine.

5.1. First Test

The results obtained from the simulation for the two methods are shown in Figures 7–12. In the case of each of the two proposed methods, the measured values of reactive and active power follow the reference values very well (see Figures 9 and 10). On the other hand, for the two techniques, the active and reactive powers perfectly track their references values, for which the TOSM controller introduces better performances in terms of settling time, overshoot damping, and steady-state error decreasing. In Figures 9 and 10, both controllers introduce a good tracking performance with a remarkable superiority of the proposed TOSM controller in terms of rise time, settling time, and steady-state error, which all decrease. The electric current has the form of active power and is related to the wind speed, as seen in Figure 12.

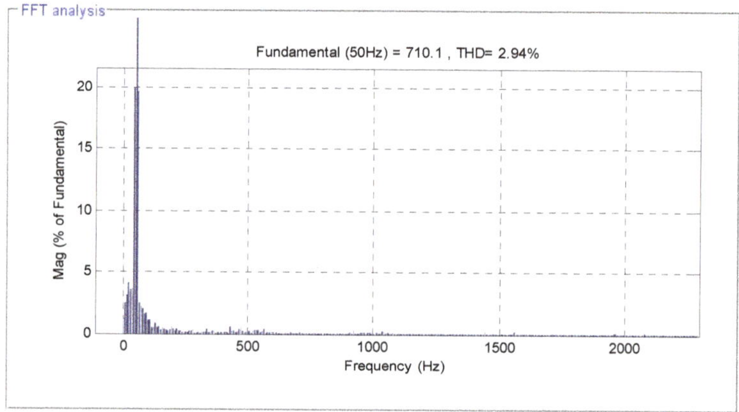

Figure 7. THD (DFOC-PI).

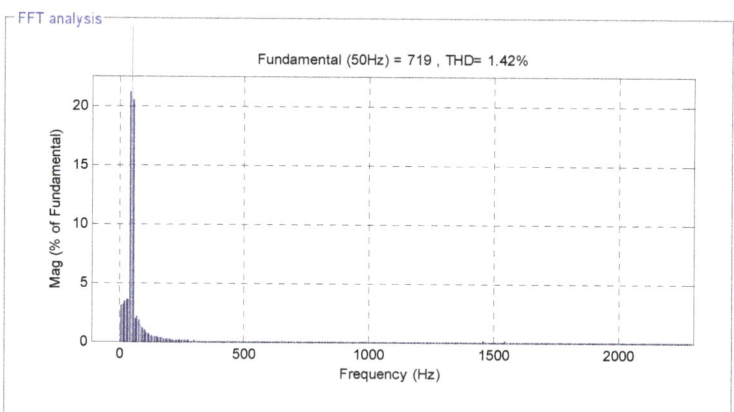

Figure 8. THD (DFOC-TOSM).

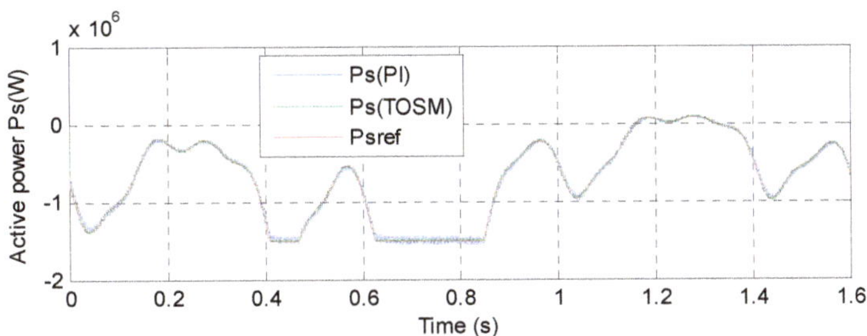

Figure 9. Active power.

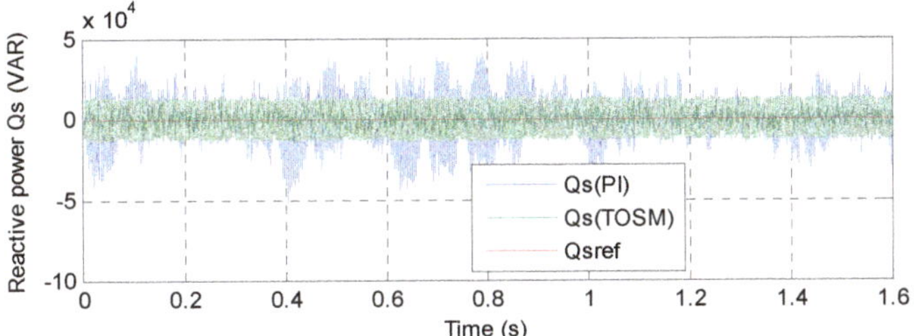

Figure 10. Reactive power.

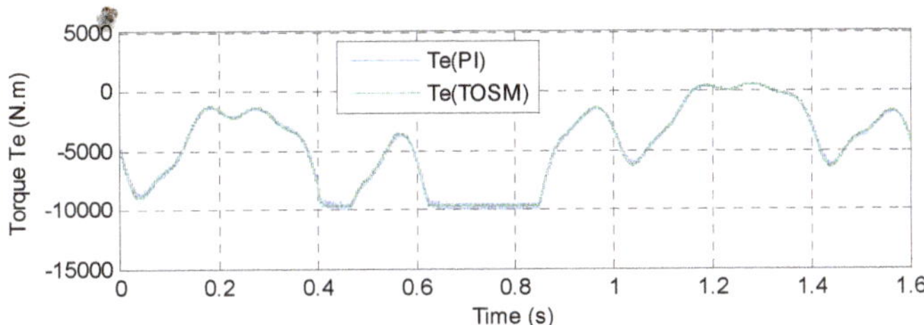

Figure 11. Torque.

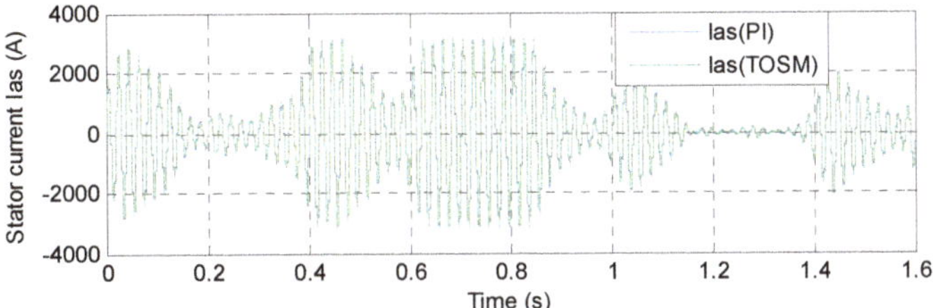

Figure 12. Current.

The torque resulting from the two proposed methods is shown in Figure 11. From this figure, we find that the value of the torque is related to the values of active power and wind speed. To further compare the characteristics of these two techniques, the stator currents harmonic spectra are compared.

Figures 7 and 8 show the THD of stator current of the AG-CRWP system for both DFOC strategies. Here, the THD value is reduced for the DFOC-TOSM (1.42%) when compared to that of the DFOC-PI (2.94%). The THD reduction can be estimated at 51.71%.

The zoom in the current, active power, and torque is shown in Figures 13–15, respectively. The DFOC-TOSM strategy reduces the undulations in torque, reactive power, current, and active power compared to that of the DFOC-PI strategy, and the reduction rate is estimated at 84%, 94.5%, 78%, and 90%, respectively, compared to that of the classical method (DARPC). Based on the results above, the DFOC-TOSM strategy proved its efficiency in minimizing undulations and chattering phenomena.

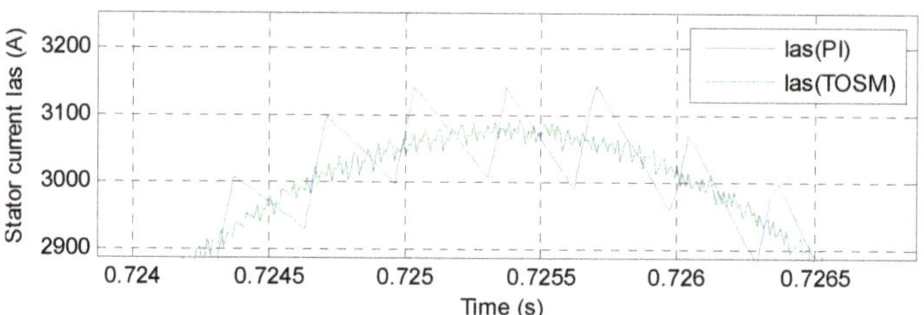

Figure 13. Zoom (Current).

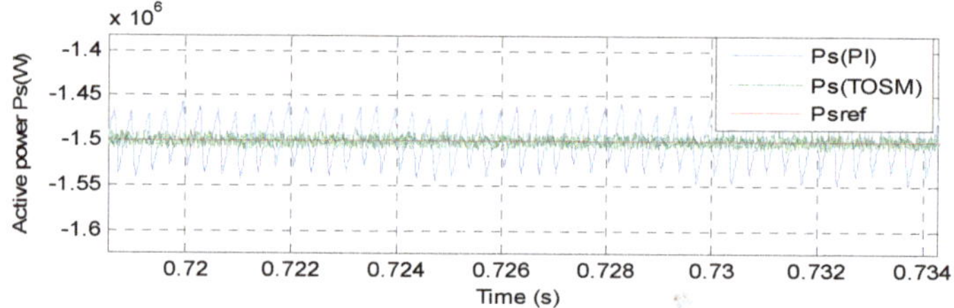

Figure 14. Zoom (Active power).

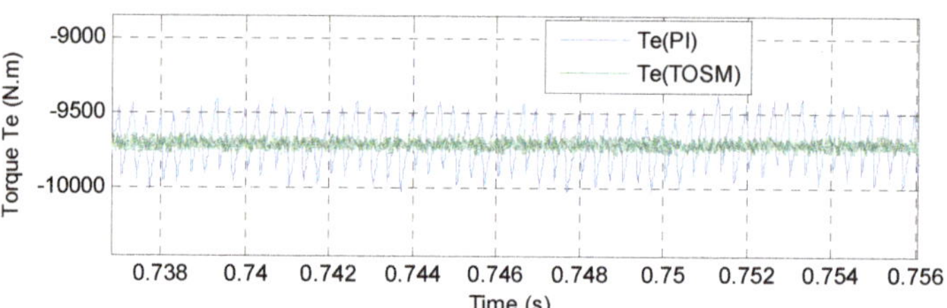

Figure 15. Zoom (Torque).

5.2. Second Test

The aim of this test is to find out which method is not affected by changing the following parameters: Rs, Rr, Ls, M, and Lr. Figures 16–21 show the simulation results for variations in inductance and resistance values. Note that there is a change in reactive power, torque, active power, and current because both active power and torque are related to changing values of parameters. The classical technique was heavily affected by the change of parameter values compared to that of the designed technique (Figures 22–24), and this is evident in the value of THD (Figures 16 and 17). The ripple reduction can be estimated at 73.94%. Parameter changes have a clear effect on the dynamic performances of active power, current, torque, and stator reactive power when using the PI controller. In addition, the decoupling between the reactive and active powers is not ensured in this case. Instead, the designed DFOC based TOSM controller is robust against parameter variations, and decoupling is perfectly ensured.

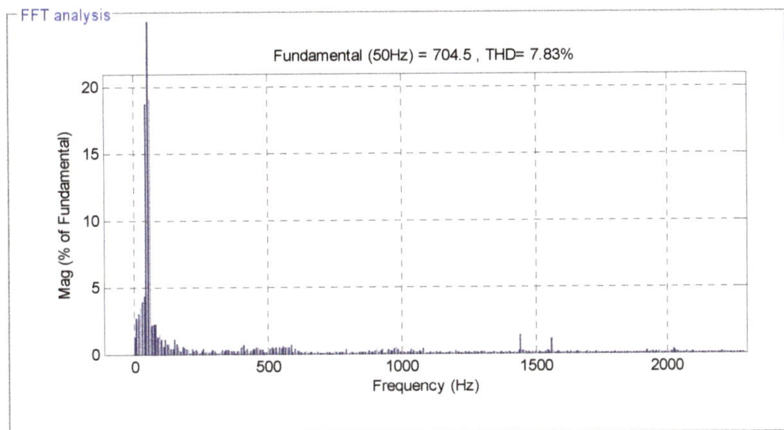

Figure 16. THD (PI-DFOC).

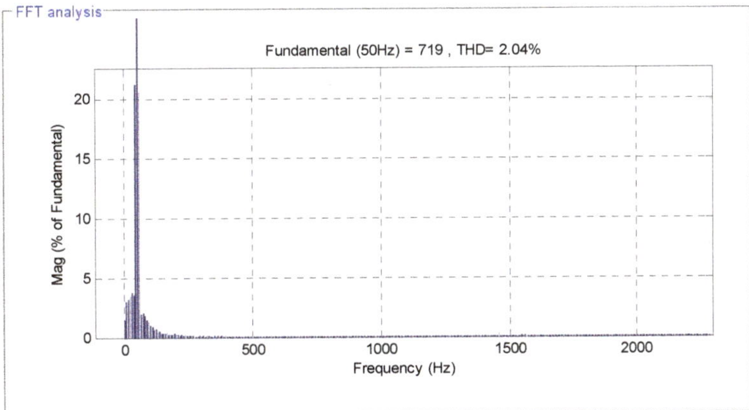

Figure 17. THD (TOSM-DFOC).

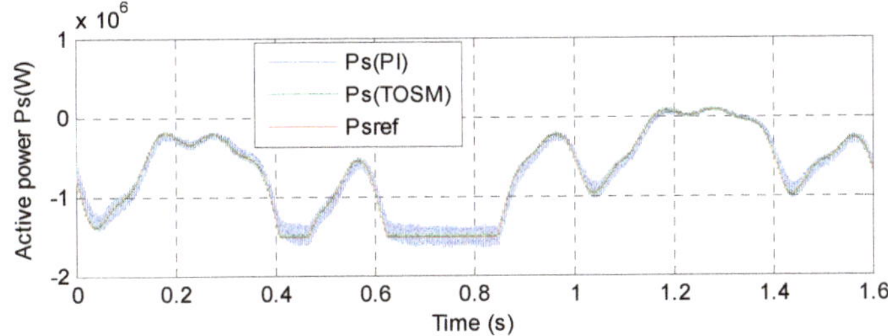

Figure 18. Active power.

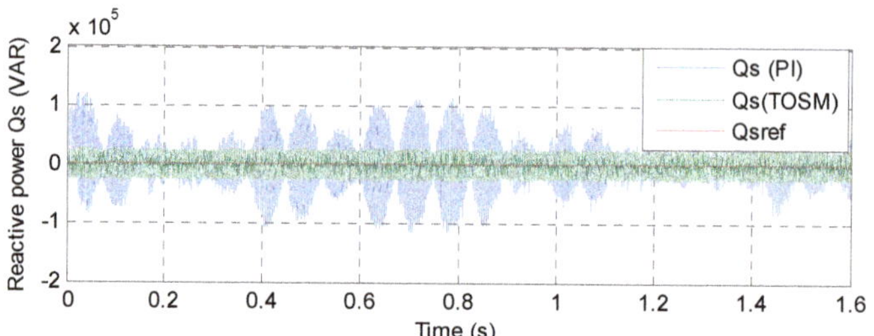

Figure 19. Reactive power.

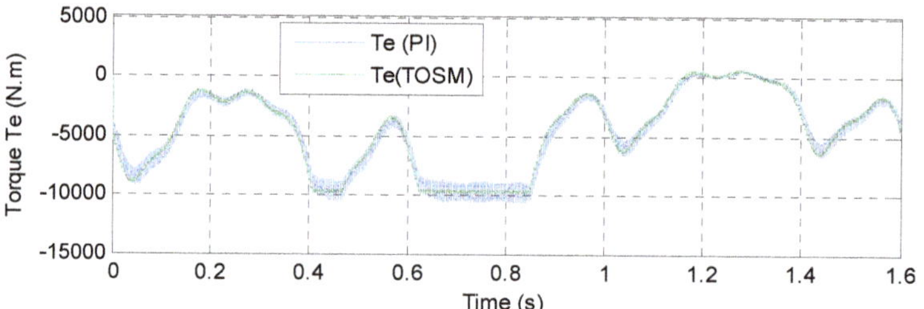

Figure 20. Torque.

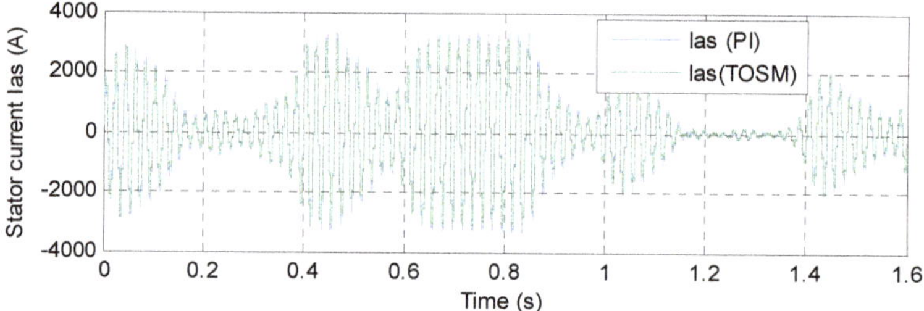

Figure 21. Current.

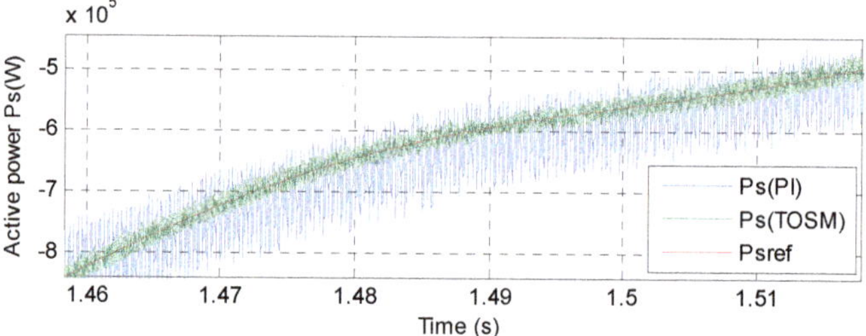

Figure 22. Zoom (Active power).

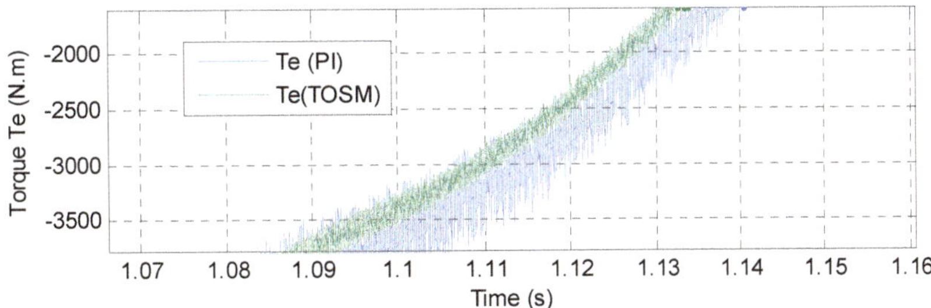

Figure 23. Zoom (Torque).

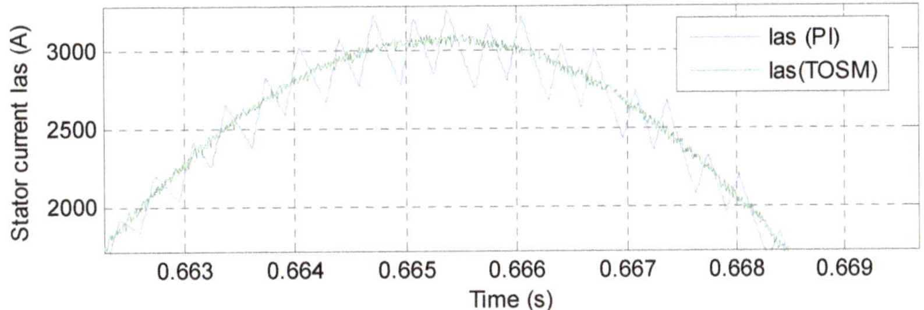

Figure 24. Zoom (Current).

Thus, the DFOC with designed TOSM algorithms is more robust than the classical DFOC technique.

The obtained results can be summarized in Table 2. Table 2 presents a comparative study between the proposed method and the classical method in all respects. Through this table, we note that the proposed method gave very satisfactory results in terms of reducing the torque and flux fluctuations. Also, the proposed method improved the dynamic response compared to that of the classical method. On the other hand, the proposed method provided better results in terms of current, reactive power, torque and active power fluctuations, reactive and active power tracking, and the quality of the produced current. The proposed DFOC-TOSM technique is more robust than that of the traditional DFOC using PI controllers except for the dynamic response, rise time, overshoot, and settling time, which is faster in the proposed technique than that of the DFOC technique. However, the proposed technique is complicated relative to the calculations to be performed compared to that of the DFOC control technique.

A comparative study between the proposed method in this paper with some published works in terms of the value of THD is presented in Table 3. Through this table, the proposed method provided a very good THD value of the current compared to that of the rest of the other controls. Tables 2 and 3 show that the proposed method is more robust and can be used to improve the quality of current and energy produced from the wind power generation system.

Table 2. Comparative results obtained from DFOC-TOSM with DFOC control.

Criteria	Control Techniques	
	DFOC	**DFOC-TOSM**
Reactive and active power tracking	Well	Excellent
THD (%)	2.94	1.42
Reactive and active power ripples	Acceptable	Excellent
Dynamic response (s)	Medium	Fast
Reactive power: ripples (VAR)	Around 20,000	Around 1100
Torque: ripple (N.m)	Around 500	Around 80
Settling time (ms)	High	Medium
Overshoot (%)	Remarkable $\approx$ 19%	Neglected $\approx$ 1.5%
Sensitivity to parameter change	High	Medium
Active power: ripples (W)	Around 10,000	Around 1000
Simplicity of converter and filter design	Simple	Simple
Rise Time (s)	High	Medium
Stator current ripple (VAR)	Around 100	Around 22
Simplicity of calculations	Simple	Rather complicated
Improvement of transient performance	Good	Excellent
Quality of stator current	Acceptable	Excellent

Table 3. Comparative results with other methods.

References	Techniques	THD (%)
Ref. [6]	Fuzzy SMC technique	3.05
Ref. [38]	Direct torque command	2.95
Ref. [64]	FOC with hysteresis current controller	3.70
Ref. [66]	Virtual flux DPC method	4.19
	DPC	4.88
Ref. [67]	Integral SMC technique	9.71
	Multi-resonant-based sliding mode controller (MRSMC)	3.14
Ref. [68]	SOSMC method	3.13
Ref. [69]	DPC control with STA controller	1.66
Designed strategies	DFOC-PI	2.94
	DFOC-TOSM	1.42

6. Conclusions

In this study, a new TOSM strategy was designed to command the generated reactive and active power from the ASG based on the variable speed CRWP. The designed strategy aims to ameliorate the command characteristics of the algorithm that are based on the proposed TOSM methods by minimizing active power, torque, current, and reactive power undulations under variable speed CRWP system.

The proposed TOSM command was used to define the attractive command section of the classic SMC strategy.

The designed strategy was compared with that of the classical PI controller. The obtained results illustrated the effectiveness of the designed TOSM strategy, even in the presence of time-varying reference trajectory, ASG parameter variations, and changes in speed wind. Current, active power, torque, and reactive power undulations were largely minimized, and response time was improved using the designed TOSM strategy. Moreover, robustness, stability, and high-decoupling between the command axes were ensured.

The simulation results showed that the introduced TOSM-based DFOC control scheme is high performance and has a robustness to uncertainties and parameter mismatches as well as the attenuation of chattering phenomena in comparison with that of the DFOC control scheme-based conventional PI controllers. The designed control technique improved the active power ripple, current ripple, torque ripple, reactive power ripple, and THD value of stator current by about 90%, 78%, 84%, 94%, and 51.71%, respectively, compared to that of the classic DFOC strategy. Finally, the designed TOSM strategy may be a good solution to improve the DFOC strategy effectiveness applied for power systems, ensuring high quality of active power and stator current.

To summarize, the main findings of this research are as follows:

- A novel nonlinear controller was presented and confirmed with numerical simulation.
- A new DFOC strategy based on the proposed TOSM algorithm was presented and confirmed with numerical simulation.
- A robust command technique was designed to minimize the ripples of current, active power, electromagnetic torque, and reactive power.
- The THD value of stator current was reduced by the proposed strategy.

In future work, to improve the quality of the voltage, torque, current, and active power, the asynchronous generator will be controlled using another combination of intelligent algorithms, such as type-two fuzzy logic and neural algorithm used for DFIG in [34]. The ratio between the obtained performance and the complexity of the control will be evaluated.

Author Contributions: Validation, N.B.; conceptualization, H.B.; software, H.B.; methodology, H.B.; investigation, H.B.; resources, H.B. and N.B.; project administration, N.B.; data curation, N.B.; writing—original draft preparation, H.B.; supervision, N.B.; visualization, N.B.; formal analysis, N.B.; funding acquisition, N.B.; writing—review and editing, N.B. and H.B. All authors have read and agreed to the published version of the manuscript.

Funding: This research received no external funding.

Institutional Review Board Statement: Not applicable.

Informed Consent Statement: Not applicable.

Data Availability Statement: Not applicable.

Conflicts of Interest: The authors declare no conflict of interest.

References

1. Xiong, L.; Li, P.; Li, H.; Wang, J. Sliding Mode Control of DFIG Wind Turbines with a Fast Exponential Reaching Law. *Energies* **2017**, *10*, 1788. [CrossRef]
2. Levant, A. Robust exact differentiation via sliding mode technique. *Automatica* **1998**, *34*, 379–384. [CrossRef]
3. Rezaei, M.M. A nonlinear maximum power point tracking technique for DFIG-based wind energy conversion systems. *Eng. Sci. Technol. Int. J.* **2018**, *21*, 901–908. [CrossRef]
4. Hu, J.; Nian, H.; Hu, B.; He, Y.; Zhu, Z.Q. Direct active and reactive power regulation of DFIG using sliding-mode control approach. *IEEE Trans. Energy Convers.* **2010**, *25*, 1028–1039. [CrossRef]
5. Benbouhenni, H. Sliding mode with neural network regulateur for DFIG using two-level NPWM strategy. *Iran. J. Electr. Electron. Eng.* **2019**, *15*, 411–419.
6. Boudjema, Z.; Meroufel, A.; Djerriri, Y.; Bounadja, E. Fuzzy sliding mode control of a doubly fed induction generator for energy conversion. *Carpathian J. Electron. Comput. Eng.* **2013**, *6*, 7–14.
7. Habib, B. ANFIS-sliding mode control of a DFIG supplied by a two-level SVPWM technique for wind energy conversion system. *Int. J. Appl. Power Eng.* **2020**, *9*, 36–47. [CrossRef]
8. Abdolhadi, H.Z.; Markadeh, G.A.; Boroujeni, S.T. Sliding mode and terminal sliding mode control of cascaded doubly fed induction generator. *Iran. J. Electr. Electron. Eng.* **2021**, *17*, 1955.
9. Xiao, L.; Zhang, L.; Gao, F.; Qian, J. Robust Fault-Tolerant Synergetic Control for Dual Three-Phase PMSM Drives Considering Speed Sensor Fault. *IEEE Access* **2020**, *8*, 78912–78922. [CrossRef]
10. Hamid, C.; Aziz, D.; Seif Eddine, C.; Othmane, Z.; Mohammed, T.; Hasnae, E. Integral sliding mode control for DFIG based WECS with MPPT based on artificial neural network under a real wind profile. *Energy Rep.* **2021**, *7*, 4809–4824. [CrossRef]

11. Du, H.; Chen, X.; Wen, G.; Yu, X.; Lü, J. Discrete-Time Fast Terminal Sliding Mode Control for Permanent Magnet Linear Motor. *IEEE Trans. Ind. Electron.* **2018**, *65*, 9916–9927. [CrossRef]
12. Mobayen, S.; Bayat, F.; Lai, C.-C.; Taheri, A.; Fekih, A. Adaptive Global Sliding Mode Controller Design for Perturbed DC-DC Buck Converters. *Energies* **2021**, *14*, 1249. [CrossRef]
13. Shehata, E.G. Sliding mode direct power control of RSC for DFIGs driven by variable speed wind turbines. *Alex. Eng. J.* **2015**, *54*, 1067–1075. [CrossRef]
14. Haibo, L.; Heping, W.; Junlei, S. Attitude control for QTR using exponential nonsingular terminal sliding mode control. *J. Syst. Eng. Electron.* **2019**, *30*, 191–200. [CrossRef]
15. Wang, Z.; Li, S.; Li, Q. Discrete-Time Fast Terminal Sliding Mode Control Design for DC–DC Buck Converters with Mismatched Disturbances. *IEEE Trans. Ind. Inform.* **2020**, *16*, 1204–1213. [CrossRef]
16. Lin, X.; Shi, X.; Li, S. Adaptive Tracking Control for Spacecraft Formation Flying System via Modified Fast Integral Terminal Sliding Mode Surface. *IEEE Access* **2020**, *8*, 198357–198367. [CrossRef]
17. Yu, X.; Man, Z. Model reference adaptive control systems with terminal sliding modes. *Int. J. Control.* **1996**, *64*, 1165–1176. [CrossRef]
18. Feng, Y.; Yu, X.; Han, F. On nonsingular terminal sliding-mode control of nonlinear systems. *Automatica* **2013**, *49*, 1715–1722. [CrossRef]
19. Morshed, M.J.; Fekih, A. A terminal sliding mode approach for the rotor side converter of a DFIG-based wind energy system. In Proceedings of the 2018 IEEE Conference on Control Technology and Applications (CCTA), Copenhagen, Denmark, 21–25 April 2018; pp. 1736–1740. [CrossRef]
20. Sami, I.; Ullah, S.; Ali, Z.; Ullah, N.; Ro, J.-S. A Super Twisting Fractional Order Terminal Sliding Mode Control for DFIG-Based Wind Energy Conversion System. *Energies* **2020**, *13*, 2158. [CrossRef]
21. Morshed, M.J.; Fekih, A. Second order integral terminal sliding mode control for voltage sag mitigation in DFIG-based wind turbines. In Proceedings of the 2017 IEEE Conference on Control Technology and Applications (CCTA), Mauna Lani, HI, USA, 27–30 August 2017; pp. 614–619. [CrossRef]
22. Levant, A. Sliding order and sliding accuracy in sliding mode control. *Int. J. Control.* **1993**, *58*, 1247–1263. [CrossRef]
23. Levant, A. Higher-order sliding modes, differentiation and output-feedback control. *Int. J. Control.* **2003**, *76*, 924–941. [CrossRef]
24. Shah, A.P.; Mehta, A.J. Direct power control of DFIG using super-twisting algorithm based on second-order sliding mode control. In Proceedings of the 2016 14th International Workshop on Variable Structure Systems (VSS), Nanjing, China, 1–4 June 2016; pp. 136–141. Available online: https://ieeexplore.ieee.org/document/7506905 (accessed on 1 September 2021). [CrossRef]
25. Alhato, M.M.; Ibrahim, M.N.; Rezk, H.; Bouallègue, S. An Enhanced DC-Link Voltage Response for Wind-Driven Doubly Fed Induction Generator Using Adaptive Fuzzy Extended State Observer and Sliding Mode Control. *Mathematics* **2021**, *9*, 963. [CrossRef]
26. Habib, B.; Boudjema, Z.; Belaidi, A. Direct power control with NSTSM algorithm for DFIG using SVPWM technique. *Iran. J. Electr. Electron. Eng.* **2021**, *17*, 1–11.
27. Habib, B.; Boudjema, Z.; Belaidi, A. DPC based on ANFIS super-twisting sliding mode algorithm of a doubly-fed induction generator for wind energy system. *J. Eur. Des. Systèmes Autom.* **2020**, *53*, 69–80.
28. Farid, B.; Tarek, B.; Sebti, B. Fuzzy super twisting algorithm dual direct torque control of doubly fed induction machine. *Int. J. Electr. Comput. Eng.* **2021**, *11*, 3782–3790. [CrossRef]
29. Mazen Alhato, M.; Bouallègue, S.; Rezk, H. Modeling and Performance Improvement of Direct Power Control of Doubly-Fed Induction Generator Based Wind Turbine through Second-Order Sliding Mode Control Approach. *Mathematics* **2020**, *8*, 2012. [CrossRef]
30. Nasiri, M.; Mobayen, S.; Zhu, Q.M. Super-twisting sliding mode control for gear less PMSG-based wind turbine. *Complexity* **2019**, *2019*, 1–15. [CrossRef]
31. Venu, G.; Kalyani, S.T. Design of fractional order based super twisting algorithm for BLDC motor. In Proceedings of the 2019 3rd International Conference on Trends in Electronics and Informatics (ICOEI), Tirunelveli, India, 23–25 April 2019; pp. 271–277. [CrossRef]
32. Rakhtala, S.M.; Casavola, A. Real time voltage control based on a cascaded super twisting algorithm structure for DC-DC converters. *IEEE Trans. Ind. Electron.* **2021**, 1. [CrossRef]
33. Sriprang, S.; Nahid-Mobarakeh, B.; Takorabet, N.; Pierfederici, S.; Bizon, N.; Kuman, P.; Thounthong, P. Permanent Magnet Synchronous Motor Dynamic Modeling with State Observer-based Parameter Estimation for AC Servomotor Drive Application. *Appl. Sci. Eng. Prog.* **2019**, *12*, 286–297. [CrossRef]
34. Benbouhenni, H. Intelligent super twisting high order sliding mode controller of dual-rotor wind power systems with direct attack based on doubly-fed induction generators. *J. Electr. Eng. Electron. Control. Comput. Sci.* **2021**, *7*, 1–8. Available online: https://jeeeccs.net/index.php/journal/article/view/219. (accessed on 1 September 2021).
35. Shah, A.P.; Mehta, A.J. Direct power control of grid-connected DFIG using variable gain super-twisting sliding mode controller for wind energy optimization. In Proceedings of the IECON 2017—43rd Annual Conference of the IEEE Industrial Electronics Society, Beijing, China, 5–8 November 2017; pp. 2448–2454. [CrossRef]
36. Kelkoul, B.; Boumediene, A. Stability analysis and study between classical sliding mode control (SMC) and super twisting algorithm (STA) for doubly fed induction generator (DFIG) under wind turbine. *Energy* **2021**, *214*, 118871. [CrossRef]

37. Bouyekni, A.; Taleb, R.; Boudjema, Z.; Kahal, H. A second-order continuous sliding mode based on DFIG for wind-turbine-driven DFIG. *Elektrotehniški Vestn.* **2018**, *85*, 29–36.
38. Boudjema, Z.; Taleb, R.; Djerriri, Y.; Yahdou, A. A novel direct torque control using second order continuous sliding mode of a doubly fed induction generator for a wind energy conversion system. *Turk. J. Electr. Eng. Comput. Sci.* **2017**, *25*, 965–975. [CrossRef]
39. Benbouhenni, H.; Bizon, N. A Synergetic Sliding Mode Controller Applied to Direct Field-Oriented Control of Induction Generator-Based Variable Speed Dual-Rotor Wind Turbines. *Energies* **2021**, *14*, 4437. [CrossRef]
40. Benbouhenni, H.; Bizon, N. Terminal Synergetic Control for Direct Active and Reactive Powers in Asynchronous Generator-Based Dual-Rotor Wind Power Systems. *Electronics* **2021**, *10*, 1880. [CrossRef]
41. Astolfi, D. Wind Turbine Operation Curves Modelling Techniques. *Electronics* **2021**, *10*, 269. [CrossRef]
42. Beik, O.; Al-Adsani, A.S. Active and Passive Control of a Dual Rotor Wind Turbine Generator for DC Grids. *IEEE Access* **2021**, *9*, 1987–1995. [CrossRef]
43. Luo, X.; Niu, S. A Novel Contra-Rotating Power Split Transmission System for Wind Power Generation and Its Dual MPPT Control Strategy. *IEEE Trans. Power Electron.* **2017**, *32*, 6924–6935. [CrossRef]
44. Zhao, Y.; Teng, D.; Li, D.; Zhao, X. Comparative Research on Four-Phase Dual Armature-Winding Wound-Field Doubly Salient Generator with Distributed Field Magnetomotive Forces for High-Reliability Application. *IEEE Access* **2021**, *9*, 12579–12591. [CrossRef]
45. Zhao, W.; Lipo, T.A.; Kwon, B. A Novel Dual-Rotor, Axial Field, Fault-Tolerant Flux-Switching Permanent Magnet Machine with High-Torque Performance. *IEEE Trans. Magn.* **2015**, *51*, 1–4. [CrossRef]
46. Fukami, T.; Momiyama, M.; Shima, K.; Hanaoka, R.; Takata, S. Steady-State Analysis of a Dual-Winding Reluctance Generator with a Multiple-Barrier Rotor. *IEEE Trans. Energy Convers.* **2008**, *23*, 492–498. [CrossRef]
47. Wang, Y.; Niu, S.; Fu, W. A Novel Dual-Rotor Bidirectional Flux-Modulation PM Generator for Stand-Alone DC Power Supply. *IEEE Trans. Ind. Electron.* **2019**, *66*, 818–828. [CrossRef]
48. Yahdou, A.; Djilali, A.B.; Boudjema, Z.; Mehedi, F. Improved vector control of a counter-rotating wind turbine system using adaptive backstepping sliding mode. *J. Eur. Des. Syst. Autom.* **2020**, *53*, 645–651.
49. Anthony, M.; Prasad, V.; Kannadasan, R.; Mekhilef, S.; Alsharif, M.H.; Kim, M.-K.; Jahid, A.; Aly, A.A. Autonomous Fuzzy Controller Design for the Utilization of Hybrid PV-Wind Energy Resources in Demand Side Management Environment. *Electronics* **2021**, *10*, 1618. [CrossRef]
50. Huangfu, Y.; Xu, J.; Zhao, D.; Liu, Y.; Gao, F. A Novel Battery State of Charge Estimation Method Based on a Super-Twisting Sliding Mode Observer. *Energies* **2018**, *11*, 1211. [CrossRef]
51. Ha, L.N.N.T.; Hong, S.K. Robust Dynamic Sliding Mode Control-Based PID–Super Twisting Algorithm and Disturbance Observer for Second-Order Nonlinear Systems: Application to UAVs. *Electronics* **2019**, *8*, 760. [CrossRef]
52. Shi, D.; Wu, Z.; Chou, W. Super-Twisting Extended State Observer and Sliding Mode Controller for Quad rotor UAV Attitude System in Presence of Wind Gust and Actuator Faults. *Electronics* **2018**, *7*, 128. [CrossRef]
53. Listwan, J. Application of super-twisting sliding mode controllers in direct field-oriented control system of six-phase induction motor: Experimental studies. *Power Electron. Drives* **2018**, *3*, 23–34. [CrossRef]
54. Shen, X.; Liu, J.; Marquez, A.; Luo, W.; Leon, J.I.; Vazquez, S.; Franquelo, L.G. A High-Gain Observer-Based Adaptive Super-Twisting Algorithm for DC-Link Voltage Control of NPC Converters. *Energies* **2020**, *13*, 1110. [CrossRef]
55. Chen, S.; Zhang, X.; Wu, X.; Tan, G.; Chen, X. Sensorless Control for IPMSM Based on Adaptive Super-Twisting Sliding-Mode Observer and Improved Phase-Locked Loop. *Energies* **2019**, *12*, 1225. [CrossRef]
56. Uddin, W.; Zeb, K.; Adil Khan, M.; Ishfaq, M.; Khan, I.; Islam, S.U.; Kim, H.-J.; Park, G.S.; Lee, C. Control of Output and Circulating Current of Modular Multilevel Converter Using a Sliding Mode Approach. *Energies* **2019**, *12*, 4084. [CrossRef]
57. Dendouga, A. Conventional and Second Order Sliding Mode Control of Permanent Magnet Synchronous Motor Fed by Direct Matrix Converter: Comparative Study. *Energies* **2020**, *13*, 5093. [CrossRef]
58. Zeb, K.; Busarello, T.D.C.; Ul Islam, S.; Uddin, W.; Raghavendra, K.V.G.; Khan, M.A.; Kim, H.-J. Design of Super Twisting Sliding Mode Controller for a Three-Phase Grid-connected Photovoltaic System under Normal and Abnormal Conditions. *Energies* **2020**, *13*, 3773. [CrossRef]
59. Kafi, M.R.; Hamida, M.A.; Chaoui, H.; Belkacemi, R. Sliding Mode Self-Sensing Control of Synchronous Machine Using Super Twisting Interconnected Observers. *Energies* **2020**, *13*, 4199. [CrossRef]
60. Liu, Y.; Fang, J.; Tan, K.; Huang, B.; He, W. Sliding Mode Observer with Adaptive Parameter Estimation for Sensorless Control of IPMSM. *Energies* **2020**, *13*, 5991. [CrossRef]
61. Valenzuela, F.A.; Ramírez, R.; Martínez, F.; Morfín, O.A.; Castañeda, C.E. Super-Twisting Algorithm Applied to Velocity Control of DC Motor without Mechanical Sensors Dependence. *Energies* **2020**, *13*, 6041. [CrossRef]
62. Azzouz, T.; Abdelhamid, B. Performance of PI controller for control of active and reactive power in DFIG operating in a grid-connected variable speed wind energy conversion system. *Front. Energy* **2014**, *8*, 371–378. [CrossRef]
63. Karima, B.; Akkila, B. Output power control of a variable wind energy conversion system. *Rev. Sci. Tech. Electrotech. Energ.* **2017**, *62*, 197–202.
64. Amrane, F.; Chaiba, A. A novel direct power control for grid-connected doubly fed induction generator based on hybrid artificial intelligent control with space vector modulation. *Rev. Sci. Tech. Electrotech. Energ.* **2016**, *61*, 263–268.

65. Amrane, F.; Chaiba, A.; Badr Eddine, B.; Saad, M. Design and implementation of high performance field oriented control for grid-connected doubly fed induction generator via hysteresis rotor current controller. *Rev. Roum. Rev. Sci. Tech. Electrotech. Energ.* **2016**, *61*, 319–324.
66. Yusoff, N.A.; Razali, A.M.; Karim, K.A.; Sutikno, T.; Jidin, A. A Concept of Virtual-Flux Direct Power Control of Three-Phase AC-DC Converter. *Int. J. Power Electron. Drive Syst.* **2017**, *8*, 1776–1784. [CrossRef]
67. Quan, Y.; Hang, L.; He, Y.; Zhang, Y. Multi-Resonant-Based Sliding Mode Control of DFIG-Based Wind System under Unbalanced and Harmonic Network Conditions. *Appl. Sci.* **2019**, *9*, 1124. [CrossRef]
68. Yahdou, A.; Hemici, B.; Boudjema, Z. Second order sliding mode control of a dual-rotor wind turbine system by employing a matrix converter. *J. Electr. Eng.* **2016**, *16*, 1–11.
69. Yaichi, I.; Semmah, A.; Wira, P.; Djeriri, Y. Super-twisting sliding mode control of a doubly-fed induction generator based on the SVM strategy. *Period. Polytech. Electr. Eng. Comput. Sci.* **2019**, *63*, 178–190. [CrossRef]

Article

Small-Signal Model and Stability Control for Grid-Connected PV Inverter to a Weak Grid

Antoine Musengimana, Haoyu Li *, Xuemei Zheng and Yanxue Yu

Department of Electrical and Automation Engineering, Harbin Institute of Technology, Harbin 150001, China; 17BF06046@hit.edu.cn (A.M.); xmzheng@hit.edu.cn (X.Z.); yuyanxueppkz@163.com (Y.Y.)
* Correspondence: lihy@hit.edu.cn

Abstract: This paper presents a small signal stability analysis to assess the stability issues facing PV (photovoltaic) inverters connected to a weak grid. It is revealed that the cause of the transient instabilities, either high-frequency or low-frequency oscillations, is dominated by the outer control loops and the grid strength. However, most challenging oscillations are low-frequency oscillations induced by coupling interaction between the outer loop controller and PLL (Phase-Locked Loop) when the inverter is connected to a weak grid. Therefore, the paper proposes a low-frequency damping methodology in order to enhance the high system integration, while maintaining the stability of the system. The control method uses a DC link voltage error to modulate the reference reactive current. The proposed control reduces the low-frequency coupling between the DVC (DC link voltage controller), AVC (AC voltage controller) and PLL (Phase-locked loop). According to this study's results, the performance capability of the grid-connected PV inverter is improved and flexibility in the outer loop controller design is enhanced. The control strategy proposed in this paper is tested using the PLECS simulation software (Plexim GmbH, Zurich Switzerland) and the results are compared with the conventional method.

Keywords: VSC (voltage source converter); PLL (Phase-Locked Loop); weak grid; small signal stability; eigenvalues

Citation: Musengimana, A.; Li, H.; Zheng, X.; Yu, Y. Small-Signal Model and Stability Control for Grid-Connected PV Inverter to a Weak Grid. *Energies* **2021**, *14*, 3907. https://doi.org/10.3390/en14133907

Academic Editor: Frede Blaabjerg

Received: 10 May 2021
Accepted: 23 June 2021
Published: 29 June 2021

Publisher's Note: MDPI stays neutral with regard to jurisdictional claims in published maps and institutional affiliations.

1. Introduction

High penetration of PV systems into AC systems reduces the grid strength, i.e., the AC system short circuit ratio (SCR) is reduced. The reduction in the SCR is due to a high impedance of long transmission lines connecting the PV systems, which are located far away from load centers [1]. In addition, the reduction in the SCR may also be due to the increased level of penetration of the inertialess PV system [2]. The grid strength is described by the SCR, where a weak grid is defined by the SCR < 3 [2]. With the increasing level of PV system penetration into the grid, the system stability becomes a challenging issue, as the grid becomes weak [3–5]. PV systems are connected to the grid through VSCs [6], which are controlled to inject a clean current to the grid.

The main method used to control VSCs is based on vector control in the synchronous reference frame (*dq*) [7–9]. Vector control consists of cascaded control loops containing a DC link voltage controller (DVC), a PCC (point of common coupling) voltage controller (AVC), and an inner current loop controller (CC) [10]. The DVC regulates the terminal voltage to follow the reference level by generating the active current reference, while the AVC regulates the PCC voltage to produce the reactive current reference. Furthermore, the CC regulates the current flowing into the grid to follow the reference set by the outer loop controllers. Apart from control loops, a PLL is necessary for the synchronization process of the VSC to the grid [11].

Stability issues encountered in distributed PV systems connected to a weak grid could be due to the physical dynamics of the PV system, to the control loop interactions or to the combination of both. To investigate the causes of the unwanted oscillations in power systems with high penetration VSC interfaced sources, many methods have been developed. References [1,12,13] proposed a reduced small-signal model to investigate the effects of outer control loops, grid dynamics and PLL on the system stability. The model only addresses the causes of low-frequency oscillations encountered when the inverter interacts with a weak grid. It has been found that the PLL, outer control loop (which includes the PCC voltage controller and DC link voltage controller), and the grid strength are the main cause of low-frequency instability issues [4,14]. The same results have been revealed in [5,15], when a full order or accurate small-signal model of the system as called by the author was used. These forms of instabilities are quite challenging and limit the integration of solar energy into the grid system. Therefore, in order to cope with the instabilities caused by the high integration of PV systems into a weak grid, different control methods are generated.

An improved PLL system has been proposed [16,17]. The proposed PLL uses a virtual impedance to compensate for the voltage drop across the line impedance in order to extend the operational capability of the system. The proposed control method yields good performance. However, such control strategies do not decouple the control loop interactions. The impact of control loop interactions was addressed in [7], where a multivariable controller was turned using H_inf. Nevertheless, the outer loop controllers were turned without considering the effects of PLL. An alternative robust vector current controller was proposed considering the effects of PLLs [3]. In reference [5], the author proposed a control method to damp the low-frequency oscillations, where the measured PCC voltage is used to update the d-axis current reference, while in [4], the author uses the measured grid current and PCC terminal voltage to set the DC link voltage reference level. All the above-mentioned methods aim to stabilize the VSC system interacting with a weak grid.

All models presented in the above sections lack a deep understanding of the degree of system control loop interactions. In addition, high-frequency instabilities that may be due to the control loop and grid interaction have not been touched. Therefore, in order to address the aforementioned problems, this paper proposes a simplified full order model that reflects on high and low-frequency instabilities.

For a weak grid, the ac terminal voltage is sensitive to the change in active power, which is controlled through the DC-link voltage controller. As a result, the DC-link controller, the ac-voltage controller that regulates the ac terminal voltage and the PLL are dynamically coupled. The resonance oscillations may take place due to those loops coupling. Thus, a decoupling methodology is required to limit the interaction between the DC-link controller, AVC controller and the PLL. Therefore, a simple compensation method that aims to damp the low-frequency oscillations induced by the outer control loop, the PLL and the grid dynamics, is proposed.

The proposed method uses the DC-link voltage error deviation to generate the additional reactive current reference. The obtained compensating reactive current is added to the reactive current reference from the PCC voltage control loop. By updating the reactive current reference with respect to the DC-link voltage, the instability issues induced by the control loops coupling are overcome. Further, the proposed control strategy also benefits from easy implementation and good robustness against AVC, DVC and PLL bandwidth variation, as well as the grid strength variation. The performance of the proposed method is evaluated by using the small-signal stability analysis based on the derived small-signal model. At last, but not the least, the comparative evaluation of the proposed control method with the conventional control method is performed.

The rest of the paper is structured as follows: in Section 2, the configuration of PV generation connected to a weak AC grid is described and its small-signal model, including PV generator, control loop dynamics, VSC power stage and grid model is derived. The stability analysis of conventional control method is presented in Section 3 and a proposed low-frequency oscillations damping method is presented in Section 4. Section 5 presents a comparative stability analysis of the conventional method and the proposed method, while in Section 6, the simulations results are presented. Finally, the conclusions are drawn in Section 7.

2. Structure of Single-Stage Grid-Connected PV Inverter and Control System

The schematic diagram of the grid-connected single-stage 3-phase 2-level PV inverter system is depicted in Figure 1a and the control system is shown in Figure 1b. The PV inverter contains a PV generator that converts solar irradiance into electrical power. The power generated by the PV generator is routed to the VSC through a DC-link capacitor (C_{dc}) that plays a major role in power balancing. The switching harmonics in the output current produced by the VSC are curtailed by the LCL filter, which consists of converter side inductor L_f with parasite resistance R_f, the capacitor C_f and the grid side inductor L_c. The filter components are designed based on reference [18]. In addition, the output current of LCL flows to the grid through the grid impedance consisting of R_g and L_g that characterize the grid strength. The control system consists of the maximum power point tracking (MPPT), the DC-link voltage controller that regulates the DC voltage at the reference level while maintaining the AC and DC power balanced, the inner CC loop implemented in *dq* reference frame and a synchronous reference frame PLL (SRF–PLL) used to synchronize the grid-connected inverter to the grid.

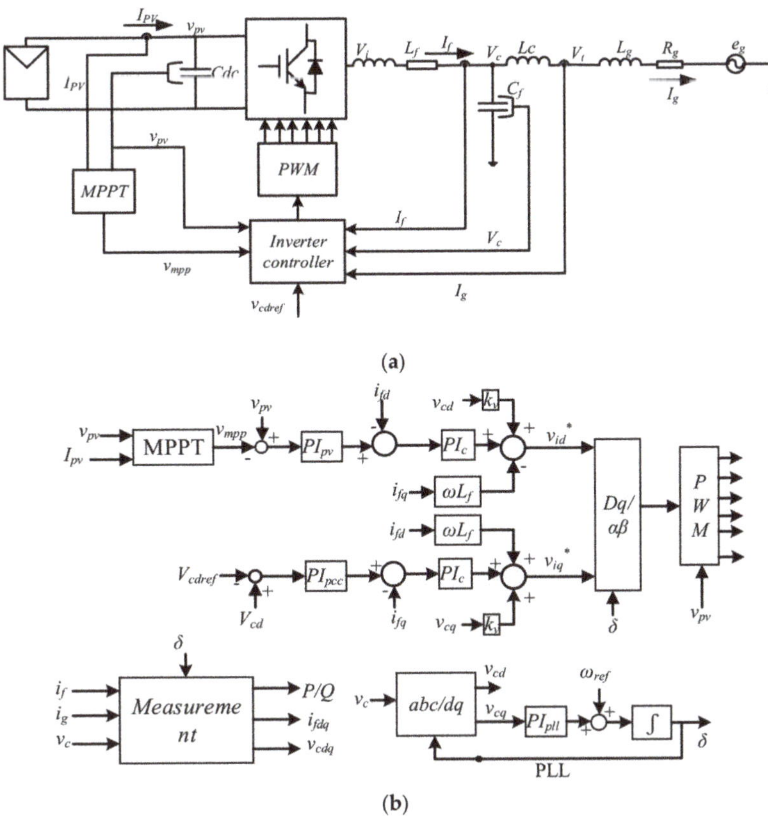

Figure 1. (**a**) Single-line representation of single-stage grid-connected PV inverter and (**b**) control system for grid-connected inverter.

3. Characteristics of PV Generator and its Mathematical Model

The main building block of a PV generator is a PV cell that converts solar energy to electrical energy. Due to the low power generated by PV cells, they are combined in series and parallel configurations to form a PV module, which can be also parallel and serially configured to form PV panels to increase power. Furthermore, PV arrays are combined in series and parallel configurations to generate the required power and produce the required DC link voltage. In compact and simple ideal form representation, a PV generator is modeled using a single diode model, due to its simplicity and accuracy. The model presented in Figure 2 contains the current source representing the photonic current generated by the PV cell, where that current mainly depends on solar irradiance and the cell temperature, the parasite diode (D), parallel resistor (R_p) and series resistance (R_p).

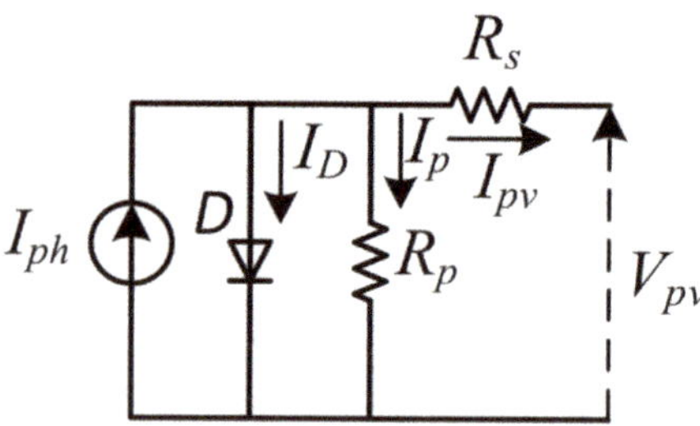

Figure 2. PV array equivalent circuit.

Mathematically, the simplified ideal PV module used to capture solar energy and convert it into electrical can be modeled by its V-I characteristics as expressed as in Equation (1) [19–21] when the parasite resistances are ignored.

$$I_{pv}(v_{pv}, T) = I_{ph} - I_{rs}\left(e^{(qv_{pv}/kaT)} - 1\right) \tag{1}$$

where the PV characteristics parameters in Equation (1) such as: $q = (1.602 \times 10^{-19}\ \text{C})$ is the electron charge, $K = (1.38 \times 10^{-23})$ is the Boltzmann constant, T is the module temperature, I_{ph} is the photon current, I_{rs} is the saturation current and A is the ideality factor. n_s and n_p represent the number of the interconnected PV module in series and parallel, respectively. The photonic current of the PV generator is determined as indicated in Equation (2), and it is linearly dependent on the solar insolation.

The module temperature affects the photonics current as well.

$$I_{ph} = (I_{sc} + K_t(T - T_{ref}))\frac{G}{G_{ref}} \tag{2}$$

where T_{ref} is the module reference temperature, I_{sc} is the module short-circuit current at the reference temperature, $G(\text{W/m}^2)$ is the solar insolation and K_t is the module temperature coefficient.

The output power extracted from the PV array can be obtained using Equation (3), where the generated power depends on the solar insolation and the terminal PV voltage.

Figure 3 shows the output power of the PV generator in the function of the PV terminal voltage at a different level of solar insolation.

$$P_{pv}(G, T, V_{pv}) = v_{pv}I_{pv} = n_p I_{ph} v_{pv} - n_p I_{rs} v_{pv}\left(e^{(v_{pv}Q/n_s KAT)} - 1\right) \tag{3}$$

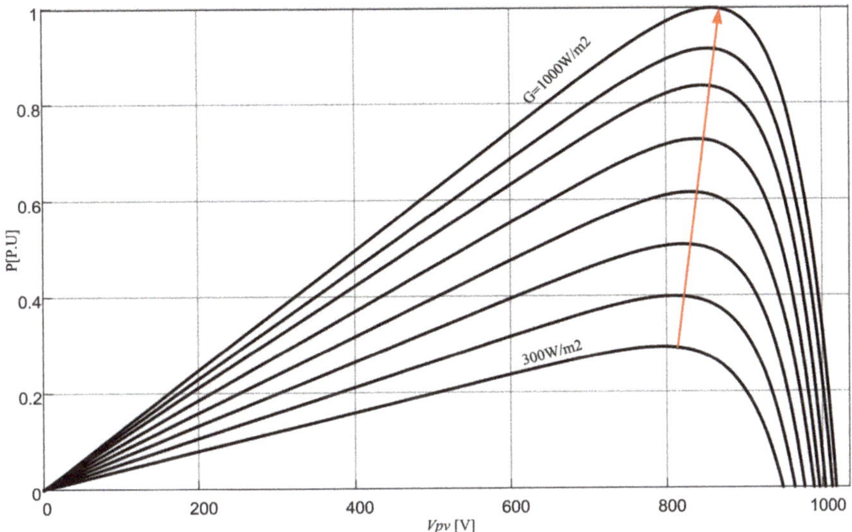

Figure 3. Ppv-v_{pv} characteristics curves of the PV generator at different solar insolation.

In this paper, a Mitsubishi PV array of a model "PVUD190MF5" is considered [22]. The output power of the PV system under different solar irradiation as a function of the DC terminal voltage, with series-connected modules of n_s = 33 and the parallel-connected string of n_p = 16 which can generate 100 KW when working at MPPT, is shown in Figure 3.

From Figure 3, the output power of a PV generator at any value of the solar insolation and module temperature varies nonlinearly as a function of the terminal voltage, where it exhibits two zero power points, with one corresponding to the open-circuit voltage and the other to the short-circuit voltage, respectively, and one single maximum power point. It is desirable to operate the grid-connected PV generator to the maximum power point for harvesting the maximum possible power available. Therefore, to guaranty that the PV generator outputs the possible maximum output power under variations of the environmental conditions, the PV terminal voltage is regulated by the MPPT algorithm [23], to remain closer to the maximum power point voltage (v_{mpp}). The incremental conductance and P&O (perturb and observe) are the most used due to their accuracy and simplicity in implementation. In this paper, a P&O is adopted.

4. Small-Signal Model of Grid-Connected PV Inverter

Considering both the power part and control system as represented in Figure 1, the small-signal mathematical model can be derived for stability analysis. The system is considered to be a 3-phase balanced system. All system variables and control loops are represented in dq reference frame. During steady-state operation, the grid dq reference frame is oriented with the converter dq rotating frame, whereas that is not the case when the converter operating is in transient mode. Therefore, the mismatch between the grid reference frame and the converter reference frame is modeled using the grid angle generated by PLL. Note that, as the control loop does not use the grid current, the Lc inductor may be taken as the grid impedance in combination with L_g. For simplicity, the variable in the grid dq frame will be superscripted by "g" while the variable in the converter frame is maintained as it is; the subscript "0" represents the steady-state value of that variable and Δ represents small signal variations around the operating point.

4.1. Frame Transformation

The input variables in grid dq reference frame are transformed into converter reference using Equation (4a), whereas the output variables of the converter in converter dq frame are transformed back into grid dq reference frame through Equation (4b)

$$x = x^g e^{-j\delta} \tag{4a}$$

$$x^g = x e^{j\delta} \tag{4b}$$

where $x = x_d + jx_q$ represents the variable in converter dq reference frame, $x^g = x_d{}^g + jx_q{}^g$ represents the variable in the grid reference frame, and δ is the grid angle output of PLL.

In linearized form around steady-state operating point, the expressions (4a) and (4b) are obtained in Equation (5a–d)

$$\Delta x_d = \Delta x_d^g + x_{q0}\Delta\delta \tag{5a}$$

$$\Delta x_q = \Delta x_q^g - x_{d0}\Delta\delta \tag{5b}$$

$$\Delta x_d^g = \Delta x_d - x_{q0}\Delta\delta \tag{5c}$$

$$\Delta x_q^g = \Delta x_q + x_{d0}\Delta\delta \tag{5d}$$

The capacitor filter dynamics and grid inductor current through $Lg + L_c$ are the only ones modeled in grid dq frame, while the remaining are modeled in the converter dq frame.

4.2. Phase-Locked Loop (PLL)

The conventional second-order SRF–PLL in [11] is used to measure the grid angle. As can be seen in Figure 1b, it consists of abc/dq transformation, a PI regulator and VCO (voltage controlled oscillator) represented by the integrator. The PI_{pll} controller regulates q-axis voltage to zero and outputs the grid rotation frequency deviation with respect to nominal frequency. Since the q-axis voltage is regulated to zero, the d-axis voltage always aligns with the capacitor voltage vector. Afterward, the sum of the angular frequency deviation and the reference angular frequency are integrated to generate the grid angle.

The mathematical equations representing the PLL dynamics are expressed in (6a) and (6b).

$$\begin{cases} \frac{d\delta}{dt} = \omega_{ref} + \omega_{pll} \\[2mm] \frac{d\gamma_{pll}}{dt} = v_{cq} \end{cases} \tag{6a}$$

$$\omega_{pll} = k_{p\omega}v_{cq} + k_{i\omega}\gamma_{pll} \tag{6b}$$

where ω_{pll} is the rotational frequency deviation, $k_{i\omega}$ and $k_{p\omega}$ are the phase-locked PI integral constant and proportional constant of PI_{pll}, respectively, γ_{pll} is the PLL integral state and ω_{ref} is the reference angular frequency.

Linearizing the Equation (6a,b), a state-space representation of PLL is obtained in (7).

$$\begin{cases} \frac{dx_{pll}}{dt} = A_{pll}x_{pll} + B_{pll}\Delta v_{cdq} \\[2mm] \Delta\omega_{pll} = C_\omega x_{pll} + D_\omega \Delta v_{cdq} \\[2mm] \Delta\delta = C_\delta x_{pll} \end{cases} \tag{7}$$

where $x_{pll} = [\delta, \gamma_{pll}]^T$, A_{pll} and B_{pll} are the state matrix and state input matrix, C_ω and D_ω are the angular frequency output matrix and disturbance input matrix, respectively, and C_δ is the synchronous angle output matrix Due to space limitation, the state-space matrices are not presented here.

4.3. DC Bus Voltage Control (DVC) and AC Voltage Control (AVC)

The outer control loops are responsible for generating the inner current references that flow into the converter. Both AVC and DVC are implemented using the classical PI controller. The DVC processes the error between the reference DC-link voltage, which is generated by the MPPT and the measured DC-link voltage to generate the d-axis current reference. The DC-link stability is guaranteed by maintaining the power generated by the PV generator equal to the power flowing to the grid. The dynamics of the MPT in this model are omitted and their impact on the stability of the system remains a topic for future study. On the other side, the AVC generates the reference reactive inductor current by regulating the d-axis component of the capacitor filter voltage to follow its reference for voltage support purpose.

The mathematical dynamics expressing the AVC and DVC are shown in Equation (8).

$$\begin{cases} \frac{d}{dt}x_1 = v_{pv} - v_{dcref} \\ \frac{d}{dt}x_2 = v_{sd} - v_{dref} \end{cases}$$
$$i_{dref} = k_{pdc}(v_{pv} - v_{mpp}) + k_{idc}x_1$$
$$i_{qref} = k_{pv}(v_{cd} - v_{cdref}) + k_{iv}x_2 \tag{8}$$

where x_1 and x_2 are the state variable of DVC and AVC, respectively; K_{pdc} and k_{pv} are the proportional constants of DVC and AVC PI-controllers, respectively; K_{idc} and k_{iv} are the integral constants of DVC and AVC PI-controllers, respectively; and i_{dref} and i_{qref} are the d-axis and q-axis current references, respectively.

After linearization of equations in (8), the state space equations of outer control-loops results in (9).

$$\frac{d}{dt}x_{12} = A_{out}x_{12} + B_{out}\Delta v_{pv} + B_{out1}\Delta v_{cd} + B_{out2}x_{ref}$$
$$\Delta i_{dqref} = C_{out}x_{12} + D_{out}\Delta v_{pv} + D_{out1}\Delta v_{cd} + D_{out2}x_{ref} \tag{9}$$

where A_{out}, B_{out}, B_{out1}, B_{out2}, C_{out}, D_{out}, D_{out1} and D_{out2} are the outer control loops state matrices and $x_{ref} = [v_{mpp}, v_{cdref}]^T$.

4.4. Inner Current Control Loop (CC)

The inner current controller, as shown in Figure 1b, also uses a PI-controller to regulate the current flowing through the filter inductor current. The reactive power is controlled by the q-axis current, whereas the active power is regulated through the d-axis current. The current controller tracks the error difference between i_{dqref} and the measured filter i_{fdq} current error to generate the modulating signal. The dq-axis feed-forward decoupling and capacitor voltage feed-forward through gain k_v are used for improving the system performance [24]. The inner loop current controller dynamics are obtained in Equation (10) and after linearization, a state-space model representation is illustrated in Equation (11).

Where γ_1 and γ_2 are the state variable of the current controller, A_c, B_c, B_{c1}, B_{c2}, C_c, D_c, D_{c1}, D_{c2} and D_{c3} are the matrices of the state-space model of the current controller and $v_{idq}{}^*$ are the output variables of the current controller. We can note that, as the current controller bandwidth was chosen to be 1/5 of the switching frequency, the delay introduced by the modulator may be omitted. This results in $v_{idq}{}^* = v_{idq}$.

$$\begin{cases} \frac{d}{dt}\gamma_1 = i_{dref} - i_{fd} \\ \frac{d}{dt}\gamma_2 = i_{qref} - i_{fq} \end{cases}$$
$$v_{id}^* = k_{pc}(i_{dref} - i_{fd}) + k_{ic}\gamma_1 - \omega_0 L_f i_{fq} + k_v v_{cd}$$
$$v_{iq}^* = k_{pc}(i_{qref} - i_{fq}) + k_{ic}\gamma_2 + \omega_0 L_f i_{fd} + k_v v_{cq} \tag{10}$$

$$\frac{d}{dt}\gamma_{12} = A_c\gamma_{12} + B_{c1}\Delta i_{fdq} + B_{c2}\Delta i_{dqref}$$

$$\Delta v^{*}_{idq} = C_c\gamma_{12} + D_c\Delta i_{dqref} + D_{c1}\Delta v_{cdq} + D_{c2}\Delta i_{fdq} + D_{c3}x_{ref} \tag{11}$$

4.5. Converter Filter Inductor Dynamics

As shown in Figure 1a, the dynamic model of the inverter side filter inductor in dq reference frame can be represented in Equation (12), and its linearized state-space model in Equation (13).

$$\begin{cases} L_f\frac{di_{fd}}{dt} = v_{id} - v_{cd} - R_f i_{fd} + \omega_0 L_f i_{fq} \\ L_f\frac{di_{fq}}{dt} = v_{iq} - v_{cq} - R_f i_{fq} - \omega_0 L_f i_{fd} \end{cases} \tag{12}$$

$$\frac{d}{dt}\Delta i_{fdq} = A_f\Delta i_{fdq} + B_f\Delta v_{idq} + B_{f1}\Delta v_{cdq} \tag{13}$$

where A_f, B_f and B_{f1} are the matrices of the filter inductor state-space model.

4.6. DC-Link Capacitor Dynamics

The dynamics rate of change of DC-link voltage is expressed mathematically by the PV power and grid power imbalance as shown in Equation (14).

$$C_{dc}v_{pv}\frac{dv_{pv}}{dt} = P_{pv} - P_{in} \tag{14}$$

where P_{pv} and P_{in} are the PV power and input power to the inverter, respectively.

The power generated by the PV generator depends on the solar insolation and temperature (see Equation (3)). The input power to the VSC can be expressed as in (15).

$$P_{in} = v_{id}i_{fd} + v_{iq}i_{fq} \tag{15}$$

The linearized DC-link dynamics is shown in (16).

$$C_{dc}v_{pv0}\frac{d\Delta v_{pv}}{dt} = I_{pv0}\Delta v_{pv} + v_{pv0}\Delta i_{pv} - 1.5v_{id0}\Delta i_{fd} - 1.5v_{iq0}\Delta i_{fq} - 1.5I_{fd0}\Delta v_{id} - 1.5I_{fq0}\Delta v_{iq} \tag{16}$$

To include the PV generator dynamics in the model, the expression (17), is obtained from Equation (3).

$$\Delta i_{pv} = G_d\Delta v_{pv} \tag{17}$$

where G_d is the conductance of the PV generator.

The state-space model of the DC link is obtained in (18) from (16) and (17).

$$\frac{d\Delta v_{pv}}{dt} = A_{pv}\Delta v_{pv} + B_{pv}\Delta v_{idq} + B_{pv1}\Delta i_{fdq} \tag{18}$$

where A_{pv}, B_{pv} and B_{pv1} are the matrices of DC-link state-space model.

4.7. Grid Dynamics Model

The grid dynamics consist of C_f, L_c and L_g. Applying the KVL, the equations describing the grid dynamics can be obtained as in Equation (19).

$$\begin{cases} C_f\frac{d}{dt}v^{g}_{cdq} = i^{g}_{fdq} - i^{g}_{gdq} + j\omega C_f v^{g}_{cdq} \\ L_t\frac{d}{dt}i^{g}_{gdq} = v_{cdq} - R_g i^{g}_{gdq} + j\omega L_t i^{g}_{gdq} - e_{dq} \end{cases} \tag{19}$$

where $L_t = L_g + L_c$, $v_{cdq}{}^{g}$ is the capacitor filter voltage in grid dq frame, $i_{fdq}{}^{g}$ is the filter inductor current in grid dq frame, $i_{gdq}{}^{g}$ is the grid side inductor current in grid dq frame and e_{dq} is the grid terminal voltage in dq frame.

Linearizing the equations in (19) around the steady-state operating point and taking into account the angular frequency dynamics, the state-space model of the grid side is obtained in (20).

$$\frac{d}{dt}x_g = A_g x_g + B_g \Delta i^g_{fdq} + B_{g1}\Delta e_{dq} + B_{g2}\Delta \omega_{pll} \tag{20}$$

where $x_g = [v_{cdq}{}^g, i_{gdq}{}^g]^\mathrm{T}$, and A_g, B_g, B_{g1} and B_{g2} are matrices of the state-space model of grid dynamics.

4.8. Overall State-Space Model

The overall small-signal representation of the grid-connected PV inverter shown in Figure 1 is illustrated in Figure 4. It is obtained by combining the converter and the grid system model as shown in Figure 4. The linearized aggregate state-space model of the overall system with 13 states variables and 2 input variables, is expressed in Equation (21). Where A_{sys} is the state matrix of the system, B_{sys} is the state input matrix and C_{sys} is the system output matrix. The proposed system model can easily be extended for multiple inverters operating in parallel, without requiring virtual resistance to define the PCC voltage as in [25]. However, in this study, a single inverter connected to a weak grid is considered and the stability analysis of parallel grid-connected PV inverters remains a topic for future study.

$$\begin{cases} \dfrac{dx_{sys}}{dt} = A_{sys}x_{sys} + B_{sys}u_{sys} \\[2mm] y_{sys} = C_{sys}x_{sys} \end{cases} \tag{21}$$

where

$$x_{sys} = \left\{\ v_{dc}\quad x_1\quad x_2\quad \gamma_1\quad \gamma_2\quad i_{fd}\quad i_{fq}\quad \delta\quad \gamma_{pll}\quad v^g_{cd}\quad v^g_{cd}\quad i^g_{gd}\quad i^g_{gq}\ \right\}\ and\ u_{sys} = \left\{\ v_{pv}\quad v_{cdref}\ \right\}$$

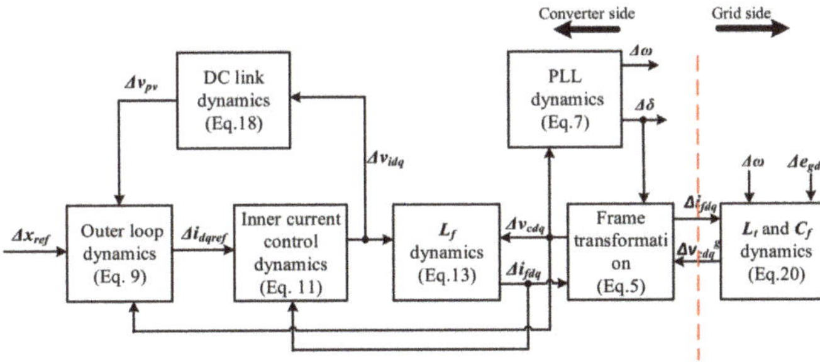

Figure 4. Overall small-signal model representation of the proposed system.

5. Stability Analysis

To analyze the stability of the PV system based on the derived small-signal model, the study system parameters in Figure 1 are defined in Table 1. The nominal power of the PV generator is 100 kVA. The inner current loop has been designed with a bandwidth equal to 1/5 of the switching frequency (f_{sw}). The AVC loop bandwidth was chosen as 20 Hz and a DVC bandwidth of 50 Hz was used. The AVC, DVC and the CC have been designed based on reference [26] for the specified above bandwidth. The LCL filter values, the parameters of the current controller, DC-link voltage controller, AVC loop controller and PLL controller are defined in Table 1. The parameters of the PLL controller were obtained as in [11] for a bandwidth of PLL equal to 10 Hz.

Table 1. Steady-state parameters of the system.

Parameter	Value	Parameter	Value
Vdc, (V)	850	$Kidc$	1401.5
Lf, Lg (mH)	0.75, 4.5	kic	471
Rf, Rg (Ω)	0.1, 0.4	Kpc	4.7
Cf, Cdc, (μF)	100, 8500	Kiv	0.0141
SN, KVA	100	Kvv	88.8889
fsw, KHz	5	G (W/m^2)	1000
Eg (V)	380	Kid	500
fg, Hz	50	kfv	0.75
$vcref$ (V)	400	$ki\omega$	12.1372
$Kpdc$	3.1540	$Kp\omega$	0.2731

The analysis of the outer control loops interaction when the system operates under weak grid conditions is presented. The impact of control parameters variation on the stability of the system is analyzed based on the eigenvalues sensitivities.

5.1. Control Loop Interaction

From the state-space model, the DC link voltage closed-loop and the ac capacitor voltage closed-loop transfer functions are derived. The bode plots of the transfer functions of the closed-loop are shown in Figure 5. The frequency response of the DC voltage closed-loop is shown in Figure 5a when the AVC bandwidth (BW_pcc) is 5 Hz, 10 Hz, 25 Hz and 35 Hz with the DC link controller bandwidth set to 50 Hz, while the PLL bandwidth is 10 Hz.

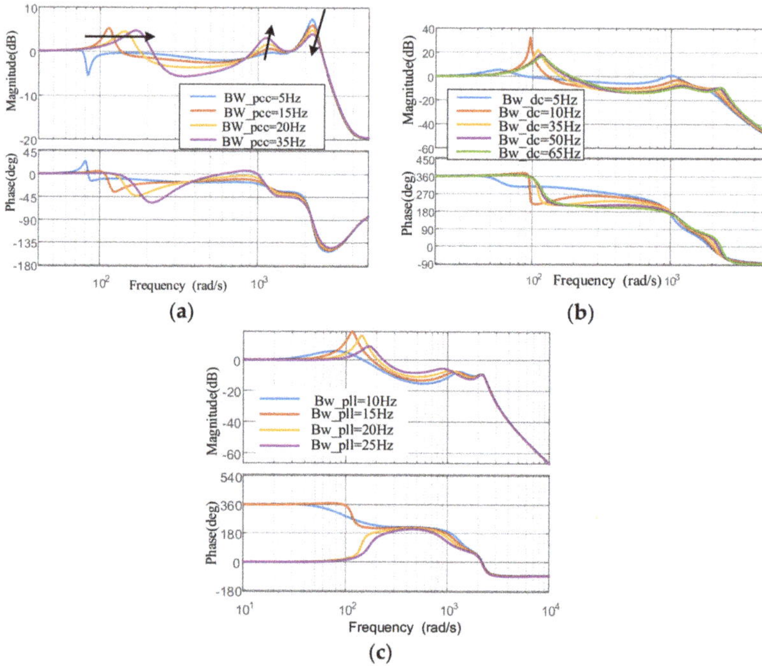

Figure 5. Frequency response of closed loop transfer function(**a**), v_{pv}/v_{mpp} (**b**) and (**c**) v_{cd}/v_{cdref}.

As shown in Figure 5a, the AVC has a significant impact on DC-link stability. For low AVC bandwidth; for example, BW_pcc = 5 Hz, the system is unstable due to the peak resonance effects with negative damping ratio induced in low-frequency range. As the AVC bandwidth increases, the low-frequency oscillations are damped and the system becomes stable. This may be due to the fact that when the inverter interacts with the weak grid, the AVC is necessary to stabilize the system. However, increasing the AVC bandwidth impacts the high-frequency oscillations. The high-frequency resonance effects appear to be closer to 1100 rad/s and shift toward 1000 rad/s as the BW_pcc increases.

The high-frequency effects are due to the resonance induced by the LCL filter. As can be seen from Figure 5, the high-frequency instabilities dominate the DC link channel compared to the ac voltage link channel, as there may be a non-zero amplification gain. Therefore, the AVC bandwidth is selected large enough to limit low-frequency oscillations and must not be high to avoid high-frequency resonances. Figure 5b shows the AVC closed-loop response with the variation of the DVC controller bandwidth. The lower frequency resonances are amplified when the DVC bandwidth increases closer to the PLL bandwidth and is damped for a DVC bandwidth greater than the PLL bandwidth. The PLL and DVC interactions are portrayed in Figure 5b.

In the same way, the impact of the PLL bandwidth on the closed-loop AVC is shown in Figure 5c. For a PLL bandwidth (BW_pll) closer to the AVC bandwidth, for example when BW_pll = 20 Hz and Bw_pll = 25 Hz, the closed-loop system has a negative phase margin, as can be seen in Figure 5c. As a result, the system becomes unstable. The PLL bandwidth in this case is selected lower than the AVC bandwidth, where a 10 Hz is selected.

5.2. Eigenvalues Analysis

The eigenvalue method is used to assess the system's stability based on the small-signal model shown in Figure 4. Table 2 shows the eigenvalues of the overall system at the steady-state operating point with system values in Table 1. The corresponding oscillation frequencies and damping ratios for each mode are also presented in Table 3.

Table 2. Eigenvalues, damping and oscillation frequency of the system at steady-state values.

Modes	Eigenvalues	Damping	Oscillation Frequency
λ_1	-6852	1	0
λ_2	-5702.3	1	0
$\lambda_{3,4}$	$-212.74 \pm 2215.2i$	0.095595	352.56
$\lambda_{5,6}$	$-275.5 \pm 1170.9i$	0.22903	186.36
$\lambda_{7,8}$	$-225.74 \pm 62.654i$	0.96358	9.9718
$\lambda_{9,10}$	$-8.0449 \pm 114.33i$	0.070194	18.196
λ_{11}	-68.643	1	0
λ_{12}	-128.64	1	0
λ_{13}	-101	1	0

Table 3. Eigenvalues of the system and the states participation factor.

States	Modes												
	λ_1	λ_2	λ_3	λ_4	λ_5	λ_6	λ_7	λ_8	λ_9	λ_{10}	λ_{11}	λ_{12}	λ_{13}
V_{dc}	0.16	0	0.18	0.18	0.09	0.09	0.07	0.07	0.18	0.175	0.038	0.03	0.043
x_1	0.01	0	0.02	0.02	0.02	0.02	0.07	0.07	0.64	0.643	0.063	0.075	0.085
x_2	0	0	0.03	0.03	0.1	0.1	0.6	0.6	0.47	0.466	0.828	0.012	0.659
γ_1	0.02	0	0	0	0	0	0	0	0.08	0.081	0.004	0.848	0.028
γ_2	0	0.03	0	0	0.03	0.03	0.04	0.04	0.22	0.216	0.226	0.008	1.131
i_{fd}	0.89	0.01	0.07	0.07	0.02	0.02	0.03	0.03	0.02	0.017	0.015	0.001	0.006
i_{fq}	0	1.13	0.01	0.01	0.03	0.03	0.01	0.01	0.04	0.042	0.007	0.007	0.044
v_{cd}	0.02	0	0.45	0.45	0.07	0.07	0.01	0.01	0.01	0.01	0.004	0.001	0.003
V_{cq}	0	0.11	0.07	0.07	0.48	0.48	0.01	0.01	0.03	0.031	0.015	0	0.017
i_{gd}	0	0	0.24	0.24	0.02	0.02	0.12	0.12	0.25	0.25	0.206	0	0.234
i_{gq}	0	0.01	0.05	0.05	0.55	0.55	0.19	0.19	0.22	0.221	0.168	0.02	0.128
δ	0	0	0	0	0.04	0.04	0.46	0.46	0.25	0.251	0.804	0	0.576
x_{pll}	0	0	0	0	0.01	0.01	0.22	0.22	0.09	0.089	1.676	0	0.536

Since all the eigenvalues have a negative real part, the system is stable for the system and control parameters in Table 1. As can be observed from Table 2, there exist four groups of complex conjugate eigenvalues that determine the dynamics of the system. Among them, complex conjugates eigenvalues $\lambda_{9,10}$ are closer to the imaginary axis, with a lower damping ratio and a small oscillation frequency of 18.19 Hz. Therefore, eigenvalues $\lambda_{9,10}$ are the critical eigenvalues to determine the system's stability. The eigenvalues $\lambda_{7,8}$ have the lowest oscillation frequency of 10 Hz. However, they do not affect the system's dynamics as they have a higher damping ratio. The remaining complex eigenvalues $\lambda_{3,4}$ and $\lambda_{5,6}$ are associated to the *LCL* filter, with oscillation frequency closer to the resonance frequency of the *LCL*, as can be noted, from Table 2.

From Table 3, the participation factor of each state variable in a particular mode corresponding to that eigenvalues are shown. The highlighted participation factors correspond to the dominant state with a participation factor greater than 0.15 for each mode. From Table 3, it can be observed that the states corresponding to L_f and C_f have no impact on critical eigenvalues. The v_{cd} and i_{gd} states affect the eigenvalues $\lambda_{3,4}$. Thus, the change in DVC and AVC controller parameters will affect the mode $\lambda_{3,4}$.

On the other hand, the v_{cq} and i_{gq} states affect the eigenvalues $\lambda_{5,6}$; meaning that they are sensitive to PLL parameter changes. The states that affect the critical eigenvalues are the states corresponding to DVC, AVC, L_g and PLL. Therefore, if there is any change in the control parameters of DVC, AVC and PLL, the system's stability is affected. In addition, any variation in line inductance or line resistance affects the robustness of the system, as they are crucial in determining the strength of the grid. Moreover, the states corresponding to the q-axis current control also impact the critical eigenvalues. This may be due to the fact that the d-axis' ac voltage is regulated through the q-axis' current and the q-current determines the voltage drop across the line inductor if the grid Xg/Rg ratio is very high. To understand how all the states enumerated above affect the system's robustness, a root locus is used to plot the sensitivity for the eigenvalues over the change of control parameters and grid strength. Only the eigenvalues closer to the imaginary axis will be shown on the root locus.

6. Proposed Damping Method for Low-Frequency Resonance Oscillations

As described in the sections above, the coupling effects between the control loops consisting of the AVC, DVC and PLL interacting with the weak grid deteriorate the robustness of the system. Therefore, a control method to damp the low-frequency oscillations by reducing the coupling effects between DC-link voltage loop and AVC loop is proposed. For a PV-inverter connected to a weak grid, the PCC voltage is sensitive to the active power flowing to the grid controlled by d-axis current through DC-link voltage regulation.

Conventionally, the PCC voltage is regulated through q-axis current control. As the PCC voltage is sensitive to both active and reactive currents when the inverter interacts with a weak grid, a control method that considers the effects of both d-axis and q-axis current on PCC voltage to enhance inverter robust performance is proposed. The proposed compensation method, as shown in Figure 6, modifies the q-axis current reference by using the DC voltage deviation error. The controller K_{damp} expressed in (22) processes the DC voltage error and generates the compensating additional reactive current reference.

$$k_{damp} = k_{id}\frac{1}{s+a} \tag{22}$$

where K_{id} is the integral constant and a is the pole of the integrator, which is chosen to be closer to zero.

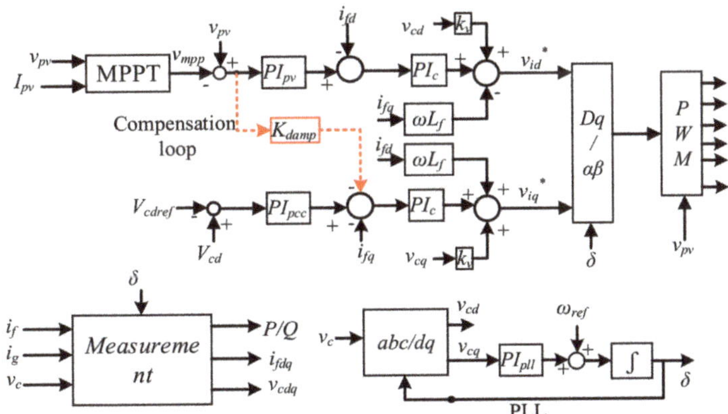

Figure 6. Control system for grid-connected PV inverter with proposed compensation loop.

The performance of the proposed control method can be analyzed based on the eigenvalues loci generated from the augmented state-space model developed in Figure 5 by the proposed damping loop dynamics. A comparison of the proposed control method and the conventional control is performed.

The outer loop dynamics in Equation (8) can be rewritten as in Equation (22). Where x_3 is the state variable of the damping loop.

$$\begin{cases} \frac{d}{dt}x_1 = v_{pv} - v_{dcref} \\ \frac{d}{dt}x_2 = v_{sd} - v_{dref} \\ \frac{d}{dt}x_3 = -ax_3 + v_{pv} - v_{dcref} \end{cases} \tag{23}$$
$$i_{dref} = k_{pdc}(v_{pv} - v_{mpp}) + k_{idc}x_1$$
$$i_{qref} = k_{pv}(v_{cd} - v_{cdref}) + k_{iv}x_2 - k_{id}x_3$$

7. Comparative Stability Analysis of Conventional and Proposed Control

Please note that the direction of the eigenvalues change from the black point corresponding to the low control variable and the eigenvalues of the posed control method are represented by the red "*" while the blue "*" represent the eigenvalues of the conventional control method. Additionally, the shaded area represents the unstable region. If any system eigenvalue is located in the shaded region, the system becomes unstable.

7.1. Eigenvalues of the Proposed System

Table 4 shows the eigenvalues of the proposed control method. By comparison with the eigenvalues of the conventional method in Table 2, one can observe that the critical mode $\lambda_{9,10}$ eigenvalues are highly damped, where the damping ratio of the proposed control is 0.638 and 0.07 for the conventional control. Consequently, the damping ratio of higher frequency eigenvalues remains approximately the same for both control methods. In addition, there is an introduction of the added eigenvalue closer to zero in the proposed method. The location of the added eigenvalue is determined by the pole of the integrator in the proposed damping method.

Table 4. Eigenvalues of the proposed control method.

Modes	Eigenvalues	Damping	Oscillation Frequency
λ_1	$-6847.8 + 0i$	1	0
λ_2	$-5701.9 + 0i$	1	0
$\lambda_{3,4}$	$-205.92 \pm 2227.6i$	0.092	354.53
$\lambda_{5,6}$	$-281.65 \pm 1159.5i$	0.236	184.54
$\lambda_{7,8}$	$-153.51 \pm 219.85i$	0.57	34.99
$\lambda_{9,10}$	$-70.76 \pm 85.231i$	0.638	13.565
$\lambda_{11,12}$	$-115.05 \pm 26.254i$	0.974	4.178
λ_{13}	-93.191	1	0
λ_{14}	-0.056512	1	0

7.2. Impact of Capacitor Voltage Feed-Forward Constant

The role of the voltage feed-forward control is to suppress the grid harmonics' effects [24,27,28]. The voltage feed-forward control may affect the system's stability when the inverter is connected to the weak grid, due to the unwanted positive feedback loop it causes [24]. The impact of voltage feed-forward gain on the system's stability is shown in Figure 7.

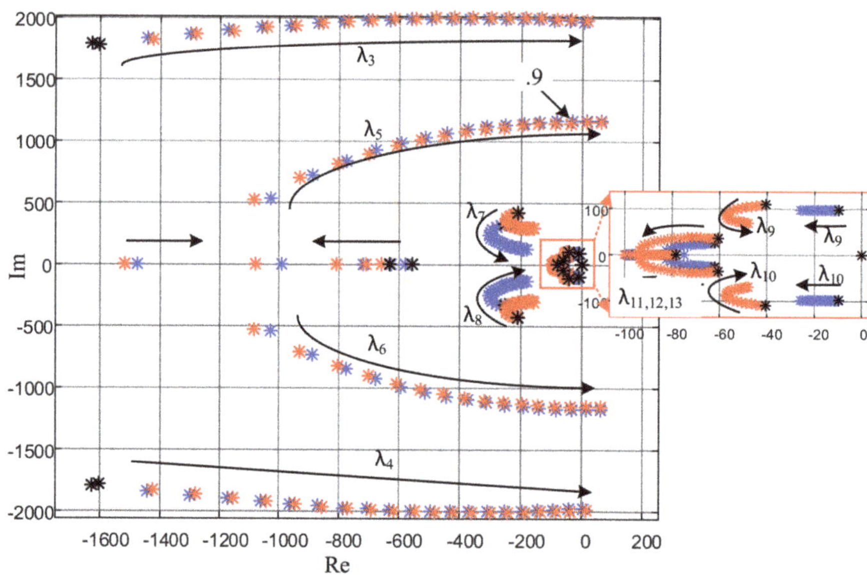

Figure 7. Sensitivity of system eigenvalues under variation of voltage feed-forward gain.

Figure 7 shows the eigenvalues locus of the system when the voltage feed-forward gain (K_v) changes from 0 to 1 for a grid strength of SCR = 1.1 (L_t = 4.5 mH and R_t = 0.4). As can be observed, the gain K_v affects both high-frequency eigenvalues and low-frequency eigenvalues. The eigenvalues move as indicated in Figure 7, where the movement of the eigenvalues is in the direction of the arrows as the feed-forward coefficient increases.

For the conventional control method, it can be seen that the critical eigenvalues $\lambda_{9,10}$ are closer to the imaginary axis compared to the critical eigenvalues of the proposed control method, and the higher damping ratio of the eigenvalues $\lambda_{9,10}$ is guaranteed for the proposed control. Since eigenvalues $\lambda_{3,4}$ and $\lambda_{5,6}$ move into the unstable region as the value of the feed-forward coefficient becomes greater than 0.9, higher frequency instabilities could occur. To guaranty the safe operation of the inverter interacting with the weak grid, the value of feed-forward should be less than or equal to 0.9. The proposed system and the conventional control method perform the same in higher frequency mode.

7.3. Impact of Grid Impedance

Figure 8 shows the movement of the system's eigenvalues when the grid inductance increases from 2.5 mH (SCR = 2.16) to 5.5 mH (SCR = 0.97). The impact of the grid inductor on the system's eigenvalues can be observed in Figure 8. The eigenvalues $\lambda_{9,10}$ are the only ones that determine the system's stability because they can move into the unstable region, while other eigenvalues remain in the stable region with high damping ratios. The eigenvalues $\lambda_{9,10}$, when the grid inductance changes, sweep toward the unstable region. As can be observed in Figure 8, the eigenvalues $\lambda_{9,10}$ become unstable when the value of the Lg > 5.35 mH (SCR = 1) for the conventional method and eigenvalues remain in stable region, far away from the imaginary axis when the proposed method is used. Thus the system's stability is worsened when the PV inverter is interfaced to a weak grid while the conventional method is in use. Consequently, the critical eigenvalues stay in a stable region for the proposed method, as can be observed in Figure 8.

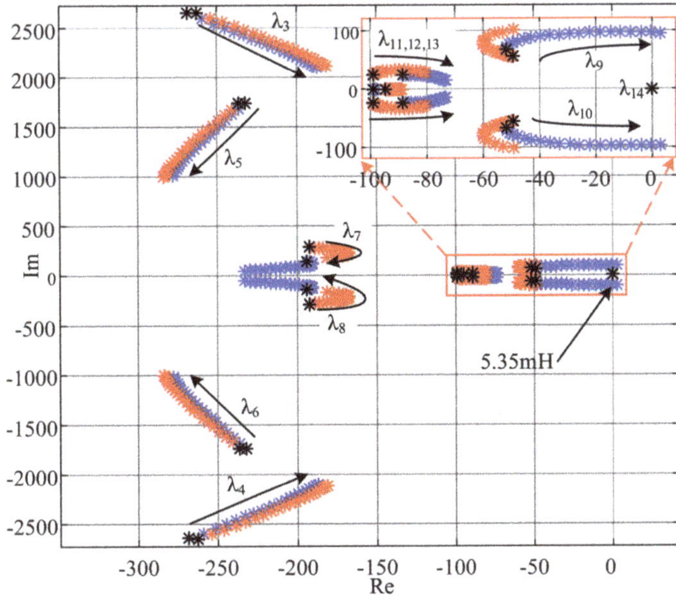

Figure 8. Sensitivity of system eigenvalues under variation of grid inductance.

7.4. Impact of AVC Control Bandwidth

Figure 9 shows the effect of the AVC controller bandwidth on system stability. As can be noted, the system remains stable for the whole range of the AVC bandwidth when it changes from 5 Hz to 40 Hz because no eigenvalue shifted in the unstable shaded region for the proposed control method. However, at low AVC bandwidth, the critical eigenvalues $\lambda_{9,10}$ for the conventional method have a positive real part. The system is stable if, and only if, the AVC bandwidth is greater than 9 Hz for the convention method. It may be observed that the increase in AVC bandwidth frequency makes the system more stable. Moreover, for a large increase of AVC bandwidth, the higher frequency eigenvalues $\lambda_{5,6}$, as can be observed in Figure 9, move toward the right side of the *s*-plan and may become unstable poles if the AVC bandwidth increases too high. Additionally, the AVC bandwidth should be lower than the DVC bandwidth to decouple their dynamics.

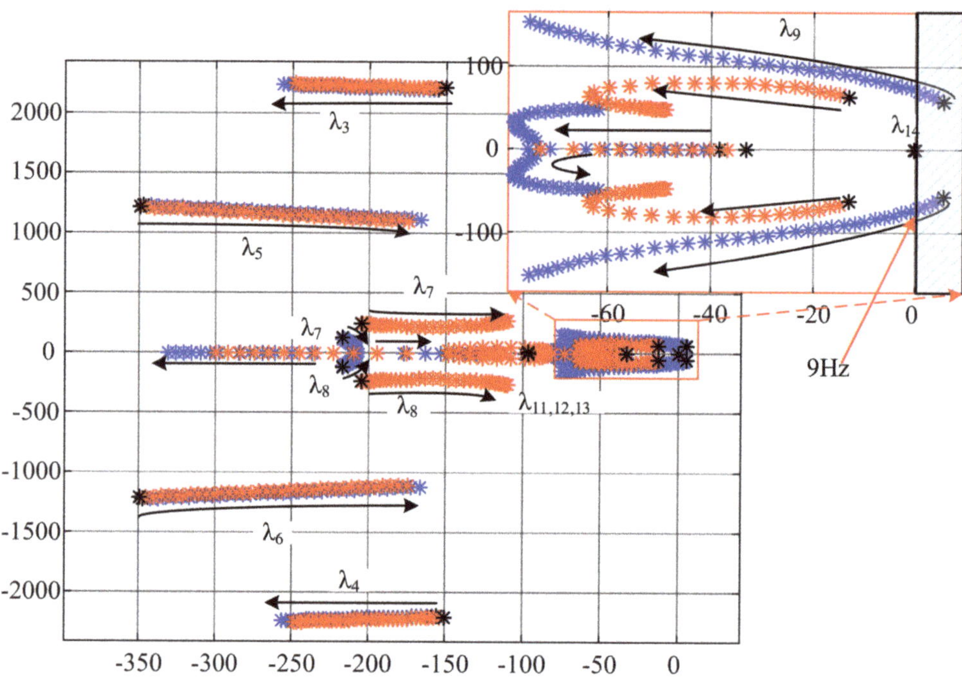

Figure 9. Sensitivity of system eigenvalues under variation of AVC control bandwidth.

7.5. Impact of DC Link Voltage Controller Bandwidth

The movement of the system eigenvalues for the DVC bandwidth controller when it changes from 5 Hz to 100 Hz can be observed in Figure 10. For the conventional method, when the bandwidth of the DVC loop is increased, the critical eigenvalues $\lambda_{9,10}$ move towards the imaginary axis as the bandwidth of the DC link controller approaches the PLL bandwidth. If the DVC bandwidth increases beyond the PLL bandwidth, the eigenvalues $\lambda_{9,10}$ return to the left side of the *s*-plan. Nonetheless, for a proposed damping method, the critical eigenvalues $\lambda_{9,10}$ move far away from the imaginary axis compared to the conventional method, thus, resulting in an improved system dynamic response. When comparing to the AVC, the increase of VDC bandwidth causes the eigenvalues $\lambda_{2,3}$ to move toward the imaginary axis. It may result in negatively damped eigenvalues if the DVC bandwidth is increased high enough, closer to the resonance frequency of the LCL filter.

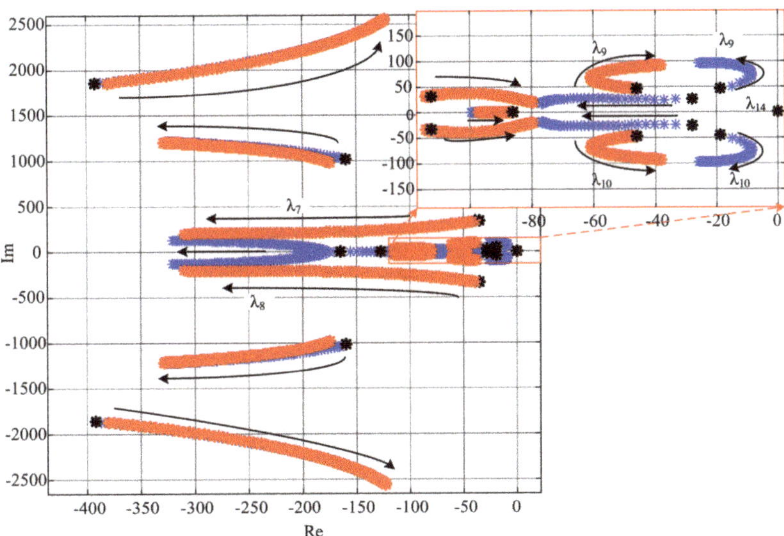

Figure 10. Sensitivity of system eigenvalues under variation of DVC control bandwidth.

7.6. Impact of PLL Controller Bandwidth

Figure 11 illustrates the effect of the PLL bandwidth on the system's stability when the PLL bandwidth changes from 5 HZ to 25 Hz. It can be noted that, while changing the PLL bandwidth, the critical eigenvalues $\lambda_{9,10}$ move towards the imaginary axis, and when the PLL bandwidth reaches 16 Hz, the real part of the critical eigenvalues becomes positive. Therefore, the system becomes unstable.

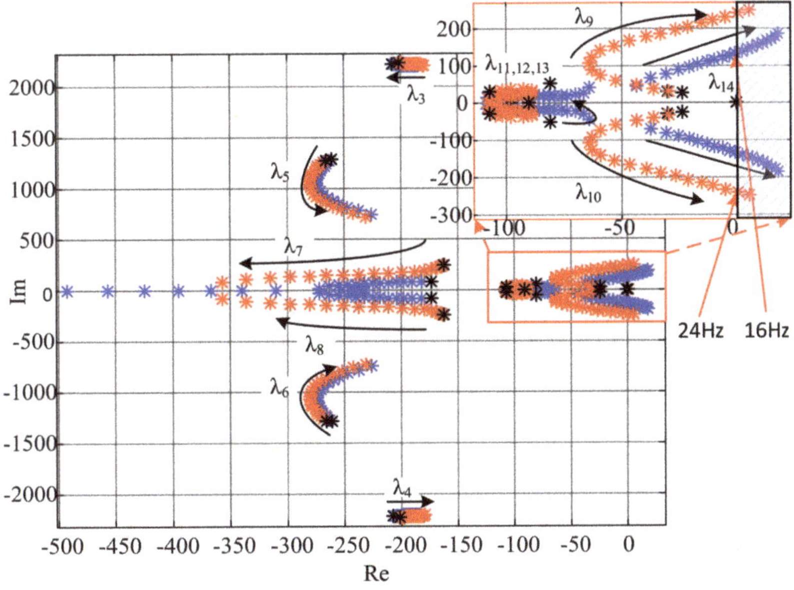

Figure 11. Sensitivity of system eigenvalues under variation of PLL control bandwidth.

On the other hand, for the proposed control, the critical eigenvalues of the system move as shown in Figure 11. They firstly move towards the left side of the *s*-plane, then return towards the imaginary axis and become negatively damped eigenvalues at 24 Hz. With the proposed control, the PLL bandwidth changes over a broad range while maintaining system stability compared to the conventional control method.

7.7. Impact of the Solar Insolation of the Stability

Figure 12 shows the eigenvalue-loci of the system when changing the values of solar irradiance from 500 W/m^2 to 1200 W/m^2. The dominant eigenvalues move as shown in Figure 12. As can be seen from Figure 12, for the proposed control loop, the system remains stable in the whole variation range of the solar insolation; whereas, for the conventional control method, the system becomes quickly unstable. For solar insolation greater than 1200 W/m^2 the conventional control method yields a negative damping ratio due to the presence of the positive critical eigenvalues, $\lambda_{9,10}$.

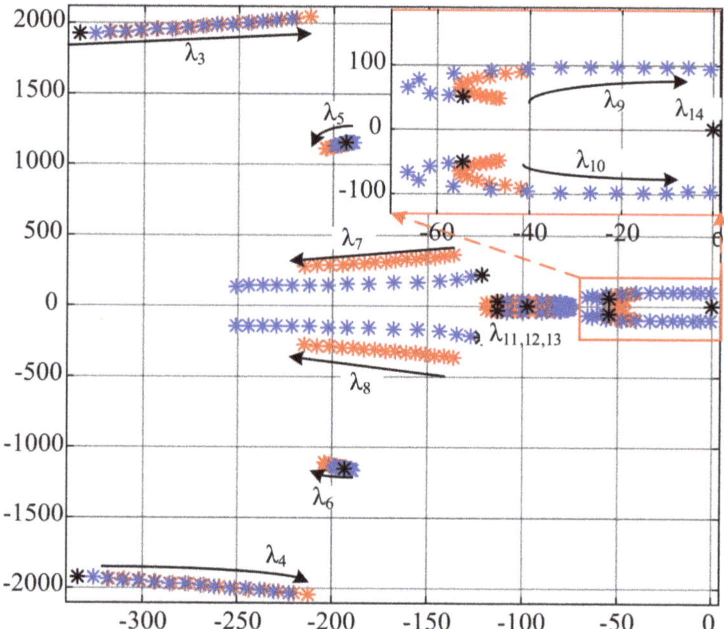

Figure 12. Eigenvalues of the system when solar insolation values change.

7.8. Impact of the Damping Loop Integrating Constant

Figure 13 shows the sensitivity of the system eigenvalues when the value of k_{id} changes from 0 to 1000. The eigenvalues move as indicated, starting from the black "*", which corresponds to $k_{id} = 0$. As can be observed, the damping ratios of critical eigenvalues increase as k_{id} increases. With the proposed control method, the additional eigenvalue appears, where its position from the origin is defined by the value of a. A value of $k_{id} = 500$ is selected in this paper. The other system eigenvalues have a high damping ratio, as they are located far from the imaginary axis as the value of k_{id} increases.

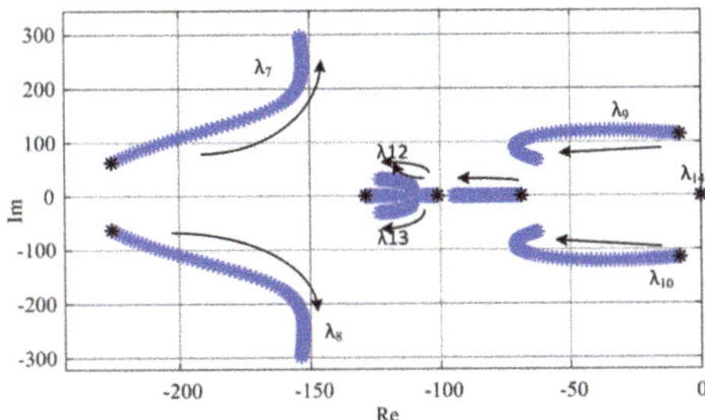

Figure 13. Sensitivity of system eigenvalues with change in k_{id}.

8. Simulation Results

To verify the stability analysis carried out in the previous section, time-domain simulations were performed in PLECS simulation software. The steady-state operating data are shown in Table 1.

The transient's response of the DC-link voltage under variation of the solar insolation, system strength, and the outer loop control bandwidth were analyzed. In addition, the grid angle measured by the PLL, and the capacitor *d*-axis voltage response were also analyzed. It is easier to say that, the transients in DC link voltage are reflected to the active power while the transients in the filter capacitor *d*-axis are reflected in reactive power when the converter operates under a strong grid. When the inverter is subjected to the disturbances and operate under weak grid conditions, the coupling effects that exist between the terminal capacitor voltage, DC-link power port and the grid dynamics induce large oscillations in reactive power compared to the oscillation in active power.

Figure 14a shows the ac voltage angle response following a step change of PLL bandwidth at time t = 2 s. Figure 14b shows the DC-link capacitor voltage and *d*-axis terminal ac voltage when the PLL bandwidth frequency is subjected to a step change from 5 Hz to 20 Hz at time t = 2 s. The growing oscillations resonating at a frequency of 24.8 Hz in both DC link voltage and ac *d*-axis voltage are yielded when the PV inverter is controlled by the conventional method. The instabilities are due to the presence of positive eigenvalues caused by the PLL bandwidth increase, as described in previous sections. However, when the proposed control system is activated, the system performance is well damped, regardless of the step-change in frequency at 2 s. Therefore, the robustness of the system over any change in the PLL bandwidth is improved as presented in the analysis.

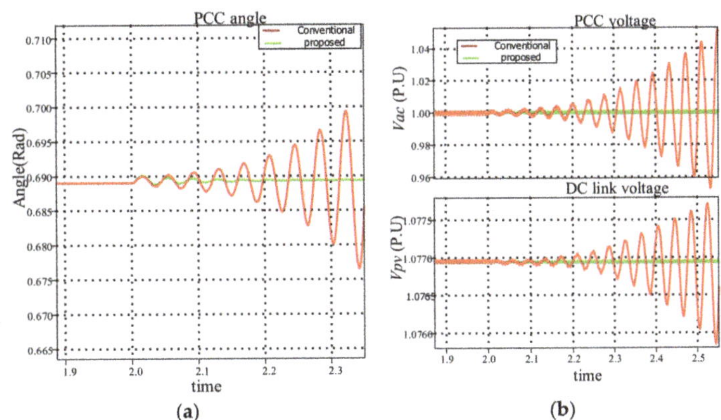

Figure 14. (**a**) grid angle and (**b**)v_{dc} and v_{cd} response when BW_PLL change from 5 Hz to 25 Hz.

Figure 15a,b shows the transients' response of the output DC link voltage, *d*-axis ac voltage and the grid voltage angle for a step-change in AVC control bandwidth. As can be observed from Figure 15, the system becomes unstable when the AVC bandwidth frequency changes from 20 Hz to 5 Hz, if the system is under conventional control method. Growing undamped oscillations that oscillate at 9.22 Hz are induced. Compared to the conventional control method, the proposed control system maintains the system's stability, even for a low AVC control bandwidth, as can be observed in the results presented in Figure 15. With the proposed control, flexibility in designing the AVC controller is achieved. This is because the system is robustly stable over a large range variation of the AVC controller bandwidth.

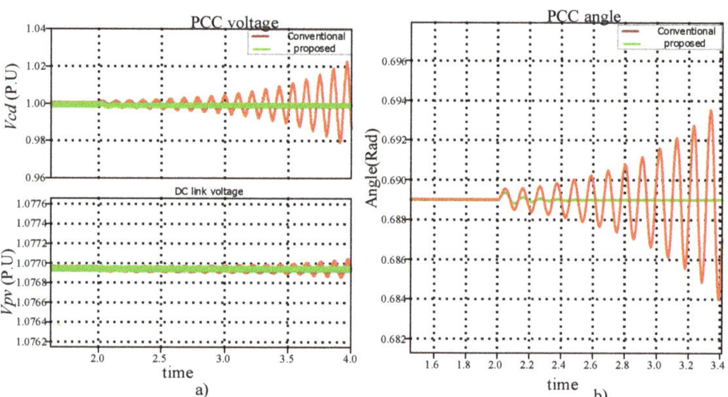

Figure 15. (**a**) v_{dc} and v_{cd} and (**b**) grid angle response when AVC bandwidth change from 20 Hz to 5 Hz.

Figure 16a shows the output response of the grid angle when the grid strength changes, and Figure 16b shows the DC-link voltage and ac *d*-axis voltage. The grid inductor changes from L_g = 4.5 mH to L_g = 5.15 mH at 1 s. Based on Figure 16a,b, it is observed that, for a step change of grid inductance that characterizes the grid strength, the induced oscillations die away with time in both control methods. However, for the proposed method, the system oscillations are rapidly damped compared to the traditional method. For both output voltage and grid angle, the steady-state is attained at time 1.3 s for the proposed control, whereas for the conventional method, the steady-state is reached at time 1.8 s. Furthermore, the transient overshoot for the proposed control is reduced compared to the conventional method.

For the convention method, the peak to peak overshot of 0.1 radians, 0.2 P.U and 0.03 P.U for grid angle, *d*-axis ac voltage and DC-link voltage, respectively, are obtained. The peak to peak overshoot when the oscillation damping control method is applied becomes 0.13 radians, 0.125 P.U and 0.0075 P.U for grid angle, *d*-axis ac voltage and DC-link voltage, respectively. The same performance can be observed in Figure 17b showing the output reactive power and active power under step change of grid inductance. Higher reactive power overshoots of 0.08 P.U for undamped system and 0.05 for the proposed control are yielded.

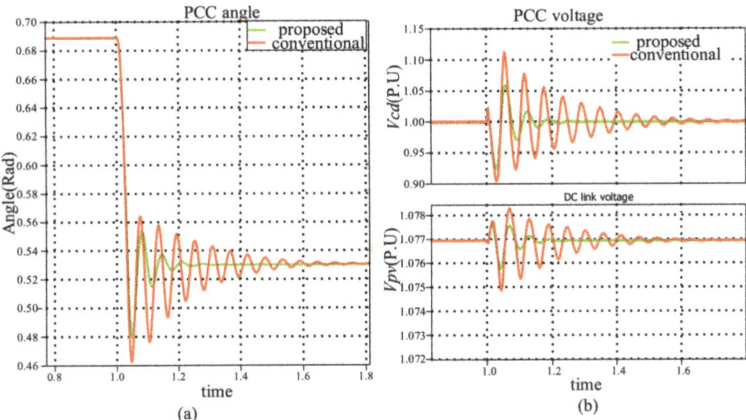

Figure 16. (**a**) PCC voltage angle and (**b**) v_{dc} and v_{cd} response when grid inductor change from 4.5 mH to 5.15 mH.

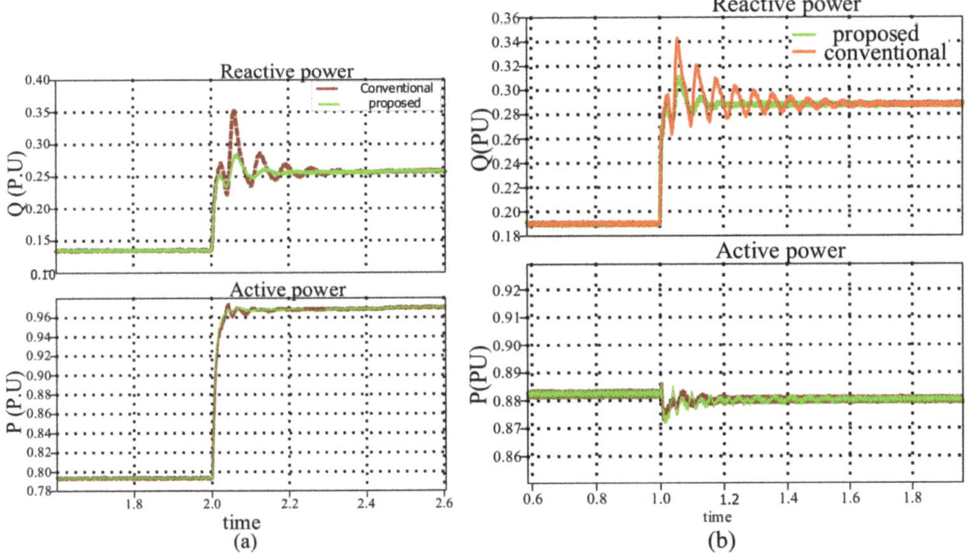

Figure 17. Reactive and active power: (**a**) step change in solar insolation and (**b**) grid inductor variation.

Apart from control and the system parameter, another factor affecting the stability of the system is the power rating of the system. The power rating depends on the solar insolation, which can change depending on weather conditions. Figure 17b shows the output power response, Figure 18a shows the grid voltage angle and Figure 18b shows the DC-link and ac d-axis voltage for solar insolation changes from 900 W/m^2 to 1100 W/m^2. Following that step change of solar insolation, the DC-link voltage changes from 1.07 P.U to 1.083 PU, whereas the active power changes from 0.785 to 0.97 PU. Correspondingly, the reactive power changes from 0.14 to 0.26 PU. Based on the results shown in Figure 17, a highly damped response is produced while the proposed method is used. It can be observed that the system remains stable in both control methods as the oscillations resulting from a step-change in solar insolation die away. However, the settling time for the proposed method is shortened to 0.2 s from 0.6 s (for the convention method). In addition, similar to the analysis carried above, the transient overshoots of the proposed method are very low compared to the response of the undamped system. For the damped system, the reactive power overshoot is 0.01 PU, while the undamped system yields 0.09 PU, the ac capacitor voltage angle overshoot is 0.15 radians and 0.075 radians for conventional and proposed control method, respectively. In addition, the overshoot of *d*-axis ac voltage is 0.15 P.U and 0.05 for the conventional and proposed methods, respectively.

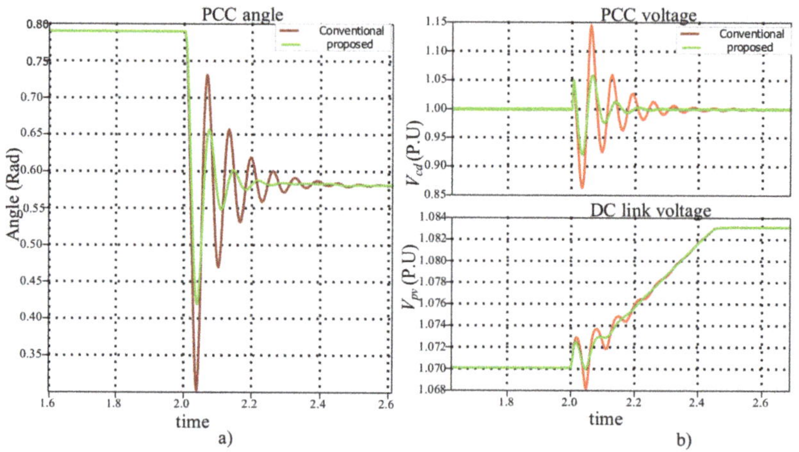

Figure 18. (**a**) PCC voltage angle and (**b**) *Vpv* and *Vcd* step change in solar insolation.

9. Conclusions

This paper firstly analyzed the stability of a single-stage PV inverter connected to a weak grid. The small-signal stability analysis was performed to evaluate the conventional control method's performance and highlighted possible sources of instabilities. The main sources of instabilities were found to be the grid dynamics, the outer control loops and the PLL. The interaction of outer control loops of the inverter and the weak grid could worsen the system's stability, thus limiting the level of integration of the grid-connected PV inverter. In order to boost the performances of the grid-connected PV inverter, a control methodology that could limit the control loops interaction was proposed. The proposed control method employs DC-link voltage error to modulate the reference reactive power current flowing into the grid.

The results of simulation in PLECS were presented to verify the performance of the proposed control methodology over the conventional method. According to the simulation results, the proposed system achieves improved system responses compared to the conventional control method. For the proposed control method, the enhanced oscillations damping and reduced transients overshot are guaranteed for a large variation range of the DC-link, PCC voltage and PLL controller bandwidth. Furthermore, the system's robustness is improved, even when the inverter is connected to a very weak grid, thanks to the proposed controller. The simulation results confirm the robust performance of the derived small-signal model and the proposed low frequencies damping method.

Author Contributions: Conceptualization, A.M., H.L. and X.Z.; methodology, A.M. and H.L.; software, A.M.; validation, A.M., Y.Y. and X.Z.; formal analysis, A.M.; investigation, A.M.; resources, H.L. and X.Z.; data curation, A.M.; writing—original draft preparation, A.M.; writing—review and editing, A.M. and X.Z.; visualization, A.M.; supervision, H.L.; project administration, H.L.; funding acquisition, H.L. All authors have read and agreed to the published version of the manuscript.

Funding: This research received no external funding.

Conflicts of Interest: The authors declare no conflict of interest.

References

1. Huang, Y.; Yuan, X.; Hu, J.; Zhou, P. Modeling of VSC Connected to Weak Grid for Stability Analysis of DC-Link Voltage Control. *IEEE J. Emerg. Sel. Top. Power Electron.* **2015**, *3*, 1193–1204. [CrossRef]
2. IEEE. *IEEE Guide for Planning DC Links Terminating at AC Locations Having Low Short-Circuit Capacities*; IEEE Std 1204-1997; IEEE: New York, NY, USA, 1997; pp. 1–216.
3. Davari, M.; Mohamed, Y.A.-R.I. Robust Vector Control of a Very Weak-Grid-Connected Voltage-Source Converter Considering the Phase-Locked Loop Dynamics. *IEEE Trans. Power Electron.* **2017**, *32*, 977–994. [CrossRef]
4. Li, Y.; Fan, L.; Miao, Z. Stability Control for Wind in Weak Grids. *IEEE Trans. Sustain. Energy* **2019**, *10*, 2094–2103. [CrossRef]
5. Rezaee, S.; Radwan, A.; Moallem, M.; Wang, J. Voltage Source Converters Connected to Very Weak Grids: Accurate Dynamic Modeling, Small-Signal Analysis, and Stability Improvement. *IEEE Access* **2020**, *8*, 201120–201133. [CrossRef]
6. Yazdani, A. Voltage-Sourced Converters in Power Systems. In *Modeling, Control, and Applications*; John Wiley & Sons, Inc.: Hoboken, NJ, USA, 2010.
7. Egea-Alvarez, A.; Fekriasl, S.; Hassan, F.; Gomis-Bellmunt, O. Advanced Vector Control for Voltage Source Converters Connected to Weak Grids. *IEEE Trans. Power Syst.* **2015**, *30*, 3072–3081. [CrossRef]
8. Durrant, M.; Werner, H.; Abbott, K. Model of a VSC HVDC terminal attached to a weak AC system. In Proceedings of the 2003 IEEE Conference on Control Applications, Istanbul, Turkey, 23–25 June 2003; Volume 1, pp. 178–182.
9. Imhof, M.C. Voltage Source Converter Based HVDC- Modelling and Coordinated Control to Enhance Power System Stability. Ph.D. Thesis, ETH Zurich, EEH—Power Systems Laboratory, Zurich, Switzerland, 2015.
10. Zhang, H.; Harnefors, L.; Wang, X.; Hasler, J.-P.; Nee, H.-P. SISO Transfer Functions for Stability Analysis of Grid-Connected Voltage-Source Converters. *IEEE Trans. Ind. Appl.* **2018**, *55*, 2931–2941. [CrossRef]
11. Se-Kyo, C. A phase tracking system for three phase utility interface inverters. *IEEE Trans. Power Electron.* **2000**, *15*, 431–438. [CrossRef]
12. Fan, L. Modeling Type-4 Wind in Weak Grids. *IEEE Trans. Sustain. Energy* **2019**, *10*, 853–864. [CrossRef]
13. Huang, Y.; Yuan, X.; Hu, J.; Zhou, P.; Wang, D. DC-Bus Voltage Control Stability Affected by AC-Bus Voltage Control in VSCs Connected to Weak AC Grids. *IEEE J. Emerg. Sel. Top. Power Electron.* **2016**, *4*, 445–458. [CrossRef]
14. Fan, L.; Miao, Z. Wind in Weak Grids: 4 Hz or 30 Hz Oscillations? *IEEE Trans. Power Syst.* **2018**, *33*, 5803–5804. [CrossRef]
15. Kalcon, G.O.; Adam, G.P.; Anaya-Lara, O.; Lo, S.; Uhlen, K. Small-Signal Stability Analysis of Multi-Terminal VSC-Based DC Transmission Systems. *IEEE Trans. Power Syst.* **2012**, *27*, 1818–1830. [CrossRef]
16. Suul, J.; Molinas, M.; D'Arco, S.; Rodriguez, P. Extended stability range of weak grids with Voltage Source Converters through impedance-conditioned grid synchronization. In Proceedings of the 11th IET International Conference on AC and DC Power Transmission, Birmingham, UK, 10–12 February 2015; pp. 1–10.
17. Jia, Q.; Yan, G.; Cai, Y.; Li, Y.; Zhang, J. Small-signal stability analysis of photovoltaic generation connected to weak AC grid. *J. Mod. Power Syst. Clean Energy* **2018**, *7*, 254–267. [CrossRef]
18. Shen, L.; Asher, G.; Wheeler, P.; Bozhko, S.; Wheeler, P. Optimal LCL filter design for 3-phase Space Vector PWM rectifiers on variable frequency aircraft power system. In Proceedings of the 2013 15th European Conference on Power Electronics and Applications (EPE), Lille, France, 3–5 September 2013; pp. 1–8.
19. Villalva, M.G.; Gazoli, J.R.; Filho, E.R. Comprehensive Approach to Modeling and Simulation of Photovoltaic Arrays. *IEEE Trans. Power Electron.* **2009**, *24*, 1198–1208. [CrossRef]
20. Yazdani, A.; Di Fazio, A.R.; Ghoddami, H.; Russo, M.; Kazerani, M.; Jatskevich, J.; Strunz, K.; Leva, S.; Martinez, J.A. Modeling Guidelines and a Benchmark for Power System Simulation Studies of Three-Phase Single-Stage Photovoltaic Systems. *IEEE Trans. Power Deliv.* **2011**, *26*, 1247–1264. [CrossRef]
21. Mahmoud, Y.; Xiao, W.; Zeineldin, H.H. A Simple Approach to Modeling and Simulation of Photovoltaic Modules. *IEEE Trans. Sustain. Energy* **2011**, *3*, 185–186. [CrossRef]
22. Zhang, L.; Harnefors, L.; Nee, H.-P. Power-Synchronization Control of Grid-Connected Voltage-Source Converters. *IEEE Trans. Power Syst.* **2010**, *25*, 809–820. [CrossRef]
23. Sumathi, S.; Kumar, L.A.; Surekha, P. Soft Computing Techniques in Solar PV. In *Solar PV and Wind Energy Conversion Systems: An Introduction to Theory, Modeling with MATLAB/SIMULINK, and the Role of Soft Computing Techniques*; Sumathi, S., Ashok Kumar, L., Surekha, P., Eds.; Springer International Publishing: Cham, Switzerland, 2015; pp. 145–245.
24. Xu, J.; Qian, Q.; Xie, S.; Zhang, B. Grid-Voltage Feedforward Based Control for Grid-Connected LCL-Filtered Inverter with High Robustness and Low Grid Current Distortion in Weak Grid. In Proceedings of the 2016 IEEE Applied Power Electronics Conference and Exposition (APEC), Long Beach, CA, USA, 20–24 March 2016; pp. 1919–1925.
25. Pogaku, N.; Prodanovic, M.; Green, T. Modeling, Analysis and Testing of Autonomous Operation of an Inverter-Based Microgrid. *IEEE Trans. Power Electron.* **2007**, *22*, 613–625. [CrossRef]
26. Pradhan, J.K.; Ghosh, A.; Bhende, C. Small-signal modeling and multivariable PI control design of VSC-HVDC transmission link. *Electr. Power Syst. Res.* **2017**, *144*, 115–126. [CrossRef]

27. Wang, X.; Ruan, X.; Liu, S.; Tse, C.K. Full Feedforward of Grid Voltage for Grid-Connected Inverter With LCL Filter to Suppress Current Distortion Due to Grid Voltage Harmonics. *IEEE Trans. Power Electron.* **2010**, *25*, 3119–3127. [CrossRef]
28. Xu, J.; Xie, S.; Tang, T. Improved control strategy with grid-voltage feedforward for LCL-filter-based inverter connected to weak grid. *IET Power Electron.* **2014**, *7*, 2660–2671. [CrossRef]

Article

Short Term Active Power Load Prediction on A 33/11 kV Substation Using Regression Models

Venkataramana Veeramsetty [1], Arjun Mohnot [2], Gaurav Singal [2] and Surender Reddy Salkuti [3,*]

[1] Center for Artificial Intelligence and Deep Learning, Department of Electrical and Electronics Engineering, S R Engineering College, Warangal 506371, India; dr.vvr.research@gmail.com
[2] Department of Computer Science Engineering, Bennett University, Greater Noida 201310, India; arjunmohnot@gmail.com (A.M.); gauravsingal789@gmail.com (G.S.)
[3] Department of Railroad and Electrical Engineering, Woosong University, Daejeon 34606, Korea
* Correspondence: surender@wsu.ac.kr

Abstract: Electric power load forecasting is an essential task in the power system restructured environment for successful trading of power in energy exchange and economic operation. In this paper, various regression models have been used to predict the active power load. Model optimization with dimensionality reduction has been done by observing correlation among original input features. Load data has been collected from a 33/11 kV substation near Kakathiya University in Warangal. The regression models with available load data have been trained and tested using Microsoft Azure services. Based on the results analysis it has been observed that the proposed regression models predict the demand on substation with better accuracy.

Keywords: dimensionality reduction; simple linear regression; multiple linear regression; polynomial regression; load forecasting

Citation: Veeramsetty, V.; Mohnot, A.; Singal, G.; Salkuti, S.R. Short Term Active Power Load Prediction on A 33/11 kV Substation Using Regression Models. *Energies* **2021**, *14*, 2981. https://doi.org/10.3390/en14112981

Academic Editor: Mohamed Benbouzid

Received: 25 April 2021
Accepted: 18 May 2021
Published: 21 May 2021

Publisher's Note: MDPI stays neutral with regard to jurisdictional claims in published maps and institutional affiliations.

1. Introduction

Electric power industries are seeking electric power prediction tools to forecast the load so that balance between load and generation can be maintained properly. Prediction of active power load is required for arranging regular interval activities and power firms are increasing their infrastructure [1]. Accurate load forecasting systems provide a better understanding of the dynamics of existing power systems [2]. Electric load forecasting was classified into three categories as presented in Table 1 based on time horizon [3].

Short-term active power load prediction is vital to effective power system service, such as dispatching power into the network to prevent regular power outages. Short term active power estimation is a critical prerequisite for optimal dispatch of generators in power plants [4]. Customers would be able to select a more cost-effective energy usage scheme if the short-term load forecasting methodology was more accurate. It helps the power system to reduce cost of power production and to utilize resources optimally [5].

Artificial Intelligence (AI) is an integral part of many fields, some of the main subparts of AI are machine learning and swarm intelligence. Machine learning has become an integral part in many fields like civil engineering applications [6,7], image processing [8] and time series data prediction [9]. Swarm intelligence was developed by taking inspiration from the swarming behavior of various natural systems and this is used to solve various optimization problems [10,11].

Estimation methods to predict the active power load was classified into two classes as shown in Figure 1. Prediction tools were used to estimate solar irradiation, temperature and wind speed. ARIMA time series forecast model was developed in [12] to predict the temperature in Pakistan and it also develops a linear trend model to estimate electric power consumption. Digital Elevation Models were developed in [13] to predict the solar irradiation.

Forecasting techniques can help power system operators exchange active power for the highest possible benefit by calculating active power load and energy price. Electric energy price was predicted using artificial neural networks in [14] by considering day category, hour marker, holiday index, electric load, nonconventional energy generation and natural gas price as input features.

Table 1. Active power load prediction classification.

Load Prediction Type	Time	Usage
Short term	Few hours to days	Electric power generation and transmission scheduling
Medium term	Few weeks to months	Fuel purchase scheduling
Long term	1–10 years	Establishment of power sector entities

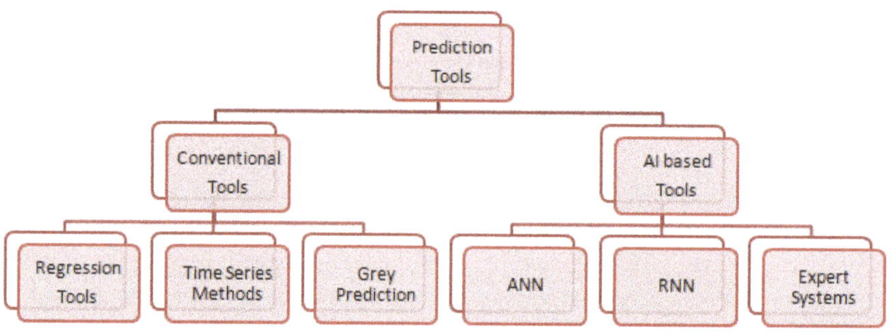

Figure 1. Prediction tools.

A new model was developed in [15] to predict the active power load. Active power load was estimated in [16] by considering day category, hour marker, holiday index, electric load, renewable energy generation using artificial neural networks and MLR model. Active power load was predicted in [17] based on load data of last four hours. Active power load was estimated in [18–20] based on load data for the last four hours and load data at the same hour for the last two days.

An ANN model was developed in [21] to forecast the half-hourly electric load demand in Tunisia. Authors have used a Levenberg–Marquardt learning algorithm to train the ANN model. Prediction of electricity demand and price one day ahead using functional models was discussed in [22]. Estimation of electric power consumption in Shanghai using grey forecasting model was discussed in [23]. All of these methods make useful advancements to load estimation, but these overlook useful elements such as dimensionality reduction, which improves model accuracy per number of model parameters.

In this paper, stochastic gradient descent optimizer [24] was used to update the parameters in the regression models. The methods described are analyzed by comparing them to previously developed models [17,18].

The main contributions of this paper are as follows:

- SLR and PR models were used to predict the active power load.
- A new approach, i.e., predict the active power load based on load at last three hours and load at one day before was used with various regression models and dimensionality reduction technique was used to reduce the complexity of the model so that overfitting problem was removed.
- Data analytic tools were used to process the data before feeding it to the model

2. Methodology

Active power load on a 33/11 kV substation has been predicted using regression models like SLR, MLR and PR. In all regression models, stochastic gradient optimizer has been used to update the parameters so that error, i.e., the difference between actual and predicted load is minimum.

2.1. Simple Linear Regression Model (SLR)

In SLR, output variable (Y_a) is related linearly with input variable (X_a). For a given input X_a, predicted output Y will be calculated using Equation (1). Gradient descent optimization method has been used to find the values of m and c for the given inputs (X_a) and corresponding outputs (Y_a), such that the total distance between all the output data points and line represented by Equation (1) is minimum, as shown in Figure 2. The main objective of gradient descent optimization method is minimization of half mean distance from all actual data points from line as shown in Equation (2). This half mean distance is also called error which represents the difference between actual output Y_a and predicted output (Y). As error which is going to minimize is a convex function, the gradient descent optimizer will perform well to reach global minimum point.

$$Y = mX_a + c \tag{1}$$

$$Min \ E = \frac{d_1 + d_2 + d_3 + d_4 + \ldots + d_s}{2n_s} = \sum_{a=1}^{n_s} \frac{d_a}{2n_s} = \sum_{a=1}^{s} \frac{\sqrt{(Y_a - Y)^2}}{2n_s} \tag{2}$$

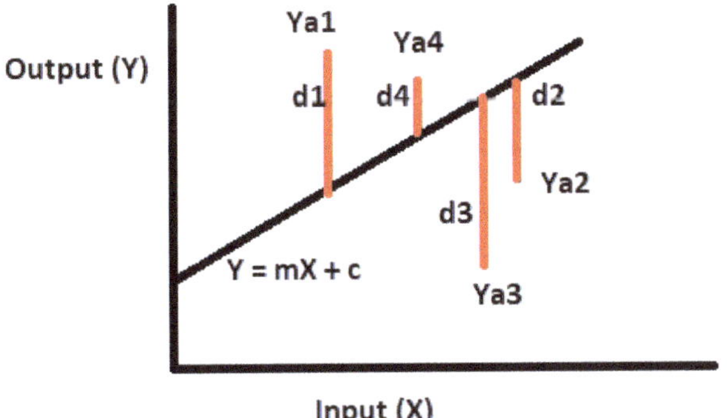

Figure 2. Distance between actual output (Ya) and predicted output (Y).

Gradient descent optimizer will update the solution such that it will reach the point where gradient is zero by choosing step size opposite to the gradient. In linear regression problem m and c are variables and step size for m, i.e., δm and step size for c, i.e., δc has been computed using Equations (3) and (4) respectively. Variables m and c will be updated using Equations (5) and (6), respectively, such that gradient will reach towards zero.

$$\delta m = -\eta \frac{\partial E}{\partial m} = \sum_{a=1}^{n_s} \eta (Y_a - mX_a - C) X_a \tag{3}$$

$$\delta c = -\eta \frac{\partial E}{\partial c} = \sum_{a=1}^{n_s} \eta (Y_a - mX_a - C) \tag{4}$$

$$m = m + \delta m \tag{5}$$

$$c = c + \delta c \tag{6}$$

SLR has been used to predict the active power load (L(D, t)) on a substation at particular hour (t) of the day (D) based on active power load (L(D−1, t)) at same time (t) but in previous day (D−1). In this scenario L(D−1, t) data points represent input (X_a), whereas L(D, t) data points represents output (Y_a). The procedure for load prediction using SLR is presented in Figure 3.

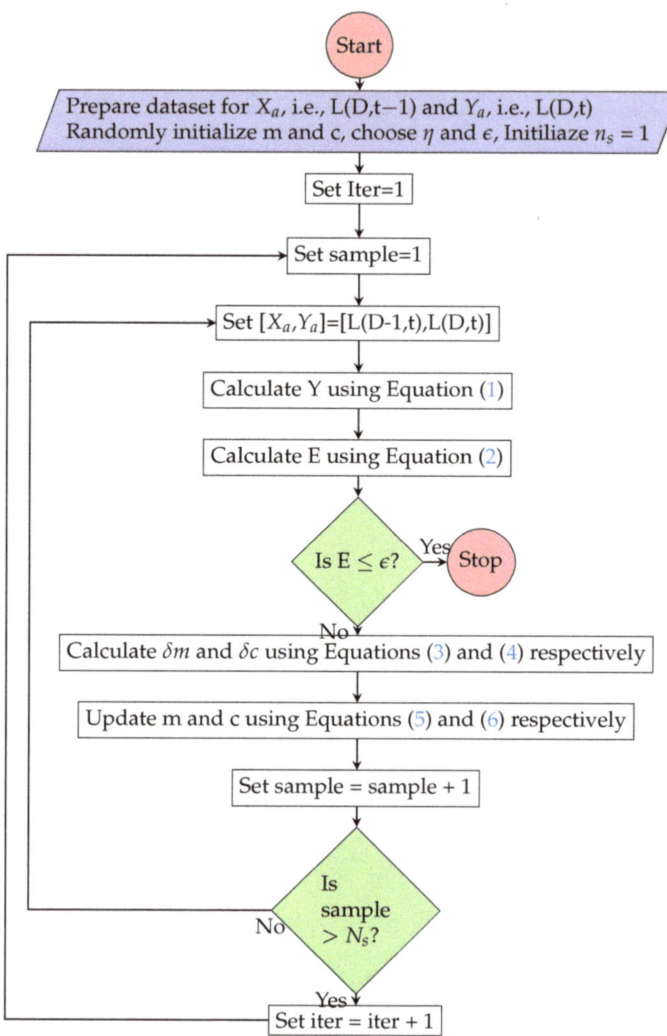

Figure 3. Simple Linear Regression model training algorithm.

2.2. Multiple Linear Regression Model (MLR)

In MLR, output variable (Y_a) is related linearly with multiple input variables like (X_1^a, X_2^a, $\cdots$, X_n^a). For a given input variables (X_1^a, X_2^a, $\cdots$, X_n^a), output Y will be predicted using Equation (7). Gradient descent optimization method has been used to find the values of (m_1, m_2, $\cdots$, m_n) and c for the given inputs (X_a) and corresponding outputs (Y_a), such that the half mean distance between all the output data points and line represented by Equation (7) is minimum as shown in Figure 4.

$$Y = m_1 X_1^a + m_2 X_2^a + ... + m_n X_n^a + c \tag{7}$$

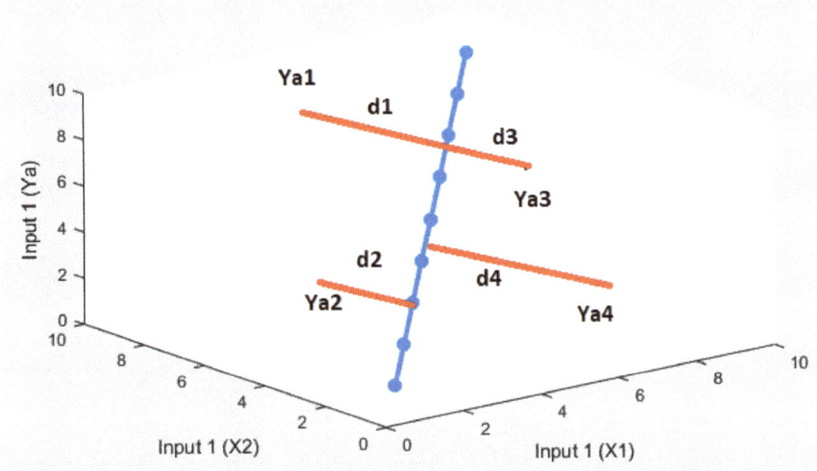

Figure 4. Distance between actual output (Ya) and predicted output (Y) for MLR.

In MLR the variables step size like $\delta m_i \in (\delta m_1, \delta m_2, \cdots, \delta m_n)$ and δc have been computed using Equations (8) and (9) respectively. Variables $\delta m_i \in (m_1, m_2, \cdots, m_n)$ and c will be updated using Equations (6) and (10) respectively such that gradient will reach toward zero.

$$\delta m_i = -\eta \frac{\partial E}{\partial m_i} = \sum_{a=1}^{n_s} \eta (Y_a - \sum_{i=1}^{n} m_i X_i^a - C) X_i^a \tag{8}$$

$$\delta c = -\eta \frac{\partial E}{\partial c} = \sum_{a=1}^{n_s} \eta (Y_a - \sum_{i=1}^{n} m_i X_i^a - C) \tag{9}$$

$$m_i = m_i + \delta m_i \tag{10}$$

MLR has been used to predict the active power load (L(D,t)) on a substation at particular hour (t) of the day (D) based on last three hours load data from time of prediction, i.e., L(D, t−1), L(D, t−2), L(D, t−3) and active power load (L(D−1, t)) at same time (t) but in previous day (D−1). In this scenario L(D, t−1), L(D, t−2), L(D, t−3) and L(D−1, t) data points represent input (X_a), whereas L(D,t) data points represents output (Y_a). The procedure for load prediction using MLR is presented in Figure 5.

2.3. Polynomial Regression Model (PR)

In PR, output variable (Y_a) is related nonlinearly with input variable (X_a). For a given input X_a, output Y will be predicted using Equation (11). Gradient descent optimization method has been used to find the values of $(m_1, m_2, \cdots, m_p)$ and c for the given input (X_a) and corresponding outputs (Y_a), such that the total distance between all the output data points and nonlinear curve represented by Equation (11) is minimum as shown in Figure 6. The main objective of gradient descent optimization method is minimization of half mean distance from all actual data points from nonlinear curve as shown in Equation (2).

$$Y = m_p X_a^p + m_{p-1} X_a^{p-1} + \ldots + m_1 X_a + c \tag{11}$$

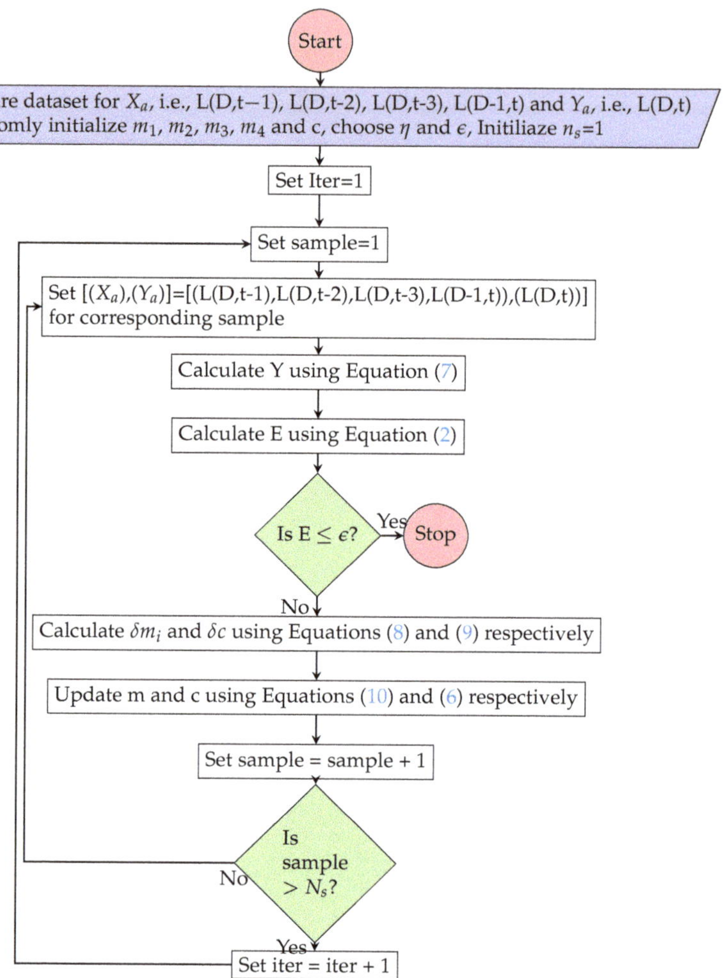

Figure 5. Multiple linear regression model training algorithm.

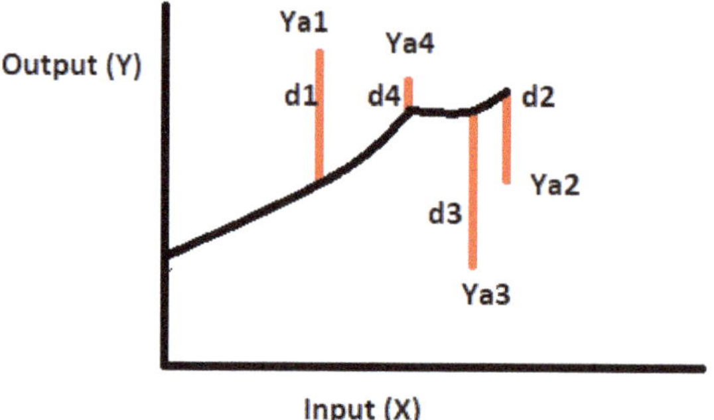

Figure 6. Distance between actual output (Ya) and predicted output (Y).

Gradient descent optimizer will update the solution such that it will reach the point where gradient is zero by choosing step size opposite to the gradient. In PR problem

$(m_1, m_2, \cdots, m_p)$ and c are variables and step size for m_i, i.e., δm_i and step size for c, i.e., δc have been computed using Equations (12) and (13) respectively. Variables δm_i and c will be updated using Equations (6) and (10), respectively, such that gradient will reach toward zero.

$$\delta m_i = -\eta \frac{\partial E}{\partial m_i} = \sum_{a=1}^{n_s} \eta \left(Y_a - \sum_{i=1}^{p} m_i X_a^i - c\right) X_a^i \tag{12}$$

$$\delta c = -\eta \frac{\partial E}{\partial c} = \sum_{a=1}^{n_s} \eta \left(Y_a - \sum_{i=1}^{p} m_i X_a^i - c\right) \tag{13}$$

PR model has been used to predict the active power load (L(D,t)) on a substation at particular hour (t) of the day (D) based on active power load (L(D−1,t)) at same time (t) but in previous day (D−1). In this scenario L(D,t) data points represents output (Y_a), whereas L(D−1,t) data points represent input (X_a). The procedure for load prediction using PR is presented in Figure 7.

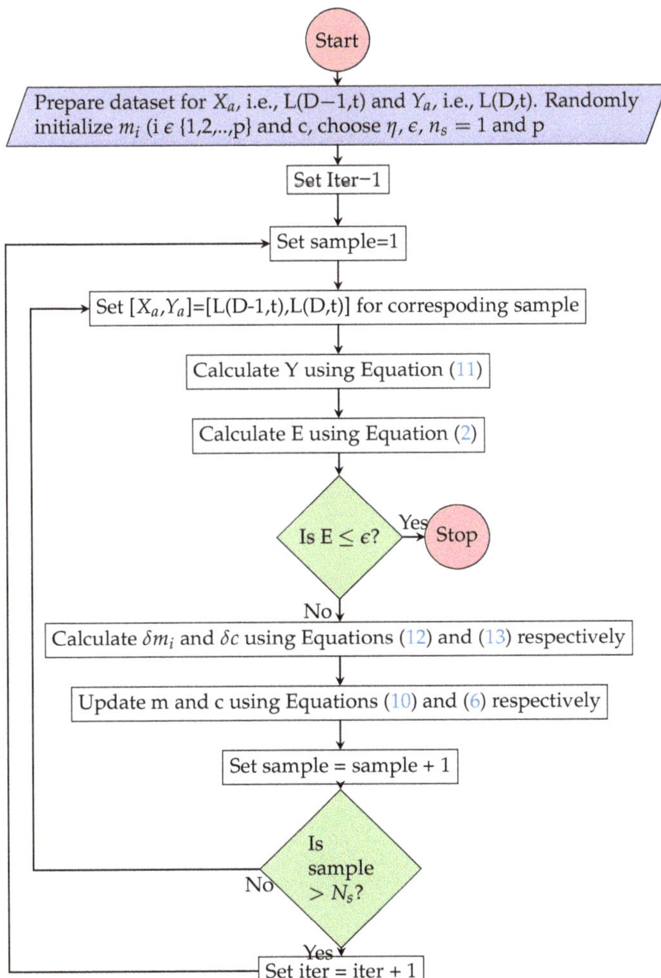

Figure 7. Polynomial Regression model training algorithm.

2.4. Dimensionality Reduction

In data science forecasting problems, there are often too many variables used to make the final estimate. These variables are also known as features. The more features there are, the more difficult it is to imagine the training set and then work on it. Occasionally, the majority of these characteristics are synonymous and therefore redundant. Features reduction algorithms come into play here. Feature reduction is the method of reducing the number of random variables taken into account by having a collection of principal variables.

Dimensionality technique based on correlation among input features was used to reduce complexity and to avoid the overfitting problem in MLR model.

2.5. Model Performance Metrics

Globally used error metrics such as MAE [25], MSE [26] and RMSE [27] as shown in Equations (14)–(16) respectively, were used to measure the performance, final decision and best model structure among simple, MLR models and PR model.

$$MAE = \frac{\sum_{i=1}^{N_s} |Y - Y_a|}{N_s} \tag{14}$$

$$MSE = \frac{\sum_{i=1}^{N_s} (Y - Y_a)^2}{N_s} \tag{15}$$

$$RMSE = \sqrt{\frac{\sum_{i=1}^{N_s} (Y - Y_a)^2}{N_s}} \tag{16}$$

The data from the electric utility, i.e., 33/11 kV substation was collected to train and test the machine learning model. The variation in predicted output with respect to each input feature is shown in Figure 8, the data samples are crammed together in a line. By observing the data distribution, we confidently begin with regression models rather than more complex advanced models that provide high nonlinear mapping between input and outputs, which were not necessary for this data. However, to improve prediction accuracy, some nonlinearity was added to the regression model in the form of a PR model, and multiple inputs were also tried. The proposed regression models are mathematically simple and have few model parameters (light weight model), allowing for quick computation and storage of the deployable model and also having good accuracy as the model considered, which is more suitable for the distribution of substation load data.

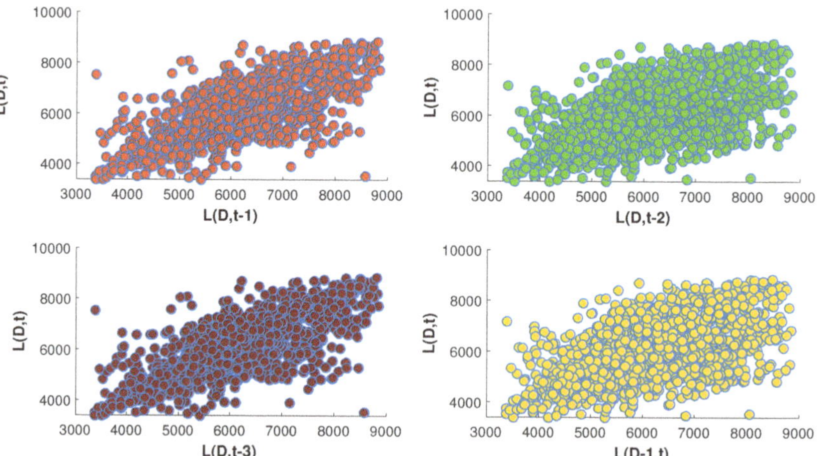

Figure 8. Impact of each input feature on output variable.

3. Results

Historical load data to train and test the models was considered from [28]. Date processing techniques for observing the data distribution and outliers and for data normalization have been used before using this data to train and test the regression model. Stochastic gradient descent optimizer has been used train the proposed regression models. The proposed regression models have been implemented and tested in cloud computing environment using Microsoft Azure Notebook [29].

3.1. Simple Linear Regression

A total of 2160 samples have been considered in the dataset; out of these, 1728 samples have been used for training and the remaining 432 samples have been used for testing.

3.1.1. Data Analysis

Statistical information of the active power load dataset is presented in Table 2. Scattering of data available in the dataset is presented in Figure 9. From the data scattering it has been observed that linear regression model can predict the load with good accuracy.

Table 2. SLR: statistical information of active power load dataset.

Parameter	Input	Output
count	2160	2160
mean	6037	6028
std	1066	1068
min	3378	3378
25%	5263	5260
50%	5950	5935
75%	6747	6739
max	8842	8842

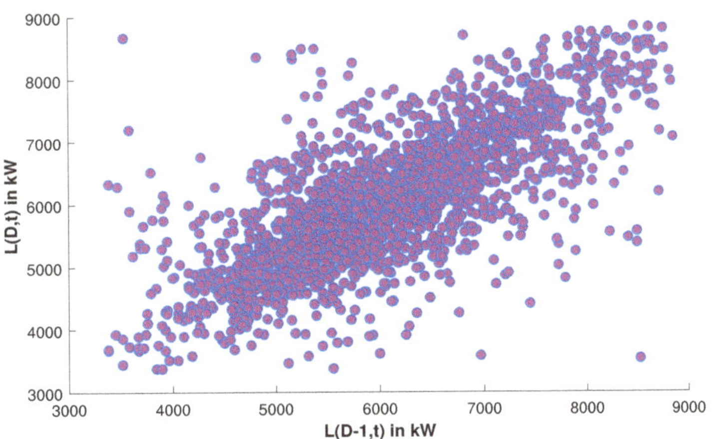

Figure 9. SLR: Scattering of data available in the dataset.

Box plot and histogram plot have been used to observe the input and output data distribution. Input and output data histograms are presented in Figure 10, and it has been observed that both input and output data samples follow the normal distribution.

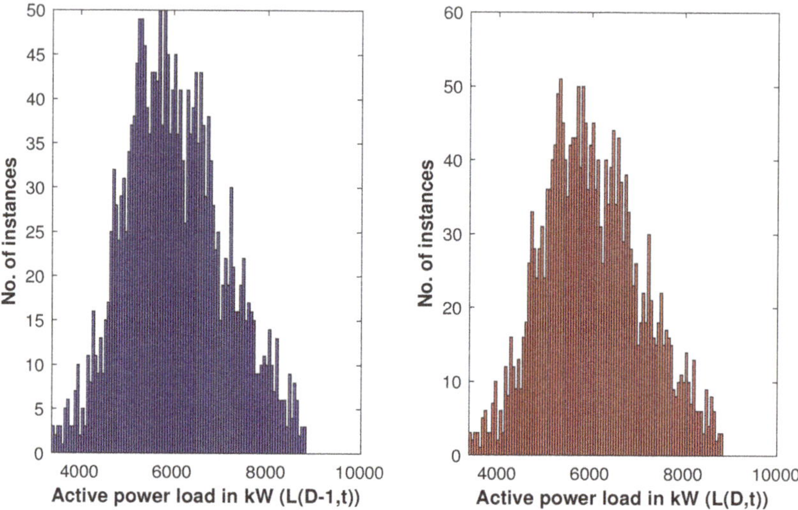

Figure 10. SLR: Data observation using histogram plot.

The box plot shown in Figure 11 is used to identify the outliers in the data and confirms that there are no outliers in the active power load dataset.

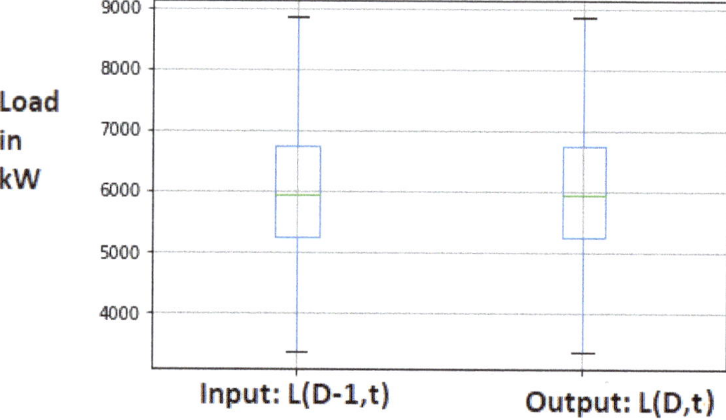

Figure 11. SLR: Data observation using box plot.

3.1.2. Simple Regression Model Performance Analysis

SLR model was trained with 1728 load data samples to find the optimum m and c values such that half mean distance from actual load data points to the line represented by optimum m and c values. The optimum m and c values after training using stochastic gradient descent optimizer are 0.7472 and 0.1173 respectively. The training performance of the model was measured in terms of MAE, MSE and RMSE and the respective values are 0.0973, 0.01742 and 0.132.

The distribution of actual load Y_a and load predicted using Equation (1) with training load data samples is presented in Figure 12 and similarly distribution of actual load Y_a and predicted load with testing load data samples is presented in Figure 13.

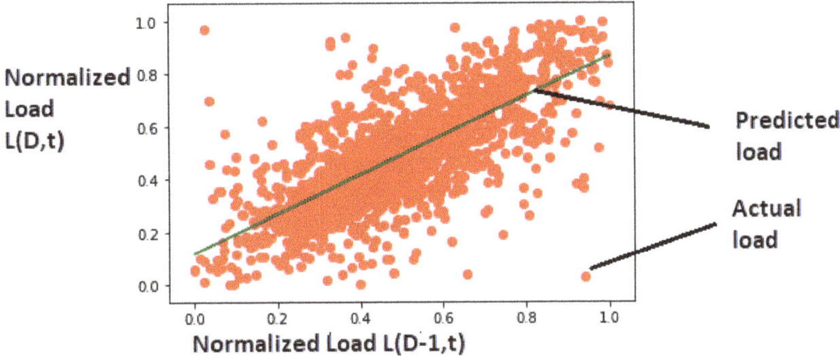

Figure 12. SLR: Distribution of actual load Y_a and predicted load with training dataset and with m = 0.7472 and c = 0.1173.

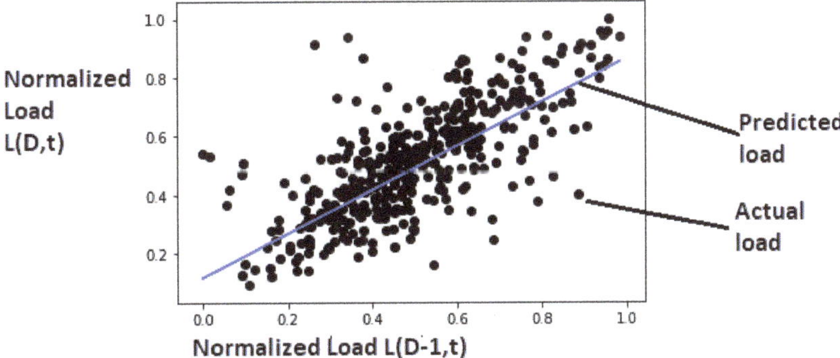

Figure 13. SLR: Distribution of actual load Y_a and predicted load with testing dataset and with m = 0.7472 and c = 0.1173.

The performance of the model was observed using MAE, MSE and RMSE and the respective values with testing dataset were 0.0939, 0.0163 and 0.1277. The comparison of actual load available in testing dataset and load predicted with trained SLR model is presented in Figure 14.

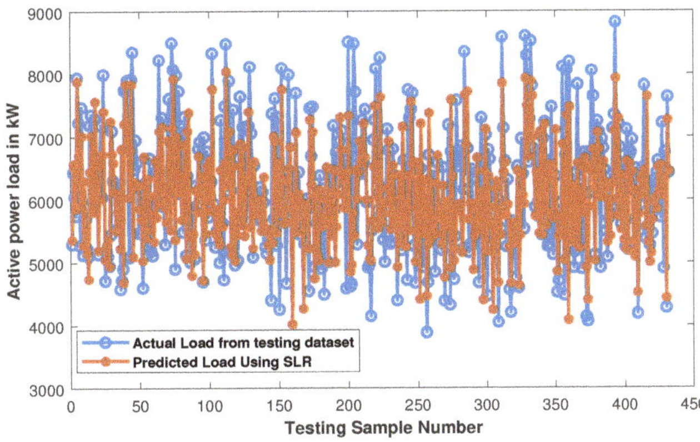

Figure 14. SLR: Comparison of actual load and predicted load with trained SLR model.

3.2. Multiple Linear Regression

A total of 2160 samples have been considered in the dataset; out of these, 1728 samples have been used for training and the remaining 432 samples have been used for testing of MLR model.

3.2.1. Data Analysis

Statistical information in terms of mean, standard deviation and inter quartile range of the active power load dataset is presented in Table 3. Dataset contains a total of four input parameters L(D,t−1), L(D,t−2), L(D,t−3) and L(D−1,t) and one output parameter L(D,t).

Table 3. MLR: Statistical information of active power load dataset.

Parameter	L(D,t−1)	L(D,t−2)	L(D,t−3)	L(D−1,t)	L(D,t)
count	2160	2160	2160	2160	2160
mean	6028	6028	6029	6037	6028
std	1068	1068	1068	1066	1068
min	3378	3378	3378	3378	3378
25%	5260	5260	5260	5263	5260
50%	5933	5935	5936	5950	5935
75%	6739	6739	6739	6747	6739
max	8842	8842	8842	8842	8842

Box plot and histogram plot have been used to observe the input and output data distribution. Input and output data histograms are presented in Figure 15, and it has been observed that both input and output data samples follow the normal distribution.

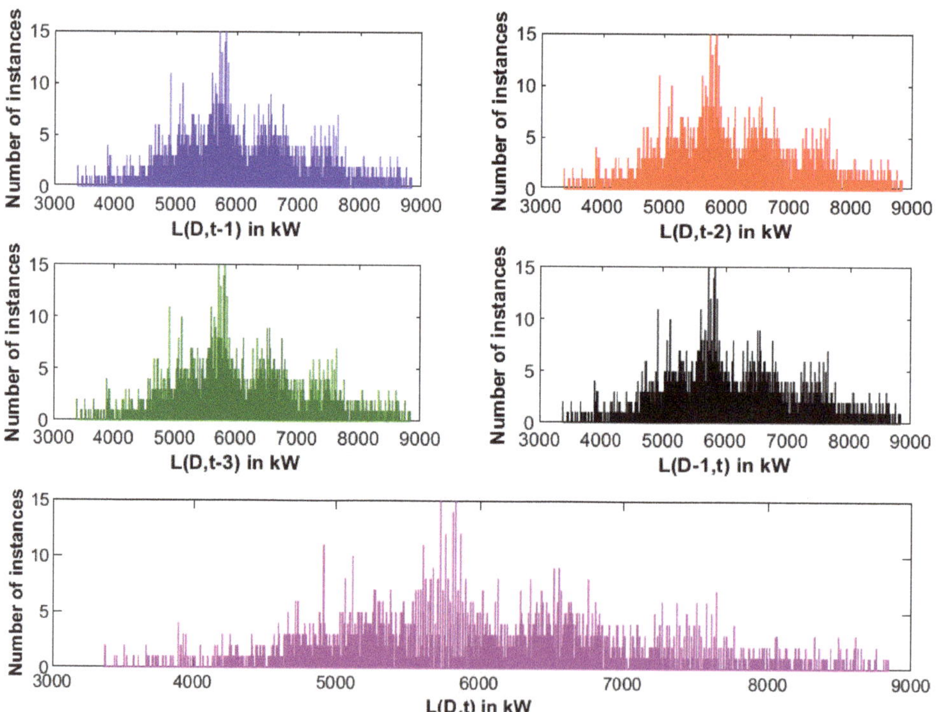

Figure 15. MLR: Data observation using histogram plot.

The box plot shown in Figure 16 is used to identify the abnormal data samples and it confirms that there are no outliers in the active power load dataset for MLR model.

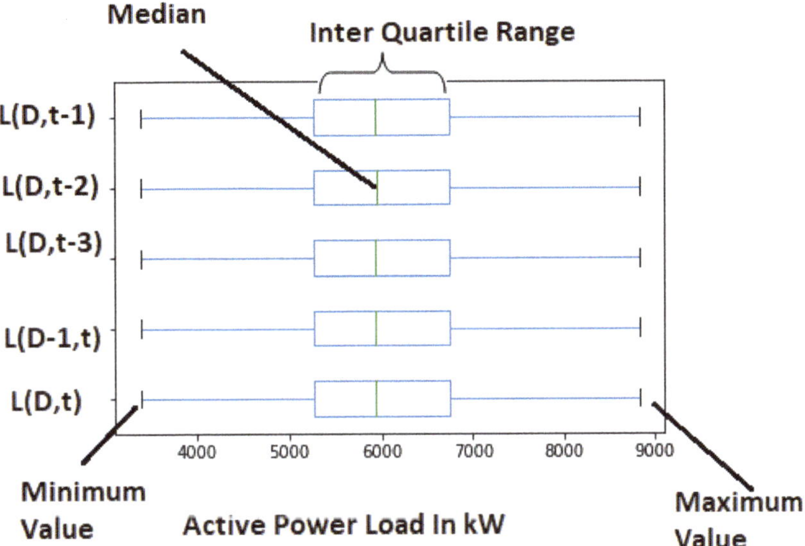

Figure 16. MLR: Data observation using box plot.

3.2.2. Multiple Regression Model Performance Analysis

MLR model was trained with 1728 load data samples to find the optimum m_4, m_3, m_2, m_1, and c values such that half mean distance from actual load data points to the line represented by optimum m_1, m_2, m_3, m_4 and c values. The optimum m_1, m_2, m_3, m_4 and c values after training using stochastic gradient descent optimizer are presented in Table 4. It has been observed that load at one hour before and one day before has positive and high impact on the predicted load. Similarly, load at two and three hours before has negative and low impact on the predicted load. The training performance of the MLR model was measured in terms of MAE, MSE and RMSE and the respective values are 0.0723, 0.0105 and 0.1026.

Table 4. MLR: coefficients and intercept information.

m_1	m_2	m_3	m_4	c
0.618733	−0.00289	−0.148361	0.374255	0.07541947

The performance of the model was observed using MAE, MSE and RMSE and the respective values with testing dataset were 0.0766, 0.0119 and 0.1092. The comparison of actual load available in testing dataset and load predicted with trained MLR model is presented in Figure 17.

3.2.3. MLR with Dimensionality Reduction (DR)

In MLR model load at particular hour (t) of the day (D) i.e., L(D,t) was predicted based on load data on last three hours from time of predicting and load at 24 h before. This means that a total of four input features were considered. Correlation among four input features were computed and presented in Table 5 and this information was used for dimensionality reduction, i.e., to reduce the number of input features and complexity of the model. As presented in Table 5, L(D,t−2) and L(D,t−3) have more than 75% correlation with L(D,t−1) and L(D,t−2), respectively. Hence feature L(D,t−2) was removed from input features and only L(D,t−1), L(D,t−3) and L(D−1,t) were considered as input features to predict the load L(D,t) using MLR model.

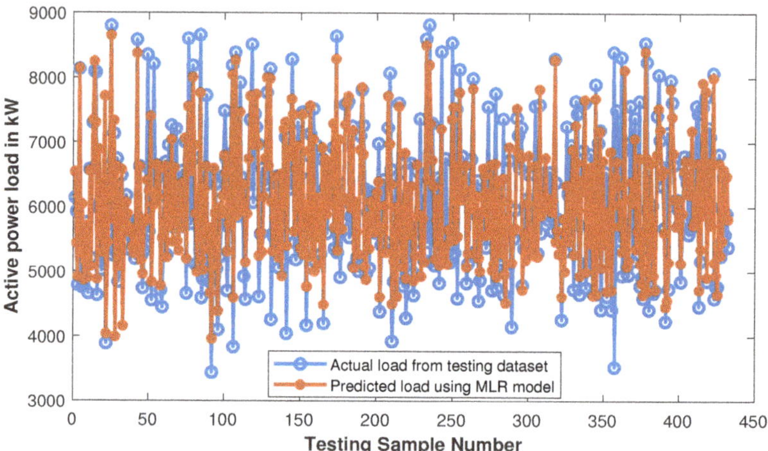

Figure 17. MLR: comparison of actual load and predicted load with trained MLR model.

Table 5. MLR: coefficients and intercept information.

Features	L(D,t−1)	L(D,t−2)	L(D,t−3)	L(D−1,t)
L(D,t−1)	1	0.783398	0.547211	0.660699
L(D,t−2)	0.783398	1	0.783336	0.472053
L(D,t−3)	0.547211	0.783336	1	0.256744
L(D−1,t)	0.660699	0.472053	0.256744	1

MLR model was trained with 1728 load data samples having three input features, i.e., L(D,t−1), L(D,t−3) and L(D−1,t) to predict the load L(D,t). The optimum m_1, m_2, m_3 and c values after training using stochastic gradient descent optimizer are presented in Table 6. It has been observed that load at one hour before and one day before has positive and high impact on the predicted load. Similarly, load three hours before has negative and low impact on the predicted load.

Table 6. MLR with dimensionality reduction: coefficients and intercept information.

m_1	m_2	m_3	c
0.596832	−0.137541	0.392209	0.07235393

The behavior of the MLR model prior to and after feature reduction was presented in Table 7. The behavior of the model was almost similar prior to and after dimensionality reduction in terms of error metrics.

Table 7. MLR with dimensionality reduction: performance of the model on training dataset.

	MAE	MSE	RMSE
With DR	0.0748	0.0113	0.107
Without DR	0.0723	0.0105	0.103

The behavior of the model with dimensionality reduction was compared with the behavior of the model without dimensionality reduction in terms of MAE, MSE and RMSE on testing dataset as presented in Table 8. The performance of the model was increased after applying the dimensionality reduction. It was happened due to removal of over fitting problem by reducing the number of input features and complexity of the model. The comparison of actual load available in testing dataset and load predicted with trained MLR model after reducing number of input features is presented in Figure 18.

Table 8. MLR with dimensionality reduction: performance of the model on testing dataset.

	MAE	**MSE**	**RMSE**
With DR	0.0679	0.009	0.093
Without DR	0.0766	0.0119	0.109

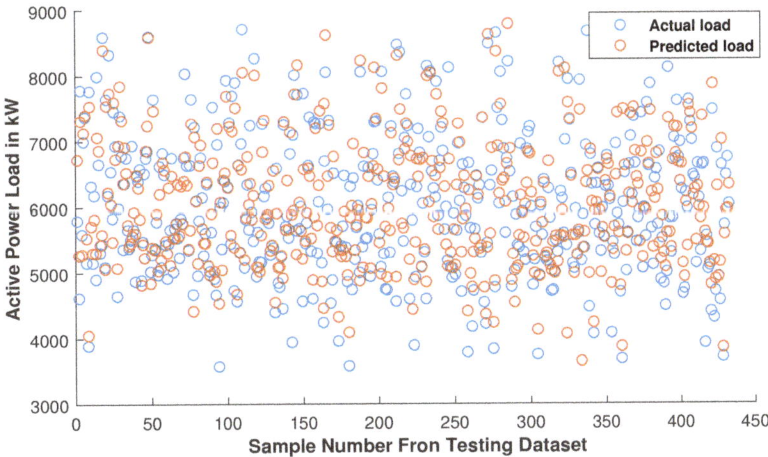

Figure 18. MLR With Dimensionality Reduction: comparison of actual load and predicted load with trained MLR model.

3.3. Polynomial Regression Model

A total of 2160 samples have been considered in the dataset; out of these, 1728 samples have been used for training and the remaining 432 samples have been used for testing of the PR model. The approach used for PR model is prediction of load at a particular hour of the day based on load at 24 h before, which is similar to the approach which was used for SLR. The same dataset was used for both simple and PR models.

PR Model Performance Analysis

PR model with different degree of polynomial (p) was trained with 1728 load data samples to find the optimum m_i where i ϵ {1,2,...,p} and c values so that half mean distance from actual load data points to the curve represented by optimum m_i and c values was minimum. Optimum m_i and c values for different degree of PR models were presented in Table 9.

Table 9. PR: coefficients and intercept information.

Degree (p)	m_1	m_2	m_3	m_4	m_5	m_6	m_7	m_8
2	0.6061	0.1345	NA	NA	NA	NA	NA	NA
3	−0.016	1.5525	−0.936	NA	NA	NA	NA	NA
4	−0.235	2.4303	−2.25139	0.65556	NA	NA	NA	NA
5	−1.458	9.886	−20.92	21.1449	−8.156	NA	NA	NA
10	3.8512	−90.59	857.68	−3885.2	10,407	17,572.45	19067.9	13006.7
15	20.437	−802.9	13357	−121,706	661,929	−2,132,575	3,267,138	2,932,709

Degree (p)	m_9	m_{10}	m_{11}	m_{12}	m_{13}	m_{14}	m_{15}	c
2	NA	NA	NA	NA	NA	NA	NA	0.15478
3	NA	NA	NA	NA	NA	NA	NA	0.22922
4	NA	NA	NA	NA	NA	NA	NA	0.24466
5	NA	NA	NA	NA	NA	NA	NA	0.30014
10	5113.5	-888.6	NA	NA	NA	NA	NA	0.25182
15	-3×10^7	7×10^7	-1×10^8	9.2×10^7	-5×10^7	18132030	-3×10^6	0.19706

The training of various PR models has been observed in terms of MAE, MSE and RMSE. Table 10 presents MAE, MSE and RMSE values during training for different PR models. It has been observed from Table 10 that the training performance metrics values, i.e., MAE, MSE and RMSE values decrease by increasing the degree of polynomial. This means that the training performance of the model is increasing with the degree of polynomial.

Table 10. PR: training performance metrics.

Polynomial Degree (p)	Training		
	MAE	MSE	RMSE
3	0.0973329	0.01732	0.131609177
4	0.0973174	0.01732	0.131591693
10	0.0970367	0.01717	0.131019329
15	0.0969165	0.01708	0.130697822
16	0.0968882	0.01708	0.130688064
18	0.0969342	0.01707	0.130644772
20	0.0970481	0.01705	0.130574147

The actual performance of various PR models has been observed with testing data and presented in Table 11. The performance metrics' values have been observed, i.e., MSE and RMSE values decrease up to the degree of polynomial 15 and beyond that increase. This means that if testing performance of the model increases with degree of polynomial up to 15 and beyond that decreases, it is due to overfitting of the model.

Table 11. PR: testing performance metrics.

Polynomial Degree (p)	Training		
	MAE	MSE	RMSE
3	0.09289	0.015969	0.12637
4	0.09276	0.015939	0.12625
10	0.09274	0.015889	0.12605
15	0.09295	0.015849	0.12589
16	0.09292	0.015853	0.12591
18	0.09308	0.015877	0.12600
20	0.09311	0.015900	0.12609

Figure 19 shows the variation of RMSE value of model with degree of polynomial on both training and testing dataset. It has been observed from the Figure 19 that training performance of the model increases with degree of polynomial. However, testing performance of the model increases with polynomial degree up to 15 but beyond that testing performance decreases due to overfitting problem. Hence, PR model with degree 15 is considered as the optimal model to deploy.

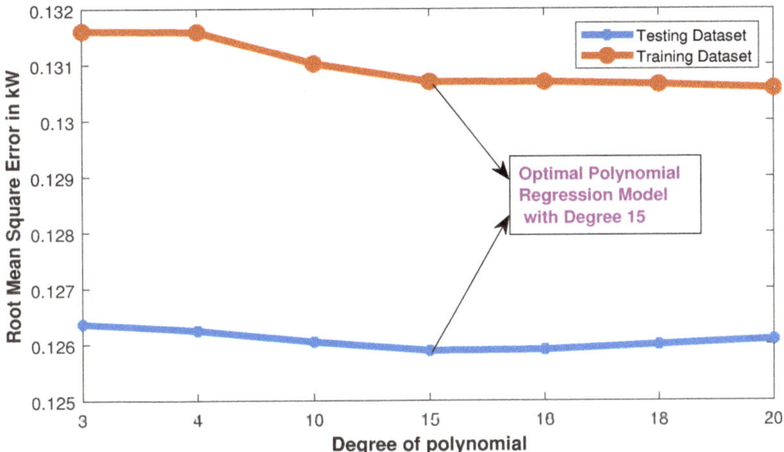

Figure 19. PR: identification of Optimal PR Model.

The comparison of actual load available in testing dataset and the load predicted with trained PR model having polynomial degree 15 is presented in Figure 20. It has been observed that most of the predicted load points are close to the actual load.

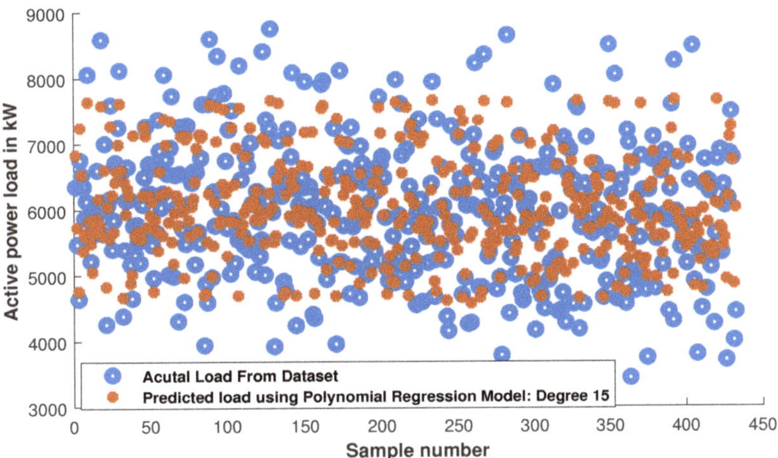

Figure 20. PR: comparison of actual load and predicted load with trained PR model.

3.4. Comparative Analysis

The proposed SLR model, MLR model with and without DR and PR model to predict the load on a 33/11 kV substation were compared in terms of MAE, MSE and RMSE as shown in Table 12. It has been observed that MLR model with DR has low MAE, MSE and RMSE values compared to simple linear and multiple linear without DR and PR models.

Which means that MLR model with DR predicted the load with good accuracy compared to simple linear and multiple linear without DR and PR models.

Table 12. Comparison of regression models' performance on testing dataset.

Regression Model	MAE	MSE	RMSE
SLR	0.0939	0.0163	0.1277
PR	0.0930	0.0158	0.1259
MLR	0.0766	0.0119	0.1092
MLR with DR	0.0679	0.009	0.093

The load on a substation varied from 3 MW to 9 MW during the observation period, and building a regression model by considering +/−0.5 MW load variation can be significant. The number of times each model failed to predict below the threshold limit is shown in Figure 21.

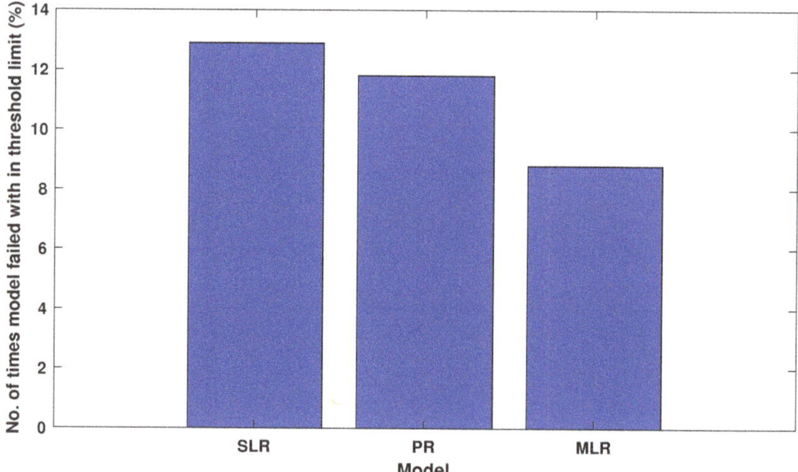

Figure 21. Model performance to predict load with in threshold limit.

The performance of the proposed models was compared with existing models available in the literature [17,18] in terms of training and testing MSE and presented in Table 13. The proposed multiple linear regression (MLR) with dimensionality reduction performs well with low mean square error 0.009 over the existing models.

Table 13. Performance comparison with existing complex models.

Model	Mean Square Error	
	Training	Testing
SLR	0.0973	0.0163
PR	0.0171	0.0158
MLR	0.0723	0.0119
MLR with DR	0.0748	0.009
[18]	0.23	0.44
[17]	0.29	1.59

In comparison with other techniques, the proposed regression models stand out on the data considered for the discussed problem in terms of speed due to light weight models, accuracy due to a suitable model chosen based on data distribution and simplicity due to a very simple model with 2 model parameters for SLR, 16 model parameters for PR and 5 model parameters for multi-variable linear regression model.

4. Conclusions

The SLR, MLR and PR models were used in this paper to estimate the active power load on a 33/11 kV substation. Dimensionality reduction based on correlation among MLR model input features was used to reduce the model's complexity and overfitting issue.

The analytical results indicate that the proposed regression models predicts active power load with good accuracy. The analytical results concluded that the proposed MLR model with dimensionality reduction performs better. This paper offers a detailed tool for network operators to efficiently exchange energy and run the network.

The methodology to predict load using regression models can be applied to other power system research areas such as LMP computation, effective trading in the energy market, power system deregulation and so on. This load prediction work can be expanded by taking into account sequential networks as well as weekdays and weekends.

Author Contributions: V.V. and S.R.S. constructed the research theories and methods; V.V. and S.R.S. developed the basic idea of the study and conducted the preliminary research; A.M. and G.S. performed the computer simulation and analyses; V.V. served as the head researcher in charge of the overall content of this study as well as modifications made. All authors have read and agreed to the published version of the manuscript.

Funding: This research was supported by S R Engineering College, Warangal, India, Woosong University, Republic of Korea and Bennett University, India.

Institutional Review Board Statement: Not applicable.

Informed Consent Statement: Not applicable.

Data Availability Statement: Active power load data used to train and test regression models is available at http://doi.org/10.17632/ycfwwyyx7d.2, accessed on 19 May 2021.

Acknowledgments: We thank S R Engineering College Warangal, India, Woosong University, South Korea and Bennett University, India for supporting us during this work. We also thank the engineers in 33/11 kV substation near Kakatiya University in Warangal for providing the historical load data.

Conflicts of Interest: The authors declare no conflict of interest.

Abbreviations

The following abbreviations are used in this manuscript:

d_a	Distance between actual and predicted outputs
DS	Dimensionality reduction
$L(D,t)$	Load at Day 'D' and at hour 't'
$L(D,t-1)$	Load at Day 'D' and at hour 't $-$ 1'
$L(D,t-2)$	Load at Day 'D' and at hour 't $-$ 2'
$L(D,t-3)$	Load at Day 'D' and at hour 't $-$ 3'
$L(D-1,t)$	Load at Day 'D $-$ 1' and at hour 't'
MAE	Mean Absolute Error
MLR	Multiple Linear Regression Model
MSE	Mean Square Error
N_a	Total number of samples
n_a	Batch size
p	Degree of polynomial
PR	Polynomial Regression Model
$RMSE$	Root Mean Square Error
SLR	Simple Linear Regression Model

X_a a^{th} sample from input dataset
X_i^a ith input parameter in a^{th} sample from input dataset
Y Predicted output using regression model
Y_a a^{th} sample from output dataset

References

1. Almeshaiei, E.; Soltan, H. A methodology for electric power load forecasting. *Alex. Eng. J.* **2011**, *50*, 137–144. [CrossRef]
2. Foldvik Eikeland, O.; Bianchi, F.M.; Apostoleris, H.; Hansen, M.; Chiou, Y.C.; Chiesa, M. Predicting Energy Demand in Semi-Remote Arctic Locations. *Energies* **2021**, *14*, 798. [CrossRef]
3. Alam, S.; Ali, M. Equation Based New Methods for Residential Load Forecasting. *Energies* **2020**, *13*, 6378. [CrossRef]
4. Mi, J.; Fan, L.; Duan, X.; Qiu, Y. Short-term power load forecasting method based on improved exponential smoothing grey model. *Math. Probl. Eng.* **2018**, *2018*. [CrossRef]
5. Hu, R.; Wen, S.; Zeng, Z.; Huang, T. A short-term power load forecasting model based on the generalized regression neural network with decreasing step fruit fly optimization algorithm. *Neurocomputing* **2017**, *221*, 24–31. [CrossRef]
6. Kumar, B.A.; Sangeetha, G.; Srinivas, A.; Awoyera, P.; Gobinath, R.; Ramana, V.V. Models for Predictions of Mechanical Properties of Low-Density Self-compacting Concrete Prepared from Mineral Admixtures and Pumice Stone. In *Soft Computing for Problem Solving*; Springer: New York, NY, USA, 2020; pp. 677–690.
7. Awoyera, P.; Akinmusuru, J.; Krishna, A.S.; Gobinath, R.; Arunkumar, B.; Sangeetha, G. Model Development for Strength Properties of Laterized Concrete Using Artificial Neural Network Principles. In *Soft Computing for Problem Solving*; Springer: New York, NY, USA, 2020; pp. 197–207.
8. Veeramsetty, V.; Singal, G.; Badal, T. Coinnet: platform independent application to recognize Indian currency notes using deep learning techniques. *Multimed. Tools Appl.* **2020**, *79*, 22569–22594. [CrossRef]
9. Bassi, S.; Gomekar, A.; Murthy, A.V. A learning algorithm for time series based on statistical features. *Int. J. Adv. Eng. Sci. Appl. Math.* **2019**, *11*, 230–235. [CrossRef]
10. Lakshmi, G.N.; Jayalaxmi, A.; Veeramsetty, V. Optimal placement of distributed generation using firefly algorithm. In *IOP Conference Series: Materials Science and Engineering*; IOP Publishing: Bristol, UK, 2020; Volume 981, p. 042060.
11. Veeramsetty, V.; Chintham, V.; D.M., V.K. Locational marginal price computation in radial distribution system using Self Adaptive Levy Flight based JAYA Algorithm and game theory. *Int. J. Emerg. Electr. Power Syst.* **2021**, *22*, 215–231.
12. Ali, M.; Iqbal, M.J.; Sharif, M. Relationship between extreme temperature and electricity demand in Pakistan. *Int. J. Energy Environ. Eng.* **2013**, *4*, 36. [CrossRef]
13. Mirmasoudi, S.; Byrne, J.; Kroebel, R.; Johnson, D.; MacDonald, R. A novel time-effective model for daily distributed solar radiation estimates across variable terrain. *Int. J. Energy Environ. Eng.* **2018**, *9*, 383–398. [CrossRef]
14. Panapakidis, I.P.; Dagoumas, A.S. Day-ahead electricity price forecasting via the application of artificial neural network based models. *Appl. Energy* **2016**, *172*, 132–151. [CrossRef]
15. Amral, N.; Ozveren, C.; King, D. Short term load forecasting using multiple linear regression. In Proceedings of the 2007 42nd International Universities Power Engineering Conference, Brighton, UK, 4–6 September 2007; pp. 1192–1198.
16. Kumar, S.; Mishra, S.; Gupta, S. Short term load forecasting using ANN and multiple linear regression. In Proceedings of the 2016 Second International Conference on Computational Intelligence & Communication Technology (CICT), Ghaziabad, India, 12–13 February 2016; pp. 184–186.
17. Shaloudegi, K.; Madinehi, N.; Hosseinian, S.; Abyaneh, H.A. A novel policy for locational marginal price calculation in distribution systems based on loss reduction allocation using game theory. *IEEE Trans. Power Syst.* **2012**, *27*, 811–820. [CrossRef]
18. Veeramsetty, V.; Chintham, V.; Vinod Kumar, D. Proportional nucleolus game theory–based locational marginal price computation for loss and emission reduction in a radial distribution system. *Int. Trans. Electr. Energy Syst.* **2018**, *28*, e2573. [CrossRef]
19. Veeramsetty, V.; Chintham, V.; DM, V.K. LMP computation at DG buses in radial distribution system. *Int. J. Energy Sect. Manag.* **2018**, *12*, 364–385. [CrossRef]
20. Veeramsetty, V.; Deshmukh, R. Electric power load forecasting on a 33/11 kV substation using artificial neural networks. *SN Appl. Sci.* **2020**, *2*, 855. [CrossRef]
21. Houimli, R.; Zmami, M.; Ben-Salha, O. Short-term electric load forecasting in Tunisia using artificial neural networks. *Energy Syst.* **2019**, *11*, 357–375. [CrossRef]
22. Lisi, F.; Shah, I. Forecasting next-day electricity demand and prices based on functional models. *Energy Syst.* **2019**, *11*, 947–979. [CrossRef]
23. Li, K.; Zhang, T. A novel grey forecasting model and its application in forecasting the energy consumption in Shanghai. *Energy Syst.* **2019**, *12*, 357–372. [CrossRef]
24. Shamir, O.; Zhang, T. Stochastic gradient descent for non-smooth optimization: Convergence results and optimal averaging schemes. In Proceedings of the 30th International Conference on Machine Learning, Atlanta, GA, USA, 16–21 June 2013; pp. 71–79.
25. Parol, M.; Piotrowski, P.; Kapler, P.; Piotrowski, M. Forecasting of 10-Second Power Demand of Highly Variable Loads for Microgrid Operation Control. *Energies* **2021**, *14*, 1290. [CrossRef]

26. Zheng, J.; Zhang, L.; Chen, J.; Wu, G.; Ni, S.; Hu, Z.; Weng, C.; Chen, Z. Multiple-Load Forecasting for Integrated Energy System Based on Copula-DBiLSTM. *Energies* **2021**, *14*, 2188. [CrossRef]
27. Jin, X.B.; Zheng, W.Z.; Kong, J.L.; Wang, X.Y.; Bai, Y.T.; Su, T.L.; Lin, S. Deep-Learning Forecasting Method for Electric Power Load via Attention-Based Encoder-Decoder with Bayesian Optimization. *Energies* **2021**, *14*, 1596. [CrossRef]
28. Venkataramana, V. Active Power Load Dataset. 2020. Available online: http://dx.doi.org/10.17632/ycfwwyyx7d.2 (accessed on 19 May 2021).
29. Barga, R.; Fontama, V.; Tok, W.H.; Cabrera-Cordon, L. *Predictive Analytics with Microsoft Azure Machine Learning*; Springer: New York, NY, USA, 2015.

Article

PV String-Level Isolated DC–DC Power Optimizer with Wide Voltage Range

Hyungjin Kim, Gibum Yu, Jaehoon Kim and Sewan Choi *

Department of Electrical & Information Engineering, Seoul National University of Science and Technology, Seoul 01811, Korea; jin1001@seoultech.ac.kr (H.K.); kibum7583@seoultech.ac.kr (G.Y.); habak@seoultech.ac.kr (J.K.)
* Correspondence: schoi@seoultech.ac.kr

Abstract: This paper proposes a photovoltaic (PV) string-level isolated DC–DC power optimizer with wide voltage range. A hybrid control scheme in which pulse frequency modulation (PFM) control and pulse width modulation (PWM) control are combined with a variable switching frequency is employed to regulate the wide PV voltage range. By adjusting the switching frequency in the above region during the PWM control process, the circulating current period can be eliminated and the turn-on period of the bidirectional switch of the dual-bridge LLC (DBLLC) resonant converter is reduced compared to that with a conventional PWM control scheme with a fixed switching frequency, resulting in better switching and conduction loss. Soft start-up control under a no-load condition is proposed to charge the DC-link electrolytic capacitor from 0 V. A laboratory prototype of a 6.25 kW DBLLC resonant converter with a transformer, including integrated resonant inductance, is built and tested in order to verify the performance and theoretical claims.

Keywords: isolated DC–DC converter; photovoltaics; LLC resonant converter; dual-bridge; wide voltage range; power optimizer

Citation: Kim, H.; Yu, G.; Kim, J.; Choi, S. PV String-Level Isolated DC–DC Power Optimizer with Wide Voltage Range. *Energies* **2021**, *14*, 1889. https://doi.org/10.3390/en14071889

Academic Editor: Marek Adamowicz

Received: 14 February 2021
Accepted: 24 March 2021
Published: 29 March 2021

Publisher's Note: MDPI stays neutral with regard to jurisdictional claims in published maps and institutional affiliations.

1. Introduction

Photovoltaic (PV) converters are key components in solar systems. These converters maximize the power extracted from PV cells for delivery to the grid. The power transferred to the utility grid is processed and organized in different concepts, as illustrated in Figure 1 [1–8]. Each grid-tie concept is a series of paralleled PV panels or strings connected to a couple of power converters (DC–DC converter and DC–AC inverter) based on the power level as well as the output voltage of the PV cells. According to the state-of-the-art technologies, grid-tie PV systems are single-stage and two-stage structures. The single-stage PV systems shown in Figure 1a,b contain only a DC–AC inverter. However, the two-stage structure employs DC-DC converters with DC–AC inverter, as shown in Figure 1c,d. Single-stage PV systems are simple and promise high efficiency due to the reduced power processing stages. Under certain ambient conditions, it is not possible to guarantee a DC-link requirement for the inverter [9,10]. Therefore, it is necessary to step up or step down the voltage of a PV string/multi-string to the required DC-link voltage for grid connection. Furthermore, the single-stage architecture does not extract the maximum energy from the PV modules, as the global maximum power point (MPP) is not the MPP for all individual modules, especially under partial shading, soiling, and mismatched conditions [11–13]. The drawbacks of the single-stage design can be alleviated using a two-stage type of architecture. Therefore, PV plants now favor two-stage architectures with DC–DC converters that reduce the effects of partial shading, improve the energy yield, provide more flexibility in the plant design process, and improve the monitoring and diagnostics capabilities [14,15].

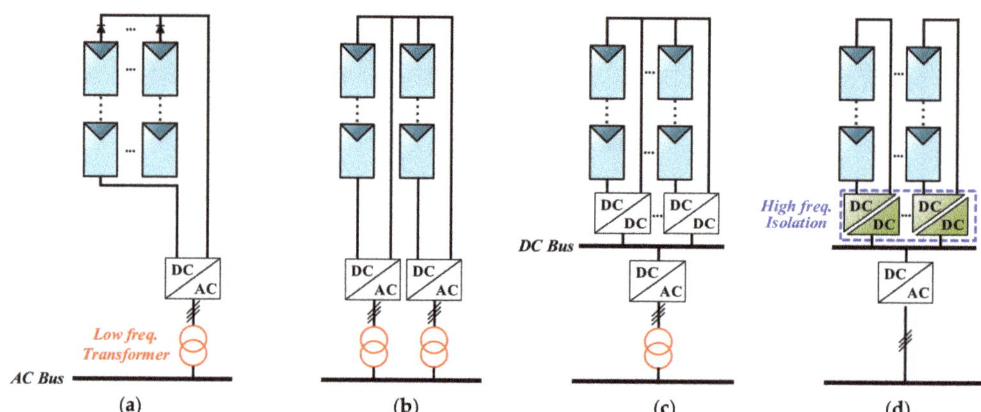

Figure 1. Grid-tie photovoltaic (PV) systems. (**a**) Centralized structure. (**b**) String structure. (**c**) Multi-string structure with non-isolated DC–DC converters. (**d**) Multi-string structure with isolated DC–DC converters.

Furthermore, in the two-stage structure, the DC–DC converter can be selectively designed as a buck, boost, or the buck–boost converter according to the PV connection. The boost converter is the best candidate if a significant step-up process is required. The basic non-isolated boost converter is the simplest solution due to its low part count and simple design. However, this converter tracks the MPP within a limited range and is associated with a limited switching frequency due to the hard switching operation. Moreover, PV systems employing a basic non-isolated boost converter necessitate a low-frequency transformer for the grid connection. Hence, the system cost and size are significantly increased. On the other hand, high-frequency isolated converters eliminate the need for a low-frequency transformer, reducing both the system size and the cost [16–18].

As a result, the isolated DC–DC converter has been considered to eliminate the low-frequency transformer [19–24]. In particular, the conventional full-bridge LLC resonant converter has an inherent entire-range zero-voltage switching (ZVS) characteristic with a low turn-off current of the primary switches [25–29]. For this reason, the resonant DC–DC converter with high-frequency isolation is widely used in various applications. However, the converter operates within a wide switching frequency range (above and below regions) under a wide PV voltage range, and low magnetizing inductance is required. The low magnetizing inductance increases the circulating current and conduction losses, especially at frequencies greater than the resonant frequency (f_r). The dual-bridge LLC (DBLLC) resonant converter presented in [30] is shown in Figure 2a. The topology operates with a fixed switching frequency (f_s) and the PV voltage V_{pv} is regulated by the mode change between the full-bridge (FB) and the half-bridge (HB) components within a half switching cycle, as shown in Figure 2b. The primary-side switches achieve ZVS turn-on while the secondary diodes are turned off with the zero-current switching (ZCS). The resonant tank of the aforementioned converter is simple given that the voltage gain is independent of the quality factor, and the magnetizing inductance has little influence on the voltage gain. Moreover, the topology utilizes high magnetizing inductance, leading to high efficiency.

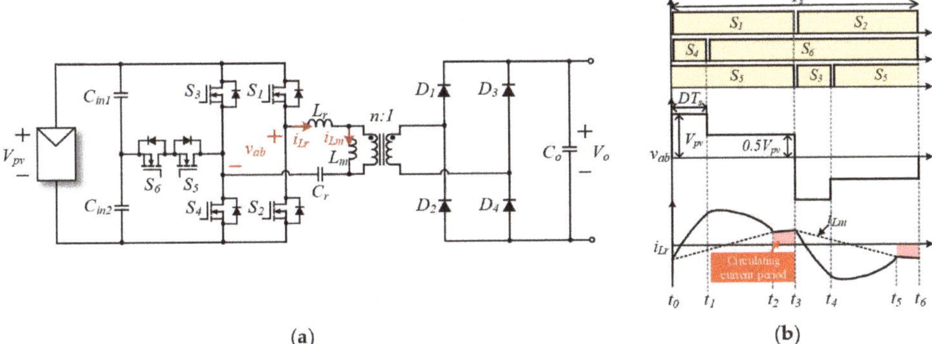

Figure 2. Dual-bridge LLC resonant converter. (**a**) Circuit configuration. (**b**) Key waveforms of the conventional pulse width modulation (PWM) control with fixed switching frequency ($f_s = f_r$).

However, in this topology, the circulating current flows for a longer duration, leading to a high root mean square (RMS) and peak current in the resonant tank. Moreover, the converter voltage gain is limited between 0.5 (HB mode) and 1 (FB mode) due to the implemented pulse width modulation (PWM) control at a fixed switching frequency. Furthermore, a soft start-up control scheme cannot be implemented as part of the traditional PWM method.

This paper proposes the PV string-level isolated power optimizer for wide voltage range. The aforementioned drawbacks of a DBLLC resonant converter are eliminated and the performance of the DBLLC resonant converter is improved. That is, the proposed power optimizer achieves wider gain and high efficiency compared to the conventional DBLLC resonant converter. Major contributions of this paper are listed as follows:

- The RMS and peak value of the resonant current are decreased by eliminating the circulating current, resulting in a reduced level of conduction loss.
- The implemented hybrid control method significantly increases the overall voltage gain. Hence, a wider voltage range is achieved.
- The current stress of the bidirectional switch is decreased because the conduction time is shorter with a minimized RMS current.
- Phase shift control is implemented for soft start-up operation, guaranteeing the initial charging of the DC-link electrolytic capacitor.

The proposed DBLLC resonant converter is presented in Section 2. In addition, design methods of the switching frequency range and transformer with integrated leakage inductance are presented in Section 3. In Section 4, the experimental results from a laboratory prototype of the proposed DBLLC resonant converter are provided to verify the theoretical analysis.

2. The Proposed Dual-Bridge LLC (DBLLC) Resonant Converter

2.1. Operation Principle and Characteristics

The dual-bridge LLC resonant converter has been considered to achieve wide voltage range [31–38]. As shown in Figure 2a, the DBLLC resonant converter has a bidirectional switch added to the conventional LLC full-bridge resonant converter for the wide PV voltage range. The DBLLC resonant converter with a fixed switching frequency in [30] operates in HB mode and/or the FB mode, providing the wide range operation with maximum gain of 2. Moreover, the DBLLC converter maintains the ZVS of all switches under entire voltage and load range. However, due to PWM control, the DBLLC converter generates a large circulating current, resulting in limitation on the efficiency.

This paper introduces the pulse frequency modulation (PFM) control operating under very wide voltage range from 300 V to 900 V based on the topology presented in [30]. In addition, in order to eliminate circulating current period, the boundary conduction mode (BCM) was introduced, thereby increasing the converter efficiency. Figure 3 shows the operation principle of the DBLLC resonant converter. The MPPT algorism is implemented by using two control methods: PFM control and PWM control. The PFM control is applied when the voltage gain M is between 0.33 and 0.5, and PWM control is applied when voltage gain M is between 0.5 and 1. Since the DBLLC resonant converter operates at the nominal MPPT voltage with PWM control, increasing the efficiency is important during the PWM control. Unlike conventional PWM control with a fixed switching frequency, the switching frequency of the proposed PWM control is varied according to the duty cycle D to increase the efficiency.

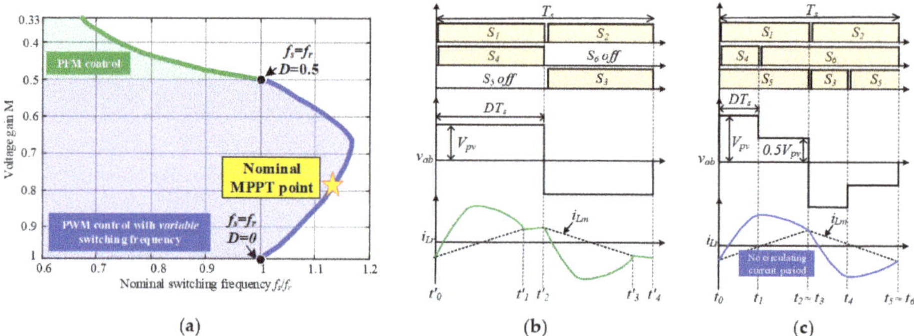

Figure 3. Operation principle of the dual-bridge LLC (DBLLC). (**a**) The proposed hybrid control scheme. (**b**) Key waveforms of the DBLLC resonant converter with Pulse Frequency Modulation (PFM) control at 300 V–450 V. (**c**) Key waveforms of the DBLLC resonant converter with the proposed PWM control scheme with a variable switching frequency of 450 V–900 V.

2.1.1. Pulse Frequency Modulation (PFM) Control

Figure 3b shows the key waveforms of the DBLLC resonant converter with PFM control for the voltage gain $M \in [0.33, 0.5]$. The converter is controlled by PFM at frequencies below the resonant frequency ($f_s < f_r$). Switches S_1 and S_2 operate in a diagonal manner with S_3 and S_4 at fixed 50% duty, while the bidirectional switch consisting of S_5 and S_6 is turned off. The voltage across the primary side of the transformer v_{ab} is denoted as V_{pv}. The operating principles of the DBLLC resonant converter in the PFM mode are not covered in this paper given that they are similar to the operating principles of a conventional full-bridge LLC resonant converter [39].

2.1.2. Pulse Width Modulation (PWM) Control with a Variable Switching Frequency

Figure 3c shows the key waveforms of the DBLLC resonant converter with PWM control of a varying switching frequency. This mode is considered within voltage gain M between 0.5 and 1. When operating in this mode, both the duty cycle D and the switching frequency are varied. Switches S_3 and S_4 operate in a diagonal manner with bidirectional switch S_5 and S_6, respectively. The duty cycles of S_3 and S_4 are also varied, while S_1 and S_2 operate at a fixed duty cycle of 0.5. The magnitude of the voltage across the resonant tank v_{ab} is V_{pv} in the FB mode and 0.5 V_{pv} in the HB mode. Because different voltages across the resonant tank lead to a circulating current period, the switching frequency is also varied in the above region such that the switching frequency f_s is higher than the resonant frequency f_r according to the duty cycle D to eliminate the circulating current period. Three operation modes during the half switching cycle are presented.

- Mode 1 $[t_0 - t_1]$;

This mode starts when S_1 and S_4 are turned on. ZVS operations of the S_1 and S_4 are achieved with the magnetizing current of the transformer. The resonant tank voltage v_{ab} equals the PV voltage V_{pv}. The series resonance between L_r and C_r starts while the voltage across the magnetizing inductor is clamped at nV_o. The equivalent circuit during this mode is shown in Figure 4a and the voltage across to the resonant tank is V_{pv}-V_o/n The resonant inductor current, resonant capacitor voltage and magnetizing current are expressed as shown below.

$$i_{Lr}(t) = (V_{pv} - \frac{V_o}{n} - v_{cr}(t_0))\frac{\sin(\omega_{r1}(t - t_0))}{Z_{r1}} + i_{Lr}(t_0)\cos(\omega_{r1}(t - t_0)) \tag{1}$$

$$v_{cr}(t) = V_{pv} - \frac{V_o}{n} - (V_{pv} - \frac{V_o}{n} - v_{cr}(t_0))\cos(\omega_{r1}(t - t_0)) + i_{Lr}(t_0)Z_{r1}\sin(\omega_{r1}(t - t_0)) \tag{2}$$

$$i_{Lm}(t) = i_{Lm}(t_0) + \frac{nV_o}{L_m}(t - t_0) \tag{3}$$

where the angular frequency between L_r and C_r $\omega_{r1} = 1/\sqrt{L_rC_r}$, resonant tank impedance of L_r and C_r $Z_{r1} = \sqrt{L_r/C_r}$ and $t_1 - t_0 = DT_s$. The power is delivered from PV to load in this mode.

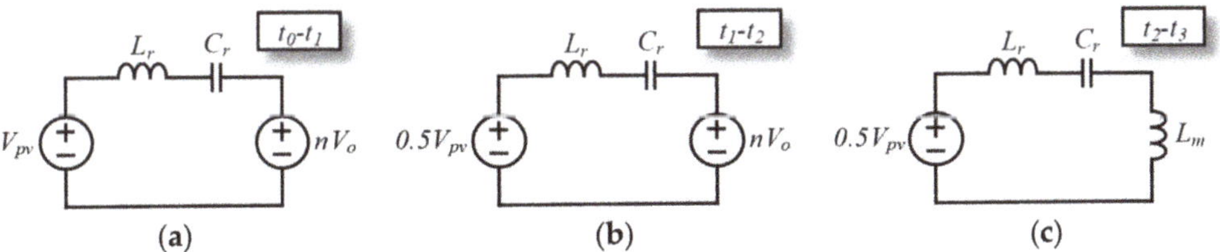

Figure 4. Equivalent circuit of the Pulse Width Modulation (PWM) control operation in the dual-bridge LLC resonant converter. (a) t_0–t_1. (b) t_1–t_2. (c) t_2–t_3.

- Mode 2 $[t_1$–$t_2]$;

When S_3 is turned off and S_6 is ZVS-turned on with resonant current, the resonant tank voltage v_{ab} equals half of the PV voltage of $0.5V_{pv}$, while the voltage across the magnetizing inductor is still clamped at nV_o. The equivalent circuit in this mode is shown in Figure 4b and the voltage across to the resonant tank is $V_{pv}/2 - V_o/n$. The resonant inductor current, resonant capacitor voltage, and magnetizing current are expressed as follows:

$$i_{Lr}(t) = (\frac{V_{pv}}{2} - \frac{V_o}{n} - v_{cr}(t_1))\frac{\sin(\omega_{r1}(t - t_1))}{Z_{r1}} + i_{Lr}(t_1)\cos(\omega_{r1}(t - t_1)) \tag{4}$$

$$v_{cr}(t) = \frac{V_{pv}}{2} - \frac{V_o}{n} - (\frac{V_{pv}}{2} - \frac{V_o}{n} - v_{cr}(t_1))\cos(\omega_{r1}(t - t_1)) + i_{Lr}(t_1)Z_{r1}\sin(\omega_{r1}(t - t_1)) \tag{5}$$

$$i_{Lm}(t) = i_{Lm}(t_1) + \frac{nV_o}{L_m}(t - t_1) \tag{6}$$

The power is delivered from PV to load in this mode. Mode 2 begins when the resonant inductor current i_{Lr} equals the magnetizing current i_{Lm}. However, in the proposed PWM control scheme with a variable switching frequency, since S_1 and S_5 are turned off when the resonant inductor current i_{Lr} equals the magnetizing current i_{Lm}, mode 3 can be eliminated.

- Mode 3 $[t_2$–$t_3]$;

In this mode, the resonant tank voltage v_{ab} still equals half of the PV voltage of $0.5V_{pv}$ and the voltage across the magnetizing inductor is no longer clamped due to the ZCS turn-off of the secondary diode. The magnetizing inductance L_m starts the resonance together with resonant inductor L_r and resonant capacitor C_r. The equivalent circuit while in this mode is shown in Figure 4b and the voltage across to the resonant tank is only $V_{pv}/2$. The resonant inductor current, resonant capacitor voltage, and magnetizing current are expressed as

$$i_{Lr}(t) = (\frac{V_{pv}}{2} - v_{cr}(t_2))\frac{\sin(\omega_{r2}(t - t_2))}{Z_{r2}} + i_{Lr}(t_2)\cos(\omega_{r2}(t - t_2)) \tag{7}$$

$$v_{cr}(t) = \frac{V_{pv}}{2} - (\frac{V_{pv}}{2} - v_{cr}(t_2))\cos(\omega_{r2}(t - t_2)) + i_{Lr}(t_2)Z_{r2}\sin(\omega_{r2}(t - t_2)) \tag{8}$$

$$i_{Lm}(t) = i_{Lr}(t) \tag{9}$$

where the angular frequency $\omega_{r2} = \sqrt{1/(L_r + L_m)C_r}$, resonant tank impedance $Z_{r2} = \sqrt{(L_r + L_m)C_r}$ containing L_m, and $t_3 - t_0 = 0.5T_s$.

In this mode, the load power is not supplied from V_{pv} to V_o, and the circulating current is conducted on the primary side. This circulating current period increases the peak and RMS values of the resonant current, resulting in high switching loss and conduction loss. In the proposed PWM control scheme with a variable frequency, this period can be eliminated by operating in the above region.

Figure 5a shows the circulating current period in the conventional PWM control scheme with a fixed switching frequency according to duty cycle period D. The resonant current i_{Lr} becomes equal to i_{Lm} at the instant t_2. The circulating current period is calculated by numerically solving Equations (1)–(4) and (6). The circulating current period is varied according to the Q factor, which is defined as $n^2 w_r L_r / R_L$ and which can be shorter with a high value of Q. Figure 5b shows the switching frequency boundary for the BCM mode when Q is 0.83. The switching frequency boundary for the BCM mode is determined when the $t_3 - t_2$ is zero. The switching frequency for the BCM mode is expressed as shown below:

$$f_{s\_BCM} = \frac{f_r}{0.5 - (t_3 - t_2)f_r} \tag{10}$$

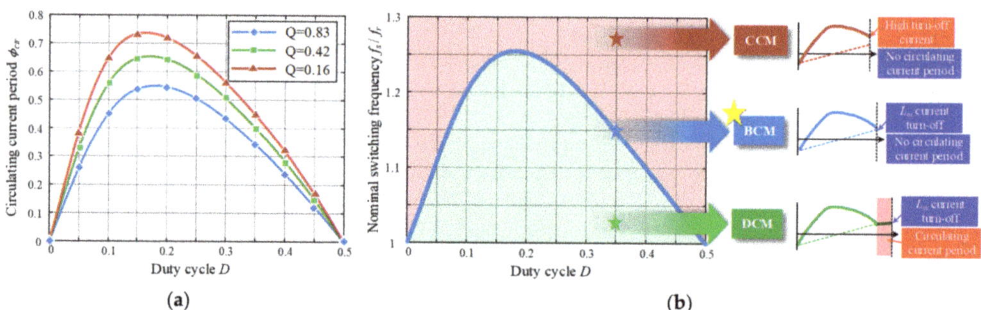

Figure 5. Selection of the switching frequency for the Boundary Conduction Mode (BCM). (**a**) Circulating current period of the PWM control scheme with a fixed switching frequency ($f_s = f_r$). (**b**) Switching frequency boundary for the BCM mode.

When the switching frequency is lower than f_{s_BCM}, the DBLLC resonant converter operates in the discontinuous conduction mode (DCM), which has a circulating current period. When the switching frequency is higher than f_{s_BCM}, the DBLLC resonant converter operates in the continuous conduction mode (CCM). Despite the fact that there is no circulating current period, the turn-off losses of S_1–S_2 and S_5–S_6 and the reverse recovery loss of the secondary diodes are increased. In the BCM mode, not only is the circulating current period eliminated, but the small turn-off current is also ensured, which results in the optimal efficiency.

Figure 6 shows the results of a steady-state trajectory comparison between the proposed PWM control method with a variable switching frequency and the conventional

PWM control with a fixed switching frequency using Equations (1)–(9). It should be noted that the proposed control scheme leads to lower magnitudes of the resonant current and resonant voltage compared to the conventional control method. Figure 6b is the resonant current comparison at time domain, which shows that the rms value of the resonant current in the proposed PWM control is smaller than that of the conventional PWM control by 15%.

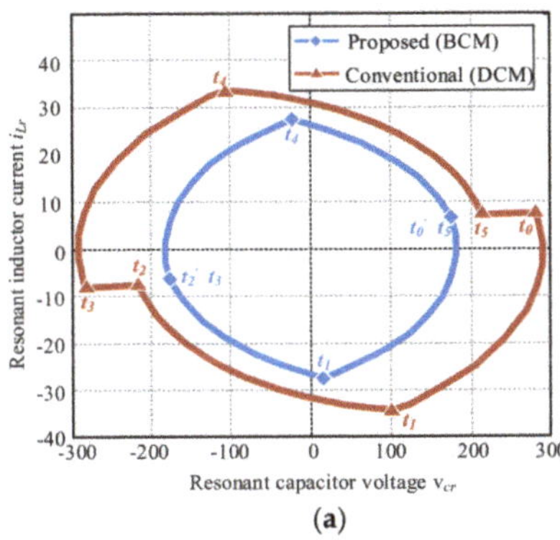

(a)

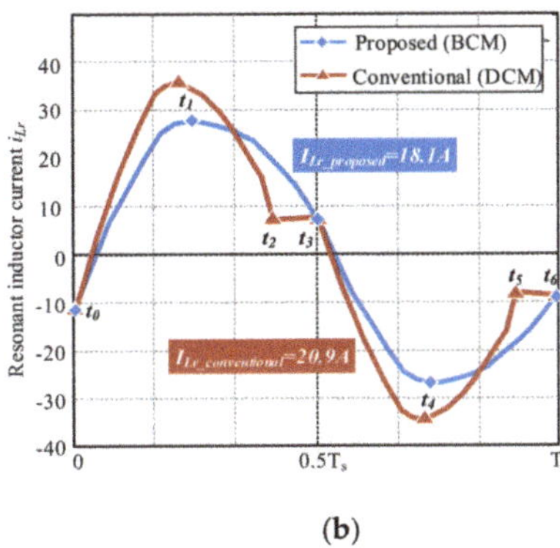

(b)

Figure 6. Steady-state waveform comparison between the proposed PWM control scheme with a variable switching frequency and the conventional PWM control scheme with a fixed switching frequency. (**a**) Steady-state trajectory; (**b**) Time domain.

Figure 7 shows a comparison of the voltage gains between the proposed PWM control and the conventional PWM control schemes. The proposed scheme requires a higher duty cycle in the entire voltage gain range than the conventional scheme. At a voltage gain of 0.65, the duty cycle of the proposed scheme is 0.23, while that of the conventional scheme is 0.3. Therefore, the period of circulating current of the proposed scheme is 0.2, which is smaller than that of the conventional scheme of 0.27. As a result, the proposed scheme eliminates the circulating current and minimizes the turn-off losses of the switches, thereby achieving high efficiency. The voltage gain of the proposed PWM control scheme is expressed using the first harmonic approximation (FHA) method as:

$$M_{PWM} = \frac{V_o}{nV_{pv}} = \sqrt{\frac{10 - 6\cos(2\pi D)/4}{\left(1 + \frac{1}{L_n} - \frac{1}{L_n \cdot f_n^2}\right)^2 + Q^2 \cdot \left(f_n - \frac{1}{f_n}\right)^2}} \tag{11}$$

where quality factor $Q = \frac{\sqrt{L_r/C_r}}{R_{ac}}$, reflected load resistance $R_{ac} = \frac{8}{\pi^2} \cdot n^2 \cdot R_o$, inductance ratio $L_n = L_m/L_r$, and nominalized switching frequency $f_n = f_s/f_r$. Since the proposed PWM control scheme operates at frequencies higher than f_r, higher voltage gain is achieved compared to that of the conventional PWM control scheme.

Therefore, a longer turn-on period of S_3–S_4 and a shorter turn-on period of the bidirectional switch S_5 and S_6 are required to regulate the same PV voltage with the conventional PWM control method. Since operating the bidirectional switch instead of operating S_3 or S_4 increases the conduction loss due to conduction though three switches together and the doubling of the conduction period during one switching cycle, as shown in Figure 7, the proposed PWM control scheme can release the current stress of the bidirectional switch S_5 and S_6.

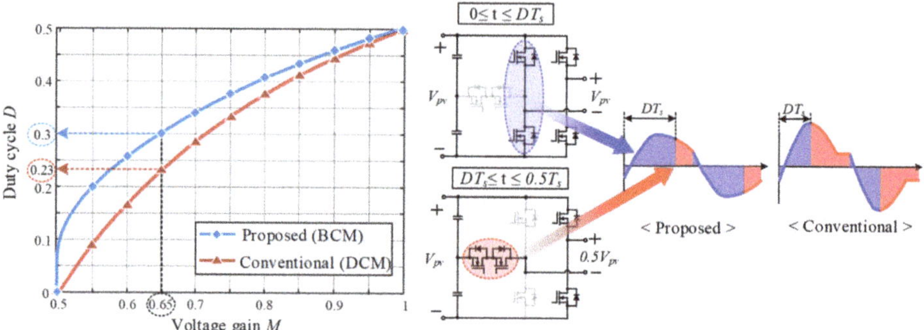

Figure 7. Voltage gain comparison between the proposed PWM control scheme and the conventional PWM control scheme.

2.2. Start-up Control of the Proposed Dual-Bridge LLC (DBLLC) Resonant Converter

If the system power of the inverter is supplied from the DC-link side, an isolated DC–DC converter should maintain the enough DC-link voltage to supply the system power of the inverter, as shown in Figure 8a, and soft-start-up control during which the DC-link voltage is increased from 0 V is required. This system can eliminate the initial charging circuit of the DC-link. However, because the minimum voltage gain is 0.5 in the DBLLC resonant converter and the voltage gain is increased when under a virtual no-load condition due to the parasitic capacitor of the secondary diode, soft start-up control with the conventional PWM scheme or the PFM control scheme is not possible.

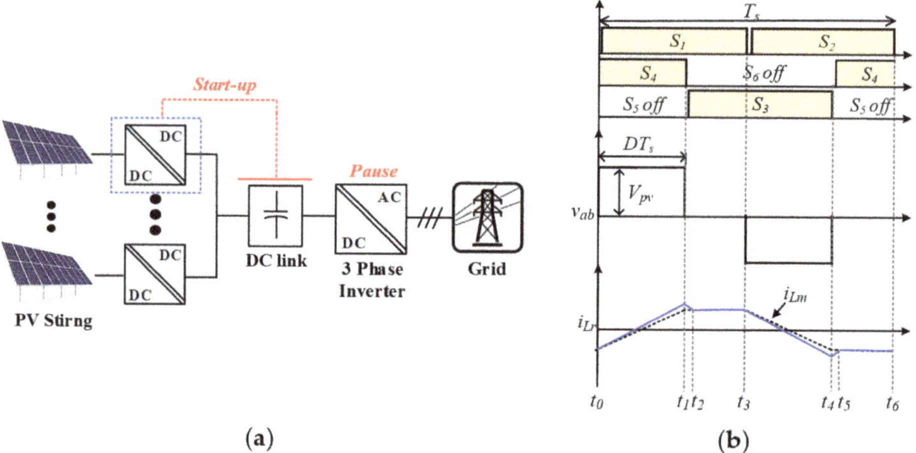

Figure 8. Selection of the switching frequency for the Boundary Conduction Mode (BCM) mode. (**a**) Circulating current period of the PWM control scheme with a fixed switching frequency ($f_s = f_r$). (**b**) The switching frequency boundary for the Boundary Conduction Mode (BCM) mode.

During the start-up process, the DBLLC resonant converter regulates the output voltage by means of phase-shift (PS) control between S_1–S_2 and S_3–S_4 while the bidirectional switch S_5 and S_6 is turned off, as shown in Figure 8b. The magnitude of the resonant tank voltage v_{ab} is the PV voltage v_{pv} during duty cycle D and is 0 V during 0.5-D. Moreover, the DC-link voltage can be increased from 0 V by increasing the duty cycle gradually from 0.

2.3. Proposed Control Scheme

Some MPPT algorithms for tracking the PV string voltage more effectively are suggested in [40,41], and the well-known perturb and observe (P&O) algorithm is applied in this paper to validate only the converter performance. When the inverter regulates the DC-link voltage, the proposed DBLLC resonant converter controls the PV string voltage by means of the perturb and observe (P&O) MPPT control, as shown in Figure 9. In this paper [42], the PV module data are listed in Table A1 of the Appendix A. The PV voltage is given as from 300 V to 900 V, and the output voltage of the converter is fixed at 900 V, as shown in Table 1. In order to operate the wide PV voltage range, the transformer turn ratio is designed to be 0.5. Thus, when the PV voltage is larger than 450 V, the converter performs PWM control to regulate the output voltage to 900 V. However, when the PV voltage is less than 450 V, it operates in PFM control to boost the output voltage to 900 V due to low PV voltage. Accordingly, when K is lower than 0.5, the duty cycle of S_3 and S_4 is the control variable and the switching frequency is varied according to f_{s_BCM}, which is calculated using Equation (10). Therefore, the DBLLC converter operates in the BCM that eliminates the circulating current period, thereby achieving high efficiency. Meanwhile, when K is higher than 0.5, the switching frequency is the control variable to operate in PFM control and the duty cycle of S_3 and S_4 is fixed at 0.5. Since the voltage gain is linear between the PWM control scheme and the PFM control scheme, there is no transient state during a mode change. When the inverter is in a pause state, one of the DBLLC resonant converters controls the DC-link voltage via PS control. The duty cycle D is increased slowly from 0 to charge the DC-link capacitor.

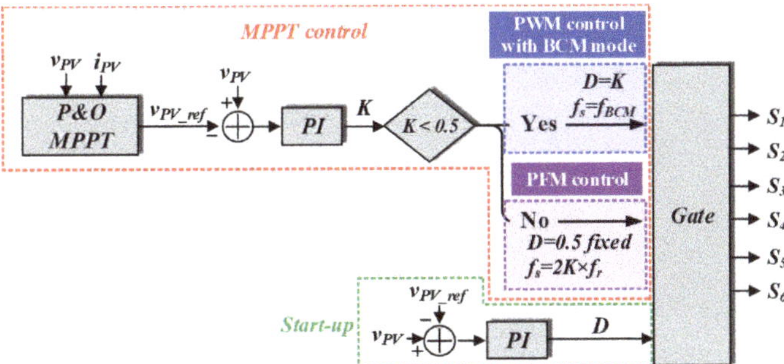

Figure 9. Proposed control scheme for Dual-Bridge LLC (DBLLC) converter with wide photovoltaic (PV) voltage range.

Table 1. System specification and designed parameters of the proposed Dual-Bridge LLC (DBLLC) resonant converter.

Specification/Parameter	Symbol	Values
PV voltage	V_{pv}	300 V–900 V
Output voltage	V_o	900 V
Maximum power	P_o	6.25 kW
Primary-side switches	S_1–S_6	C2M0025120D (1200 V, 90 A)
Secondary-side diode	D_1–D_4	FFSH30120A-D (1200 V, 30 A)
Transformer turn ratio	n	0.5
Magnetizing inductance	L_m	120 µH
Resonant inductance	L_r	22.3 µH
Resonant capacitance	C_r	60 nH
Resonant frequency	f_r	120 kHz
Input capacitance	C_{pv1}, C_{pv2}	40 µF
Output capacitance	C_o	10 µF
Switching frequency range	f_s	70 kHz–160 kHz

3. Design Procedure

3.1. Selecting the Switching Frequency Range

The resonant tank parameters are designed considering both the switching frequency range and the magnetizing inductance value. When the resonant frequency is 120 kHz, the resonant capacitor value is determined by the resonant inductance value. Figure 10 shows the voltage gain of the PFM control scheme with different resonant inductance L_r values.

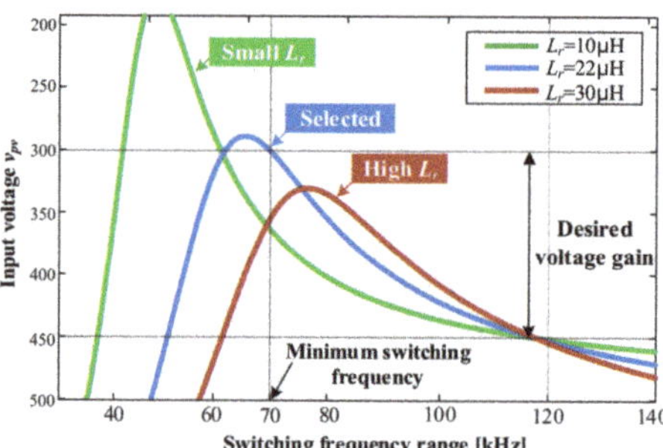

Figure 10. Voltage gain of the PFM control with different resonant inductance values when V_o = 900 V and L_m = 120 μH.

When the output voltage is 900 V and the magnetizing inductance value is fixed at 120 μH, the voltage gain of the PFM control scheme is expressed using the first harmonic approximation (FHA) method, as shown below.

$$M_{PFM} = \frac{1}{\sqrt{\left(1 + \frac{1}{L_n} - \frac{1}{L_n \cdot f_n^2}\right)^2 + Q^2 \cdot \left(f_n - \frac{1}{f_n}\right)^2}} \tag{12}$$

where quality factor $Q = \frac{\sqrt{L_r/C_r}}{R_{ac}}$, reflected load resistance $R_{ac} = \frac{8}{\pi^2} \cdot n^2 \cdot R_o$, inductance ratio $L_n = L_m/L_r$, and nominalized switching frequency $f_n = f_s/f_r$. With a small L_r value (green line), a wide switching frequency range is required to satisfy the requirements of the desired voltage gain due to the flat gain curve. Because the minimum switching frequency is limited to 70 kHz, considering the volume of the transformer, a small magnetizing inductance value design is required to increase the voltage gain, which increases the conduction loss and the switching loss. The maximum switching frequency for the BCM mode is increased due to the low Q factor. Although the voltage gain curve is sharp with a high L_r value (red line), the desired voltage gain in this case as well cannot be acquired due to the reduced peak voltage gain. Therefore, a design with a small magnetizing inductance value is required to increase the voltage gain. A resonant inductance value of 22 μH is selected considering the maximum switching frequency for the BCM mode and the magnetizing inductance value.

3.2. Soft Switching

The magnetizing current of the transformer discharges the output capacitor of the primary switches, while ZVS operation of S_3 and S_4 is ensured when using the PWM control scheme due to the high turn-off current of the resonant current. Therefore, the peak value of the magnetizing current should be high enough to discharge the output capacitor of the switches. The peak current of the magnetizing current for ZVS operation is expressed as

$$i_{Lm}(t_3) \geq i_{Coss} = 2C_{oss}\frac{V_{pv}}{t_d} \tag{13}$$

where C_{oss} is the output capacitance value of the switches and t_d is the dead-time period. The peak value of the magnetizing current is expressed as

$$i_{Lm}(t_3) = \frac{V_o T_{s\_min}}{4nL_m} \tag{14}$$

where T_{s_min} denotes a single switching cycle at the maximum switching frequency. From Equations (13) and (14), the magnetizing inductance value for the entire range of ZVS operation of the switches is expressed as

$$L_m = \frac{t_d T_{s\_min}}{8C_{oss}} = \frac{80 \cdot 10^{-9} \cdot \frac{1}{160 \cdot 10^3}}{8 \cdot 80 \cdot 10^{-12}} = 535\mu\text{H} \tag{15}$$

where V_{pv} = 900 V, the output voltage V_o = 900 V, turn ratio of the transformer n = 0.5, dead-time t_d = 80 ns, C_{oss} = 80 pF (from the CREE C2M0025120D datasheet), and one switching period is considered as the maximum operation switching frequency of 160 kHz of the proposed DBLLC resonant converter. The magnetizing inductance value in the PFM control scheme was set to 120 µH according to the design procedure in Section 3.1, a value which satisfies the Equation (15) requirement, the condition for the entire range of ZVS operation.

3.3. Transformer with Integrated Leakage Inductance

Generally, both primary and secondary windings are wound on the center leg, as shown in Figure 11a [43,44]. The L_m value can easily be adjusted by changing the air gap distance of the core; however, there is limitation to the L_k value because it can only be adjusted by changing the distance between the primary and secondary wires. This winding method has the lowest L_k value because the distance between the primary and secondary windings is narrow [45]. Therefore, for this winding method, an additional resonant inductor should be used when a sufficiently large L_k value is required to operate over a wide voltage range. An example of this would be an LLC resonant converter. Meanwhile, by placing some of the secondary turns on the outer legs of the core, as shown in Figure 11b, the leakage inductance can be increased since there is more leakage magnetic flux flowing into the air. The leakage inductance can be increased further by distributing more windings on the outer legs. The largest L_k value can be obtained by winding all of the secondary side turns on the outer leg, as shown in Figure 11c, in which case the internal L_k of the transformer can be used as the resonant inductor without using an additional inductor [46].

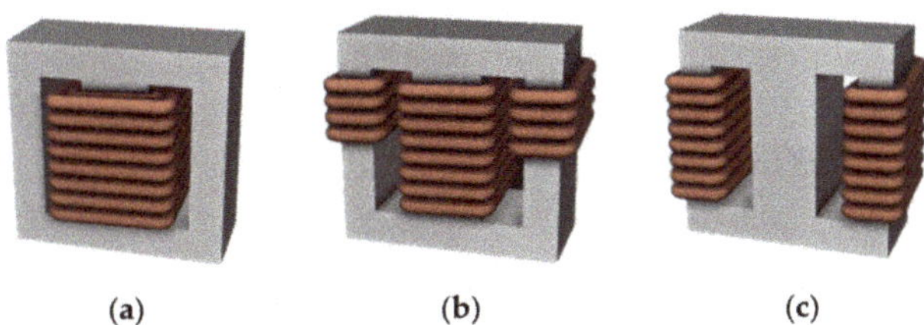

(a)   (b)   (c)

Figure 11. Secondary-side winding methods. (**a**) Winding only the center leg of the core. (**b**) Winding both the center leg and the outer legs of the core. (**c**) Winding only the outer legs of the core.

Figure 12a shows the relationship between the number of turns of the secondary winding on the outer legs, T1, and the leakage inductance value. In this paper, the winding method in Figure 11c was adopted because at least 20 µH is required for the regulation of the LLC converter. PM12/EE555S (TODA ISU Co. Ltd., Wonju, Korea) was selected as the transformer core. To minimize heat generation in the core, the maximum flux density (B_{max}) was chosen to be 0.3 T. To achieve the desired B_{max}, the required number of primary turns is 11, which is wound around the core center leg. A 3D finite element analysis (FEA) simulation was conducted on ANSYS Maxwell, and these results are shown Figure 12b.

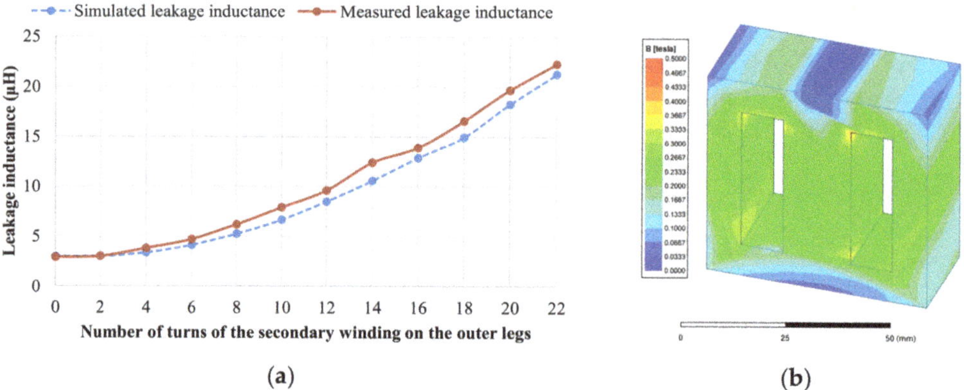

(a) (b)

Figure 12. Finite Element Analysis (FEA) simulation result. (**a**) Relationship between the number of turns of the secondary winding on the outer legs and the leakage inductance. (**b**) The magnetic flux density distributions of the core.

4. Experimental Results

In order to verify the performance of the proposed DBLLC resonant converter, a 6.25 kW laboratory prototype was built, as shown in Figure 13. The system specifications and designed parameters of the proposed DBLLC resonant converter are summarized in Table 1. The primary switches and secondary-side diodes were implemented using SiC devices (S_1–S_6: C2M0025120D from CREE, D_1–D_4: FFSH30120A from ON Semiconductor). The transformer was implemented with EE55/55 ferrite cores from Core Electronics. Values of 40 µF for input capacitors C_{pv1} and C_{pv2} and 10 µF for output capacitor C_o were selected, respectively, considering the voltage ripple. The control algorithm was implemented in a floating-point digital signal processing (DSP) platform TMS320F28377S, and a power analyzer (YOKOGAWA WT3000) was used to measure the efficiency.

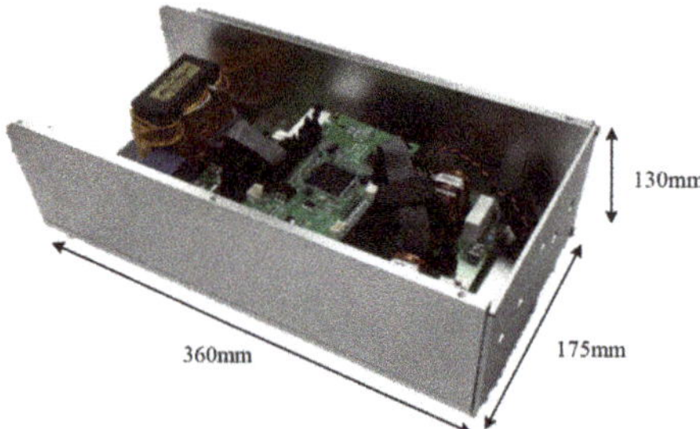

Figure 13. Prototype of the proposed 6.25 kW DBLLC resonant converter.

Figure 14 shows the experimental waveforms of the proposed hybrid control in a wide PV voltage while the output voltage is constant at 900 V. When the PV voltage is 300 V, the DBLLC resonant converter operates in the FB mode, and the switching frequency of 84 kHz is lower than the resonant frequency. The amplitude of the resonant tank voltage v_{ab} is equal to the PV voltage, and the PV voltage is regulated using the PFM control scheme. When the PV voltage is 450 V, the DBLLC resonant converter operates in the FB mode, and the switching frequency is near to the resonant frequency of 120 kHz. The amplitude of the resonant tank voltage v_{ab} is still equal to the PV voltage. When the PV voltage is equal to 720 V, the DBLLC resonant converter operates with the proposed PWM control scheme, and the switching frequency is varied in the above region in order to eliminate the circulating current period. The FB mode and the HB mode are varied to regulate the PV voltage. When V_{pv} is 900 V, the DBLLC resonant converter operates in the HB mode and the switching frequency is near to the resonant frequency of 120 kHz. The amplitude of the resonant tank voltage v_{ab} is equal to half of the PV voltage. As a result, the peak and RMS values of the resonant tank current i_{Lr} are able to be minimized by the proposed PWM control scheme with a variable switching frequency. Furthermore, a wide PV voltage range (300 V–900 V) is regulated by the PFM control scheme.

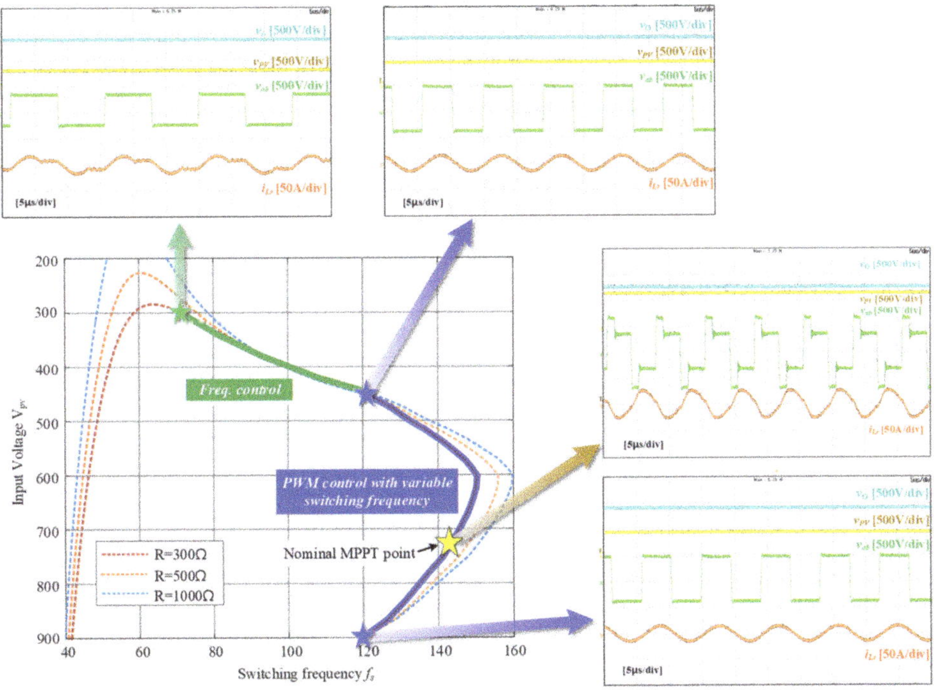

Figure 14. Experimental waveforms of the proposed hybrid control in a wide PV voltage range.

Figure 15 shows the results of a comparison of the experimental waveforms between the proposed PWM control scheme and the conventional PWM control scheme in terms of the peak current value and the RMS value of the resonant current when V_{pv} = 720 V, V_o = 900 V, and P_0 = 6.25 kW. In the proposed PWM control scheme, the circulating current period is eliminated by operating in the above region, as shown in Figure 15a, and the peak value and RMS value of the resonant current are 29.3 A and 18.1 A, respectively. In the conventional PWM control scheme shown in Figure 15b, the peak value and RMS value of the resonant current are 34.2 A and 20.9 A, respectively, due to the circulating current period. The proposed PWM control scheme provides a smaller peak value and a smaller RMS value of the resonant current by 17% and 15%, respectively, resulting in less turn-off loss and conduction loss.

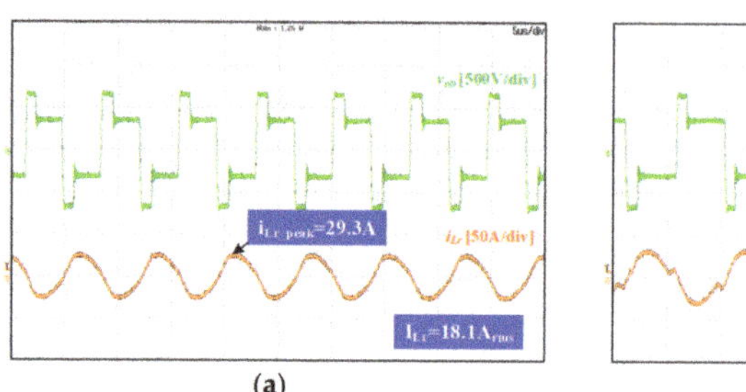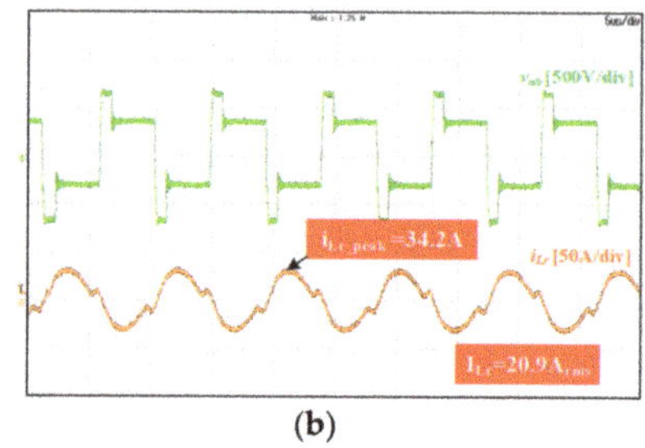

(a)   (b)

Figure 15. Experimental waveforms of the PWM control scheme: (**a**) proposed variable switching frequency; (**b**) conventional fixed switching frequency.

Figure 16 shows the experimental waveforms of the proposed soft-start up and initial operation sequence. This method is used to smoothly charge the bulky electrolytic capacitors of the DC-link to decrease inrush current. Therefore, the resistive pre-charge circuits of the inverter were eliminated. In period (a), the DBLLC resonant converter regulates the DC-link voltage gradually from 0 V to supply the system power such as in the inverter condition using the PS control scheme. Then, the inverter regulates the DC-link voltage during period (b), while the DBLLC resonant converter is paused. In period (c), the DBLLC resonant converter regulates the PV voltage with the proposed hybrid control scheme to track the MPPT point.

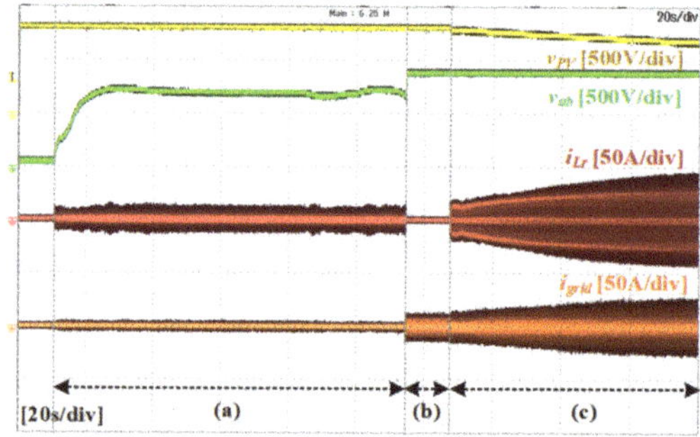

Figure 16. Experimental waveforms of the soft-start up and initial operation sequence.

Figure 17 shows a comparison of the measured efficiency rates between the proposed PWM control and the conventional PWM control schemes when $V_{PV} = 720$ V and $V_o = 900$ V. The maximum and California energy commission (CEC) efficiency of the proposed PWM control are 98.23% and 98.04%, respectively, values greater than those by the conventional PWM control by 0.1% and 0.2%, respectively. This is because the switches losses are reduced by eliminating the circulating current period by the proposed PWM control. Figure 18 shows the results of a loss analysis when $P_o = 6.25$ kW with PWM control. Despite the fact that the switching loss is higher due to the increased switching frequency, the conduction losses of the primary switches of the proposed PWM control scheme are smaller due to the elimination of the circulating current period and the reduced turn-on period of the bidirectional switch.

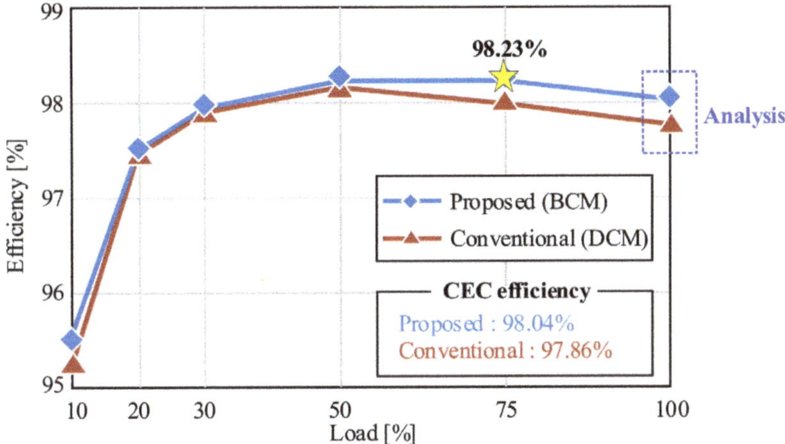

Figure 17. Measured efficiency comparison between the proposed PWM control scheme and the conventional PWM control scheme when V_{pv} = 720 V and V_o = 900 V.

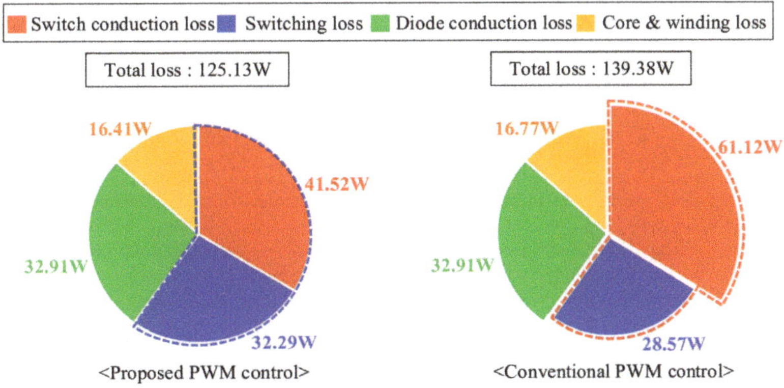

Figure 18. Loss analysis under the full load condition with PWM control.

5. Conclusions

This paper proposes an isolated DC–DC power optimizer for a wide PV string voltage range, and this wide PV voltage range (300 V–900 V) is regulated by the proposed hybrid control combining the PFM control scheme and PWM control scheme with a variable switching frequency. By operating in the above region of the LLC resonant converter with PWM control, the circulating current period is eliminated and the peak and RMS values of the resonant current are decreased to 17% and 15% under the rated power, respectively, compared to the conventional PWM control scheme with a fixed switching frequency. The current stress of the bidirectional switches is also decreased due to the reduced turn-on cycle. To charge the DC-link capacitor, the start-up control is proposed and validated with experimental results. The leakage inductance integrated transformer with a distributed secondary turn ratio is designed for high efficiency and a high power density. From a 6.25 kW laboratory prototype, it is shown that the maximum efficiency and CEC efficiency are 98.22% and 98.04%, respectively, which are increased by 0.1% and 0.2%, respectively, compared to those of the conventional PWM control scheme with a fixed switching frequency. It should be noted that the proposed hybrid control in a DBLLC resonant converter provides higher efficiency and a wider voltage range without extra components.

Author Contributions: The authors contributed to the paper as follows: Study conception and idea: H.K. and S.C.; design procedure: G.Y., experiment and results: J.K.; draft manuscript preparation: H.K., G.Y., J.K. and S.C. All authors have read and agreed to the published version of the manuscript.

Funding: This research was funded by the Korea government (MSIT).

Institutional Review Board Statement: Not applicable.

Informed Consent Statement: Not applicable.

Data Availability Statement: Not applicable.

Acknowledgments: This work was supported by the National Research Foundation of Korea(NRF) grant funded by the Korea government(MSIT). (No. 2020R1A2C2006301).

Conflicts of Interest: The authors declare no conflict of interest.

Appendix A

Table A1. Specification of the one PV Module [42].

Parameter	Values
Total maximum power	400 W
MPP voltage	40.6 V
MPP current	9.86 A
Open circuit voltage	49.3 V
Short circuit current	10.47 A
Number of series	16

References

1. Yang, Y.; Blaabjerg, F. Overview of Single-phase Grid-connected Photovoltaic Systems. *Electr. Power Compon. Syst.* **2015**, *43*, 1352–1363. [CrossRef]
2. Choi, H.; Ciobotaru, M.; Jang, M.; Agelidis, V.G. Performance of Medium-Voltage DC-Bus PV System Architecture Utilizing High-Gain DC–DC Converter. *IEEE Trans. Sustain. Energy* **2015**, *6*, 464–473. [CrossRef]
3. Kouro, S.; Leon, J.I.; Vinnikov, D.; Franquelo, L.G. Grid-Connected Photovoltaic Systems: An Overview of Recent Research and Emerging PV Converter Technology. *IEEE Ind. Electron. Mag.* **2015**, *9*, 47–61. [CrossRef]
4. Teodorescu, R.; Liserre, M.; Rodriguez, P. *Grid Converters for Photovoltaic and Wind Power Systems*; Wiley: New York, NY, USA, 2011.
5. Blaabjerg, F.; Iov, F.; Kerekes, T.; Teodorescu, R. Trends in power electronics and control of renewable energy systems. In Proceedings of the 14th International Power Electronics and Motion Control Conference EPE-PEMC, Ohrid, Republic of Macedonia, 6–8 September 2010; pp. K-1–K-19. [CrossRef]
6. Rivera, S.; Kouro, S.; Wu, B.; Leon, J.I.; Rodríguez, J.; Franquelo, L.G. Cascaded H-bridge multilevel converter multistring topology for large scale photovoltaic systems. In Proceedings of the 2011 IEEE International Symposium on Industrial Electronics, Gdansk, Poland, 27–30 June 2011; pp. 1837–1844. [CrossRef]
7. Acharya, A.B.; Ricco, M.; Sera, D.; Teoderscu, R.; Norum, L.E. Performance Analysis of Medium-Voltage Grid Integration of PV Plant Using Modular Multilevel Converter. *IEEE Trans. Energy Convers.* **2019**, *34*, 1731–1740. [CrossRef]
8. Shi, Y.; Li, R.; Xue, Y.; Li, H. High-Frequency-Link-Based Grid-Tied PV System With Small DC-Link Capacitor and Low-Frequency Ripple-Free Maximum Power Point Tracking. *IEEE Trans. Power Electron.* **2016**, *31*, 328–339. [CrossRef]
9. Abramovitz, A.; Zhao, B.; Smedley, K.M. High-Gain Single-Stage Boosting Inverter for Photovoltaic Applications. *Ieee Trans. Power Electron.* **2016**, *31*, 3550–3558. [CrossRef]
10. Sathyan, S.; Suryawanshi, H.M.; Shitole, A.B.; Ballal, M.S.; Borghate, V.B. Soft-Switched Interleaved DC/DC Converter as Front-End of Multi-Inverter Structure for Micro Grid Applications. *IEEE Trans. Power Electron.* **2018**, *33*, 7645–7655. [CrossRef]
11. Sharma, P.; Agarwal, V. Exact Maximum Power Point Tracking of Grid-Connected Partially Shaded PV Source Using Current Compensation Concept. *IEEE Trans. Power Electron.* **2014**, *29*, 4684–4692. [CrossRef]
12. Rani, B.I.; Ilango, G.S.; Nagamani, C. Enhanced Power Generation From PV Array Under Partial Shading Conditions by Shade Dispersion Using Su Do Ku Configuration. *IEEE Trans. Sustain. Energy* **2013**, *4*, 594–601. [CrossRef]
13. Patel, H.; Agarwal, V. MATLAB-Based Modeling to Study the Effects of Partial Shading on PV Array Characteristics. *IEEE Trans. Energy Convers.* **2008**, *23*, 302–310. [CrossRef]

14. Zapata, J.W.; Kouro, S.; Carrasco, G.; Renaudineau, H.; Meynard, T.A. Analysis of Partial Power DC–DC Converters for Two-Stage Photovoltaic Systems. *IEEE J. Emerg. Sel. Top. Power Electron.* **2019**, *7*, 591–603. [CrossRef]
15. Sangwongwanich, A.; Yang, Y.; Blaabjerg, F. High-Performance Constant Power Generation in Grid-Connected PV Systems. *IEEE Trans. Power Electron.* **2016**, *31*, 1822–1825. [CrossRef]
16. Vázquez, N.; Rosas, M.; Hernández, C.; Vázquez, E.; Perez-Pinal, F.J. A New Common-Mode Transformerless Photovoltaic Inverter. *IEEE Trans. Ind. Electron.* **2015**, *62*, 6381–6391. [CrossRef]
17. Chen, L.; Amirahmadi, A.; Zhang, Q.; Kutkut, N.; Batarseh, I. Design and Implementation of Three-Phase Two-Stage Grid-Connected Module Integrated Converter. *IEEE Trans. Power Electron.* **2014**, *29*, 3881–3892. [CrossRef]
18. Calais, M.; Myrzik, J.; Spooner, T.; Agelidis, V.G. Inverters for single-phase grid connected photovoltaic systems-an overview. In Proceedings of the 2002 IEEE 33rd Annual IEEE Power Electronics Specialists Conference. Proceedings (Cat. No.02CH37289), Cairns, QLD, Australia, 23–27 June 2002; Volume 4, pp. 1995–2000. [CrossRef]
19. Zhu, H.; Zhang, D.; Athab, H.S.; Wu, B.; Gu, Y. PV Isolated Three-Port Converter and Energy-Balancing Control Method for PV-Battery Power Supply Applications. *IEEE Trans. Ind. Electron.* **2015**, *62*, 3595–3606. [CrossRef]
20. Vakacharla, V.R.; Rathore, A.K. Isolated Soft Switching Current Fed LCC-T Resonant DC–DC Converter for PV/Fuel Cell Applications. *IEEE Trans. Ind. Electron.* **2019**, *66*, 6947–6958. [CrossRef]
21. De Oliveira, R.N.M.; Mazza, L.C.d.S.; Filho, H.M.d.; Oliveira, D.d.S. A Three-Port Isolated Three-Phase Current-Fed DC–DC Converter Feasible to PV and Storage Energy System Connection on a DC Distribution Grid. *IEEE Trans. Ind. Appl.* **2019**, *55*, 4910–4919. [CrossRef]
22. Prasanna, U.R.; Rathore, A.K. Analysis, Design, and Experimental Results of a Novel Soft-Switching Snubberless Current-Fed Half-Bridge Front-End Converter-Based PV Inverter. *IEEE Trans. Power Electron.* **2013**, *28*, 3219–3230. [CrossRef]
23. Liu, Y.; Abu-Rub, H.; Ge, B. Front-End Isolated Quasi-Z-Source DC–DC Converter Modules in Series for High-Power Photovoltaic Systems—Part II: Control, Dynamic Model, and Downscaled Verification. *IEEE Trans. Ind. Electron.* **2017**, *64*, 359–368. [CrossRef]
24. Buticchi, G.; Barater, D.; Costa, L.F.; Liserre, M. A PV-Inspired Low-Common-Mode Dual-Active-Bridge Converter for Aerospace Applications. *IEEE Trans. Power Electron.* **2018**, *33*, 10467–10477. [CrossRef]
25. Kundu, U.; Yenduri, K.; Sensarma, P. Accurate ZVS Analysis for Magnetic Design and Efficiency Improvement of Full-Bridge LLC Resonant Converter. *IEEE Trans. Power Electron.* **2017**, *32*, 1703–1706. [CrossRef]
26. Kathiresan, R.; Das, P.; Reindl, T.; Panda, S.K. A Novel ZVS DC–DC Full-Bridge Converter With Hold-Up Time Operation. *IEEE Trans. Ind. Electron.* **2017**, *64*, 4491–4500. [CrossRef]
27. Wang, H.; Dusmez, S.; Khaligh, A. Design and Analysis of a Full-Bridge LLC-Based PEV Charger Optimized for Wide Battery Voltage Range. *IEEE Trans. Veh. Technol.* **2014**, *63*, 1603–1613. [CrossRef]
28. Wang, H.; Li, Z. A PWM LLC Type Resonant Converter Adapted to Wide Output Range in PEV Charging Applications. *IEEE Trans. Power Electron.* **2018**, *33*, 3791–3801. [CrossRef]
29. Awasthi, A.; Bagawade, S.; Jain, P. Analysis of a Hybrid Variable Frequency-Duty Cycle Modulated low-Q LLC Resonant Converter for Improving Light-Load Efficiency for Wide Input Voltage Range. *IEEE Trans. Power Electron.* **2020**. [CrossRef]
30. Sun, X.; Li, X.; Shen, Y.; Wang, B.; Guo, X. Dual-Bridge LLC Resonant Converter With Fixed-Frequency PWM Control for Wide Input Applications. *IEEE Trans. Power Electron.* **2017**, *32*, 69–80. [CrossRef]
31. Li, X.; Bhat, A.K.S. Analysis and Design of High-Frequency Isolated Dual-Bridge Series Resonant DC/DC Converter. *IEEE Trans. Power Electron.* **2010**, *25*, 850–862. [CrossRef]
32. Li, X. A LLC-Type Dual-Bridge Resonant Converter: Analysis, Design, Simulation, and Experimental Results. *IEEE Trans. Power Electron.* **2014**, *29*, 4313–4321. [CrossRef]
33. Wu, H.; Chen, L.; Xing, Y. Secondary-Side Phase-Shift-Controlled Dual-Transformer-Based Asymmetrical Dual-Bridge Converter with Wide Voltage Gain. *IEEE Trans. Power Electron.* **2015**, *30*, 5381–5392. [CrossRef]
34. Hu, S.; Li, X. Performance Evaluation of a Semi-Dual-Bridge Resonant DC/DC Converter With Secondary Phase-Shifted Control. *IEEE Trans. Power Electron.* **2017**, *32*, 7727–7738. [CrossRef]
35. Wu, J.; Li, Y.; Sun, X.; Liu, F. A New Dual-Bridge Series Resonant DC–DC Converter With Dual Tank. *IEEE Trans. Power Electron.* **2018**, *33*, 3884–3897. [CrossRef]
36. Zhou, S.; Li, X.; Zhong, Z.; Zhang, X. Wide ZVS operation of a semi-dual-bridge resonant converter under variable-frequency phase-shift control. *Iet Power Electron.* **2020**, *13*, 1746–1755. [CrossRef]
37. Wu, F.; Wang, Z.; Luo, S. Buck-Boost Three-Level Semi-Dual-Bridge Resonant Isolated DC-DC Converter. *IEEE J. Emerg. Sel. Top. Power Electron.* **2020**. [CrossRef]
38. Wei, Y.; Luo, Q.; Du, X.; Altin, N.; Nasiri, A.; Alonso, J.M. A Dual Half-Bridge LLC Resonant Converter With Magnetic Control for Battery Charger Application. *IEEE Trans. Power Electron.* **2020**, *35*, 2196–2207. [CrossRef]
39. Kazimierczuk, M.K.; Czarkowski, D. *Resonant Power Converters*; Wiley: New York, NY, USA, 2012.
40. Olivares-Rodríguez, C.; Castillo-Calzadilla, T.; Kamara-Esteban, O. Bio-inspired Approximation to MPPT Under Real Irradiation Conditions. In *IDC 2018. Studies in Computational Intelligence*; Springer: Cham, Switzerland, 2018; Volume 798.
41. De Brito, M.A.G.; Galotto, L.; Sampaio, L.P.; Melo, G.d.A.e.; Canesin, C.A. Evaluation of the Main MPPT Techniques for Photovoltaic Applications. *IEEE Trans. Ind. Electron.* **2013**, *60*, 1156–1167. [CrossRef]
42. Available online: https://www.lg.com/us/business/solar-panels/lg-lg400n2w-v5 (accessed on 26 March 2021).

43. Naradhipa, A.M.; Kim, S.; Yang, D.; Choi, S.; Yeo, I.; Lee, Y. Power Density Optimization of 700 kHz GaN-Based Auxiliary Power Module for Electric Vehicles. *IEEE Trans. Power Electron.* **2021**, *36*, 5610–5621. [CrossRef]
44. Ouyang, Z.; Thomsen, O.C.; Andersen, M.A.E. Optimal Design and Tradeoff Analysis of Planar Transformer in High-Power DC–DC Converters. *IEEE Trans. Ind. Electron.* **2012**, *59*, 2800–2810. [CrossRef]
45. McLyman, C.W.T. *Transformer and Inductor Design Handbook*; Marcel Dekker: New York, NY, USA, 1988; Chapter 17.
46. Li, S.; Min, Q.; Rong, E.; Zhang, R.; Du, X.; Lu, S. A Magnetic Integration Half-Turn Planar Transformer and its Analysis for LLC Resonant DC-DC Converters. *IEEE Access* **2019**, *7*, 128408–128418. [CrossRef]

Article

Design and Fabrication of Solar Thermal Energy Storage System Using Potash Alum as a PCM

Muhammad Suleman Malik [1], Naveed Iftikhar [2], Abdul Wadood [3,*], Muhammad Omer Khan [4], Muhammad Usman Asghar [4], Shahbaz Khan [3], Tahir Khurshaid [5,*], Ki-Chai Kim [5,*], Zabdur Rehman [6] and S. Tauqeer ul Islam Rizvi [7]

[1]   Electrical Construction Division, Tribal Areas Electric Supply Company (TESCO), Peshawar 25000, Pakistan; Suleman2113@gmail.com

[2]   Department of Mechanical Engineering, Ghulam Ishaq Khan Institute of Science & Technology, Swabi, Khyber Pakhtunkhwa 23640, Pakistan; naveed_iftikhar@hotmail.com

[3]   Department of Electrical Engineering, Air University, Islamabad, Kamra Campus, Kamra 43570, Pakistan; shahbaz@au.edu.pk

[4]   Department of Electrical Engineering & Technology, Ripah International University, Faisalabad Campus, Punjab 38000, Pakistan; omerkhan@riphahfsd.edu.pk (M.O.K.); usman.asghar@riphahfsd.edu.pk (M.U.A.)

[5]   Department of Electrical Engineering, Yeungnam University, 280, Daehak-Ro, Gyeongsangbuk do 38541, Korea

[6]   Department of Mechanical Engineering, Air University, Islamabad, Kamra Campus, Kamra 43570, Pakistan; zabdurrehman@au.edu.pk

[7]   Department of Aeropsace Engineering, Air University, Islamabad, Kamra Campus, Kamra 43570, Pakistan; tauqeerulislam@aack.au.edu.pk

*   Correspondence: wadood@au.edu.pk (A.W.); tahir@ynu.ac.kr (T.K.); kckim@ynu.ac.kr (K.-C.K.)

Received: 12 October 2020; Accepted: 22 November 2020; Published: 24 November 2020

Abstract: Renewable energy resources like solar energy, wind energy, hydro energy, photovoltaic etc. are gaining much importance due to the day by day depletion of conventional resources. Owing to the lower efficiencies of renewable energy resources, much attention has been paid to improving them. The concept of utilizing phase change materials (PCMs) has attracted wide attention in recent years. This is due to their ability to extract thermal energy when used in collaboration with photovoltaic (PV), thus improving the photoelectric conversion efficiency. In this paper, the objective is to design and fabricate a novel thermal energy storage system using phase change material. An investigation on the characteristics of Potash Alum as a phase change material due to its low cost, easy availability and its usage as an energy storage for the indoor purposes are taken into account. The use of a latent heat storage system using phase change materials (PCMs) is an effective way of storing thermal energy and has the advantage of high-energy storage density and the isothermal nature of the storage process. In the current study, potash alum was identified as a phase change material combined with renewable energy sources, that can be efficiently and effectively used in storing thermal energy at compartively lower temperatures that can later be used in daily life heating requirements.A parabolic dish which acts of a heat collector is used to track and reflects solar radiation at a single point on a receiver tank. Heat transfer from the solar collector to the storage tank is done by using a circulating heat transfer fluid with the help of a pump. The experimental results show that this system is capable of successfully storing and utilizing thermal energy on indoor scale such as cooking, heating and those applications where temperature is below 92 °C.

Keywords: thermal energy storage; parabolic dish; latent heat; phase change material; heat transfer fluid

---

1. Introduction

Renewable energy is playing a vital role in the clean energy generation and avoiding hazardous and negative effects of pollutions in our environment. The use of different renewable energy sources like hydroelectric, wind, solar, tidal is increasing day by day with the addition of thousands of megawatts to the grid systems. Photovoltaic (PV) technology, which converts solar irradiance directly into electricity, has made tremendous progress on the scientific as well as the commercial scale. Still the research and development continues to push its efficiency along with lowering its cost [1,2]. On the other hand, photovoltaics are already used but the problem is that of the system durability and low expenditure because panels don't work the same over time. Similarly, storage batteries are also very expensive and not long lasting. However, along with other barriers, the prime barrier is that of intermittent sunlight which is available for only a portion of a day. Hence, incorporating efficient energy storage systems along with renewable energy sources is becoming essential with time [3,4]. The traditional mechanism for the solar energy storage is first converting solar energy into electrical energy through photovoltaic panels and then storing it in a batteries which are expensive [5]. In recent years, few researchers have proposed the usage of phase change materials (PCMs) as an alternative method for the storage of solar energy [6]. In this method, the solar energy is converted into thermal energy using PCMs and then stored into storage tank which acts as a thermal battery [7,8]. In thermal energy storage, the useful energy is transferred to the storage medium and stored in the form of latent and sensible heat during the phase transition process with a minimum rise in temperature. Among the two, latent heat storage is more attractive than the sensible heat storage system because of its high temperature swing and relatively small size [9].

In latent heat storage system, radiation from the sun fall on a parabolic concentrator which directs the incident radiation toward the base of the receiver tank resulting in the temperature rise of the heat transfer fluid (HTF), hence storing energy in the form of sensible heat for a specific period of time. The receiver tank is connected to a storage tank by rubber pipes and HTF is circulated in them at a constant flow rate with the help of a small pump. As the hot HTF flows in the storage tank, the heat is supplied to the PCM situated inside the storage tank which changes its state [10–12]. The bulk amount of heat (latent heat) used to change the phase of PCM can be stored by using effective insulation. The storage tank is also connected to a heating application by using pipes. Valves are used to direct the flow of HTF within the receiver tank and storage tank to store the energy during the availability of the solar energy and from storage tank to the heating application when the stored energy is to be used.

Several types of solar energy storage devices using different PCMs like paraffin, perlite, metal foam and beeswax [13–15] are used. The selection of suitable PCM for a particular application requires a lot of factors to be taken into consideration. These PCMs are classified into organic and inorganic, with the later PCMs are costly and not readily available [16–18]. Bushnell and Sohi designed a modular phase change heat exchanger with pentaerythritol as PCM for thermal storage and tested it in an oven with circulating heat transfer oil with a pattern of electrical heating to stimulate the concentrating solar collector. The comparative analysis of thermal energy retention times and cooking extraction times with efficiencies were assessed with reported non-modular heat exchanger. The results indicated that foods were cooked at temperatures between 95–97 degrees [19]. Sharma and his team used commercial grade acetamide as a latent heat storage material with significant values of melting point and latent heat of fusion, designed a cylindrical PCM storage unit for box-type solar cookers and to utilized it for late evening cooking. This unit provides higher heat transfer between PCM and cooking and consumes less time for cooking. It is proved that storage of solar energy doesn't affect the performance of solar cookers for late evening cooking, thus having a PCM melting temperature between 105–110 °C is needed and this design can be implemented for late-evening cooking as well. The only thing to be considered for this design is to identify a material with appropriate melting point, concentration and quantity [20]. Buddhi used acetanilide with a melting point of 118.9 °C and latent heat of fusion as 222 kj/kg as a PCM for night cooking [21]. Thus the proper significance and the assessment of the materials with capability of increasing melting point and higher latent heat of fusion can be tried by further experiments.

Schmerse et al. [22] studied PCM materials in building design. They concluded that PCM incorporated in the construcition of buildings can save up to 27% of the energy consumption annually.

Nems and Puertas [23] experimentally studied a dual PCM for heat storage, they made a I–D model for dual PCM and validated their results using experimental results. Leang et al. [24] performed design and optimization of composite walls by integrating a PCM in their design. Their study concluded that the higher the latent heat, the lower the heating demand and greater the thermal comfort of the interior of the house. Pasupathi et al. [25] studied a hybrid PCM that included nano-particles for the purpose of a thermal energy storage system. Their study concluded that utilizing a hybrid PCM material can increase the performance of PCM if the mass fraction of the nanomaterial is 1.0. In all the aforementioned studies, different PCM materials were employed for thermal energy storage. There is no specific studies that employed potash alum as PCM with a renewable energy source for thermal energy storage systems. In the current study, potash alum is used as PCM because it is cheap and readily available on the market.

The aim of this study was to design and study a latent heat thermal energy storage system which can store solar energy for a reasonable amount of time. This stored energy is environmentally friendly and can be used for any indoor heating purpose such as cooking, water and room heating applications in the absence of sunlight. The experimental results of the designed system show that the system is capable of storing enough energy during sunshine hours which can later be used for any heating application in the absence of sunlight.

This research article has been organized as follows: In Section 2 the system design is discussed and analyzed. Simulations are performed and results are presented in Section 3. Section 4 is about the fabrication of the system, whereas, Section 5 throws light on the experimental results and discussion

2. Design and Analysis

2.1. Design Requirements

Heat required for the purpose of cooking through heater can be calculated using Equation (1):

$$Q = P_h \times t \tag{1}$$

The heat losses cannot be ignored in every real system. Total heat required incorporating heat loss can be calculated using Equation (2):

$$Q_T = Q + Q_l \tag{2}$$

The required design parameters are summarized in Table 1.

Table 1. Design requirement for cooking.

S. No	Description, Symbol (Units)	Values
1	Temperature requirement, T (°C)	95–97
2	Cooking time, t (minutes)	55
3	Power of electric Heater, P_h (W)	1000
4	Required heat, Q (kJ)	3300
5	Heat losses, Q_l (kJ)	100
6	Total Heat, Q_T (kJ)	3400
7	Steel pipe inner diameter (inches)	1
8	Steel pipe length (inches)	30

For the purpose of encapsulating the PCM, several tests were done on small one inch and half inch pipes as PCM encapsulating materials since they are effective and can be cheaply manufactured.

2.2. Selection of Potash Alum as PCM and Heat Transfer Fluid

It is possible to find materials with a heat of fusion and melting temperature in the desired range (184 kJ/kg, 92 °C) but a material has to exhibit certain properties to become a feasible PCM. The properties and characteristics required for selection of a phase change material desired for heat storage are shown in Table 2.

Table 2. Properties of Phase Change Materials [26,27].

Thermal Properties	Physical Properties	Kinetic Properties	Chemical Properties	Economics
Suitable phase-transition temperature	Favorable phase equilibrium	No super cooling	Long-term chemical stability	Abundant
High latent heat of transition	High density	Sufficient crystallization rate	Compatibility with materials of construction	Available
Good heat transfer	Small volume change		No toxicity	Cost effective
	Low vapor pressure		No fire hazard	

Based upon the above properties and special consideration to economics and thermal stability of the PCM, potash alum, which is readily available in the market, was selected as a suitable PCM for latent thermal energy storage systems [28]. Its melting temperature is about 92 °C, which is why it is suitable for cooking or for heating water which is near to our target range of 95–97 °C. The latent heat of potash alum is about 184 kJ/kg, which is very high compared to rest of PCMs. Potash alum is a PCM with exceptional stability behavior [29]. Table 3 shows the characteristics of the potash alum and its suitability as a phase change material. Thermal stability is an extremely important characteristic to select a PCM as the material has to withstand several heating and cooling cycles in order to ensure the efficient functioning of the thermal system. Therefore several tests were carried out with eutectic salt mixtures and potash alum to check which one is the most thermally stable. Firstly, the eutectic salt Al_2O_3 (66%) and NaCl (34%) were encapsulated, which melted twice and by the third cycle it was rock solid. Secondly, potsh alum was encapsulated, melted and solidified 30 times and every time the material melted easily and did not show degradation which demonstrates the good stability characteristics of potash alum.

Table 3. Design parameters of the storage tank, PCM and HTF.

S.No	Description, Symbol (Units)	Values
1	Heat of fusion of potash alum (PCM), h_m (kJ/kg)	184
2	Required mass of potash alum (PCM), m_{pcm} (kg)	18.4
3	Specific heat of potash alum in the liquid phase (melted), C_{lp} (kJ/kg.K)	2.76
4	Specific heat of potash alum in liquid phase, C_{sp} (kJ/kg.K)	1.38
5	Melted fraction, a_m	1
6	Melting point of potash alum, T_m (°C)	92
7	Initial temperature, T_i (°C)	10
8	Final temperature, T_f (°C)	140
9	Total heat storage capacity, Q_{pcm} (kJ)	4591
10	Density of potash alum in the liquid phase (melted), ρ_i (kg/m^3)	1300
11	Volume of potash alum, V (m^3)	0.0079
12	Outer diameter of cylinder, d_{oo} (m)	0.201
13	Inner diameter of cylinder, d_{ii} (m)	0.2
14	Inner diameter of tube, d_{it} (m)	0.0254
15	Number of tubes, n	89
16	Length of PCM tube, L_{pcmt} (m)	0.74
17	Length of storage tank, L_{st} (m)	0.77
18	Inner radius of storage tank inlet, r_{ii} (m)	0.00635
19	Inner inlet area, A_i (m^2)	0.000125
20	Volume flow rate, q' (m^3/s)	1.235×10^{-5}
21	Mass flow rate, m' (kg/s)	0.013
22	Velocity of HTF at the inlet, u (m/s)	0.1
23	Viscosity of HTF (at 20 °C), μ_{htf} (Pa · S)	2.1671×10^{-3}
24	Reynold's number, Re	605
25	Required flow rate of pump, (m^3/s)	2.5×10^{-5}

The selection criteria for heat transfer fluid is based on the specific heat density, boiling point, decomposition point, viscosity and availability of the fluid [30,31]. Among the other heat transfer fluids, water has a boiling point of 100 °C at 1 atm and highest specific heat (4.184 kJ/kg-K). However, adding ethylene glycol to water increases its melting point significantly and reduces its specific heat slightly [32], so a mixture of 30% ethylene glycol and 70% water was selected as a heat transfer fluid in this study.

2.3. Modeling and Design of Storage Tank

The mass of phase change material required for the storage of heat is calculated using Equation (3). Where Q_T is the value of heat required for cooking and h_m stands for the heat of fusion of potash alum:

$$m_{pcm} = \frac{Q_T}{h_m} \tag{3}$$

The total heat storage capacity of a latent heat system in the concrete case of solid-liquid transformation incorporating sensible heat can be found by knowing the values of the mass of phase change material (m_{pcm}), specific heat of the potash alum in the liquid phase (Melted) (C_{lp}), the specific heat of the potash alum in the liquid phase (C_{sp}), melted fraction (a_m), heat of fusion of potash alum (h_m), melting point of potash alum (T_m), initial temperature (T_i) and final temperature (T_f) using Equation (4):

$$Q_{pcm} = m_{pcm} \times \left((a_m \times h_m) + \left(C_{sp} \times (T_m - T_i) \right) + \left(C_{lp} \times (T_f - T_m) \right) \right) \tag{4}$$

The increase in energy occurs due to $C_{sp} \times (T_m - T_i) + C_{lp} \times (T_f - T_m)$, which represents the sensible heat. This sensible heat is lost very quickly, so the useful energy to be considered is the latent energy of the PCM which is to be calculated from the PCM mass. This means that the PCM material considered in the study has sufficient capacity to store the required amount of heat. The required volume of the potash alum (PCM) can be calculated using Equation (5):

$$V_{pcm} = \frac{m}{\rho_l} \tag{5}$$

The length of the PCM tubes (L_{pcmt}) can be calculated using Equation (6) where V_{pcm} stands for PCM volume, d_{it} for inner diameter of tubes and n for number of tubes:

$$L_{pcmt} = \frac{(V \times 4)}{(\pi \times n \times d_{it}{}^2)} \tag{6}$$

The length of the storage tank (L_{st}) can be found from the length of PCM tube using Equation (7):

$$L_{st} = L_{pcmt} + 0.03 \tag{7}$$

The inner inlet area of the storage tank is calculated using Equation (8), where r_{ii} stands for the radius of the storage tank:

$$A_i = \pi \times r_{ii}{}^2 \tag{8}$$

The mass flow rate of HTF is calculated using Equation (9), where ρ_{HTF} stands for the density of HTF and $\dot{q}$ stands for the volume flow rate:

$$\dot{m} = \rho_{HTF} \times \dot{q} \tag{9}$$

The velocity of the HTF at the inlet is calculated using Equation (10):

$$u = \frac{\dot{q}}{A_i} \tag{10}$$

The Reynold's number for the HTF is calculated using Equation (11) which specifies the flow pattern of a fluid. A low Reynold's number indicates laminar flow, whereas a high Reynold's number indicates turbulent flow. μ_{HTF} stands for the velocity of the HTF:

$$Re = \frac{\rho_{HTF} \times u \times 2 \times r_{ii}}{\mu_{HTF}} \tag{11}$$

The results calculated above for the design parameters of the storage tank, PCM and HTF are summarized in Table 3.

In order to get an idea how much time it takes to melt the PCM material within the tank (charging time) and also how much time can the energy be stored within the storage tank (storing time) we use two separate simulation studies.

A 2D model of the latent energy storage tank is constructed in COMSOL, a software used for simulation-based modeling and designing physical and mechanical structures. The COMSOL software is used to analyze the temperature and phase change results of PCM material after a specific amount of charging and storing time. The geometry of the 2D model is shown in the Figure 1. The HTF enter through the inlet and exits through outlet, and transfer heat throughout the storage tank container which contains a cross section of 10 PCM tubes as shown in Figure 1.

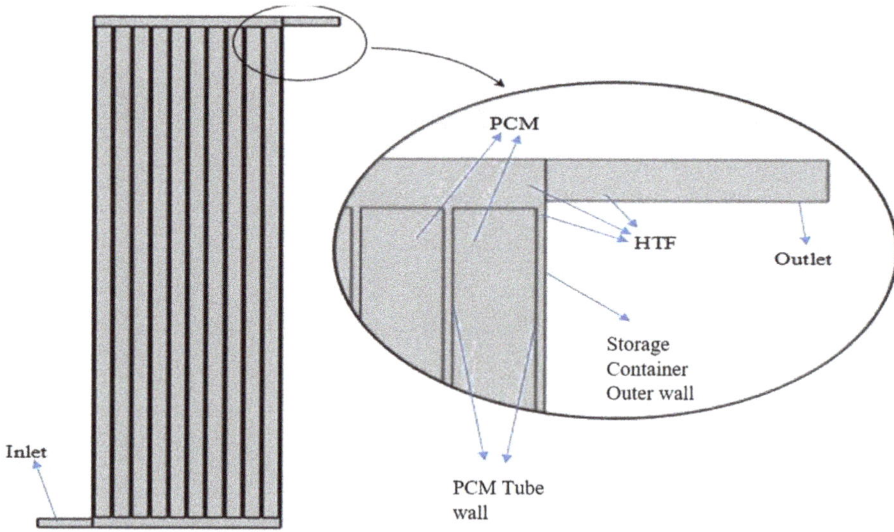

Figure 1. Energy storage tank modeling in COMSOL.

3. Simulation Results

For charging time simulation, laminar flow is considered, incorporating gravitational effects and insulation effects on the walls of the storage tank. At the inlet, the HTF flows with a velocity of 0.1 m/s normal to the inlet boundary. Temperature boundary condition is specified as 105 °C (378 K). In the simulation, it has been assumed that the temperature of the HTF has reached the maximum value of 105 °C. The outflow boundary condition was specified at the outlet. After 5 h of simulation time for HTF flow in the storage tank, the results are observed. Figure 2a,b illustrate the temperature contour along with liquid level indicator plots for charging the storage tank at various time steps.

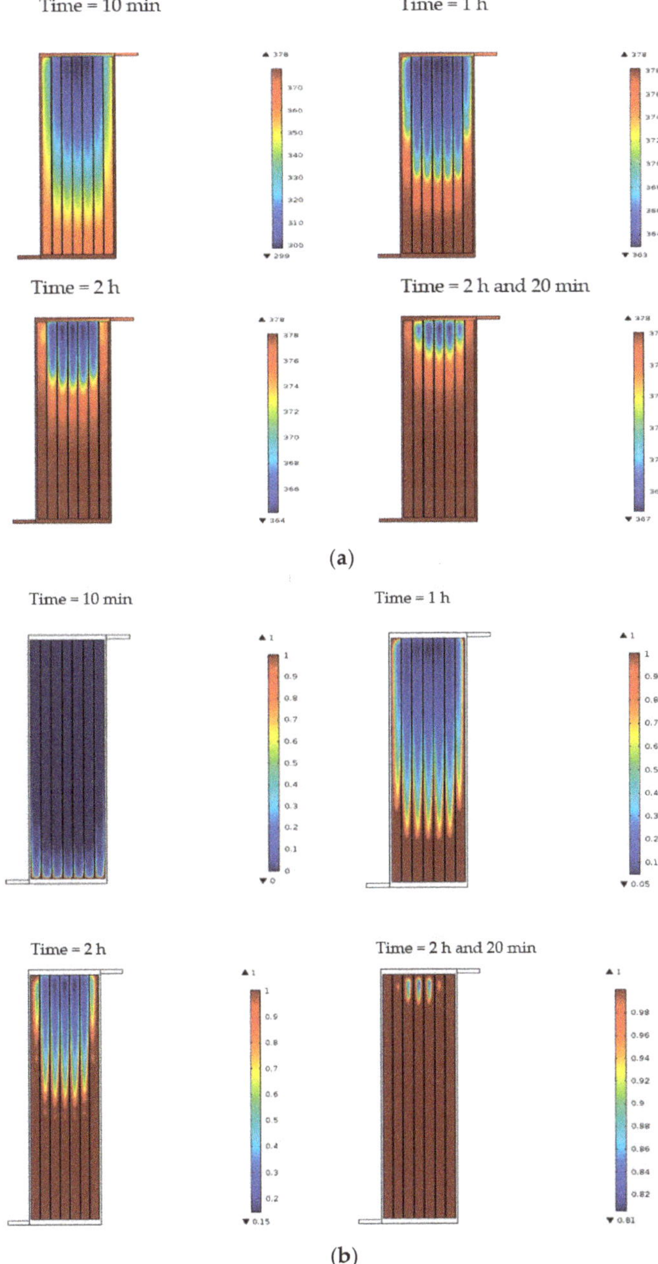

Figure 2. (a) Temperature for 10 min, 1 h, 2 h and 2 h and 20 min; (b) Liquid level temperature for 10 min, 1 h, 2 h and 2 h and 20 min.

The liquid level indicator shows how much portion of the PCM is converted into liquid so a scale of 0 to 1 is defined where 0 stands for the PCM to be solid while 1 stands for the liquid state. Any value between 0 and 1 is partially liquid and partially solid state of the PCM.

From the temperature contours it is observed that, once the HTF reaches a temperature of 105 °C (378 K) it takes about 140 min to completely melt the PCM as shown in the last contour in Figure 2. It is also observed that the liquid level indicator shows a small portion of the PCM which isn't fully melted while the temperature contour shows that the temperature is nearly uniform at 104 °C (377 K) everywhere in the tube except the small solid PCM region. Figure 3 shows the relation of average temperature in the PCM region with respect to time showing the melting behavior of the PCM.

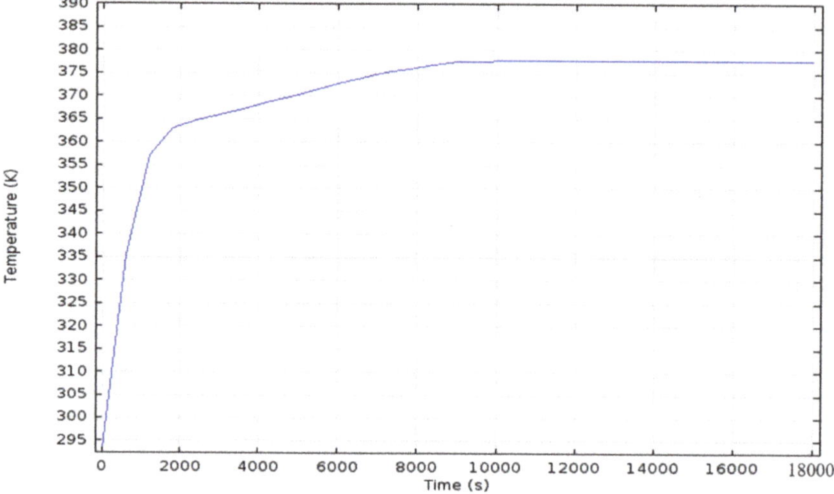

Figure 3. Temperature vs. Time of melting of the PCM.

Simulation for Charging Time

Using the same geometry as Figure 1 for our storage time simulation, it is considered that the PCM was already melted and is at a temperature of 105 °C (378 K). The inlet and outlet boundary conditions are replaced by walls because during storage the valves will be closed so fluid will not flow. Convective heat flux condition is given to the outside boundary of the storage tank with ambient air temperature taken to be 20 °C (293 K). After 12 h of simulation time for airflow over the charged storage tank the results are observed. Figure 4 illustrates the temperature contour along with liquid level indicator plots for storing heat in the storage tank for various time steps.

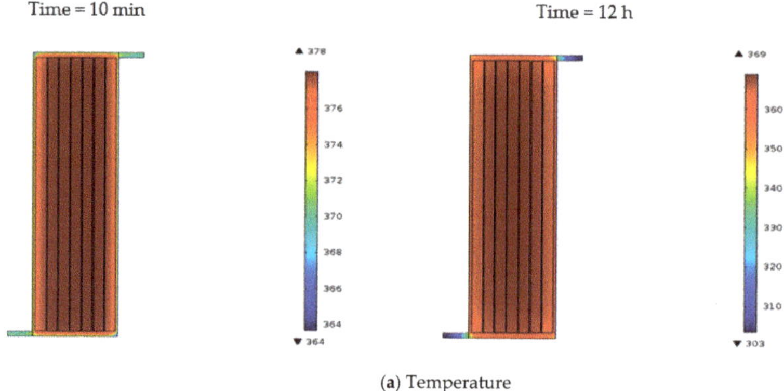

(a) Temperature

Figure 4. *Cont.*

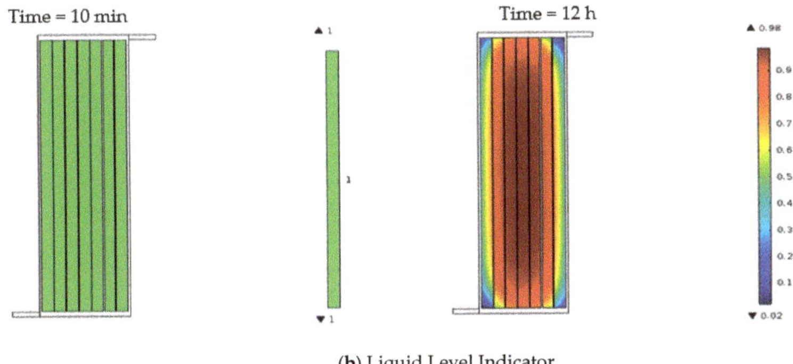

(b) Liquid Level Indicator

Figure 4. Storage time by: (**a**) temperature and (**b**) liquid level for 10 min and 12 h.

As it can be observed that, it takes about 12 h to re-solidify the PCM up to some extent. The storing capacity for 8 cm thick insulation is very high. The temperature plot for storage time is shown in Figure 5.

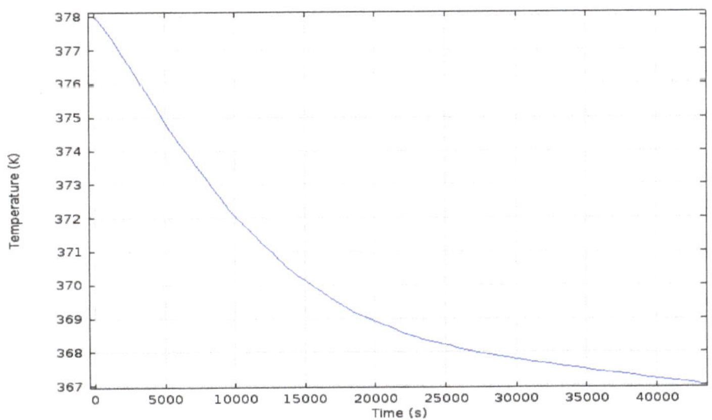

Figure 5. Temperature vs. storage time to resolidify the PCM.

4. Fabrication

4.1. Prototype Modeling of Solar Concentrator

The prototype design was carried out with CAD/CAM software Creo parametric, which contemplates the parabola and focus characteristics, as well as the dimensions adjusted to the estimated size from the above calculations. As shown in Figure 6, the parabolic concentrator support has four support points; each support point has a piece to level the structure to the surface. There are four wheels attached to these four support points. The parabolic dish concentrator can rotate only in one axis with the help of screws. The whole structure (stand dish assembly) can be rotated in the other axis with the help of the support wheels

381

Figure 6. CAD/CAM Model of the solar concentrator.

4.2. Design of Solar Concentrator

The parabolic dish is made up of fiber glass and small pieces of mirror are attached to the inner surface of the concentrator, which reflects the solar radiation at a single point known as the focal point. At the focal point, the receiver tank is placed which contains the heat transfer fluid which is heated up through the sunlight directed toward it. This heat transfer fluid is moved from receiver tank to the storage tank with the help of pump until the PCM is melted. Figure 7 shows the solar concentrator that was manufactured and utilized in this study.

Figure 7. Manufactured solar concentrator.

4.3. PCM Tubes

Steel tubes are used for encapsulating the PCM. These PCM tubes are filled by melted PCM and welded from both sides as shown in Figure 8. The dimensional aspect of the PCM tubes has already been summarized in Table 1.

Figure 8. PCM steel tubes used for encapsulating the PCM.

4.4. Storage Tank

The storage tank consists of two cylindrical tanks made of steel as shown in Figure 9. The internal tank also known as PCM tank contains the PCM tubes while the other tank is used as an external tank. There is styrofoam and glass wool insulation between the internal tank and external tanks.

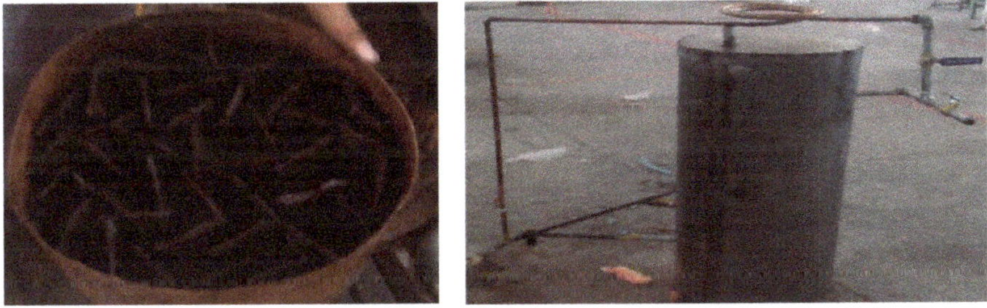

Figure 9. Fabrication of the storage tank.

4.5. Assembled Design

Flexible rubber pipes are used to assemble the receiver tank and storage tank, so that the concentrator can rotate freely on two axes. A copper coil which utilizes the stored energy is connected to storage tank by steel pipes. In order to circulate the HTF in the system, a 12 Volt pump of the type originally used to spray water on the windshield in cars is used. Five ball valves are used to direct the flow of the HTF into the two thermal circuits. Two dial temperature sensors are attached to the receiver tank and at outlet of the storage tank to measure the temperature in the respective tanks. The flow chart of charging and discharging of the PCM is shown in Figure 10a,b, respectively. Figure 11 shows the design of the assembled solar energy storage system

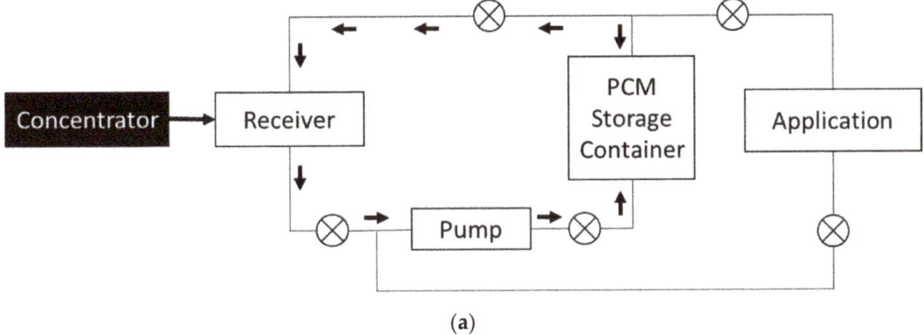

(a)

Figure 10. *Cont.*

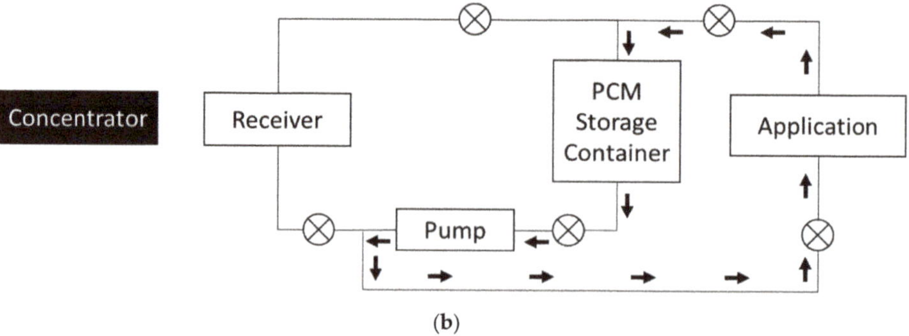

(**b**)

Figure 10. (**a**) Flow chart of charging PCM; (**b**) Flow chart of discharging PCM.

Figure 11. Completely assembled prototype of the solar thermal energy storage system.

5. Experimental Results and Discussion

In this section, the experimental results for the solar thermal energy storage system based upon the assembled experimental setup are discussed. The experimental results are compared and discussed with simulated results.

5.1. Experimental Results

Figure 12 shows the temperature behavior of the HTF. The following results were obtained for charging the PCM storage tank.

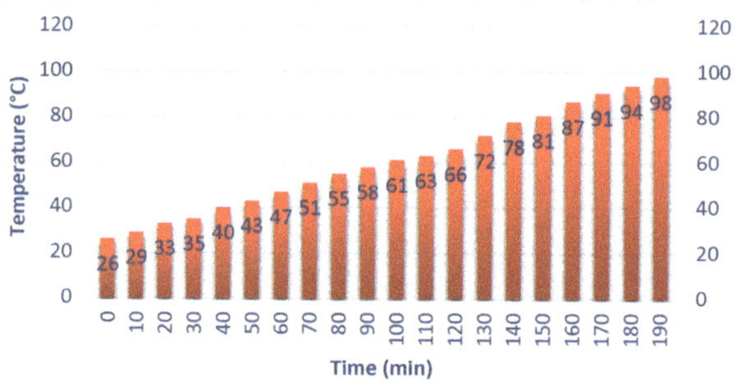

Figure 12. Plot of HTF temperature vs charging time.

The results show a linear trend as the temperature of the HTF follows a linear pattern as time advances and reaches a value of 98 °C. At this point, the temperature of the HTF approaches closer to its boiling point, i.e., 104.4 °C, so the process is halted for a while to regain the temperature of 98 °C. The temperature is maintained by this process for about 2 h and 50 min in order to completely melt the PCM.

Figure 13 shows the results for the storing time of the solar thermal energy storage system. The charged storage tank was left in the open air after sunset and the recorded temperature relation with respect to time is given as follows.

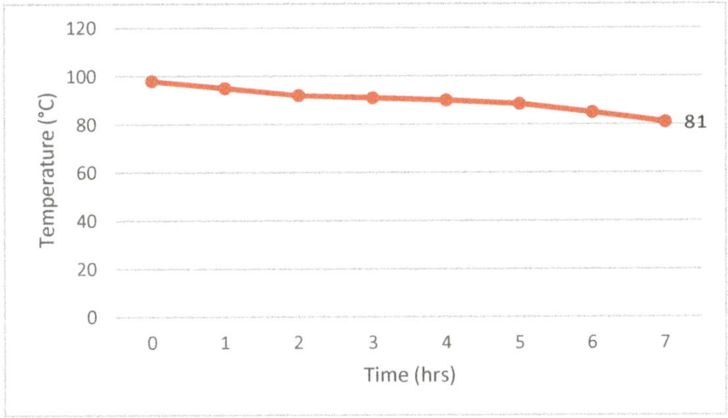

Figure 13. Plot of temperature vs. storage time.

Figure 14 shows that initially the temperature of the PCM at 98 °C steadily decreases to 92 °C and then remains constant for about 3 h at said temperature. The temperature remains constant at this value due to the latent heat of the PCM being released at 92 °C and then the temperature steadily decreases to about 81 °C after 7 h. This means that the solar thermal energy storage system has the capacity to store energy for about 7 h after sunset. The discharge time of the system was found from another experiment by first charging the storage tank at 97 °C and then storing the energy for about 1 h, and then by putting water in a pot placed above the heating coil.

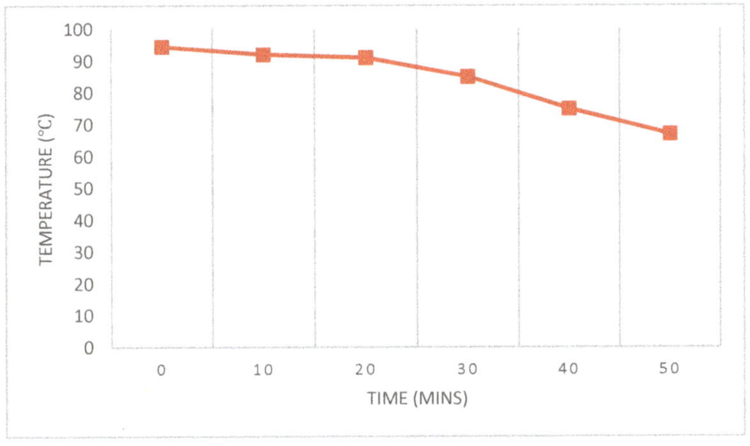

Figure 14. Plot of temperature vs. discharge time after 1 h.

Figure 14 shows that the stored energy can be utilized for 40 min before discharging the system completely as the temperature falls below 70 °C after this time and the remaining available heat energy is not enough for cooking.

5.2. Analysis and Discussion

The comparison of the experimental and simulation results highlights that both the results are in conformance for the charging duration of the analysis but differ slightly when it comes to storage analysis. The difference in the results occurs because of the manufacturing defects with the insulation of the storage tank as the storage tank was not tightly insulated. This defect results in a significant heat loss affecting the storage capacity of the storage tank to reduce to 7 h as compared to 12 h as derived from the simulation results. However the experimentally measured charging time (2.83 h) is within reasonable error (5%) of the simulated charging time (2.66 h). From the results, it was analyzed that the discharging time of 40 min is reasonable to green vegetables and some pulses.

6. Conclusions

This paper presented an assessment of the design and fabrication of a thermal energy storage system using potash alum as a phase change material. Our preference of potash alum as a phase changing material over other materials is due to its easy availability at a lower price along with better thermal energy storage compatibilities for low temperature cooking and heating purposes. The prime importance for thermal energy storage performance is the coupling between the heat transfer fluid characteristics and storage performance. In this study, a mixture of ethylene glycol and water is used as heat transfer fluid. The experimental setup was developed and the results were compared with the simulation results performed in COMSOL. The propsed setup shows that thermal energy can be stored for up to 7 h at a maximum temperatue of 92 °C and minimum temperature of 81 °C after sunset, which is a very beneficial output of the proposed system. The simulation results were consistent with experimental results and confirm that the proposed thermal energy storage system could be successfully used efficiently for cooking and heating purposes for applications which require temperatures of less than 92 °C. i.e., low temperature cooking and water heating can be done using this setup. The application of this system is not limited to indoor cooking, and it can be used for several other applications requiring temperatures below 92 °C, for example, it can be used for heating water for indoor applications and also used as an indoor heater, etc.

As future work we can mention the following aspects: (1) We used two temperature sensors in the system, but no pressure sensor to measure the pressure in the system. The system should have a feedback loop from the pressure and temperature sensor to stop the pump from pumping the high temperature HTF; (2) The selection of the HTF can be improved; (3) The insulation of the tank can be improved by using a good manufacturing process and materials to enhance the storage capacity of the storage tank; (4) Simulation of discharging can be added by linking in to the already stored energy in the concentrator to an application; (5) The charging simulation can be improved to incorporate even more multiphysics and (6) a better concentrator design which can be linked to the phase change simulation of the concentrator discussed in this paper can be implemented. For example sunlight incidence upon a concentrator can be simulated using particle ray tracing simulation and concentrated upon a receiver tank containing a HTF which heats up. A pump can be added from the receiver to the concentrator and back to the receiver to complete the loop Design improvements can be further achieved by using different variables for various parameters in the simulation and the optimized results can be found based upon the best results for charging, storing and discharging times. A few improvements in the design of the concentrator and storage tank could yield a commercially viable product.

Author Contributions: M.S.M. and N.I., Conceptualization, invistgation, methodology, writing draft, validation Data curation, formal analysis. A.W. & S.K. writing review, investagation, funding acquisition, supervision, review and editing. M.O.K. and M.U.A. investagation, review and editing, resources, editing, validation. T.K. and K.-C.K., funding acquisition, resources. Z.R., experimental anaysis, revised the draft, investigation and S.T.u.I.R. investigate, review & Editing & experimental anaysis. All authors have read and agreed to the published version of the manuscript.

Funding: This work was supported by the Basic Science Research program through the National Research Foundation of Korea (NRF) and funded by the ministry of Education under grant NRF-2019R1D1A1A09058357.

Conflicts of Interest: The authors declare no conflict of interest.

References

1. Li, M.-J.; Tao, W.-Q. Review of methodologies and polices for evaluation of energy efficiency in high energy-consuming industry. *Appl. Energy* **2017**, *187*, 203–215. [CrossRef]
2. Ascione, F. Energy conservation and renewable technologies for buildings to face the impact of the climate change and minimize the use of cooling. *Sol. Energy* **2017**, *154*, 34–100. [CrossRef]
3. Kittner, N.; Lill, F.; Kammen, D.M. Energy storage deployment and innovation for the clean energy transition. *Nat. Energy* **2017**, *2*, 17125. [CrossRef]
4. Prasad, A.A.; Taylor, R.A.; Kay, M. Assessment of solar and wind resource synergy in Australia. *Appl. Energy* **2017**, *190*, 354–367. [CrossRef]
5. Li, Q.; Liu, Y.; Guo, S.; Zhou, H. Solar energy storage in the rechargeable batteries. *Nano Today* **2017**, *16*, 46–60. [CrossRef]
6. Nazir, H.; Batool, M.; Osorio, F.J.B.; Isaza-Ruiz, M.; Xu, X.; Vignarooban, K.; Phelan, P.; Kannan, A.M. Recent developments in phase change materials for energy storage applications: A review. *Int. J. Heat Mass Transf.* **2019**, *129*, 491–523. [CrossRef]
7. Pelay, U.; Luo, L.; Fan, Y.; Stitou, D.; Rood, M. Thermal energy storage systems for concentrated solar power plants. *Renew. Sustain. Energy Rev.* **2017**, *79*, 82–100. [CrossRef]
8. Esteves, L.; Magalhães, A.; Ferreira, V.M.; Pinho, C. Test of Two Phase Change Materials for Thermal Energy Storage: Determination of the Global Heat Transfer Coefficient. *ChemEngineering* **2018**, *2*, 10. [CrossRef]
9. Zeinelabdein, R.; Omer, S.; Gan, G. Critical review of latent heat storage systems for free cooling in buildings. *Renew. Sustain. Energy Rev.* **2018**, *82*, 2843–2868. [CrossRef]
10. Patankar, S. *Numerical Heat Transfer and Fluid Flow*; CRC Press: Boca Raton, FL, USA, 2018.
11. González-Roubaud, E.; Pérez-Osorio, D.; Prieto, C. Review of commercial thermal energy storage in concentrated solar power plants: Steam vs. molten salts. *Renew. Sustain. Energy Rev.* **2017**, *80*, 133–148. [CrossRef]
12. Alva, G.; Liu, L.; Huang, X.; Fang, G. Thermal energy storage materials and systems for solar energy applications. *Renew. Sustain. Energy Rev.* **2017**, *68*, 693–706. [CrossRef]
13. Zhang, P.; Meng, Z.; Zhu, H.; Wang, Y.L.; Peng, S. Melting heat transfer characteristics of a composite phase change material fabricated by paraffin and metal foam. *Appl. Energy* **2017**, *185*, 1971–1983. [CrossRef]
14. Amin, M.; Putra, N.; Kosasih, E.A.; Prawiro, E.; Luanto, R.A.; Mahlia, T. Thermal properties of beeswax/graphene phase change material as energy storage for building applications. *Appl. Therm. Eng.* **2017**, *112*, 273–280. [CrossRef]
15. Karaipekli, A.; Biçer, A.; Sari, A.; Tyagi, V.V. Thermal characteristics of expanded perlite/paraffin composite phase change material with enhanced thermal conductivity using carbon nanotubes. *Energy Convers. Manag.* **2017**, *134*, 373–381. [CrossRef]
16. Zeng, J.-L.; Cao, Z.; Yang, D.W.; Sun, L.-X.; Zhang, L. Thermal conductivity enhancement of Ag nanowires on an organic phase change material. *J. Therm. Anal. Calorim.* **2009**, *101*, 385–389. [CrossRef]
17. Shukla, A.; Buddhi, D.; Sawhney, R. Thermal cycling test of few selected inorganic and organic phase change materials. *Renew. Energy* **2008**, *33*, 2606–2614. [CrossRef]
18. Singh, R.; Sadeghi, S.; Shabani, B. Thermal Conductivity Enhancement of Phase Change Materials for Low-Temperature Thermal Energy Storage Applications. *Energies* **2018**, *12*, 75. [CrossRef]
19. Bushnell, D.L. Performance studies of a solar energy storing heat exchanger. *Sol. Energy* **1988**, *41*, 503–512. [CrossRef]

20. Sharma, S.; Buddhi, D.; Sawhney, R.; Sharma, A. Design, development and performance evaluation of a latent heat storage unit for evening cooking in a solar cooker. *Energy Convers. Manag.* **2000**, *41*, 1497–1508. [CrossRef]

21. Buddhi, D.; Sharma, S.; Sharma, A. Thermal performance evaluation of a latent heat storage unit for late evening cooking in a solar cooker having three reflectors. *Energy Convers. Manag.* **2003**, *44*, 809–817. [CrossRef]

22. Schmerse, E.; Ikutegbe, C.A.; Auckaili, A.; Farid, M.M. Using PCM in Two Proposed Residential Buildings in Christchurch, New Zealand. *Energies* **2020**, *13*, 6025. [CrossRef]

23. Nemś, A.; Puertas, A.M. Model for the Discharging of a Dual PCM Heat Storage Tank and its Experimental Validation. *Energies* **2020**, *13*, 5687. [CrossRef]

24. Leang, E.; Tittelein, P.; Zalewski, L.; Lassue, S. Design Optimization of a Composite Solar Wall Integrating a PCM in a Individual House: Heating Demand and Thermal Comfort Considerations. *Energies* **2020**, *13*, 5640. [CrossRef]

25. Pasupathi, M.K.; Alagar, K.; Mm, M.; Aritra, G. Characterization of Hybrid-nano/Paraffin Organic Phase Change Material for Thermal Energy Storage Applications in Solar Thermal Systems. *Energies* **2020**, *13*, 5079. [CrossRef]

26. Wei, G.; Wang, G.; Xu, C.; Ju, X.; Xing, L.; Du, X.; Yang, Y. Selection principles and thermophysical properties of high temperature phase change materials for thermal energy storage: A review. *Renew. Sustain. Energy Rev.* **2018**, *81*, 1771–1786. [CrossRef]

27. Qiu, S.; Solomon, L.; Fang, M. Study of Material Compatibility for a Thermal Energy Storage System with Phase Change Material. *Energies* **2018**, *11*, 572. [CrossRef]

28. Palomba, V.; Brancato, V.; Palomba, G.; Borsacchi, S.; Forte, C.; Freni, A.; Frazzica, A. Latent Thermal Storage for Solar Cooling Applications: Materials Characterization and Numerical Optimization of Finned Storage Configurations. *Heat Transf. Eng.* **2018**, *40*, 1033–1048. [CrossRef]

29. Prabhu, A.N.; Shetty, P.K.; Vali, I.; Kumar, H.; Udupa, S. Gamma and neutron irradiation effects on the structural and optical properties of potash alum crystals. *Nucl. Instrum. Methods Phys. Res. Sect. B Beam Interact. Mater. Atoms* **2017**, *413*, 37–41. [CrossRef]

30. Benoit, H.; Spreafico, L.; Gauthier, D.J.; Flamant, G. Review of heat transfer fluids in tube-receivers used in concentrating solar thermal systems: Properties and heat transfer coefficients. *Renew. Sustain. Energy Rev.* **2016**, *55*, 298–315. [CrossRef]

31. Lorenzin, N.; Abánades, A. A review on the application of liquid metals as heat transfer fluid in Concentrated Solar Power technologies. *Int. J. Hydrog. Energy* **2016**, *41*, 6990–6995. [CrossRef]

32. Li, H.; He, Y.; Hu, Y.; Jiang, B.; Huang, Y. Thermophysical and natural convection characteristics of ethylene glycol and water mixture based ZnO nanofluids. *Int. J. Heat Mass Transf.* **2015**, *91*, 385–389. [CrossRef]

Article

Robust Decentralized Tracking Voltage Control for Islanded Microgrids by Invariant Ellipsoids

Hisham M. Soliman [1], Ehab Bayoumi [2], Amer Al-Hinai [1,*] and Mostafa Soliman [3]

[1]   Department of Electrical and Computer Engineering, Sultan Qaboos University, Muscat 123, Oman; hsoliman1@squ.edu.om
[2]   Department of Electrical and Electronics Engineering, University of Eswatini, Private Bag 4, Kwaluseni M201, Swaziland; ehab.bayoumi@gmail.com
[3]   Department of Computer Engineering, Cairo University, Cairo 12613, Egypt; msoliman.w@gmail.com
*   Correspondence: hinai@squ.edu.om; Tel.: +968-24141356

Received: 6 October 2020; Accepted: 27 October 2020; Published: 3 November 2020

Abstract: This manuscript presents a robust tracking (servomechanism) controller for linear time-invariant (LTI) islanded (autonomous, isolated) microgrid voltage control. The studied microgrid (MG) consists of many distributed energy resources (DERs) units, each using a voltage-sourced converter (VSC) for the interface. The optimal tracker design uses the ellipsoidal approximation to the invariant sets. The MG system is decomposed into different subsystems (DERs). Each subsystem is affected by the rest of the system that is considered as a disturbance to be rejected by the controller. The proposed tracker (state feedback integral control) rejects bounded external disturbances by minimizing the invariant ellipsoids of the MG dynamics. A condition to design decentralized controllers is derived in the form of linear matrix inequalities. The proposed controller is characterized by rapid transient response, and zero error in the steady state. A robustness analysis of the control strategy (against load changes, load unbalances, etc.) is carried out. A MATLAB/SimPowerSystems (R2017b, MathWorks, Natick, MA, USA) simulation of the case study confirm the robustness of the proposed controller.

Keywords: islanded mode; microgrid; decentralized control; robust tracking; invariant set

1. Introduction

Distributed energy resources (DERs), such as photovoltaic arrays (PVs) and wind, are now connected to the power grids to address environmental issues and emissions of global warming gas [1]. The loads represented by constant impedances are connected to DGs [2–14] for the operation of MG at the distribution networks. Stability analysis is one of the most critical issues for inverter-interfaced MG [4–8]. Attaining precise power-sharing while controlling the voltage magnitude and frequency in an autonomous MG is the main DERs control objective [4]. Researchers have used various centralized and decentralized control strategies to improve MG dynamic performance [8–12]. The benefits and drawbacks of the schemes are summarized in [8]. In [8,9], different centralized control schemes are proposed to control multiple parallel inverters, maximize the DERs output power, and optimize the power exchange between the MG and the main grid. Centralized control in remote areas with a long distance between inverters is impractical and expensive due to the requirement of building a reliable communication link [9]. To avoid the utilization of expensive communication networks, decentralized systems are utilized [10,11]. Distributed control that lies in between the centralized and decentralized approaches in terms of complexity, price, and effectiveness [12–15]. It allows communication between the subsystems. However, it faces the problems of packet loss, communication delay, and quantization errors.

MGs can operate either connected to the utility main grid or isolated form it. The MG operates in islanded mode when it is disconnected from the grid and continues to provide power to local loads. The voltage and frequency are no longer dictated by the grid, and they can have values dictated by the MG's DER units. Because of the load variations and the DERs' output power intermittency, an islanded MG may encounter problems of reliability, robustness and power quality. Novel optimal and robust control strategies are required to minimize these problems.

Voltage source converters (VSC) are widely utilized to interface the MG to the grid. VSCs perform power flow conversion and control. Several control strategies for the autonomous operation of VSC-based DER units are suggested. The frequency/power and voltage/reactive-power droop control [16] are among the most widely used control techniques. Considering an active load to the autonomous MG is given in [17]. A control strategy is proposed in [18] which ensures robust stability despite parametric uncertainties due to load variations using Kharitonov's theorem. Reference [18] does not consider neither multi-DG micro-grids nor decentralized control. This strategy is based on a low order proportional integral controller, which only uses its d-axis part in dq-frame to regulate the load voltage. A more robust control strategy is suggested in [19,20], using a servomechanism controller. Nevertheless, this is a high-order controller that is more complex than the one in [18]. Note that [19] considers only a single-DG islanded system. The decentralized control in [20] has higher dynamics than the proposed one in this manuscript.

This paper presents the decentralized control for each DER. The suggested controller uses only its local information. Decentralization is tackled by decomposing the global MG system into subsystems (DERs). The dynamic effect of the rest of the system on a particular DER is considered as an external disturbance. The proposed controller has to achieve fast response + zero steady state error in addition to rejecting the external disturbance. Note that the designer faces two challenges: (1) a modeling problem and (2) a control problem. The designer has to obtain the model for the global large system. The resulting centralized controller will be of large dimension, difficult to implement. In this case, the designer needs to adopt model reduction or decentralized control. This paper adopts decentralized control for each subsystem, for each only the local (subsystem) states are used.

The designed controllers are decentralized in the sense that it uses only the local information of its subsystem. Centralized versus decentralized control can be summarized as follows. The benefit of using decentralized control via local subsystem information is avoiding using a hub computer (controller) whose failure will cause loss of the global system stability. It also avoids a costly communication network (and its associated delay, packet loss) to transmit information of the whole system to the centralized controller. This is in addition to the high dimensionality of centralized control.

One of the key issues in the control theory is the rejection of external disturbance. It is studied by both the linear quadratic Gaussian optimization (where the disturbance is assumed to be random) and the H∞-optimization (where the disturbance is considered as arbitrary bounded). An alternative approach to disturbance rejection relying on the method of invariant sets and invariant ellipsoids is proposed in [21,22]. The invariant ellipsoid method is a newly born powerful method in robust control theory.

The present work formulates the problem of external disturbances (arbitrary bounded) rejection in terms of the invariant ellipsoids. The proposed feedback+ integral controller is designed by minimizing the size of the invariant ellipsoids of the dynamic system. The controller synthesis is formulated using linear matrix inequalities (LMIs) that can be effectively solved using convex optimization techniques.

The main contributions of this paper are summarized as follows:

1. An effective decentralized control is developed. The overall system is decomposed into subsystems, for each a local controller is installed. Each controller is designed to achieve fast response, zero steady-state error, and effective rejection of external disturbances that are caused by the different subsystems.
2. A new method, based on LMIs and invariant sets, is proposed to design an optimal tracking controller that effectively rejects external disturbances.

The rest of this paper is organized as follows. Section 2 describes the MG under investigation and the proposed control strategy. The MG mathematical model is provided in Section 3. Section 4 details the controller synthesis that is based on the ellipsoidal design. Verification of the performance and viability of the proposed method, based on a simulation case study. Conclusions are stated in Section 5.

Notations and Facts

R^m is the set of m × 1 vectors, $R^{r×q}$ is the set of real matrices of dimension r × q, and $(.)'$ denotes the transpose of a vector or a matrix. For a matrix P, P > 0 (< 0) means that P is a symmetric positive (negative) definite matrix. Also, the shorthand $\begin{bmatrix} M & N \\ * & L \end{bmatrix}$ means $\begin{bmatrix} M & N \\ N' & L \end{bmatrix}$. Similarly, $(M + N + *)$ means $(M + N + M' + N')$. Matrices are denoted by capital letters, vectors are denoted by small letters, and scalers are denoted by small Greek letters. Finally, 0 and I denote the zero matrices and the identity matrix, respectively.

Fact 1: The time varying uncertainty $\Delta(t)$ can be removed using the fact $M\Delta(t)N + * < \epsilon MM' + \epsilon^{-1}N'N$

Fact 2: (Schur complements). Given a matrix M composed of

$$M = \begin{bmatrix} M_1 & M_3 \\ * & M_2 \end{bmatrix}$$

where $M_1 = M_1'$, $M_2 = M_2' > 0$, then $M > 0$ if and only if

$$M_1 - M_3 M_2^{-1} M_3' > 0$$

Fact 2 is used to linearize a special class of nonlinear matrix inequalities.

2. Microgrid Study System

2.1. Microgrid Description

The MG under study is shown in Figure 1a. A control strategy and power management system (PMS) for several-distributed energy resources (DER) islanded microgrids is proposed in [20]. DERs can also be viewed as distributed generators (DGs). In this system, the PMS defines the voltage set points at the DER terminals and communicates them (via a low bandwidth communication medium) to the local controllers (LCs) of DER units. The DER terminal voltages are robustly regulated by the LCs in a decentralized manner. The MG frequency is controlled using an open-loop control scheme using the DER units' internal crystal oscillators. To ensure the same frequency, these oscillators are synchronized by a common timing signal received from a global positioning system (GPS), Figure 1b.

The MG has of three DER units rated at 0.6 kV. Their power ratings are 1.6, 1.2, and 0.8 MVA. The DER units represent photovoltaic PV generating stations. Each DER unit is modelled using a 1.5 kV DC voltage source, a voltage sourced converter (VSC), and a series RL filter. A 0.6 kV/13.8 kV step-up transformer is used to interface each DER to the grid at its corresponding point of coupling (PC) bus. The three busses in Figure 1b has their local loads and they are connected using two sections of the 13.8 kV distribution line. The utility grid is described by an AC voltage source behind series R and L elements. The MG can be operated in grid-connected or islanded mode based on the circuit breaker CBg status. The parameters of the study system are given in Table 1.

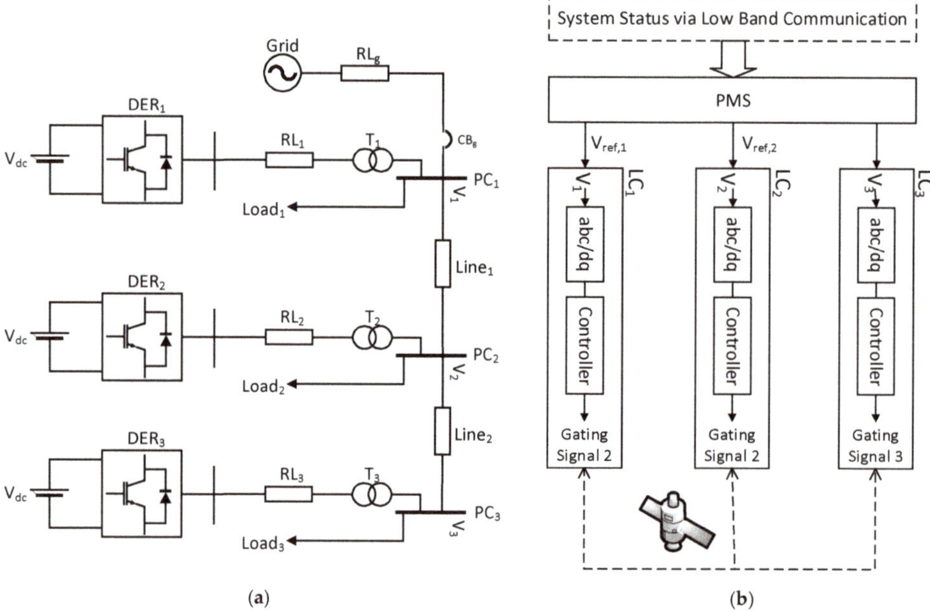

Figure 1. Case-study system (**a**) The studied microgrid (MG) and (**b**) power management system (PMS) and non-droop control system.

Table 1. Parameters of the microgrid.

Base Values							
S_{base} = 1.6 MVA			$V_{base, low}$ = 0.6 kV			$V_{base, high}$ = 13.8 kV	
Transformers							
0.6/13.8 kV			Δ/Y_g			X_T = 8%	
Load Parameters							
	Load$_1$			Load$_2$		Load$_3$	
R_1	350 Ω	2.94 pu	R_2	375 Ω	3.15 pu	R_3	400 Ω	3.36 pu
X_{L1}	41.8 Ω	0.35 pu	X_{L2}	37.7 Ω	0.32 pu	X_{L3}	45.2 Ω	0.38 pu
X_{C1}	44.2 Ω	0.37 pu	X_{C2}	40.8 Ω	0.34 pu	X_{C3}	48.2 Ω	0.41 pu
R_{l1}	2 Ω	0.02 pu	R_{l2}	2 Ω	0.02 pu	R_{l3}	2 Ω	0.02 pu
Line Parameters							
R 0.34 Ω / km 0.0029 pu Line$_1$ 5 km							
X 0.31 Ω / km 0.0026 pu Line 10 km							
Filter Parameters (Based on DER$_i$ Ratings)							
X_f = 15% Quality Factor = 50							

In a low-voltage distribution system, the values line parameters are difficult to acquire, especially when the devices are installed in an ad hoc manner by individual users. Reference [23] proposes a novel method for automated impedance estimation based on practically available parameters at the terminal nodes.

2.2. Mathematical Model of The Microgrid

To design robust linear controllers, a linearized model of the microgrid of Figure 1a using a synchronously rotating dq-frame, which is based on the fundamental frequency of the system, is required. The linearized state space model of the MG is derived based on the single-line diagram shown in Figure 2. Each DER unit is described by a three-phase voltage source inverter with a series RL branch. Loads are represented by an equivalent parallel RLC circuit. Distribution lines are modeled using lumped series RL elements.

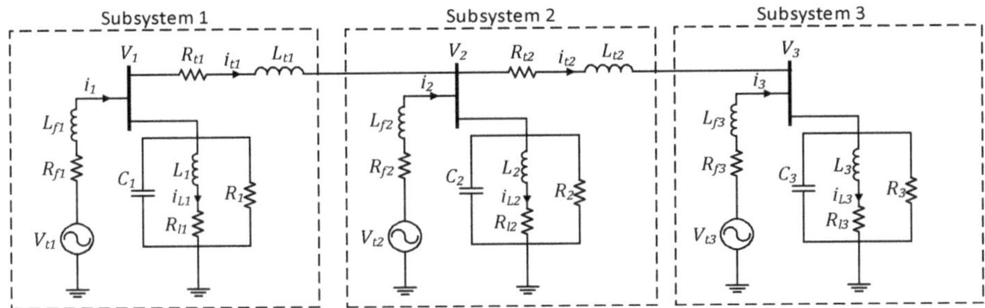

Figure 2. Single-line diagram of the microgrid.

The MG in Figure 2 is partitioned into three subsystems. Subsystem 1 model, in the abc reference frame, is

$$i_{1,abc} = i_{t1,abc} + C_1 \dot{v}_{1,abc} + i_{L1,abc} + \frac{v_{1,abc}}{R_1}$$

$$v_{t1,abc} = L_{f1} \frac{di_{1,abc}}{dt} + R_{f1} i_{t1,abc} + v_{1,abc}$$

$$v_{1,abc} = L_1 \frac{di_{L1,abc}}{dt} + R_{l1} i_{L1,abc} \tag{1}$$

$$v_{1,abc} = L_{t1} \frac{di_{t1,abc}}{dt} + R_{t1} i_{t1,abc} + v_{2,abc}$$

where x_{abc} is a 3×1 vector. The coefficients of differential Equation (1) are time varying, difficult to handle. To simplify the controller design, the *abc* frame of system (1) is transformed to the *dq* frame using Park's transformation. This results in differential equations with constant coefficients, easy to work with. Assuming a three-wire system and using the Park's transformation in Equation (3), the mathematical model (1) can be rewritten in a synchronously rotating *dq*-frame. In Equation (3),

$$\theta(t) = \int_0^t w(\tau)d\tau + \theta_0 \tag{2}$$

is the phase angle and w is the angular frequency of the crystal oscillator internal to DER. The resulting state equation in the *dq* frame is given in Equation (4).

$$f_{dq} = \frac{2}{3} \begin{bmatrix} \cos\theta & \cos\left(\theta - \frac{2}{3}\pi\right) & \cos\left(\theta - \frac{4}{3}\pi\right) \\ -\sin\theta & -\sin\left(\theta - \frac{2}{3}\pi\right) & -\sin\left(\theta - \frac{4}{3}\pi\right) \\ \frac{1}{\sqrt{2}} & \frac{1}{\sqrt{2}} & \frac{1}{\sqrt{2}} \end{bmatrix} \tag{3}$$

$$\dot{V}_{1,dq} = \frac{1}{C_1}I_{1,dq} - \frac{1}{C_1}I_{t1,dq} - \frac{1}{C_1}I_{L1,dq} - \frac{V_{1,dq}}{C_1 R_1} - jwV_{1,dq}$$

$$\dot{I}_{1,dq} = \frac{1}{L_{f1}}V_{t1,dq} - \frac{R_{f1}}{L_{f1}}I_{1,dq} - \frac{1}{L_{f1}}V_{1,dq} - jwI_{1,dq}$$

$$\dot{I}_{L1,dq} = \frac{1}{L_1}V_{1,dq} - \frac{R_{l1}}{L_1}I_{L1,dq} - jwI_{L1,dq}$$

$$\dot{I}_{t1,dq} = \frac{1}{L_{t1}}V_{1,dq} - \frac{R_{t1}}{L_{t1}}I_{t1,dq} - \frac{1}{L_{t1}}V_{2,dq} - jwI_{t1,dq}$$

(4)

Likewise, the subsystems 2, and 3 *dq*-models are both derived and used to write the overall system's state-space model

$$\dot{x} = Ax + Bu, y = Cx \tag{5}$$

where

$$x' = [V_{1,d}, V_{1,q}, I_{1,d}, I_{1,q}, I_{L1,d}, I_{L1,q}, I_{t1,d}, I_{t1,q}, V_{2,d}, V_{2,q}, I_{2,d}, I_{2,q}, I_{L2,d}, I_{L2,q}, I_{t2,d}, I_{t2,q},$$
$$V_{3,d}, V_{3,q}, I_{3,d}, I_{3,q}, I_{L3,d}, I_{L3,q}]$$

$$u' = [V_{t1,d}, V_{t1,q}, V_{t2,d}, V_{t2,q}, V_{t3,d}, V_{t3,q}], \quad y' = [V_{1,d}, V_{1,q}, V_{2,d}, V_{2,q}, V_{3,d}, V_{3,q}]$$

$A \in R^{22 \times 22}$, $B \in R^{22 \times 6}$ and $C \in R^{6 \times 22}$ are the state matrices as given in the Appendix A.

3. Decentralized Tracking Control

In control system design, the output has to follow the input. If the input is constant, the control problem is termed as a regulator problem. If the input is time varying, it is called a servomechanism (tracking) problem. A decentralized servomechanism controller for the system (5) is designed in this section. System (5) represents an interconnected composite system composed of three subsystems. Each subsystem can be controlled by using only local controllers about each subsystem [3]. Splitting the matrix $A = (A_{ij}, i, j = 1, 2, 3)$ in Equation (5) into a two parts: diagonal, A_d, and off diagonal, D, one gets,

$$\dot{x} = A_d x + Bu + Dx, y = Cx \tag{6}$$

where

$$A_d = \begin{bmatrix} A_{11} & 0 & 0 \\ 0 & A_{22} & 0 \\ 0 & 0 & A_{33} \end{bmatrix}, B = \begin{bmatrix} B_1 & 0 & 0 \\ 0 & B_2 & 0 \\ 0 & 0 & B_3 \end{bmatrix}, D = \begin{bmatrix} 0 & A_{12} & A_{13} \\ A_{21} & 0 & A_{23} \\ A_{31} & A_{32} & 0 \end{bmatrix}$$

$$C = \begin{bmatrix} C_1 & 0 & 0 \\ 0 & C_2 & 0 \\ 0 & 0 & C_3 \end{bmatrix}$$

In Equation (6), the effect of interconnecting the rest of the system on a specific subsystem is considered to be an external disturbance, Dx, to be rejected by the controller proposed. The vector x is assumed to be an external bounded disturbance w.

The following structure is considered to implement a decentralized controller, consisting of three control agents, for the MG under study.

$$u = Kx + K_I \xi, \dot{\xi} = r - Cx \tag{7}$$

where

$$K = \begin{bmatrix} K_1 & 0 & 0 \\ 0 & K_2 & 0 \\ 0 & 0 & K_3 \end{bmatrix}, K_I = \begin{bmatrix} K_{I1} & 0 & 0 \\ 0 & K_{I2} & 0 \\ 0 & 0 & K_{I3} \end{bmatrix}$$

so that the augmented closed-loop system is described by

$$\begin{bmatrix} \dot{x} \\ \dot{\xi} \end{bmatrix} = \begin{bmatrix} A_d + BK & BK_I \\ -C & 0 \end{bmatrix} \begin{bmatrix} x \\ \xi \end{bmatrix} + \begin{bmatrix} D \\ 0 \end{bmatrix} w + \begin{bmatrix} 0 \\ I \end{bmatrix} r, \tag{8}$$

Note that the proposed controller (7–8) is a decentralized one equipped with an integral part to eliminate the steady state errors thus the desired reference voltage is tracked. The microgrid dynamics and the suggested control is modelled in Figure 3.

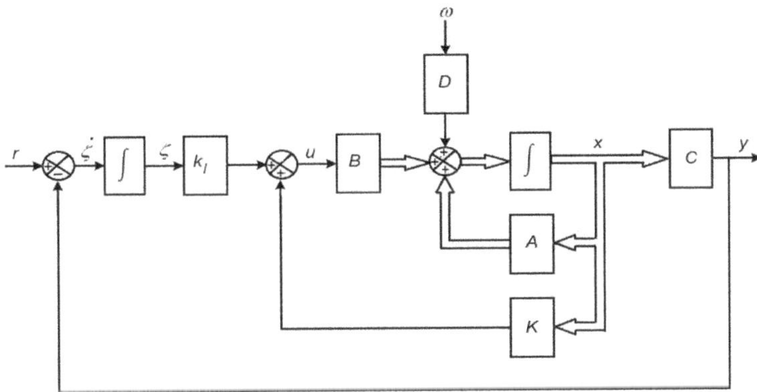

Figure 3. Servo system.

The problem of MG voltage control is the design of a decentralized controller in which the output voltage tracks the reference voltage. The tracker must be robust against the load variations and must minimize the interconnection effect of the rest of the system on the output voltage of a particular subsystem. This is termed disturbance rejection control.

The proposed controller relies on using the invariant (or attracting as will be seen) ellipsoid's concept. A summary is described as follows:

3.1. Attracting Ellipsoid

Consider a continuous-time state space model of a linear dynamical system:

$$\dot{x} = Ax + Dw, \; z = C_z x \tag{9}$$

where $x(t) \in R^n$ is the system state vector, $z(t) \in R^l$ is the system output to be optimized(minimized). It is assumed that the output for feedback y is equal to the output to be optimized, z. Hence $C = C_z$.

The external disturbances $w(t) \in R^m$ are bounded i.e., subject to the constraint

$$\|w(t)\| \le 1, \forall t \ge 0 \tag{10}$$

For the vector $(.)$, $\|(.)\|$ denotes the Euclidean norm of $(.)$. The control target is to minimize the impact of disturbance Dw on the output z. The constraint (10) can be always satisfied by properly scaling the matrix D. The disturbances $w(t)$ is considered to be L∞ bounded external disturbance. It is assumed that (9) is stable, the pair (A, D) is controllable, and that C has maximum-rank. The goal is to determine the system's family of attracting ellipsoids.

An ellipsoid E with origin at the center can be written as

$$E = \{x : x'P^{-1}x \le 1\}, \quad P > 0 \tag{11}$$

where P is a symmetric positive definite matrix called the matrix of the ellipsoid.

The ellipsoid E is termed state invariant if for any initial state $x(0)$ lies *inside* E, the trajectory $x(t)$ stays inside the ellipsoid for $t > 0$ for all admissible disturbances (10). When the initial state $x(0)$ lies outside the ellipsoid, the trajectory $x(t)$ is attracted to the ellipsoid for $t > 0$. This is achieved if the ellipsoid E represented a Lyapunov function

$$V(x) = x'P^{-1}x, \; P > 0 \tag{12}$$

does not increase outside (including the boundary, $V(x) \geq 1$) of this ellipsoid, that is, if

$$\dot{V}(x) \leq 0 \; for \; all \; x(t) \; subject \; to \; V(x) \geq 1 \tag{13}$$

Reference [22] shows that the ellipsoid E is invariant and attracting for system (8) if and only if

$$AP + PA' + \alpha P + \frac{1}{\alpha}DD' \leq 0 \tag{14}$$

When the system is subjected to the family of bounded disturbance in Equation (10), the trajectories of the system relies inside the ellipsoid. To reduce the disturbance impact on the system states and outputs, the volume of the attracting ellipsoid must be minimized. Thus, an objective function must be formulated to represent ellipsoid volume.

In this paper, the trace function in Equation (15) is selected as the objective function. The $tr(.)$ is defined as the sum of the diagonal elements of $(.)$. It corresponds to the sum of the squares of the semi-axes of the state-invariant ellipsoid. The linearity of the trace function offers a significant advantage (the optimization problem becomes convex, easy to handle) compared to other functions that can be used to calculate the volume of the ellipsoid. With this choice, the optimization problem can be easily cast into a standard semi-definite program (SDP) (optimization of a linear function subject to LMI constraints) that can be easily solved using convex optimization solvers that are available using the MATLAB-LMI(R2017b, MathWorks, Natick, MA, USA) toolbox.

$$f(P) = tr(P) \tag{15}$$

It is more important to minimize the disturbance impact on the controlled outputs z rather that the whole state vector x. This can be easily achieved by substituting the ellipsoid of Equation (11) by $E_z = z'P^{-1}z$, $z = Cx$. Thus, if the attracting ellipsoid in Equation (10) bounds the x trajectory, then the output z is contained in the ellipsoid E_z. Minimizing the volume of E_z will minimize the MG voltage variations due to external disturbances.

The optimal state feedback controller design using attracting ellipsoids is discussed next.

3.2. Attracting Ellipsoid Design of State-Feedback Plus Integral Tracker

The state feedback regulator is discussed first. In this case, the reference is assumed to be constant. The main control objectives are to drive all the states and system outputs to the origin, ensure fast transient response and optimize the disturbance rejection performance. The results of this subsection are, then, extended to address a state feedback tracker where the reference is time varying.

Regulator Attracting Ellipsoid Design

Consider the linear system

$$\begin{aligned} \dot{x} &= Ax + Bu + Dw \\ z &= Cx + B_2u \end{aligned} \tag{16}$$

where $x \in R^n$ is the system state, $z \in R^l$ is the controlled output, $u \in R^p$ is the control signal, and $w \in R^m$ is the external disturbance satisfying the constraint (10). The goal is to calculate the gain K of a linear proportional state feedback controller,

$$u = Kx \tag{17}$$

That guarantees stability of the closed-loop system and optimizes the disturbance rejection performance by minimizing the trace of the attracting output ellipsoid. It is worth noting that the inclusion of term $B_2 u$ in Equation (16) prevents large values of control effort.

By substituting Equation (17) in Equation (16), the closed loop system is

$$\dot{x} = (A + BK)x + Dw, \; z = (C + B_2 K)x \tag{18}$$

The following theorem provides a semidefinite program formulation for the state feedback controller synthesis problem.

Theorem 1. *Let the external disturbances be $L\infty$-bounded and the pair (A, B) controllable for the system (16). Then, the problem of designing a state feedback controller by state (17) that optimally rejects the external disturbance, in the sense of the trace that is output-attracting ellipsoid, is equivalent to*
Minimize $tr\!\left[CPC' + B_2 Z B_2'\right]$,
subject to the following constraints:

$$(AP + BY + *) + \alpha P + \frac{1}{\alpha} DD' \leq 0, \alpha > 0 \tag{19}$$

$$\begin{bmatrix} Z & Y \\ * & P \end{bmatrix} \geq 0, \; P > 0 \tag{20}$$

where $Y = KP$. The minimization is carried out with respect to the variables α, $P = P' \in R^{n \times n}$, $Y \in R^{p \times n}$ and $Z = Z' \in R^{p \times p}$ [22].

Note that Equation (19) is nonlinear matrix equation (due to the product term $\alpha P, \frac{1}{\alpha}$). If the scalar $\alpha > 0$ is fixed, the optimization problem in Theorem 1 becomes convex which can be solved efficiently using convex optimization algorithms.

Theorem 1 does not consider a reference input in the regulator formulation. Thus, reference tracking cannot be achieved. In the next subsection, Theorem 1 is extended to offer both good dismissal of disturbance and good reference tracking.

3.3. Tracker Attracting Ellipsoid Design

It can be verified that the MG model (5) has no integrator (type 0 plant), and therefore, non-zero steady error will occur for step changes in the reference voltage. To ensure offset-free tracking of the voltage reference, the integrator must be inserted in the forward path of the loop. In a state feedback controller setup, the integrator is added as shown Figure 3 (which facilitates the Simulink modeling (R2017b, MathWorks, Natick, MA, USA).

From Figure 3, we obtain

$$\dot{x} = A_d x + Bu + Dw, y = Cx, u = Kx + k_i \xi, \dot{\xi} = r - Cx \tag{21}$$

Therefore, the augmented system is

$$\begin{bmatrix} \dot{x} \\ \dot{\xi} \end{bmatrix} = \begin{bmatrix} A & 0 \\ -C & 0 \end{bmatrix} \begin{bmatrix} x \\ \xi \end{bmatrix} + \begin{bmatrix} B \\ 0 \end{bmatrix} u + \begin{bmatrix} D \\ 0 \end{bmatrix} w + \begin{bmatrix} 0 \\ I \end{bmatrix} r, z = [C \; 0] \begin{bmatrix} x \\ \xi \end{bmatrix} + B_2 u \tag{22}$$

or

$$\begin{bmatrix} \dot{x} \\ \dot{\xi} \end{bmatrix} = \hat{A} \begin{bmatrix} x \\ \xi \end{bmatrix} + \hat{B}u + \hat{D}w + \begin{bmatrix} 0 \\ I \end{bmatrix} r, z = \hat{C} \begin{bmatrix} x \\ \xi \end{bmatrix} + B_2 u \tag{23}$$

where the augmented matrices are

$$\hat{A} = \begin{bmatrix} A & 0 \\ -C & 0 \end{bmatrix}, \hat{B} = \begin{bmatrix} B \\ 0 \end{bmatrix}, \hat{D} = \begin{bmatrix} D \\ 0 \end{bmatrix} \hat{C} = [C \ 0]$$

The ellipsoidal design of regulators, theorem 1, can be generalized to the trackers design. The proposed controller has to be decentralized and robust against load variation (uncertainty) of the MG. For this, Equation (23) is modified to

$$\begin{bmatrix} \dot{x} \\ \dot{\xi} \end{bmatrix} = (\hat{A} + \Delta\hat{A}) \begin{bmatrix} x \\ \xi \end{bmatrix} + \hat{B}u + (\hat{D} + \Delta\hat{D})w + \begin{bmatrix} 0 \\ I \end{bmatrix} r, z = \hat{C} \begin{bmatrix} x \\ \xi \end{bmatrix} + B_2 u \tag{24}$$

The system time varying uncertainties $\Delta\hat{A}(t), \Delta\hat{D}(t)$ have the norm-bounded form

$$\Delta\hat{A} = M\Delta_A N, \Delta\hat{D} = F\Delta_D H, \|\Delta_A(t)\| \le 1, \|\Delta_D(t)\| \le 1 \tag{25}$$

The following theorem is developed.

Theorem 2. *Consider system in Equation (24) with the controllable pair $(\hat{A}, \hat{B})$, and L∞-bounded external disturbances. Then, the decentralized robust state feedback with integral controller $u = [K \ k_i] \begin{bmatrix} x \\ \xi \end{bmatrix}$, which rejects optimally the external disturbances (in the sense of the trace that is output-invariant to the ellipsoid) is equivalent to minimize $tr[\hat{C}\hat{P}\hat{C}' + B_2 Z B_2']$ subject to the following constraints:*

$$\begin{bmatrix} (\hat{A}\hat{P} + \hat{B}\hat{Y} + *) + \alpha\hat{P} + \epsilon MM' + \rho FF' & \hat{D} & \hat{P}N' & 0 \\ * & -\alpha I & 0 & H' \\ * & * & -\epsilon I & 0 \\ * & * & * & -\rho I \end{bmatrix} \le 0, \quad \alpha, \epsilon, \rho > 0 \tag{26}$$

$$\begin{bmatrix} Z & \hat{Y} \\ * & \hat{P} \end{bmatrix} \ge 0, \hat{P} > 0$$

where $\hat{Y} = \hat{K}\hat{P}, \hat{K} = [K \ k_i]$, and the minimization is carried out with respect to the variables $\alpha, \epsilon, \rho, P = P' \in R^{(n+l).(n+l)}, \hat{Y} \in R^{p.(n+l)}$ and $Z = Z' \in R^{p \times p}$.

Proof of Theorem 2. For the tracking problem, the matrices in Equation (19) are replaced by the augmented system matrices. Using Fact 2, the following can be obtained:

$$\begin{bmatrix} (\hat{A}\hat{P} + \hat{B}\hat{Y} + *) + \alpha\hat{P} & \hat{D} \\ * & -\alpha I \end{bmatrix} \le 0, \alpha > 0 \tag{27}$$

To cope with load variation (uncertainty) in the MG, $\hat{A}$ is replaced with $(\hat{A} + \Delta\hat{A})$, and $\hat{D}$ with $(\hat{D} + \Delta\hat{D})$ in Equation (27). The resultant system is:

$$\begin{bmatrix} (\{\hat{A} + \Delta\hat{A}\}\hat{P} + \hat{B}\hat{Y} + *) + \alpha\hat{P} & \hat{D} + \Delta\hat{D} \\ * & -\alpha I \end{bmatrix} \le 0, \alpha > 0$$

Substituting for the norm-bounded uncertainty in Equation (25), one gets

$$\begin{bmatrix} \left(\{\hat{A} + M\Delta N'\}\hat{P} + \hat{B}\hat{Y} + *\right) + \alpha\hat{P} & \hat{D} + F\Delta_D H \\ * & -\alpha I \end{bmatrix} \le 0, \ \alpha > 0$$

Separating the uncertainty terms, we obtain

$$\begin{bmatrix} \left(\hat{A}\hat{P} + \hat{B}\hat{Y} + *\right) + \alpha\hat{P} & \hat{D} \\ * & -\alpha I \end{bmatrix} + \left(\begin{bmatrix} M \\ 0 \end{bmatrix}\Delta_A\begin{bmatrix} N\hat{P} & 0 \end{bmatrix} + *\right) + \left(\begin{bmatrix} F \\ 0 \end{bmatrix}\Delta_D\begin{bmatrix} 0 & H \end{bmatrix} + *\right) \le 0 \qquad (28)$$

Using Fact 1 to eliminate $\Delta_A(t)$, $\Delta_D(t)$, Equation (28) is satisfied if the following matrix equation is satisfied:

$$\begin{bmatrix} \left(\hat{A}\hat{P} + \hat{B}\hat{Y} + *\right) + \alpha\hat{P} & \hat{D} \\ * & -\alpha I \end{bmatrix} + \epsilon\begin{bmatrix} M \\ 0 \end{bmatrix}\begin{bmatrix} M \\ 0 \end{bmatrix}' + \epsilon^{-1}\begin{bmatrix} \hat{P}N' \\ 0 \end{bmatrix}\begin{bmatrix} \hat{P}N' \\ 0 \end{bmatrix}'$$
$$\rho\begin{bmatrix} F \\ 0 \end{bmatrix}\begin{bmatrix} F \\ 0 \end{bmatrix}' + \rho^{-1}\begin{bmatrix} 0 \\ H' \end{bmatrix}\begin{bmatrix} 0 \\ H' \end{bmatrix}' \le 0, \quad \alpha, \epsilon, \rho > 0$$

Using Fact 2, theorem is proved. $\square$

Decomposing global MG system into N-subsystems (DGs), the above theorem can be applied for each subsystem by adding a subscript i to every variable.

For $\pm$ 10% changes in load resistance, the norm-bounded uncertainty in Equation (25) for the subsystems DERS are shown in Table 2. Note that the load resistance does not appear in the off-diagonal matrices $A_{ij}, j \ne i$ (the disturbance), hence the uncertainty matrices of the disturbance (25) F, and $H = 0$.

Table 2. Norm-bounded uncertainties in $\Delta A_{i,j}$.

DER #	M	N
1	[2.8994,0,0,0,0,0,0,0;0,2.8994,0,0,0,0,0,0]',	[1.836,0,0,0,0,0,0,0;0,1.836,0,0,0,0,0,0]
2	[2.6912,0,0,0,0,0,0,0;0,2.6912,0,0,0,0,0,0]'	[1.7042,0,0,0,0,0,0,0;0,1.7042,0,0,0,0,0,0];
3	[2.8323,0,0,0,0,0;0,2.8323,0,0,0,0]'	[1.7935,0,0,0,0,0;0,1.7935,0,0,0,0]

The following optimal decentralized robust controller parameters are obtained by solving the LMIs of Theorem 2, with B_2 selected as a unit matrix.

Decentralized robust tracker for DER1:

$$K1 = [-0.062127, 0.16641, -16.023, 0.0019958, -4.0591, 5.2101, 5.7804, 2.7372;$$
$$-0.16641, -0.062127, -0.0019958, -16.023, -5.2101, -4.0591, 2.7372, 5.7804]$$
$$K1, I = [292.7000 - 161.8750; 161.8750\ 292.7000]$$

Decentralized robust tracker for DER2:

$$K2 = [-0.23452, 0.38322, -34.774, -0.00019014, -4.1782, 16.381, 13.573, -4.5376;$$
$$-0.38322, -0.23452, 0.00019014, -34.774, -16.381, -4.1782, 4.5376, 13.573]$$
$$K2, I = [372.8000 - 120.3425; 120.3425\ 372.8000]$$

Decentralized robust tracker for DER3:

$$K3 = [-0.94669, 0.37784, -79.496, 0.00025575, 48.315, 9.9538;$$
$$-0.37784, -0.94669, -0.00025575, -79.496, -9.9538, 48.315]$$
$$K3, I = [223.39 - 95.798; 95.798\ 223.39]$$

4. Simulation Results

The proposed control system performance is verified and tested using the MATLAB/SimPowerSystem Toolbox. The selected islanded PWM-inverter microgrid consisting of three-DGs as shown in Figure 2. It is implicit that each DG maintains a local load. The parameters of all DGs, transmission lines and local loads are presented in Table 1. In Figures 4–9, the dynamic performance of the proposed designed controllers is assessed via several test scenarios, including voltage tracking, load change and load unbalance.

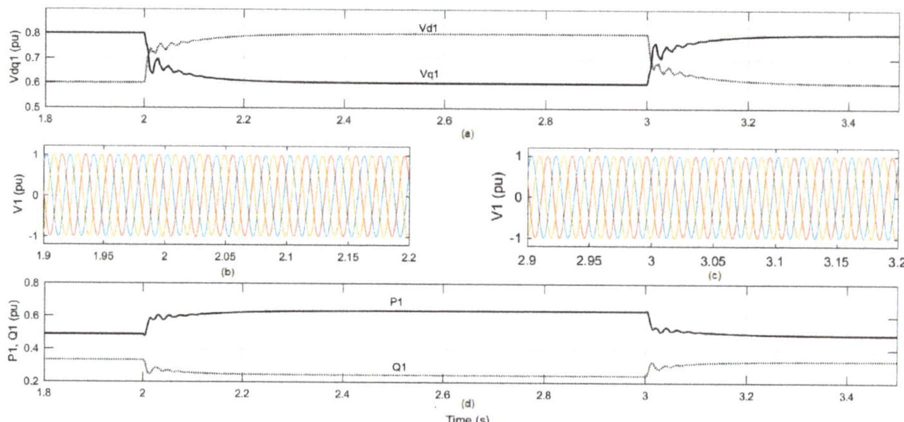

Figure 4. Proposed controllers' performance of DG 1 as a result of two-step changes in the voltages set point. (**a**) dq-components of the load voltage at point of coupling (PC) 1. (**b,c**) Instantaneous load voltages of PC 1. (**d**) Output active and reactive power of DG 1.

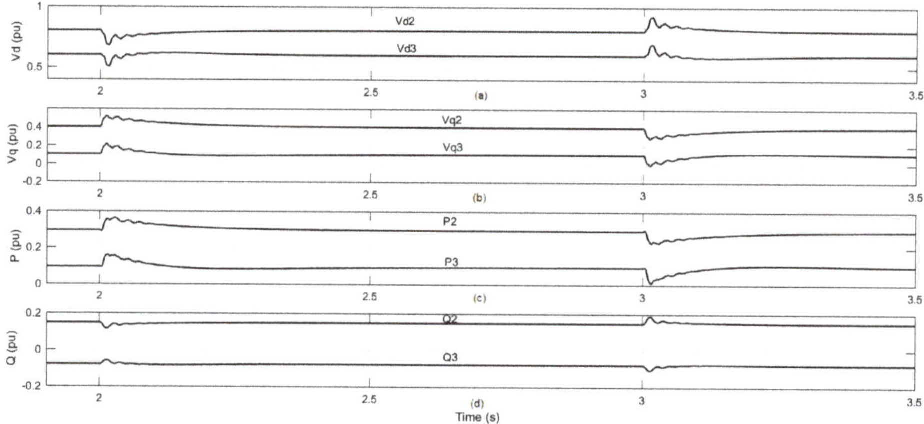

Figure 5. Dynamic performances of DGs 2 and 3 as a result of two-step changes in V_{dq1} set point. (**a**) d-component of the load voltages at PCs. (**b**) q-component of the load voltages at PCs. (**c**) Output active power of DGs. (**d**) Output reactive power of DGs.

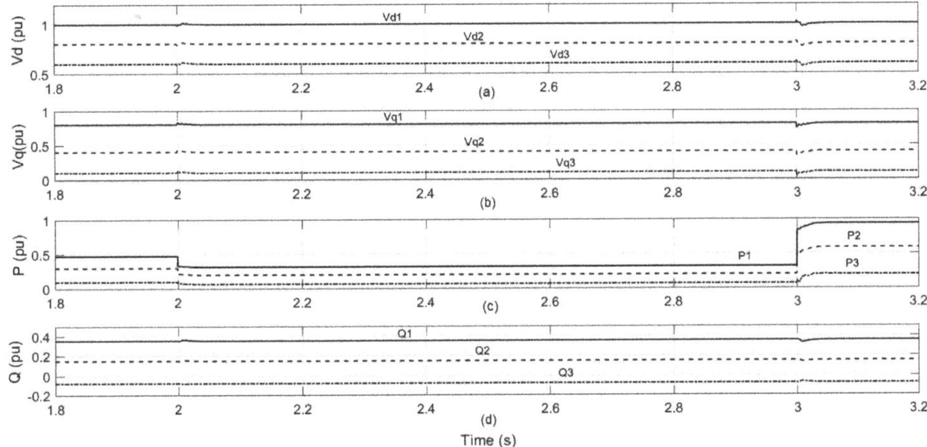

Figure 6. Dynamic performances of DG 1, DG2, and DG 3 as a result of step changes in the load resistance in three-phase system. (**a**) d-component of the load voltages at three DGs. (**b**) q-component of the load voltages at three-DGs. (**c**) Output active power of three DGs. (**d**) Output reactive power of three DGs.

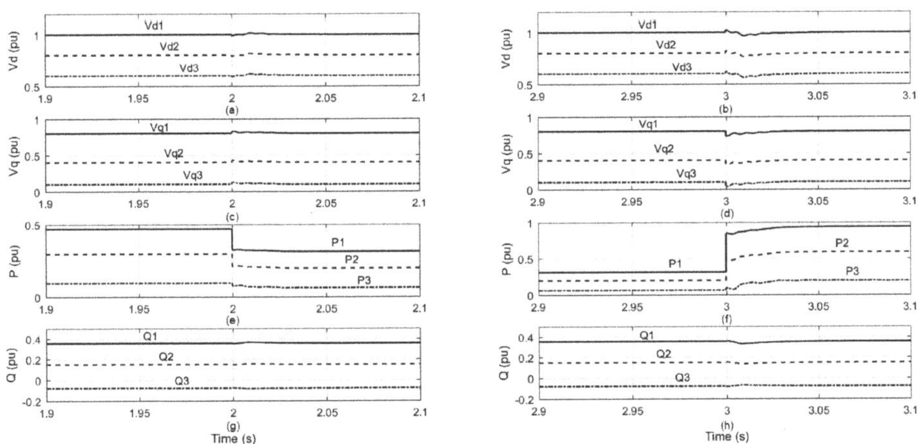

Figure 7. Dynamic performances of DG 1, DG2, and DG 3 as a result of step changes in the load resistance in three-phase system. (**a,b**) d-component of the load voltages at three DGs. (**c,d**) q-component of the load voltages at three DGs. (**e,f**) Output active power of three DGs. (**g,h**) Output reactive power of three DGs.

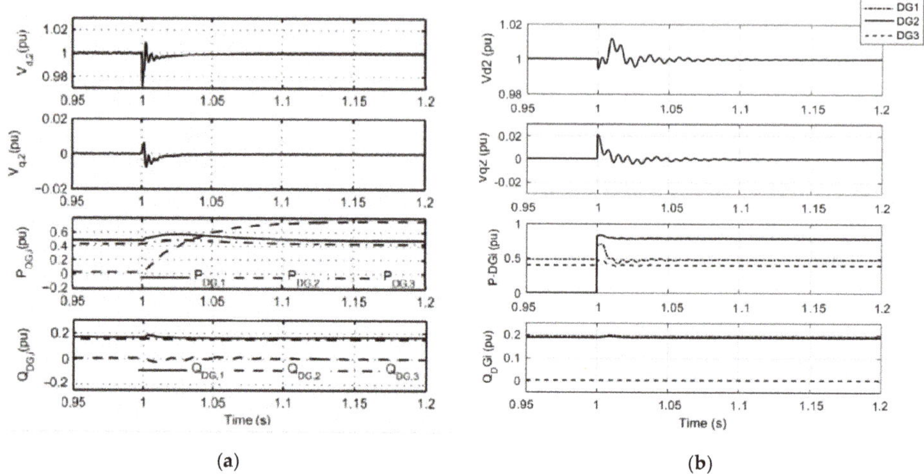

Figure 8. Response for an increase in real power of load2. (**a**) Control of Reproduced from [20], the name of the publisher: IEEE Trans. Power Deliv [20] (**b**) Proposed control.

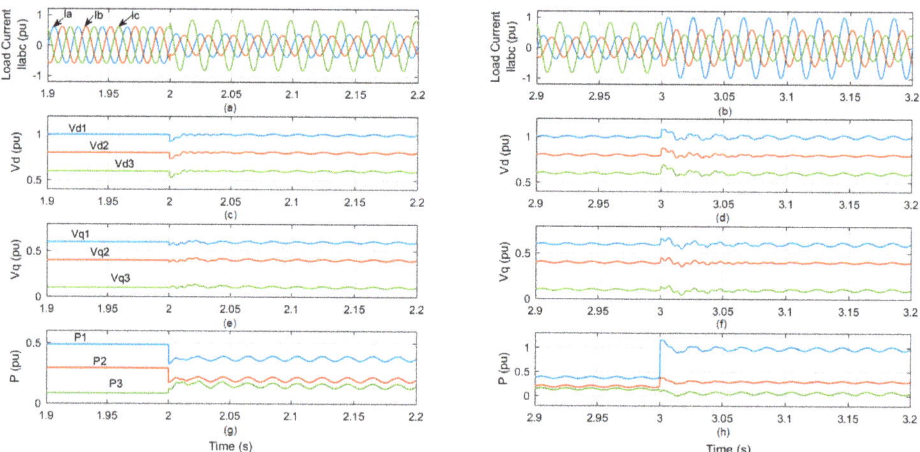

Figure 9. Dynamic performances of DG 1, DG 2, and DG 3 during islanded micro-grid and unbalance load condition. (**a**) load currents, unbalance at t = 2 s, (**b**) load currents, unbalance at t = 3 s, (**c**) V_d at PCC, unbalance at t = 2 s, (**d**) V_d at PCC, unbalance at t = 3 s, (**e**) V_q at PCC, unbalance at t = 2 s, (**f**) V_q at PCC, unbalance at t = 3 s, (**g**) active power, unbalance at t = 2 s, (**h**) active power, unbalance at t = 3 s.

4.1. Scenario 1: Voltage Tracking

The microgrid system shown in Figure 2 which has three DGs. Each DG delivers the active and reactive power for its particular local loads corresponding to the information/set points obtained from PMS shown in Figure 1a. The voltage values of the dq components of the DGs' references are listed in Table 1. The tracking response is checked by step changes in the reference voltage. The d and q components of the reference voltage for DG1 are respectively changed in steps as follows: (i) from 0.6 and 0.8 pu to 0.8 and 0.6 pu at t = 2 s. and (ii) at t = 3 s, from 0.8 and 0.6 pu to 0.6 and 0.8 pu.

The dynamic responses of DG 1 as a result of new reference voltages are shown in Figure 4. Figure 4a shows the d and q components of the load voltage of DG 1. The proposed controller successfully adjusts the load voltage in less than 0.2 s with zero steady state error. Figure 4b,c respectively, shows the instantaneous load voltages of PCC 1 during the two-step changes. Figure 5d represents the output active and reactive power of DG 1. Moreover, the dq voltages and the active and reactive power of the other two-DGs (DG2 and DG3) are depicted in Figure 5. The results show that there is a short transient period (approximately 0.2 s) in the load voltages and the active and reactive powers at PC2 and PC3 as a result of the two-step changes in the voltage set points of DG 1.

4.2. Scenario 2: Load Change

Although the proposed controllers are designed to reject any disturbance up to ±10% of load changes; in this scenario, we will examine the robustness of the proposed controller by applying load changes to the microgrid system more than the design range. The local load of any DG in Figure 2 is modeled by a three-phase parallel RLC network whose parameters are given in Table 1.

The load resistances R at any of the three-DGs in the three phases are equally changed from 100% to 150% at t = 2 s and changed again from 150% to 50% at t = 3 s. The results illustrated in Figure 6 show the robustness of the controller and how it copes with respect to the load changes even if it is more than the designed range. Figure 7 demonstrates that the designed controllers are too fast where it took less than 0.05 s to reject the significant changes.

4.3. Comparison with Other Controller

For an increase in real power of load2 at t = 1 s from 0 to 0.8 p.u, the response using the control in [20] and the proposed one is shown respectively in Figure 8a,b.

The comparison can be summarized in Table 3.

Table 3. Comparison with the control in [20]. Reproduced from [20], the name of the publisher: IEEE Trans. Power Deliv.

Case	Response with Control in [20]	Response with Proposed Control
Vd2	Faster, but has higher undershoot	Slower, but has less undershoot
Vq2	Faster, less undershoot	Slower, higher undershoot
P_{DGi}	Slower	Faster
Q_{DGi}	Noticeable disturbance	Unnoticeable disturbance

It can be concluded that the proposed control outperforms that in [20] in PDGi, and QDGi.

4.4. Scenario 3: Unbalanced Load

In this scenario, the proposed system is primarily operating with a balanced load, then the islanded micro-grid load is assumed to become unbalanced. At t = 2 s, the load becomes unbalanced by changing the resistance load in phase a, b and c to be: R_a = 130% of R_a, R_b = 150% of R_a and R_c = 60% of R_a. The dynamic performance of the three-DGs is shown in Figure 9a,c,e,g. At t = 3 s, the load is changed from unbalanced to another unbalanced by changing the resistance load in phase a, b and c to be: R_a = 50% of R_a, R_b = 100% of R_a and R_c = 140% of R_a. The dynamic performance of the three-DGs is given in Figure 9b,d,f,h. Figure 9a,b shows the abc-load current response. Figure 9c,d depicts the d-components of the PCC voltage. Figure 9e,f illustrates the q-components of the PCC voltage, and Figure 9g,h gives the output active power of the three-DGs. Figure 9 demonstrates that the proposed voltage controller is effective against the load unbalances. The *d-q* components are polluted by a small oscillation which is produced from the load side. The frequency of this oscillation is double the PCC frequency. This case shows that the proposed control can partially compensate for the effect of the unbalanced load.

5. Conclusions

This paper proposes a control strategy for an islanded, multi-DER microgrid. Each DER's local controller is designed based on a multivariable decentralized robust servomechanism approach that uses a microgrid's linear state-space model. The design is simple and based on the ellipsoidal approximation to invariant sets. The local controller for each DER considers the interaction effect of the rest of the system as an external disturbance to be rejected. The proposed controller achieves robust stability and desired performance (fast transient response, negligible interaction, and zero steady-state tracking error, despite uncertainties in the load parameters).

The performance of the proposed robust servomechanism controller is evaluated and verified using MATLAB/Simulink. The controller performance is evaluated when the study system is subjected to

- Set point step changes;
- Uncertainties in the load parameters.

The simulation results show that the proposed controller

- Provides excellent tracking of the reference signals (fast and smooth non-peaking transients);
- Robustly maintains voltage magnitude of the load despite the load parameter uncertainties.

Author Contributions: Data curation, E.B.; Formal analysis, E.B.; Funding acquisition, A.A.-H.; Methodology, H.M.S.; Project administration, A.A.-H.; Software, M.S.; Validation, M.S.; Writing—review & editing, H.M.S. and A.A.-H. All authors have read and agreed to the published version of the manuscript.

Funding: This research was funded by SQU through His Majesty Trust Fund, grant number SR/ENG/ECED/17/01.

Acknowledgments: The authors would like to acknowledge the support from Sultan Qaboos University and His Majesty Trust Fund for funding this research work from the research grant SR/ENG/ECED/17/01.

Conflicts of Interest: The authors declare no conflict of interest.

Appendix A System Equations and State Matrices

The A-matrix of (4) is given below.

$$A = \begin{bmatrix} A_{11} & A_{12} & A_{13} \\ A_{21} & A_{22} & A_{23} \\ A_{31} & A_{32} & A_{33} \end{bmatrix}$$

$$A_{11} = \begin{bmatrix} \frac{-1}{R_1 c_1} & \omega & \frac{1}{c_1} & 0 & \frac{-1}{c_1} & 0 & \frac{-1}{c_1} & 0 \\ -\omega & \frac{-1}{R_1 c_1} & 0 & \frac{1}{c_1} & 0 & \frac{-1}{c_1} & 0 & \frac{-1}{c_1} \\ \frac{-1}{L_{f1}} & 0 & \frac{-R_{f1}}{L_{f1}} & \omega & 0 & 0 & 0 & 0 \\ 0 & \frac{-1}{L_{f1}} & -\omega & \frac{-R_{f1}}{L_{f1}} & 0 & 0 & 0 & 0 \\ \frac{1}{L_1} & 0 & 0 & 0 & \frac{-R_{l1}}{L_1} & \omega & 0 & 0 \\ 0 & \frac{1}{L_1} & 0 & 0 & -\omega & \frac{-R_{l1}}{L_1} & 0 & 0 \\ \frac{1}{L_{t1}} & 0 & 0 & 0 & 0 & 0 & \frac{-R_{t1}}{L_{t1}} & \omega \\ 0 & \frac{1}{L_{t1}} & 0 & 0 & 0 & 0 & -\omega & \frac{-R_{t1}}{L_{t1}} \end{bmatrix}, \quad A_{22} = \begin{bmatrix} \frac{-1}{R_2 c_2} & \omega & \frac{1}{c_2} & 0 & \frac{-1}{c_2} & 0 & \frac{-1}{c_2} & 0 \\ -\omega & \frac{-1}{R_2 c_2} & 0 & \frac{1}{c_2} & 0 & \frac{-1}{c_2} & 0 & \frac{-1}{c_2} \\ \frac{-1}{L_{f2}} & 0 & \frac{-R_{f2}}{L_{f2}} & \omega & 0 & 0 & 0 & 0 \\ 0 & \frac{-1}{L_{f2}} & -\omega & \frac{-R_{f2}}{L_{f2}} & 0 & 0 & 0 & 0 \\ \frac{1}{L_2} & 0 & 0 & 0 & \frac{-R_{l2}}{L_2} & \omega & 0 & 0 \\ 0 & \frac{1}{L_2} & 0 & 0 & -\omega & \frac{-R_{l2}}{L_2} & 0 & 0 \\ \frac{1}{L_{t2}} & 0 & 0 & 0 & 0 & 0 & \frac{-R_{t2}}{L_{t2}} & \omega \\ 0 & \frac{1}{L_{t2}} & 0 & 0 & 0 & 0 & -\omega & \frac{-R_{t2}}{L_{t2}} \end{bmatrix},$$

$$A_{33} = \begin{bmatrix} \frac{-1}{R_3 c_3} & \omega & \frac{1}{c_3} & 0 & \frac{-1}{c_3} & 0 \\ -\omega & \frac{-1}{R_3 c_3} & 0 & \frac{1}{c_3} & 0 & \frac{-1}{c_3} \\ \frac{-1}{L_{f3}} & 0 & \frac{-R_{f3}}{L_{f3}} & \omega & 0 & 0 \\ 0 & \frac{-1}{L_{f3}} & -\omega & \frac{-R_{f3}}{L_{f3}} & 0 & 0 \\ \frac{1}{L_3} & 0 & 0 & 0 & \frac{-R_{l3}}{L_3} & \omega \\ 0 & \frac{1}{L_3} & 0 & 0 & -\omega & \frac{-R_{l3}}{L_3} \end{bmatrix},$$

$$A_{12} = \frac{-1}{L_{t1}} \times \begin{bmatrix} 0_{6\times2} & 0_{6\times6} \\ I_{2\times2} & 0_{2\times6} \end{bmatrix}, A_{13} = 0_{8\times6},$$

$$A_{12} = \frac{1}{C_2} \times \begin{bmatrix} 0_{2\times6} & I_{2\times2} \\ 0_{6\times6} & 0_{6\times2} \end{bmatrix},$$

$$A_{23} = \frac{-1}{L_{t2}} \times \begin{bmatrix} 0_{6\times2} & 0_{6\times4} \\ I_{2\times2} & 0_{2\times4} \end{bmatrix},$$

$$A_{31} = 0_{6\times8}, A_{32} = \frac{1}{C_3} \times \begin{bmatrix} 0_{2\times6} & I_{2\times2} \\ 0_{4\times6} & 0_{4\times2} \end{bmatrix},$$

Nonzero entries of $B_{22\times6}$ are $b_{3,1} = b_{4,2} = \frac{1}{L_{f1}}, b_{11,3} = b_{12,4} = \frac{1}{L_{f2}}$, and $b_{19,5} = b_{21,6} = \frac{1}{L_{f3}}$.

Non-zero entries of $C_{6\times22}$, i.e., $C_{1,1}$, $C_{2,2}$, $C_{9,3}$, $C_{10,4}$, $C_{17,5}$, and $C_{18,6}$ are unity.

References

1. Colmenar-Santos, A.; Reino-Rio, C.; Borge-Diez, D.; Collado-Fernández, E. Distributed generation: A review of factors that can contribute most to achieve a scenario of DG units embedded in the new distribution networks. *Renew. Sustain. Energy Rev.* **2016**, *59*, 1130–1148. [CrossRef]
2. Mehdi, M.; Saad, M.; Jamali, S.Z.; Kim, C.-H. Output-feedback based robust controller for uncertain DC islanded microgrid. *Trans. Inst. Meas. Control.* **2019**, *42*, 1239–1251. [CrossRef]
3. Hassan, M.A.; Abido, M.A. Optimal Design of Microgrids in Autonomous and Grid-Connected Modes Using Particle Swarm Optimization. *IEEE Trans. Power Electron.* **2010**, *26*, 755–769. [CrossRef]
4. Raju, E.S.N.; Jain, T. Robust optimal centralized controller to mitigate the small signal instability in an islanded inverter based microgrid with active and passive loads. *Int. J. Electr. Power Energy Syst.* **2017**, *90*, 225–236. [CrossRef]
5. Han, H.; Hou, X.; Yang, J.; Wu, J.; Su, M.; Guerrero, J.M. Review of Power Sharing Control Strategies for Islanding Operation of AC Microgrids. *IEEE Trans. Smart Grid* **2015**, *7*, 200–215. [CrossRef]
6. Chandorkar, M.C.; Divan, D.M.; Adapa, R. Control of parallel connected inverters in standalone AC supply systems. *IEEE Trans. Ind. Appl.* **1993**, *29*, 136–143. [CrossRef]
7. Olivares, D.E.; Mehrizi-Sani, A.; Etemadi, A.H.; Canizares, C.A.; Iravani, R.; Kazerani, M.; Hajimiragha, A.H.; Gomis-Bellmunt, O.; Saeedifard, M.; Palma-Behnke, R.; et al. Trends in Microgrid Control. *IEEE Trans. Smart Grid* **2014**, *5*, 1905–1919. [CrossRef]
8. Tsikalakis, A.G.; Hatziargyriou, N.D. Centralized Control for Optimizing Microgrids Operation. *IEEE Trans. Energy Convers.* **2008**, *23*, 241–248. [CrossRef]
9. Tan, K.T.; Peng, X.Y.; So, P.L.; Chu, Y.C.; Chen, M.Z.Q. Centralized Control for Parallel Operation of Distributed Generation Inverters in Microgrids. *IEEE Trans. Smart Grid* **2012**, *3*, 1977–1987. [CrossRef]
10. Liang, H.; Choi, B.J.; Zhuang, W.; Shen, X. Stability Enhancement of Decentralized Inverter Control Through Wireless Communications in Microgrids. *IEEE Trans. Smart Grid* **2013**, *4*, 321–331. [CrossRef]
11. Wang, Y.; Wang, X.; Chen, Z.; Blaabjerg, F. Distributed Optimal Control of Reactive Power and Voltage in Islanded Microgrids. *IEEE Trans. Ind. Appl.* **2016**, *53*, 340–349. [CrossRef]
12. Li, D.; Zhao, B.; Wu, Z.; Zhang, X.; Zhang, L. An Improved Droop Control Strategy for Low-Voltage Microgrids Based on Distributed Secondary Power Optimization Control. *Energies* **2017**, *10*, 1347. [CrossRef]
13. Alyazidi, N.; Mahmoud, M.; Abouheaf, M. Adaptive critics based cooperative control scheme for islanded Microgrids. *Neurocomputing* **2018**, *272*, 532–541. [CrossRef]
14. Bidram, A.; Nasirian, V.; Davoudi, A.; Lewis, F.L. *Cooperative Synchronization in Distributed Microgrid Control*; Springer: Berlin, Germany, 2017.
15. Mahmoud, M.S.; Al-Sunni, F.M. *Control and Optimization of Distributed Generation Systems*; Springer Science and Business Media: Berlin, Germany, 2015.
16. Bevrani, H.; Francois, B.; Ise, T. *Microgrid Dynamics and Control*; John Wiley & Sons, Inc.: Hoboken, NJ, USA, 2017.
17. Hassan, A.M.; Worku, Y.M.; Abido, A.M. Optimal Design and Real Time Implementation of Autonomous Microgrid Including Active Load. *Energies* **2018**, *11*, 1109. [CrossRef]
18. Habibi, F.; Naghshbandy, A.H.; Bevrani, H. Robust voltage controller design for an isolated Microgrid using Kharitonov's theorem and D-stability concept. *Int. J. Electr. Power Energy Syst.* **2013**, *44*, 656–665. [CrossRef]
19. Karimi, H.; Davison, E.J.; Iravani, R. Multivariable Servomechanism Controller for Autonomous Operation of a Distributed Generation Unit: Design and Performance Evaluation. *IEEE Trans. Power Syst.* **2009**, *25*, 853–865. [CrossRef]
20. Etemadi, A.; Davison, E.J.; Iravani, R. A Decentralized Robust Control Strategy for Multi-DER Microgrids—Part I: Fundamental Concepts. *IEEE Trans. Power Deliv.* **2012**, *27*, 1843–1853. [CrossRef]
21. Blanchini, F.; Miani, S. Set-Theoretic Methods in Control. *Dyn. Control.* **2015**. [CrossRef]

22. Nazin, S.A.; Polyak, B.T.; Topunov, M.V. Rejection of bounded exogenous disturbances by the method of invariant ellipsoids. *Autom. Remote Control.* **2007**, *68*, 467–486. [CrossRef]
23. Han, S.; Kodaira, D.; Han, S.; Kwon, B.; Hasegawa, Y.; Aki, H. An Automated Impedance Estimation Method in Low-Voltage Distribution Network for Coordinated Voltage Regulation. *IEEE Trans. Smart Grid* **2015**, *7*, 1–9. [CrossRef]

Publisher's Note: MDPI stays neutral with regard to jurisdictional claims in published maps and institutional affiliations.

MDPI
St. Alban-Anlage 66
4052 Basel
Switzerland
Tel. +41 61 683 77 34
Fax +41 61 302 89 18
www.mdpi.com

Energies Editorial Office
E-mail: energies@mdpi.com
www.mdpi.com/journal/energies